CONFORMAL FIELD THEORY

CONFORMAL FIELD THEORY

Sergei V. Ketov

Institut für Theoretische Physik
Universität Hannover
Germany

World Scientific
Singapore • New Jersey • London • Hong Kong

Published by

World Scientific Publishing Co. Pte. Ltd.

P O Box 128, Farrer Road, Singapore 912805

USA office: Suite 1B, 1060 Main Street, River Edge, NJ 07661

UK office: 57 Shelton Street, Covent Garden, London WC2H 9HE

First published 1995
First reprint 1997

CONFORMAL FIELD THEORY

ISBN   981-02-1608-4

Printed in Singapore by Uto-Print

To Tatiana Ketova, my wife,
for all her love and support

Preface

The aim of physics as a science is to describe the most fundamental laws of Nature. Being the most fundamental, the physical laws are simultaneously the simplest ones. To be simple does not mean however to be trivial. The simplicity of fundamental physics exhibits itself in fundamental symmetry principles. It is therefore fundamental physical symmetry principles that we should be looking for, and it is mainly a theoretical problem. It is very often in physics that underlying symmetries are hidden by the complexity of actual physical phenomena, and it is the objective of theoretical physics to uncover them.

Taking high energy physics as an example, the theoretical basis for our current understanding of properties and dynamics of elementary particles is provided by *quantum field theory* (QFT). A general QFT however seems to be too general for these purposes. It is the *gauge* QFT's that are apparently distinguished in describing fundamental physical interactions of Nature. Compared to a general QFT, quantum gauge theories are much more restrictive and possess additional symmetries. The well-known achievements of gauge theories in high energy physics during last decades perfectly illustrate the power of symmetry principles in physics. The higher an energy scale, the more symmetric become interactions among elementary particles. This is a clear sign of some (normally broken) symmetry which should be formally restored at infinite energies or vanishing masses. Another relevant example is provided by the theory of critical phenomena. It is a well-known fact that quite different physical systems at their criticality exhibit *universal* behaviour. This also indicates a presence of some underlying fundamental symmetry. It is the *conformal* symmetry that is responsible in both cases.

One important point should however be emphasized. Benefits of having conformal symmetry usually come true in *two* dimensions, where this symmetry becomes infinite-dimensional. An infinite number of conservation laws is responsible for an integrability of two-dimensional physical systems possessing conformal invariance. It is the two-dimensional *conformal field theory* (CFT) that is actually addressed in this book.

CFT is viewed as a specific QFT having conformal symmetry and its own general principles. The underlying theoretical principles of CFT were actually extracted from a careful analysis of numerous examples in statistical mechanics and field theory, as far as their critical or short-distance behaviour is concerned. These principles are not self-evident, and they do not follow from some underlying theory. In this book, after introducing the basic CFT principles in Chapter I, their meaning is explored without

vii

actually discussing their origin. This is made on purpose since CFT itself is still in a process of evolution. It may not be a good idea to extensively discuss CFT fundamentals when there is a good chance for their being changed or even replaced by more fundamental principles. Nevertheless, modern CFT has its own sound basis. Concentrating on this basis, which I believe will survive in time, gives me a chance to represent CFT in a self-contained way.

Compared to a QFT description, a CFT description of a given physical system is less informative. Nevertheless, it contains relevant universal information about high-energy scattering or short-distance (critical) behaviour. The main task of CFT is to calculate QFT correlation functions in the limit where conformal symmetry is thought of as being present. An infinite number of physical degrees of freedom presumably contributes to the correlation functions at very high energies which are hardly accessible in experiments, or near a critical point of a statistical system under consideration. Therefore, one should try to understand the conformal physics without solving a general (very complicated) interacting problem by the methods of QFT. The latter are usually of a perturbative nature and, hence, they are very restricted in applications. Instead, CFT is capable of going beyond perturbation theory, and it is just what this book is all about.

The idea of conformal symmetry is not the only one which helps. A related idea is the assumption about *locality* of CFT operators which should form a complete set of local operators satisfying some kind of algebra to be described by the operator product expansions. A good analogue is provided by non-relativistic quantum physics, where a scattering problem can be reduced to solving a non-linear differential equation. In practice, one usually needs only asymptotics of a solution, not the exact one. It may well be quite difficult to find an exact solution. Knowing asymptotics clearly gives only partial information about an exact solution. However, calculating asymptotics is usually a simpler problem, and it is sometimes possible to reformulate the problem in terms of a closed set of equations known as the '*conformal bootstrap*'. It may be very difficult to get such a picture by using the tools of QFT, if it is even possible at all.

Specific CFT models, as well as solutions to CFT bootstrap equations, may possess additional symmetries. It is one of the tasks of CFT to describe all possible extensions of conformal symmetry. This leads to remarkable connections with various branches of modern mathematics, such as affine Lie and Kač-Moody algebras and 'quantum groups'. It is a physical problem how to identify the relevant operators for a given conformally invariant physical system.

This book intends to provide an introduction to CFT from the first principles, and to reach the current frontiers of research in this area. Throughout the book I tried to keep a balance between introductory and advanced topics, sometimes omitting general proofs and substituting them by examples. This seems to be quite reasonable for describing a theory which has yet to establish itself in a strict mathematical way. Special attention was paid to making a clear distinction between the well-established mathematical results and their applications in CFT.

The monograph is organized as follows: Chapter I is devoted to the fundamentals of conformal symmetry in field theory, where a few basic examples are also considered in detail. In Chapter II, I review representations of conformal algebra, conformal boot-strap equations, fusion rules and the most important particular constructions. This Chapter plays a key role for CFT applications considered in the following Chapters. The partition functions and the Virasoro characters are introduced in Chapter III. The approach I adopted is based on free fields. Affine Lie and Kač-Moody algebras, and their role in CFT are discussed in Chapter IV. Their classification and free field representations are reviewed. An underlying topological and group-theoretical structure of the Wess-Zumino-Novikov-Witten theories possessing such symmetries is explained. (Extended) supersymmetry in CFT is discussed in Chapter V. A generalization of the group-theoretical results of Chapter IV to cosets is the topic of Chapter VI. Chapters I–VI essentially comprise the introductory part of the book. Chapters VII–XI contain more advanced material which is close to the frontiers of current research. In particular, Chapter VII is devoted to the W algebras generalizing the conformal Virasoro algebra. In Chapter VIII, I briefly discuss strings and superstrings from a viewpoint of CFT. Two-dimensional quantum gravity and topological field theories in relation to CFT are addressed in Chapter IX. Matrix models in relation to two-dimensional quantum gravity and CFT are reviewed in Chapter X. Chapter XI concludes the odyssey through CFT by addressing some mathematical relations between two-dimensional integrable models and CFT. My attitude towards conformal models as the particular integrable models becomes finally justified in this Chapter. Finally, Chapter XII plays an auxiliary role. There I collected comments about the Literature, mathematical details about Lie and affine (Kač-Moody) algebras, Riemann surfaces, BRST and BFV quantization, and 'quantum groups'. The mathematical comments are written in a pedagogical way, so that they may serve as introductory notes as well.

Almost every section contains a few exercises. Most of them are not difficult at all, but they may help for an individual's assessment. Some marginal information used in

x

the text is put into the exercises labelled by a star (*). For the reader's convenience, the list of abbreviations used in the text is supplied at the end of the book, as well as an index. Key words in the text are emphasized by different type styles.

Including all the relevant material for a wide range of CFT topics covered in this book is obviously impossible, and it was not my desire at all. Though being rather extensive, the bibliography collected at the end of the book and my comments about the Literature in section XII.1 are by no means complete. They only reflect my personal preferences in selecting the material. I apologize to the authors whose papers somehow escaped my attention or were cited inappropriately throughout the book.

This book grew out of my lectures given at the Physics Department of Maryland University in 1991-1992. A major part of the selected material is based on existing reviews and original papers. All the editorial work on the manuscript was done at the Institute of Theoretical Physics in Hannover.

It is my pleasure to thank colleagues, friends and students, as well as participants of Spring and Summer Schools at the International Center for Theoretical Physics in Trieste over the years, for numerous discussions on CFT and strings. In particular I would like to thank Nathan Berkovits, Jim Gates Jr., Marc Grisaru, Brian Dolan, Marty Halpern, Sergei Kuzenko, Dima Lebedev, Olaf Lechtenfeld, Alexei Morozov, Hermann Nicolai, Hitoshi Nishino, Yaroslav Pugai, Jens Schnittger, John Schwarz, Warren Siegel, Harald Skarke, Arkady Tseytlin, Igor Tyutin and Cumrum Vafa, who helped me to understand some particular issues. I am especially grateful to Olaf Lechtenfeld for his careful reading of the manuscript and critical remarks. The TeX typesetting system combined with the original macros and figures was used by the author in preparing the LaTeX file for printing the camera-ready manuscript. The assistance of Jens Johannesson and Thorsten Schwander in choosing the computer software is appreciated. At last but not least, this book would never appear without my wife Tatiana, supporting me and taking care of our children for all these years.

Sergei V. Ketov
Hannover, Germany
October, 1994

Contents

Chapter I

Conformal Symmetry and Fields

In this Chapter we introduce basic concepts of *conformal field theory* (CFT), study basic examples of *two-dimensional* (2d) conformal fields and calculate their correlation functions on a plane. The operator product expansion and the conformal algebra are shown to play key roles in CFT.

I.1 Conformal invariance

The basic ingredients of any field theory are a set of fields $\Phi_\Delta(x)$ depending on space-time coordinates x^μ, and the action $S[\Phi]$ to be a functional of the fields. In *quantum field theory* (QFT), one usually starts with the field theory action and then one calculates the correlation functions via the path integral approach. Our goal will be to calculate the correlation functions in CFT directly, by exploiting its symmetries. In this process, the action itself becomes redundant. In fact, more than one action may correspond to the same physical theory.

A field theory is normally classified by some set Δ of fundamental fields, which are generically infinite in number. The fundamental fields are supposed to form a basis in the field space. Though space-time is usually assumed to be equipped with the Minkowski signature, we choose to work in Euclidean space, so that our spacetime vector indices will be represented by the middle Greek letters, $\mu, \nu, \ldots = 1, 2, \ldots, d$. We find it convenient to allow the number d of Euclidean dimensions be arbitrary at the beginning, since it helps to understand the special role of two dimensions in CFT.

1

Our goal is not just to consider arbitrary field theories (there is not much to say about them) but those which are reparametrization-invariant. This means that among the fields of the theory there is a d-dimensional metric ('gravity') $g_{\mu\nu}(x)$, and the theory is invariant under the transformations

$$\delta x^\mu = \varepsilon^\mu(x) \; ,$$

$$\delta \Phi_\Delta(x) = L_\varepsilon \Phi_\Delta(x) \; ,$$

$$\delta g_{\mu\nu}(x) = \partial_\mu \varepsilon_\nu + \partial_\nu \varepsilon_\nu - \varepsilon^\lambda \partial_\lambda g_{\mu\nu} \; , \tag{1.1}$$

where $\varepsilon^\mu(x)$ are the infinitesimal parameters, and L_ε is the Lie derivative. The fields Φ_Δ are supposed to transform according to their tensor nature. The metric $g_{\mu\nu}(x)$ defines an invariant line element $ds^2 = g_{\mu\nu} dx^\mu dx^\nu$, and it transforms under a finite transformation $x^\mu \to \tilde{x}^\mu$ as a rank-2 symmetric tensor

$$g_{\mu\nu}(x) \to \tilde{g}_{\mu\nu}(\tilde{x}) = \frac{\partial x^\lambda}{\partial \tilde{x}^\mu} \frac{\partial x^\rho}{\partial \tilde{x}^\nu} g_{\lambda\rho}(x) \; . \tag{1.2}$$

The invariance under reparametrizations is not in fact even a restriction, since any field theory is usually assumed to be either Lorentz-invariant (in a flat space-time) or general coordinate invariant (in a curved space-time), and this implies some metric-dependence in the theory. [1] We assume almost everywhere in the book that our Euclidean spacetime is flat and isomorphic to $\mathbf{R}^d$, so that there exists a globally defined coordinate system in which $g_{\mu\nu} = \delta_{\mu\nu}$. [2]

After being quantized, the fields Φ_Δ become operators in the Fock space of quantum states of the theory. For the moment, there is no need to specify the quantization procedure, which can be really non-trivial only for gauge fields. In the first few Chapters we will deal only with the ordinary (non-gauge) quantum fields, whose quantization can be performed in the standard canonical way known in quantum mechanics and quantum field theory (see, e.g., any of the textbooks [1, 2, 3, 4]). The outcome of the quantization procedure can equally be represented in terms of the formal path integral describing the *vacuum expectation value* (VEV) for a product of the field operators:

$$\langle \Phi_{\Delta_1}(x_1) \Phi_{\Delta_2}(x_2) \cdots \Phi_{\Delta_M}(x_M) \rangle$$

$$= \mathcal{N}^{-1} \int \prod [\mathcal{D}\Phi_\Delta] \, \Phi_{\Delta_1}(x_1) \Phi_{\Delta_2}(x_2) \cdots \Phi_{\Delta_M}(x_M) e^{-S[\Phi_\Delta]} \; , \tag{1.3}$$

---

[1] This is not true for the topological field theories to be considered in sect. IX.3.

[2] What happens in the presence of quantum 2d gravity is discussed in Chs. IX and X.

where the functional measure is yet to be specified, and $\mathcal{N}$ is the normalization factor, $\mathcal{N} = \int \prod [\mathcal{D}\Phi_\Delta] e^{-S[\Phi_\Delta]}$. The VEV (1.3) is called a **correlator** or a **correlation function**. The quantum field theory is solved once all its correlators are calculated. One of the standard ways of doing that is to use the perturbation theory (the Wick's theorem) usually described in terms of the conventional Feynman graphs. Quite generally, there is no way to exactly calculate the correlators, unless there are powerful symmetries in the theory. The general idea is to impose such symmetries on QFT, which would severely constrain the correlation functions in such a way that they could be calculated, while being non-trivial. The CFT's are a subclass of QFT's, which are invariant under the conformal transformations. It is of interest to test the power of the conformal symmetry in quantum field theory, and to explore the consequences of this invariance. This will be the main story in our discussion. The physical examples where the conformal symmetry is particularly relevant are represented by strings in high-energy physics, and critical phenomena at the second order phase transitions in statistical physics.

By definition, the **conformal** transformations are the restricted general coordinate transformations (diffeomorphisms), $x \to \tilde{x}$, for which the metric is invariant *up to a scale factor*,

$$g_{\mu\nu}(x) \to \tilde{g}_{\mu\nu}(\tilde{x}) = \Omega(x)g_{\mu\nu} , \quad \Omega(x) \equiv e^{\omega(x)} . \tag{1.4}$$

Clearly, these transformations form a group, which is known as the *conformal* group. Since our space is flat, the reference metric can always be chosen to be flat also, $g_{\mu\nu} = \delta_{\mu\nu}$.

Our first task is to identify the conformal transformations from eq. (1.4). They clearly preserve the angle $A \cdot B / \sqrt{A^2 B^2}$ between any two vectors A^μ and B^μ, $A \cdot B = g_{\mu\nu} A^\mu B^\nu$. That's why such transformations are called conformal. They also contain the Poincaré transformations (translations and Euclidean rotations of the flat space $\mathbf{R}^d$) as a subgroup, since the latter obviously satisfy eq. (1.4) with $g_{\mu\nu} = \delta_{\mu\nu}$ and $\Omega = 1$. To determine all the conformal transformations in a flat space, consider the infinitesimal coordinate transformations $x^\mu \to x^\mu + \varepsilon^\mu(x)$ w.r.t. the flat metric $g_{\mu\nu} = \delta_{\mu\nu}$. One easily finds from eqs. (1.2) and (1.4) that

$$\delta_{\mu\nu} + \partial_\mu \varepsilon_\nu + \partial_\nu \varepsilon_\mu = \Omega \delta_{\mu\nu} = [1 + \omega(x)]\delta_{\mu\nu} , \quad \omega(x) = \frac{2}{d}\partial^\mu \varepsilon_\mu . \tag{1.5}$$

Hence, one gets the equation

$$\partial_\mu \varepsilon_\nu + \partial_\nu \varepsilon_\mu = \frac{2}{d}(\partial \cdot \varepsilon)\delta_{\mu\nu} . \tag{1.6}$$

The simplest way of finding its general solution, is to consider its corollary:

$$[\delta_{\mu\nu}\Box + (d-2)\partial_\mu\partial_\nu]\partial \cdot \varepsilon = 0 , \tag{1.7}$$

which follows after two more differentiations and one contraction of indices from eq. (1.6). From the exercise # 1 we learn that $\varepsilon(x)$ is at most quadratic in x when $d > 2$. Substituting the general second-order polynomial in x into eq. (1.6) and fixing its coefficients by that equation yield the result that the conformal algebra consists of the ordinary *translations* ($\varepsilon^\mu = a^\mu$) and *rotations* ($\varepsilon^\mu = \omega^\mu{}_\nu x^\nu, \omega_{\mu\nu} = -\omega_{\nu\mu}$), *scale* transformations ($\varepsilon^\mu = \lambda x^\mu$) and the *special conformal* transformations ($\varepsilon^\mu = b^\mu x^2 - 2x^\mu b \cdot x$). A special conformal transformation can be recognized as a composition of an inversion and a translation: $\tilde{x}^\mu/\tilde{x}^2 = x^\mu/x^2 + b^\mu$. The whole algebra is locally isomorphic to $so(d+1, 1)$. [3]

The conformal group is comprised of finite conformal transformations. They include the Poincaré group

$$x^\mu \to \tilde{x}^\mu = x^\mu + a^\mu \ ,$$

$$x^\mu \to \tilde{x}^\mu = \Lambda^\mu{}_\nu x^\nu \ , \quad \Lambda^\mu{}_\nu \in SO(d) \ , \tag{1.8}$$

and, in addition, the scale transformations (dilatations) and the special conformal transformations, resp.,

$$x^\mu \to \tilde{x}^\mu = \lambda x^\mu \ , \quad \omega = -2\ln\lambda \ ,$$

$$x^\mu \to \tilde{x}^\mu = \frac{x^\mu + b^\mu x^2}{1 + 2b \cdot x + b^2 x^2} \ , \quad \omega = 2\ln[1 + 2b \cdot x + b^2 x^2] \ . \tag{1.9}$$

It is important to realize that the d-dimensional conformal group is finite-dimensional for $d > 2$, and all its transformations are globally defined. It has been known for a long time that the conformal group is a symmetry of massless fields with dimensionless coupling constants, and that it is the maximal kinematical extension of relativistic invariance [5].

The case of two dimensions is special, as one can already see from eq. (1.7). When $g_{\mu\nu} = \delta_{\mu\nu}$ and $d = 2$, eq. (1.6) takes the form of the Cauchy-Riemann equation

$$\partial_1 \varepsilon_1 = \partial_2 \varepsilon_2 \ , \quad \partial_1 \varepsilon_2 = -\partial_2 \varepsilon_1 \ . \tag{1.10}$$

In terms of the complex coordinates and fields

$$z = x^1 + ix^2 \ , \quad \bar{z} = x^1 - ix^2 \ ,$$

$$\varepsilon^z(z, \bar{z}) = \varepsilon^1(z, \bar{z}) + i\varepsilon^2(z, \bar{z}) \ , \quad \bar{\varepsilon}^{\bar{z}}(z, \bar{z}) = \varepsilon^1(z, \bar{z}) - i\varepsilon^2(z, \bar{z}) \ , \tag{1.11}$$

resp., eq. (1.10) implies a *holomorphic* dependence: $\varepsilon^z = \varepsilon^z(z)$ and $\bar{\varepsilon}^{\bar{z}} = \bar{\varepsilon}^{\bar{z}}(\bar{z})$. Therefore, the 2d conformal transformations can be identified with the *analytic* coordinate transformations:

$$z \to f(z) \quad \bar{z} \to \bar{f}(\bar{z}) \ , \quad f'(z) \neq 0 \ . \tag{1.12}$$

---

[3]Given a flat d-dimensional space with an indefinite metric of signature (p, q), $p + q = d > 2$, the corresponding conformal algebra is isomorphic to $so(p + 1, q + 1)$ [5].

This can be seen in yet another way, when rewriting the line element in the complex coordinates on a plane as $ds^2 = dzd\bar{z}$. Under the holomorphic transformations (1.12), we get

$$dzd\bar{z} \rightarrow \left|\frac{\partial f}{\partial z}\right|^2 dzd\bar{z} \,, \tag{1.13}$$

which just means that we have a conformal transformation, in accordance to the definition (1.4), with $\omega(x) = 2\ln|\partial f/\partial z|$.

It is possible to define the infinitesimal analytic functions, parametrizing the conformal transformations, by restricting the plane to a finite region around the origin and assuming that all the singularities of the analytic functions are outside the region chosen. Since eq. (1.10) allows an arbitrary dependence on z, it is quite reasonable to assume this dependence to be *meromorphic* in general.

The suitable basis for the infinitesimal conformal transformations $z \rightarrow \tilde{z} = z + \epsilon(z)$, and $\bar{z} \rightarrow \tilde{\bar{z}} = \bar{z} + \bar{\epsilon}(\bar{z})$ is given by

$$z \rightarrow z - a_n z^{n+1} \,, \quad \bar{z} \rightarrow \bar{z} - \bar{a}_n \bar{z}^{n+1} \,, \quad n \in \mathbf{Z} \,, \tag{1.14}$$

which are generated by the operators

$$l_n = -z^{n+1}\frac{d}{dz} \,, \quad \bar{l}_n = -\bar{z}^{n+1}\frac{d}{dz} \,, \quad n \in \mathbf{Z} \,. \tag{1.15}$$

These operators satisfy an algebra which consists of the two commuting pieces,

$$[l_n, l_m] = (n-m)l_{n+m} \,, \quad [\bar{l}_n, \bar{l}_m] = (n-m)\bar{l}_{n+m} \,, \quad [l_n, \bar{l}_m] = 0 \,, \tag{1.16}$$

each one being known as the 2d local **conformal algebra**. The independence of the two algebras $\{l_n\}$ and $\{\bar{l}_n\}$ justifies the use of z and $\bar{z}$ as the independent coordinates. This literally means a *complexification* of the initial space: $\mathbf{C} \rightarrow \mathbf{C}^2$, which gives us a freedom to choose various reality conditions by making 'sections' in the complexified space. In particular, the section defined by $\bar{z} = z^*$ recovers a Euclidean plane.

A Minkowski plane can be recovered by the section $z^* = -z$, which implies

$$(z, \bar{z}) = i(\bar{\tau} + \bar{\sigma}, \bar{\tau} - \bar{\sigma}) \,, \quad ds^2 = -d\bar{\tau}^2 + d\bar{\sigma}^2 \,. \tag{1.17}$$

However, these are not the coordinates we want to associate with the 2d Minkowski space-time. First, we make the conformal transformation $z = e^\zeta$, $\bar{z} = e^{\bar{\zeta}}$, with $\zeta = \tau + i\sigma$, $\bar{\zeta} = \tau - i\sigma$, from the z-plane to a cylinder, $-\infty < \tau < +\infty$, $0 \leq \sigma \leq 2\pi$. By definition, the Minkowski space-time formulation of a field theory is obtained from its Euclidean formulation on the cylinder by the Wick rotation, $\zeta = \tau + i\sigma \to i(\tau + \sigma) \equiv i\zeta^+$, $\bar{\zeta} = \tau - i\sigma \to i(\tau - \sigma) \equiv i\zeta^-$, where the Minkowski light-cone coordinates $\zeta^\pm = \tau \pm \sigma$ have been introduced. In terms of the Minkowski *light-cone* coordinates, the line element takes the form $ds_{\rm M}^2 = d\zeta^+ d\zeta^-$, and the conformal transformations take the form of the reparametrizations of ζ^+ and $\zeta^- : \zeta'^+ = f(\zeta^+)$, $\zeta'^- = g(\zeta^-)$, which leave the light-cone invariant. The line element $ds_{\rm M}^2$ is clearly preserved by these transformations up to the scale factor $\Omega = f'(\zeta^+)g'(\zeta^-)$.

The relation between Euclidean and Minkowski space-time formulations of CFT is important, to understand the quantization in what folllows. We will frequently discuss the holomorphic dependence only, and ignore the similar anti-holomorphic dependence.

The number of generators in the 2d local conformal algebra (1.16) is *infinite*. Obviously, the infinite number of symmetries should imply severe restrictions on the conformally invariant field theories in two dimensions, and that's what our story is all about. It is also worthy to mention that the conformal transformations (1.16) we consider are neither globally well-defined nor invertible, even on the Riemann sphere $S^2 = \mathbf{C} \cup \infty$. This is because the vector fields

$$\mathcal{V}(z) = -\sum_n a_n l_n = \sum_n a_n z^{n+1} \frac{d}{dz} \, , \tag{1.18}$$

generating holomorphic transformations, are globally defined only if $a_n = 0$ for $n < -1$ *and* $n > 1$. This is necessary for the absence of singularities of $\mathcal{V}(z)$ at $z \to 0$ and $z \to \infty$ (use the conformal transformation $z = -1/w$ in the latter case!). Therefore, a global group of the well-defined and invertible conformal transformations on the Riemann sphere is generated by $\{l_{-1}, l_0, l_1\} \cup \{\bar{l}_{-1}, \bar{l}_0, \bar{l}_1\}$ only. Eq.(1.15) tells us that l_{-1} and $\bar{l}_{-1}$ can be identified with the generators of translations, $(l_0 + \bar{l}_0)$ and $i(l_0 - \bar{l}_0)$ — with the generators of dilatations and rotations, resp., and l_1 and $\bar{l}_1$ — with the generators of special conformal transformations. The corresponding finite transformations form a group known as the complex **Möbius** group :

$$z \to \tilde{z} = \frac{az + b}{cz + d} \, , \tag{1.19}$$

where a, b, c, d are complex parameters and $ad - bc = 1$. The group of transformations (1.19) parametrized by six real parameters is isomorphic to $SL(2, \mathbf{C})/\mathbf{Z}_2 \cong SO(3, 1)$. The need for the quotient $\mathbf{Z}_2$ is caused by the fact that the transformation (1.19) is not sensitive to a simultaneous change of sign for all parameters a, b, c, d. This group is sometimes referred to as the group of **projective** conformal transformations.

The lesson we learn from here is that the local conformal algebra in two dimensions cannot be fully integrated to a globally defined group, but it is the full local conformal symmetry (1.16) that will be the symmetry we want to investigate.

Representations of the global conformal algebra (after quantization) assign quantum numbers to physical states. It is quite natural to assume the existence of the *vacuum state* $|0\rangle$ among the physical states, which is invariant under the Möbius transformations and has vanishing quantum numbers. The eigenvalues h and $\bar{h}$ of the operators l_0 and $\bar{l}_0$, resp., are known as **conformal weights** of a state. Given the conformal weights of a state, its *scaling dimension* Δ and *spin s* are determined by $\Delta = h + \bar{h}$ and $s = h - \bar{h}$, in accordance to the similar assignment for the generators of dilatations and rotations.

Instead of the Riemann sphere, which is a *closed* Riemann surface of genus zero, the simplest *open* Riemann surface to be represented by an upper half plane (with the infinity attached) could equally be considered. In this case, the parameters a, b, c, d in eq. (1.19) have to be real, thus leading to the *real* Möbius group $SL(2, \mathbf{R})/\mathbf{Z}_2$. There is an obvious similarity between the 'open' case and a holomorphic part of the 'closed' case.

Exercises

I-1 ▷ Consider eqs. (1.6) and (1.7) and show that the third derivatives of $\varepsilon(x)$ must vanish when $d > 2$.

I-2 ▷ Calculate the number of generators of the conformal algebra in $d > 2$ dimensions and compare it with the dimension of the Lie algebra $so(d + 1, 1)$.

I-3 ▷ (*) Rewrite the two-dimensional conformal transformations (1.8) and (1.9) in terms of the $SL(2, \mathbf{C})$ matrices. What is the well-known group theory decomposition which is illustrated by the result?

I.2 Symmetries and currents

Let us go back to a general field theory defined in an arbitrary number $d \geq 2$ of Euclidean dimensions. The reparametrization symmetry (1.1), like any other symmetry of the field theory action S, implies the *Ward identity* for its correlation functions defined by eq. (1.3). This identity follows from the obvious equations:

$$\sum_{j=1}^{m}\left\langle \Phi_{\Delta_1}(x_1)\cdots\delta\Phi_{\Delta_j}(x_j)\cdots\Phi_{\Delta_M}(x_M)\right\rangle$$

$$=\mathcal{N}^{-1}\int\prod[\mathcal{D}\Phi_{\Delta}]\sum_{j=1}^{m}\Phi_{\Delta_1}(x_1)\cdots\delta\Phi_{\Delta_j}(x_j)\cdots\Phi_{\Delta_M}(x_M)e^{-S[\Phi_{\Delta}]}$$

$$=\tfrac{1}{2}\int d^2x\sqrt{g}\,\delta g_{m\nu}(x)\left\langle T^{\mu\nu}(x)\Phi_{\Delta_1}(x_1)\cdots\Phi_{\Delta_M}(x_M)\right\rangle\,, \tag{1.20}$$

where the *stress-energy tensor* $T^{\mu\nu}$ of the field theory has been introduced,

$$T^{\mu\nu}(x)=-\frac{2}{\sqrt{g}}\frac{\delta S}{\delta g_{\mu\nu}(x)}\,. \tag{1.21}$$

This tensor encodes the reaction of field theory to the metric deformations. In the flat space with the reference metric $g_{\mu\nu}=\delta_{\mu\nu}$, eqs. (1.1) and (1.20) imply

$$\sum_{j=1}^{m}\left\langle\Phi_{\Delta_1}(x_1)\cdots\delta\Phi_{\Delta_j}(x_j)\cdots\Phi_{\Delta_M}(x_M)\right\rangle$$

$$=\int d^2x\,\partial_\mu\varepsilon_\nu(x)\left\langle T^{\mu\nu}(x)\Phi_{\Delta_1}(x_1)\cdots\Phi_{\Delta_M}(x_M)\right\rangle\,, \tag{1.22}$$

which is just the reparametrizational Ward identity between the correlation functions we are looking for, since the stress-energy tensor itself is a composite operator in terms of the fundamental (basis) fields.

The Ward identity (1.22) is quite general, and it is valid for any reparametrization invariant field theory. It simply means that the stress-energy tensor components are the generators of the general coordinate transformations. In particular, in a flat space, the $T_{\mu\nu}$ themselves generate translations, whereas the $x_\mu T_{\lambda\rho}-x_\lambda T_{\mu\rho}$ are the generators of Euclidean rotations. These currents are both conserved, $\partial^\mu T_{\mu\nu}=0$. The latter follows from eq. (1.22) after integrating by parts and using the arbitrariness of the parameter, and it is clearly valid for all points on the plane except the distinguished ones (called *punctures*) $\{x_i\}$, where the field operators are inserted.

We actually want to study the conformal field theories having a larger symmetry then the Poincaré group. Let us consider first the consequences of the global scaling invariance $x^\mu\rightarrow\lambda x^\mu$, which is a part of the global conformal group. The associated current is given by

$$j_\mu=x^\nu T_{\nu\mu}\,. \tag{1.23}$$

The conservation of the current (1.23) is equivalent to a tracelessness condition for the stress-energy tensor, $T^\mu{}_\mu = 0$. This restriction is known as the **scale invariance** condition. It implies, in particular, the vanishing of any correlator between Φ_Δ's and $T^\mu{}_\mu$. It is easy to see that any current of the form $f^\lambda(x)T_{\lambda\nu}$ will also be conserved provided

$$\partial^\mu f^\nu + \partial^\nu f^\mu - \varphi(x)\eta^{\mu\nu} = 0 \,, \tag{1.24}$$

where $\varphi(x)$ is an arbitrary function. In case of the special conformal transformations, the associated current reads $f_\mu{}^\lambda(x)T_{\lambda\nu} = x_\mu x^\lambda T_{\lambda\nu} - x^2 T_{\mu\nu}$. Therefore, the global scale invariance *implies* global conformal invariance in any *local* field theory.

To get more physical insights into the nature of scale and conformal invariances, it is useful to remember what is known in general about the scaling theory of the second-order phase transitions in statistical mechanics, where this symmetry has been studied for a long time (see, e.g., refs. [6, 7, 8, 9, 10]). A scale-invariant statistical system near a critical point can be described by a set of field operators $\{\Phi_i\}$ transforming under the scale transformations as

$$\Phi_i(x) \rightarrow \lambda^{\Delta_i}\Phi_i(\lambda x) \,, \tag{1.25}$$

where Δ_i are their scaling (*anomalous*) dimensions. Being generically different from the canonical dimensions of free (non-interacting) fields, these dimensions are indeed 'anomalous' since the fields are supposed to be strongly interacting at criticality. There exist various techniques to calculate the *critical indices*, which are just powers of singularities in the physical quantities like free energy or magnetization near criticality [6, 7, 8]. All the critical indices are determined from the anomalous dimensions by using the standard scaling relations [6, 7, 8]. Many statistical systems in *two* dimensions can be exactly solved [6, 9], so that the spectra of basic fields and their anomalous dimensions may all be known. Therefore, in the framework of CFT, it is quite reasonable to make one more technical (and, in fact, also physical) assumption that it is always possible to select among the fundamental (basic) fields a subset of fields transforming under the global conformal transformations as

$$\Phi_i(x) \rightarrow \left|\frac{\partial\tilde{x}}{\partial x}\right|^{\Delta_j/d} \Phi_i(\tilde{x}) \,, \tag{1.26}$$

which generalizes eq. (1.25) to the whole global conformal group. Such fields are called '*quasi-primary*'. In statistical systems, the 'quasi-primary' fields correspond to the 'observable' operators like energy, spin or any other order parameter. The rest of the fields is assumed to be linear combinations of the 'quasi-primary' fields and their 'derivatives' in a generalized sense to be specified in the next section. The number of 'quasi-primary' fields in CFT can be infinite.

It was the original idea due to Polyakov [11] that the critical statistical systems should also be invariant under the *local* scale transformations with $\lambda = \lambda(x)$, which would allow us to exploit the full conformal symmetry. Since the conformal symmetry is finite-dimensional for more than two dimensions, only a limited numer of constraints for correlation functions can be obtained this way (see below). The situation is drastically changed in two dimensions, where the local conformal algebra is infinite-dimensional. From now on, we restrict ourselves to the case of *two* dimensions only. A fundamental contribution to modern CFT was due to Belavin, Polyakov and Zamolodchikov in their seminal paper [12]. Many reviews are now available (see, e.g., [13, 14, 15, 16, 17, 18, 19, 20, 21]).

On a Euclidean plane parametrized by the complex coordinates $z, \bar{z}$ with the line element $ds^2 = dzd\bar{z}$, the conservation of the stress-energy tensor $\partial^\mu T_{\mu\nu} = 0$ takes the form

$$\partial_{\bar{z}} T_{zz} + \partial_z T_{\bar{z}z} = 0 \,, \quad \partial_z T_{\bar{z}\bar{z}} + \partial_{\bar{z}} T_{z\bar{z}} = 0 \,, \tag{1.27}$$

while the scale invariance condition $T^\mu{}_\mu = 0$ yields

$$T_{z\bar{z}} = T_{\bar{z}z} = 0 \,, \tag{1.28}$$

where the corresponding metric and stress-energy tensor components have been introduced,

$$g_{z\bar{z}} = g_{\bar{z}z} = \tfrac{1}{2} \,, \quad g_{zz} = g_{\bar{z}\bar{z}} = 0 \,,$$

$$T_{zz} = \tfrac{1}{4}(T_{11} + 2iT_{12} - T_{22}) \,, \quad T_{\bar{z}\bar{z}} = \tfrac{1}{4}(T_{11} - 2iT_{12} - T_{22}) \,,$$

$$T_{z\bar{z}} = T_{\bar{z}z} = \tfrac{1}{4}(T_{11} + T_{22}) \equiv \tfrac{1}{4} T^\mu{}_\mu \,. \tag{1.29}$$

Therefore, the stress-energy tensor in any 2d CFT can be split into a holomorphic part and an anti-holomorphic part [4]

$$T_{zz} \equiv T(z) \,, \quad T_{\bar{z}\bar{z}} \equiv \bar{T}(\bar{z}) \,. \tag{1.30}$$

The holomorphic and anti-holomorphic parts of a field in Euclidean space are related by complex conjugation, whereas in Minkowski space they correspond to the *left-moving*

---

[4]In what follows, we call a 'stress-energy tensor' as a *'stress tensor'* for brevity.

(LM) and *right-moving* (RM) modes, resp., which are truly independent. The relation between the two formulations is provided via the complexification discussed in the previous section, which allows us to consider z and $\bar{z}$ as the independent variables, in particular. The Ward identity (1.22) with the independent meromorphic parameters $\varepsilon^z(z)$ and $\bar{\varepsilon}^{\bar{z}}(\bar{z})$ can now be rewritten to the form

$$\sum_{j=1}^{m} \left\langle \Phi_{\Delta_1}(z_1, \bar{z}_1) \cdots \delta\Phi_{\Delta_j}(z_j, \bar{z}_j) \cdots \Phi_{\Delta_M}(z_M, \bar{z}_M) \right\rangle$$

$$= \int d^2z \, \partial_{\bar{z}}\varepsilon^z(z, \bar{z}) \left\langle T_{zz}(z)\Phi_{\Delta_1}(z_1, \bar{z}_1) \cdots \Phi_{\Delta_M}(z_M, \bar{z}_M) \right\rangle , \qquad (1.31)$$

where only the variations w.r.t. the holomorphic coordinates were considered. A quite similar formula holds for the variations of the anti-holomorphic coordinates. We thus effectively reduced the 2d dependence of CFT to a 1-dimensional dependence: $(z, \bar{z}) \to z$. This observation is very crucial for the whole integrability of CFT, as we shall see all the time throughout the book.

The generators of infinitesimal conformal transformations can be defined in terms of $T(z)$ as

$$L_n = \oint \frac{dz}{2\pi i} \, z^{n+1} T(z) , \qquad (1.32)$$

with the contour encircling the origin. The formal operatorial equation (1.32) makes actual sense when acting on fields whose arguments are inside the integration contour. The contour shape is irrelevant because of Cauchy's theorem. The same theorem yields the statement

$$T(z) = \sum_{n \in \mathbf{Z}} L_n z^{-n-2} , \qquad (1.33)$$

and similarly for the $\bar{L}_n$ and $\bar{T}$. We now have all the power of complex calculus at our disposal.

To incorporate the standard machinery of canonical quantization into CFT, it is convenient to use the *radial quantization* techniques [22, 23, 24] on a plane. It uses the following parameterization of $\mathbf{C}^2$:

$$z = e^{\zeta} , \quad \zeta = \tau + i\sigma , \qquad (1.34)$$

in terms of the Euclidean 'world-sheet' time and space coordinates, $\tau \in \mathbf{R}$ and $0 < \sigma \leq 2\pi$, resp. [5]

---

[5]Eq. (1.34) is just the definition of the τ and σ used here. In string theory (Ch. VIII), they can be interpreted as the coordinates of the Euclidean closed string world-sheet. This gives the useful vizualization of quantization procedure.

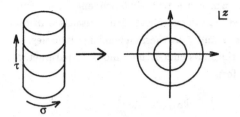

Fig. 1. The conformal map of a cylinder to a plane.

We interpret eq. (1.34) as the conformal map of a cylinder to a plane (Fig. 1). Infinite past and future, $\tau = \mp\infty$, on the cylinder are mapped into the points $z = 0, \infty$, resp., on the plane. Equal-time slices are the circles of fixed radius on the plane, whereas equal-space slices are the lines radiating from the origin. The *time translations* $\tau \to \tau + \lambda$ are the *dilatations* on $\mathbf{C}$: $z \to e^\lambda z = z + \lambda z + \ldots$, whereas the *space translations* $\sigma \to \sigma + \theta$ are the *rotations* on $\mathbf{C}$: $z \to e^{i\theta} z$. Therefore, the Hamiltonian of the system can be identified with the dilatation generator on the plane, while the Hilbert space of states comprises surfaces of constant radius. The stress-tensor components $T(z)$ and $\bar{T}(\bar{z})$ in eq. (1.30) are identified with the generators of local conformal transformations on the z-plane. In the radial quantization, an 'equal-time' surface becomes a contour on the z-plane surrounding the origin.

To convert these heuristic arguments into precise statements, consider the time evolution of any operator Φ in the Heisenberg picture with a Hamiltonian H. The dynamics is described by the operator equation of motion

$$\frac{d\Phi}{d\tau} = [H, \Phi] \;, \tag{1.35}$$

where the square brackets mean an *equal-time* commutator. Equivalently, eq. (1.35) can be rewritten to the infinitesimal form

$$\delta_\lambda \Phi = [\lambda H, \Phi] \;, \quad \text{where} \quad H = \oint \frac{dz}{2\pi i} \, z T(z) = L_0 \;, \tag{1.36}$$

since H is just the charge of the scaling current $zT(z)$ associated with the dilatational transformation and the parameter $\varepsilon(z) = \lambda z$. In the case of a general conformal transformation $z \to z + \varepsilon(z)$, one should take

$$T_\varepsilon = \oint \frac{dz}{2\pi i} \, \varepsilon(z) T(z) \tag{1.37}$$

as the conserved charge in the equation of motion $\delta_\varepsilon \Phi = [T_\varepsilon, \Phi]$. Eq. (1.37) is formal until we specify the other fields inside the integration contour. Including the analogous anti-holomorphic term, the general statement is

$$\delta_{\varepsilon,\bar{\varepsilon}}\Phi(w,\bar{w}) = \frac{1}{2\pi i} \oint \left[dz\, T(z)\varepsilon(z), \Phi(w,\bar{w}) \right] + \left[d\bar{z}\, \bar{T}(\bar{z})\bar{\varepsilon}(\bar{z}), \Phi(w,\bar{w}) \right] . \qquad (1.38)$$

This equation is still formal and ill-defined because of the operator product ambiguities, whose presence is usual in QFT. The operator ordering will be addressed in the next section.

As was noticed above, among the CFT fields one can distinguish the 'quasi-primary' fields transforming like tensors under the conformal transformations. If the form

$$\Phi \equiv \Phi_{h,\bar{h}}(z,\bar{z})dz^h d\bar{z}^{\bar{h}} \qquad (1.39)$$

is conformally invariant in two dimensions, the field $\Phi_{h,\bar{h}}(z,\bar{z})$ is called a **primary** field. The primary field transforms as

$$\tilde{\Phi}(\tilde{z},\tilde{\bar{z}}) = \Phi(z,\bar{z}) \left(\frac{dz}{d\tilde{z}}\right)^h \left(\frac{d\bar{z}}{d\tilde{\bar{z}}}\right)^{\bar{h}} , \qquad (1.40)$$

where the conformal weight $(h,\bar{h})$ is real-valued. The rest of CFT fields are called **secondary** fields. The infinitesimal form of eq. (1.40) is

$$\delta_{\varepsilon,\bar{\varepsilon}}\Phi(z,\bar{z}) = \left[(h\partial\varepsilon + \varepsilon\partial) + (\bar{h}\bar{\partial}\bar{\varepsilon} + \bar{\varepsilon}\bar{\partial}) \right] \Phi(\varepsilon,\bar{\varepsilon}) , \qquad (1.41)$$

where the shortened notation has been introduced,

$$\partial \equiv \partial_z , \qquad \bar{\partial} \equiv \partial_{\bar{z}} . \qquad (1.42)$$

Eq. (1.40) is useful to relate mode expansions of a primary field on the cylinder and on the plane. Taking, for example, a holomorphic field with $\bar{h} = 0$, its Fourier expansion on the cylinder,

$$\Phi_h(\zeta) = \sum_{n\in\mathbf{Z}} \phi_n e^{-n\zeta} , \quad \zeta = \tau + i\sigma , \qquad (1.43)$$

can be transformed into an expansion on the plane,

$$\Phi_h(z) = \sum_{n\in\mathbf{Z}} \phi_n z^{-n-h} . \qquad (1.44)$$

Eq. (1.33) is now recognized as the particular case of eq. (1.44). This means that the stress tensor has conformal dimension $h = 2$.

Let us now consider the constraints imposed by the conformal invariance on the correlation functions of the primary fields in a quantized 2d field theory. This means that its correlation functions do not change under the transformations (1.41). The two-point function $G^{(2)}(z_i, \bar{z}_i) \equiv \langle \Phi_1(z_1, \bar{z}_1)\Phi_2(z_2, \bar{z}_2)\rangle$ provides the simplest example, where an invariance under conformal transformations leads to a partial differential equation of the form

$$[(\varepsilon(z_1)\partial_{z_1} + h_1\partial\varepsilon(z_1)) + (\varepsilon(z_2)\partial_{z_2} + h_2\partial\varepsilon(z_2))$$

$$+(\bar{\varepsilon}(\bar{z}_1)\partial_{\bar{z}_1} + \bar{h}_1\bar{\partial}\bar{\varepsilon}(\bar{z}_1)) + (\bar{\varepsilon}(\bar{z}_2)\partial_{\bar{z}_2} + \bar{h}_2\bar{\partial}\bar{\varepsilon}(\bar{z}_2))]\, G^{(2)}(z_i, \bar{z}_i) = 0 \, . \tag{1.45}$$

Substituting $\varepsilon(z) = \bar{\varepsilon}(\bar{z}) = 1$ implies that $G^{(2)}$ depends on $z_{12} = z_1 - z_2$ and $\bar{z}_{12} = \bar{z}_1 - \bar{z}_2$; substituting $\varepsilon(z) = z$ and $\bar{\varepsilon}(\bar{z}) = \bar{z}$ leads to $G^{(2)} = C_{12}/(z_{12}^{h_1+h_2}\bar{z}_{12}^{\bar{h}_1+\bar{h}_2})$; finally substituting $\varepsilon(z) = z^2$ and $\bar{\varepsilon}(\bar{z}) = \bar{z}^2$ gives $h_1 = h_2 = h$, $\bar{h}_1 = \bar{h}_2 = \bar{h}$ (otherwise $G^{(2)} = 0$). Putting all together determines the form of the two-point function as

$$G^{(2)}(z_i, \bar{z}_i) = \frac{C_{12}}{z_{12}^{2h}\bar{z}_{12}^{2\bar{h}}} \, , \tag{1.46}$$

where the C_{12} is a constant to be fixed by normalizations of the fields.

The 3-point function $G^{(3)} \equiv\, < \Phi_1\Phi_2\Phi_3 >$ is determined along the similar lines,

$$G^{(3)}(z_i, \bar{z}_i) = C_{123}\frac{1}{z_{12}^{h_1+h_2-h_3} z_{23}^{h_2+h_3-h_1} z_{13}^{h_3+h_1-h_2}}$$

$$\times \frac{1}{\bar{z}_{12}^{\bar{h}_1+\bar{h}_2-\bar{h}_3} \bar{z}_{23}^{\bar{h}_2+\bar{h}_3-\bar{h}_1} \bar{z}_{13}^{\bar{h}_3+\bar{h}_1-\bar{h}_2}} \, , \tag{1.47}$$

where $z_{ij} = z_i - z_j$, and C_{123} is a constant. Hence, we can conclude that the coordinate dependence of the 2- and 3-point functions are determined by conformal invariance. The geometrical reason why it happens is because any three points z_1, z_2, z_3 can be mapped by a complex Möbius transformation into any three reference points, say $\infty, 1, 0$, where

$$\lim_{z_1\to\infty} z_1^{2h_1} \bar{z}_1^{2\bar{h}_1} G^{(3)} = C_{123} \, . \tag{1.48}$$

The coefficients like C_{123} carry a nontrivial dynamical information about the operator algebra of a given CFT. Knowledge of these coefficients is equivalent to solving CFT. The equations satisfied by these coefficients are discussed in Ch. II.

Things become more complicated for the n-point correlation functions with $n \geq 4$. The *global* conformal invariance restricts the 4-point function of the primary fields to the form [17]

$$G^{(4)}(z_i, \bar{z}_i) = f(x, \bar{x}) \prod_{i<j} z_{ij}^{-(h_i+h_j)+h/3} \prod_{i<j} \bar{z}_{ij}^{-(\bar{h}_i+\bar{h}_j)+\bar{h}/3} \, , \tag{1.49}$$

where $h = \sum_{i=1}^{4} h_i$, $\bar{h} = \sum_{i=1}^{4} \bar{h}_i$, $x = z_{12} z_{34} / z_{13} z_{24}$, and f is an arbitrary function. However, this is not a whole story. It will be shown in the next Chapter how this 4-point function, and, in fact, all the other correlation functions could, in principle, be fixed by the full conformal invariance.

Exercises

I-4 ▷ Derive eq. (1.24) and find the explicit form of functions $f^\lambda(x)$ and $\varphi(x)$ for the case of the special conformal transformations.

I-5 ▷ Calculate the two-point correlation function of quasi-primary fields in d dimensions from conformal invariance. In $d > 2$ dimensions, the higher-point correlation functions cannot be fixed by the conformal symmetry alone.

I.3 Operator product expansion

To be well-defined, eq. (1.38) requires a resolution of the *operator ordering ambiguity*. The problem arises in the radial quantization because it uses the 'equal-time' commutators at $|z| = |w|$. In general, an operatorial product $A(z)B(w)$ on a Euclidean plane is defined only for $|z| > |w|$, since this is equivalent to the time-ordering in QFT. Therefore, we can proceed in the usual way known there to resolve the operator ordering problem, namely, by using *Schwinger's time-splitting* technique [1]. The latter implies the following definition: [6]

$$[T_\varepsilon, \Phi(w, \bar{w})] \overset{\text{def}}{=} \lim_{|z| \to |w|} \left\{ \oint_{|z| > |w|} \frac{dz}{2\pi i} \, \varepsilon(z) \mathcal{R}\left(T(z)\Phi(w, \bar{w})\right) \right.$$
$$\left. - \oint_{|z| < |w|} \frac{dz}{2\pi i} \, \varepsilon(z) \mathcal{R}\left(T(z)\Phi(w, \bar{w})\right) \right\} , \tag{1.50}$$

where the *radial ordering operator* $\mathcal{R}$ has been introduced,

$$\mathcal{R}\left(A(z)B(w)\right) = \begin{cases} +A(z)B(w), & \text{if } |z| > |w| , \\ \pm B(w)A(z), & \text{if } |z| < |w| . \end{cases} \tag{1.51}$$

The sign factor in the last equation counts the statistics: it should be plus in all cases, except the one when both operators are fermionic.

We can now deform the integration contours as depictured in Fig. 2, to obtain [7]

---

[6] The anti-holomorphic part is assumed to be added everywhere, when it is necessary.

[7] We normally drop the $\mathcal{R}$-symbol and always assume that the Euclidean correlators are *radially-ordered* in what follows.

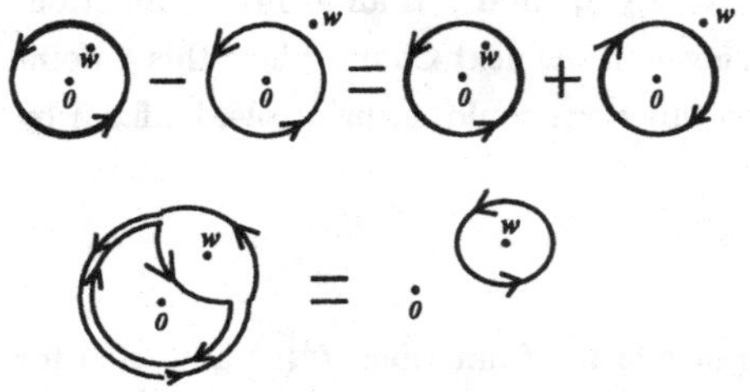

Fig. 2. For the evaluation of an equal-time
commutator on a conformal plane.

$$[T_\varepsilon, \Phi(w, \bar{w})] = \lim_{|z| \to |w|} \oint \frac{dz}{2\pi i} \varepsilon(z) T(z) \Phi(w, \bar{w}) , \qquad (1.52)$$

where the contour encircles the point w in the z-plane. This integral is clearly non-vanishing only if there is a *singularity* in the operator product

$$\lim_{z \to w} T(z) \Phi(w, \bar{w}) , \qquad (1.53)$$

whose residue contributes to the r.h.s. of eq. (1.52) by Cauchy's theorem.

In general, when two operators $A(z)$ and $B(w)$ approach each other, Wilson's **operator product expansion** (OPE) [25, 26] takes the form

$$A(z)B(w) \sim \sum_\Delta C_\Delta(z - w) O_\Delta(w) , \qquad (1.54)$$

where the $\{O_\Delta(w)\}$ are a *complete* set of local operators, and the C_Δ's are (singular) numerical coefficients. Equations like eq. (1.54) have to be understood as being valid when the product $A(z)B(w)$ is inserted into a Green's function with other elementary operators of the theory, i.e.

$$\lim_{z \to w} \left\langle \mathcal{R} \left\{ A(z)B(w) - \sum_\Delta C_\Delta(z - w) O_\Delta(w) \right\} \Phi_1(w_1) \cdots \Phi_M(w_M) \right\rangle = 0 . \qquad (1.55)$$

In CFT one can use eqs. (1.38) and (1.41) to introduce the OPE between the $T(z)$ and a *primary* field $\Phi(w, \bar{w})$ in the form

$$T(z)\Phi(w, \bar{w}) = \frac{h}{(z - w)^2} \Phi(w, \bar{w}) + \frac{1}{z - w} \partial_w \Phi(w, \bar{w})$$

$$+ \Phi^{(-2)}(w, \bar{w}) + (z - w)\Phi^{(-3)}(w, \bar{w}) + \dots , \qquad (1.56)$$

where the dots represent an infinite set of other *regular* terms depending on the new local fields called the **descendants** or the **secondary fields** w.r.t. a primary field Φ of conformal dimension h. The secondary fields are determined from the above equation (1.56) by the contour integration:

$$\Phi^{(-n)}(w, \bar{w}) = \hat{L}_{-n}(w)\Phi(w, \bar{w}) \equiv \oint_w \frac{dz}{2\pi i} (z-w)^{-n+1}T(z)\Phi(w, \bar{w}) , \qquad (1.57)$$

where the associated operators $\hat{L}_n(w)$ have been introduced. These operators appear in the formal expansion of the stress tensor $T(z)$ around the point w:

$$T(z) = \sum_{n \in \mathbf{Z}} \frac{\hat{L}_n(w)}{(z-w)^{n+2}} . \qquad (1.58)$$

One has, in particular, [8]

$$\hat{L}_0 \Phi(z, \bar{z}) = h\Phi(z, \bar{z}) , \qquad \hat{L}_{-1}\Phi(z, \bar{z}) = \partial_z\Phi(z, \bar{z}) ;$$

$$\hat{L}_n \Phi(z, \bar{z}) = 0 , \qquad n \geq 1 , \qquad (1.59)$$

where $\Phi^{(-n)} = \hat{L}_{-n}\Phi$ are the new descendant fields.

Eq. (1.59) represents only the first generation of the secondary fields resulting from the expansion of a product of one primary field Φ_h with one stress tensor T. We can also consider products of other CFT primaries with several stress tensors, $T \cdots T\Phi_h$. It is a way to obtain the complete spectrum of CFT. In accordance with the previous section, we assume that primary fields (or just *primaries*) in any 2d CFT form a complete set of local operators in the *algebraic* sense. Namely, the rest of CFT fields are assumed to be linear combinations of the primary *and* secondary fields, the latter being the 'derivatives' of the primaries in the generalized sense.

There exists an infinite *conformal family* of the secondary fields associated with a given primary field and generated by the repeated use of the $\hat{L}_{-n}$-operators:

$$\Phi_h^{\{-\vec{k}\}}(z) = \hat{L}_{-n_1}\hat{L}_{-n_2} \cdots \hat{L}_{-n_k}\Phi_h(z) , \qquad \vec{k} = (n_1, n_2, \ldots, n_k) . \qquad (1.60)$$

They are all naturally classified by their $\hat{L}_0$ quantum number,

$$\hat{L}_0 \Phi_h^{\{-\vec{k}\}}(z) = (h + |\vec{k}|)\Phi_h^{\{-\vec{k}\}}(z) , \qquad |\vec{k}| = \sum_{j=1}^k n_j . \qquad (1.61)$$

---

[8]The argument of the $\hat{L}_n$-operator is always the same as that of the field it acts on. That's why it is safe to omit the argument of this operator, while keeping explicit the argument of the field.

It is important to notice that the conformal transformation properties of the descendants are *different* from those of the primaries. In particular, the descendants are not even conformally covariant operators, since they do not satisfy the condition

$$\hat{L}_n \Phi = 0 \ \text{ for } \ n > 0 \ . \tag{1.62}$$

In what follows, we adopt the shortened notation for the OPE's, *viz.*

$$T(z)\Phi(w) \sim \frac{h}{(z-w)^2}\Phi(w) + \frac{1}{z-w}\partial_w\Phi(w) \ , $$

$$\bar{T}(\bar{z})\Phi(\bar{w}) \sim \frac{\bar{h}}{(\bar{z}-\bar{w})^2}\Phi(\bar{w}) + \frac{1}{\bar{z}-\bar{w}}\partial_{\bar{w}}\Phi(\bar{w}) \ , \tag{1.63}$$

keeping only singular terms on the r.h.s., and displaying, as a rule, only the holomorphic dependence. Only these terms are relevant for the correlation functions of the primary fields with the stress tensor. All the operatorial relations we are going to consider, and the OPE's in particular, should be understood as being inserted inside correlation functions. The latter carry all physical information about CFT, and they are therefore the only ones we need to know.

The primary fields $\{\Phi_i\}$ in CFT are conveniently normalized by taking their 2-point correlation functions to be of the form

$$\langle \Phi_i(z,\bar{z})\Phi_j(w,\bar{w}) \rangle = \delta_{ij}\frac{1}{(z-w)^{2h_i}}\frac{1}{(\bar{z}-\bar{w})^{2\bar{h}_i}} \ . \tag{1.64}$$

The OPE for a product of two primary fields takes the form

$$\Phi_i(z,\bar{z})\Phi_j(w,\bar{w}) \sim \sum_k C_{ijk}(z-w)^{h_k-h_i-h_j}(\bar{z}-\bar{w})^{\bar{h}_k-\bar{h}_i-\bar{h}_j}\Phi_k(w,\bar{w}) \ , \tag{1.65}$$

where the symmetric OPE coefficients C_{ijk} can be determined by taking the coincidence limit of any two z's in the 3-point function $\langle \Phi_i\Phi_j\Phi_k \rangle$. These coefficients coincide with the numerical factors in the 3-point functions (1.47).

The OPE's (1.63) and (1.65) contain all information about dynamics and transformation properties of the primaries in CFT, and can actually substitute for a CFT Lagrangian, which does not play a significant role in CFT. In part, eq. (1.63) can serve as the definition of a primary field. In terms of the modes defined by eqs. (1.32) and (1.44), the OPE (1.63) is equivalent to the commutator

$$[L_m, \phi_n] = (hm - m - n)\phi_{n+m} \ . \tag{1.66}$$

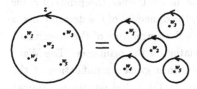

Fig. 3. The deformation of the contour in eq. (1.67).

In practice, one can use normal ordering of the operators for computing their OPE.

Eq. (1.63) can be used to calculate the correlator $\langle T(z)\Phi_1(w_1)\cdots\Phi_M(w_M)\rangle$ by using the analyticity properties of $T(z)$ as a meromorphic quadratic differential ($h = 2$) on a plane. Integrating $\varepsilon(z)T(z)$ around the z-contour as pictured in Fig. 3, and deforming the contour to a sum of the small contours encircling each of the points w_i , one finds

$$\left\langle \oint_0 \frac{dz}{2\pi i}\, \varepsilon(z)T(z)\Phi_1(w_1)\cdots\Phi_M(w_M)\right\rangle$$

$$= \sum_{j=1}^{m} \left\langle \Phi_1(w_1)\cdots\left(\oint_{w_j}\frac{dz}{2\pi i}\,\varepsilon(z)T(z)\Phi_j(w_j)\right)\cdots\Phi_M(w_M)\right\rangle$$

$$= \sum_{j=1}^{m} \langle \Phi_1(w_1)\cdots\delta_\varepsilon\Phi_j(w_j)\cdots\Phi_M(w_M)\rangle \ . \tag{1.67}$$

The known transformation law of a primary field,

$$\delta_\varepsilon\Phi(w) = \oint_w \frac{dz}{2\pi i}\,\varepsilon(z)T(z)\Phi(w) = [\varepsilon(w)\partial + h\partial\varepsilon(w)]\Phi(w) \ , \tag{1.68}$$

has to be substituted into the r.h.s. of eq. (1.67). Since it is still valid for an arbitrary parameter $\varepsilon(z)$ and an arbitrary contour in $\oint dz\, T(z)$, one gets the following **conformal Ward identity**:

$$\langle T(z)\Phi_1(w_1)\ldots\Phi_M(w_M)\rangle$$

$$= \sum_{j=1}^{M} \left[\frac{h_j}{(z-w_j)^2} + \frac{1}{z-w_j}\frac{\partial}{\partial w_j}\right] \langle \Phi_1(w_1)\ldots\Phi_M(w_M)\rangle \ . \tag{1.69}$$

Eq. (1.69) shows that the correlators are also meromorphic functions of z with singularities at the positions of inserted operators, the residues being determined by the conformal transformation laws of the operators.

Eq. (1.69) gives the *local* version of the conformal Ward identity, which is the fundamental relation in CFT and string theory [28]. The OPE's for products of the stress tensor with primary fields are equivalent to the Ward identitites in CFT.

The descendant or secondary fields are important in CFT since the primary fields alone do *not* form a basis in the field space. However, the dynamics of the descendants is completely determined by that of the primaries, since the correlators of the descendant fields can always be expressed in terms of the correlators of the primaries. The Ward identity (1.69) plays the crucial role here. It results in a great simplification of the structure of CFT, when compared to that of a general 2d QFT. Indeed, the correlators with only one descendent field can be treated as follows:

$$\Big\langle \Phi_1(w_1) \cdots \hat{L}_{-n} \Phi_j(w_j) \cdots \Phi_M(w_M) \Big\rangle$$

$$= \oint_{w_j} \frac{dz}{2\pi i} \, (z - w_j)^{-n+1} \, \langle T(z) \Phi_1(w_1) \ldots \Phi_j(w_j) \ldots \Phi_M(w_M) \rangle$$

$$= \oint_{w_j} \frac{dz}{2\pi i} \, (z - w_j)^{-n+1} \left[\frac{h_j}{(z - w_j)^2} + \frac{1}{z - w_j} \frac{\partial}{\partial w_j} \right] \langle \Phi_1(w_1) \ldots \Phi_M(w_M) \rangle \ . \qquad (1.70)$$

The correlators with several descendants are similarly reducible to the ones just considered, after simplifying a product of the stress tensors by using their (Virasoro) algebra, which will be the subject of the next section.

Exercises

I-6 ▷ (*) Derive the conformal Ward identity (1.69) from the reparametrizational Ward identity (1.31) and the primary field transformation law (1.41) in two dimensions. *Hint*: use the substitution $\varepsilon^z(z) = \frac{1}{\pi(z-w)}$ and the identity

$$\partial_{\bar{z}} \frac{1}{z - w} = \pi \delta^{(2)}(z - w) \ . \qquad (1.71)$$

I-7 ▷ Prove the identity (1.71) assuming that the delta-function $\delta^{(2)}(z - w)$ is normalized by the condition

$$\int \delta^{(2)}(z - w) f(w) = f(z) \ , \quad \int \equiv 2i \int dz \wedge d\bar{z} \ , \qquad (1.72)$$

valid for any regular function $f(z)$.

I-8 ▷ Check the violation of the conformal covariance condition (1.62) for the descendant fields.

I.4 Central charge and Virasoro algebra

By dimensional reasons and analyticity, the OPE for a product of stress tensor $T(z)$ with itself has the general form

$$T(z)T(w) = \frac{c/2}{(z-w)^4} + \frac{2}{(z-w)^2}T(w) + \frac{1}{z-w}\partial_w T(w)$$

$$+\hat{L}_{-2}T(w) + (z-w)\hat{L}_{-3}T(w) + \dots , \qquad (1.73)$$

where the constant c has been introduced. The simplest way to isolate this constant is to consider the 2-point correlation function of T's,

$$\langle T(z)T(0)\rangle = \frac{c/2}{z^4} . \qquad (1.74)$$

Eq. (1.74) can equally be justified by eq. (1.46), which is a consequence of the *global* conformal invariance. The constant c depends on which CFT the T is computed for, and is called the **central charge** or the **conformal anomaly**. Eq. (1.73) implies

$$T(z)T(w) \sim \frac{c/2}{(z-w)^4} + \frac{2}{(z-w)^2}T(w) + \frac{1}{z-w}\partial_w T(w) . \qquad (1.75)$$

This equation means that the $T(z)$ is an 'almost' primary field of conformal dimension $h = 2$. This is in fact violated by the 'anomalous' term. In other words, $T(z)$ is not a primary field, unless $c = 0$. It is not even holomorphic at the origin (exercise # I-9). Classically, one always has $c = 0$ since $T(z)$ is a tensor. Therefore, the non-vanishing central charge is a purely quantum effect.

An anomalous QFT always has internal inconsistencies, since the anomaly violates some of the basis symmetries used to build up the theory. At the same time, symmetry violations in CFT due to a non-vanishing central charge, with the anomaly 'sitting' at only one singular point (namely, at the origin) for the stress tensor, can normally be isolated. The art of dealing with CFT is, to some extent, just the art of dealing with the anomalies.

The additional and, in principle, independent constant $\bar{c}$ appears in the similar anti-holomorphic counterparts to eqs. (1.74) and (1.75). The difference $(c - \bar{c})$ is known as the local *gravitational anomaly*.

The equivalent useful representation of the OPE (1.75) is given by the commutation relations between the operatorial modes $\hat{L}_n$ defined by the Laurent expansion of $T(z)$ in

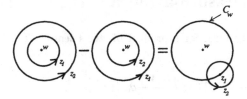

Fig. 4. For the calculation of eq. (1.77).

eq. (1.58). It follows that their action on the operator $\Phi(w)$ takes the form (cf eq. (1.57))

$$\hat{L}_m \Phi(w) = \oint \frac{dz}{2\pi i} (z - w)^{m+1} T(z) \Phi(w) . \tag{1.76}$$

The successive action of the second operator $\hat{L}_n$ to this equation yields

$$\hat{L}_n \hat{L}_m \Phi(w) = \oint_{C_2} \frac{dz_2}{2\pi i} \oint_{C_1} \frac{dz_1}{2\pi i} (z_2 - w)^{n+1}(z_1 - w)^{m+1} T(z_2) T(z_1) \Phi(w) . \tag{1.77a}$$

Similarly, one gets

$$\hat{L}_m \hat{L}_n \Phi(w) = \oint_{C_1} \frac{dz_1}{2\pi i} \oint_{C_2} \frac{dz_2}{2\pi i} (z_1 - w)^{m+1}(z_2 - w)^{n+1} T(z_1) T(z_2) \Phi(w) , \tag{1.77b}$$

The difference between the two equations is

$$[\hat{L}_n, \hat{L}_m]\Phi(w) = \oint_{C_w} \frac{dz_1}{2\pi i} (z_1 - w)^{m+1} \oint_{C_{z_1}} \frac{dz_2}{2\pi i} (z_2 - w)^{n+1} T(z_2) T(z_1) \Phi(w) , \tag{1.77c}$$

where the contours of integration are shown in Fig. 4.

Subsituting eq. (1.75) into eq. (1.77c) allows us to take the z_2-integral, since only the singular contributions from the product of the two T's are relevant there. One finds

$$[\hat{L}_n, \hat{L}_m]\Phi(w) = \oint_{C_w} \frac{dz_1}{2\pi i} (z_1 - w)^{m+1} \frac{c}{12} n(n^2 - 1)(z_1 - w)^{n-2}\Phi(w)$$

$$+ (n - m) \oint_{C_w} \frac{dz_1}{2\pi i} (z_1 - w)^{n+m+1} T(z_1) \Phi(w)$$

$$= \frac{c}{12} n(n^2 - 1)\delta_{n+m,0}\Phi(w) + (n - m)\hat{L}_{n+m}\Phi(w) . \tag{1.77d}$$

Finally, one gets an algebra

$$[\hat{L}_n, \hat{L}_m] = (n - m)\hat{L}_{n+m} + \frac{c}{12} n(n^2 - 1)\delta_{n+m,0} , \tag{1.78a}$$

which is known as the **Virasoro algebra** [29, 30, 31].

The L_n-operators introduced in eq. (1.33) are just the operators $\hat{L}_n(z)$ referred to the origin in the complex z-plane, $L_n = \hat{L}_n(0)$. They satisfy the Virasoro algebra of the form

$$[L_n, L_m] = (n - m)L_{n+m} + \frac{c}{12}n(n^2 - 1)\delta_{n+m,0} , \qquad (1.78b)$$

which can be recognized as a central extension of the 2d conformal algebra (1.16). A uniqueness of the central extension, which is fixed by the only constant c, can also be seen in the following way [32]. Quantum mechanically, after taking into account the conformal anomalies, the l_n's of eq. (1.16) become the operators L_n's in a centrally extended conformal algebra of the generic form

$$[L_n, L_m] = (n - m)L_{n+m} + c_{n,m} . \qquad (1.79)$$

The antisymmetric coefficients $c_{n,m}$ defined by this equation are constrained by the requirement for the algebra (1.79) to obey the Jacobi identity for the commutators. From the exercise # I-10 we can now learn that the only way to obey the identity is to choose

$$c_{n,m} = \frac{c}{12}n(n^2 - 1)\delta_{n+m,0} , \qquad (1.80)$$

modulo additive constant redefinitions of the generators. The coefficient c appearing in this equation is just the central charge. In the normalization (1.75) or (1.80), one finds $c = 1$ for a real scalar, and $c = 1/2$ for a *Majorana-Weyl* (MW) fermion in two dimensions (see the next section). The vanishing of the central terms for $m = 0, \pm 1$ in eq. (1.80) reflects the $SL(2, \mathbf{C})$ invariance of the ground state. Therefore, the global conformal group $SL(2, \mathbf{C})$ remains the *exact* symmetry of CFT despite the central extension (1.80). In particular, the $T(z)$ is still a primary field w.r.t. the $SL(2, \mathbf{C})$, even when $c \neq 0$.

The infinitesimal equivalent to eq. (1.75) reads

$$\delta_\varepsilon T(w) = [T_\varepsilon, T(w)] = \oint_w dz\, \varepsilon(z)T(z)T(w)$$

$$= [\varepsilon(w)\partial_w + 2\partial_w\varepsilon(w)]\, T(w) + \frac{c}{12}\partial_w^3\varepsilon(w) . \qquad (1.81)$$

It can be integrated to a finite transformation $z = f(w)$ of the form:

$$T(z) \to \tilde{T}(w) = \left(\frac{dz}{dw}\right)^2 T(z) + \frac{c}{12}S[z; w] , \qquad (1.82)$$

where the $S[z; w] \equiv S[f; w]$ is known as the **Schwartzian derivative**,

$$S[f; w] = [\partial_w f \partial_w^3 f - \tfrac{3}{2}(\partial_w^2 f)^2]/(\partial_w f)^2 . \qquad (1.83)$$

For the conformal transformation (1.34) from a cylinder to a plane, one finds, in particular, after substituting eq. (1.34) into eq. (1.83), that

$$T_{\text{cyl}}(z) = z^2 T(z) - \frac{c}{24} = \sum_{n \in \mathbf{Z}} L_n z^{-n} - \frac{c}{24} , \qquad (1.84a)$$

and, hence,

$$(L_0)_{\text{cyl}} = L_0 - \frac{c}{24} . \qquad (1.84b)$$

This relation allows us to interpret the central charge c as representing the Casimir energy.

Exercises

\# I-9 ▷ Prove the violation of analyticity for the stress tensor $T(z)$ in the origin $z = 0$ when $c \neq 0$. This implies the violation of the stress-energy conservation law in two dimensions, and justifies the use of the word 'anomaly'. *Hint*: consider the OPE (1.75) and differentiate it w.r.t. $\bar{z}$. Use the identity (1.71) and the new one,

$$\frac{1}{z - w} = \partial_z \ln|z - w| , \qquad (1.85)$$

which is a consequence of eq. (1.71) when using the explicit form of the Green's function for the 2d Laplace operator,

$$\partial_z \partial_{\bar{z}} \ln|z - w| = \pi \delta^{(2)}(z - w) . \qquad (1.86)$$

\# I-10 ▷ (*) Derive eq. (1.80) from the Jacobi identity for the commutators.

\# I-11 ▷ Use the group property (composition law) for the infinitesimal conformal transformation (1.81) of the stress tensor to derive its finite form in eq. (1.82). The Schwartzian derivative satisfies the following composition law

$$S[z; w] = (\partial_w f)^2 S[z; f] + S[f; w] . \qquad (1.87)$$

The Schwartzian derivative is the unique object of dimension two that vanishes when restricted to the global conformal group $SL(2, \mathbf{C})$.

I.5   Free bosons and fermions on a plane

The simplest example of CFT and the best illustration of the formal aspects developed so far are provided by the two-dimensional theory of a single free massless scalar boson whose action is

$$S_\Phi = \frac{1}{4\pi} \int \left(\partial\Phi\bar\partial\Phi \right) , \tag{1.88}$$

where the integration measure on a complex plane has been defined in eq. (1.72). Eq. (1.86) determines the propagator of the $\Phi(z,\bar z)$ field,

$$\langle \Phi(z,\bar z)\Phi(w,\bar w) \rangle = -\ln|z-w|^2 . \tag{1.89}$$

The general solution to the classical equation of motion splits the scalar field Φ into a sum of holomorphic and anti-holomorphic pieces,

$$\Phi(z,\bar z) = \phi(z) + \bar\phi(\bar z) . \tag{1.90}$$

These pieces are related by complex conjugation in the Euclidean case (that's why one uses complexification of the plane), but independent in the Minkowski case where they become left-movers and right-movers, resp. The decomposition (1.90) is unique modulo constants.

The propagators of the (anti)-holomorphic scalar fields read

$$\langle \phi(z)\phi(w) \rangle = -\ln(z-w) , \quad \left\langle \bar\phi(\bar z)\bar\phi(\bar w) \right\rangle = -\ln(\bar z - \bar w) . \tag{1.91}$$

Though the field $\phi(z)$ itself does *not* have a conformal dimension, the fields $\partial_z\phi(z)$ and $: e^{i\alpha\phi(z)} :$ have definite dimensions, i.e. they transform as primary fields. [9] To see this explicitly, let us first compute the leading contribution to the OPE of $\partial\phi(z)$ with itself:

$$\partial\phi(z)\partial\phi(w) = -\frac{1}{(z-w)^2} + \dots . \tag{1.92}$$

The action (1.88) gives rise to (the holomorphic part of) the stress tensor of the form [10]

$$T_\phi(z) = -\tfrac{1}{2} : \partial\phi(z)\partial\phi(z) :\equiv -\tfrac{1}{2} \lim_{w\to z} \left[\partial\phi(z)\partial\phi(w) + \frac{1}{(z-w)^2} \right] , \tag{1.93}$$

where eq. (1.92) has been used to define the normal-ordered stress tensor by subtracting the divergence in the operator product. We are now in a position to compute the OPE between $\partial\phi$ and T by using the Wick theorem:

$$T_\phi(z)\partial\phi(w) \sim \frac{\partial\phi(w)}{(z-w)^2} + \frac{1}{z-w}\partial^2\phi(w) . \tag{1.94}$$

---

[9] The normal ordering used here is just the radial ordering defined by OPE, or, equivalently, the normal ordering via the Wick theorem in terms of modes.

[10] Because of the normalization convention we use for the 2d actions, the factor of 2π should be inserted into the r.h.s. of the definition of the stress tensor in eq. (1.21).

This means that $\partial\phi$ is a primary field of dimension $h = 1$, and it is now quite legitimate to define its mode expansion as

$$i\partial\phi(z) = \sum_{n\in\mathbf{Z}} \alpha_n z^{-n-1} . \tag{1.95}$$

The proof that $: e^{i\alpha\phi(z)} :$ is a primary field of conformal dimension $h = \alpha^2/2$ goes along similar lines (exercise # 12):

$$T_\phi(z) : e^{i\alpha\phi(w)} : \sim \frac{\alpha^2/2}{(z-w)^2} : e^{i\alpha\phi(w)} : + \frac{1}{z-w}\partial : e^{i\alpha\phi(w)} : . \tag{1.96}$$

Some useful identities are

$$: e^{i\alpha\phi(z)} :: e^{i\beta\phi(w)} := \frac{1}{(z-w)^{-\alpha\beta}} : e^{i[\alpha\phi(z)+\beta\phi(w)]} : ,$$

$$\left\langle : e^{i\alpha\phi(z)} :: e^{-i\beta\phi(w)} : \right\rangle = e^{\alpha\beta<\phi(z)\phi(w)>} = \frac{\delta_{\alpha,\beta}}{(z-w)^{\alpha\beta}} . \tag{1.97}$$

Actually, only the complete (non-chiral) vertex $: e^{i\alpha\Phi(z,\bar{z})} :$ with $h = \bar{h}$ is well-defined in QFT, while the chiral correlators in CFT are to be defined as (anti-)holomorphic roots of the full (non-chiral) correlation functions.

It is a good exercise (# I-13) to find the central charge of a free boson by computing the OPE of $T_\phi = -\frac{1}{2} : \partial\phi\partial\phi :$ with itself by using the Wick rules,

$$T_\phi(z)T_\phi(w) \sim \frac{1/2}{(z-w)^4} + \frac{2}{(z-w)^2}\left[-\frac{1}{2}\left(\partial\phi(w)\right)^2\right] + \frac{1}{z-w}\partial\left[-\frac{1}{2}\left(\partial\phi(w)\right)^2\right] . \tag{1.98}$$

This means, in particular, that a single free chiral boson has $c = 1$ indeed.

A free theory of a single massless *Majorana-Weyl* (MW) fermion [11] is another simple example of the 2d CFT, described by the action

$$S_{\psi,\bar{\psi}} = -\frac{1}{4\pi}\int\left(\psi\bar\partial\psi + \bar\psi\partial\bar\psi\right) , \tag{1.99}$$

where ψ is a real chiral Grassmannian (i.e. anticommuting) MW fermion. Eq. (1.99) is the Euclidean form of the usual Dirac action for spinors since

$$\hat{\partial} \equiv \sigma_x\partial_x + \sigma_y\partial_y = \begin{pmatrix} 0 & \partial_x - i\partial_y \\ \partial_x + i\partial_y & 0 \end{pmatrix} \sim \begin{pmatrix} 0 & \partial \\ \bar\partial & 0 \end{pmatrix} ,$$

---

[11]The *Majorana* condition for a spinor in 2d *Minkowski* space is just the reality condition, whereas the *Weyl* condition means a definite chirality of a spinor w.r.t. the 2d analogue of the four-dimensional γ_5 matrix.

$$\begin{pmatrix} \psi \\ \bar{\psi} \end{pmatrix} = \tfrac{1}{2}(1+\sigma_z)\begin{pmatrix} \psi \\ 0 \end{pmatrix} + \tfrac{1}{2}(1-\sigma_z)\begin{pmatrix} 0 \\ \bar{\psi} \end{pmatrix} , \qquad (1.100)$$

where the Pauli matrices $(\sigma_x, \sigma_y, \sigma_z)$ have been introduced, with the σ_z playing the role of the two-dimensional 'γ_5'.

The stress tensor components for the action (1.99) result in

$$\begin{aligned} T_\psi(z) &= \tfrac{1}{2} : \psi(z)\partial\psi(z) : , \\ \bar{T}_{\bar{\psi}}(\bar{z}) &= \tfrac{1}{2} : \bar{\psi}(\bar{z})\bar{\partial}\bar{\psi}(\bar{z}) : . \end{aligned} \qquad (1.101)$$

It is straightforward to derive the OPE's

$$\psi(z)\psi(w) \sim -\frac{1}{z-w} , \qquad (1.102)$$

and

$$T_\psi(z)\psi(w) \sim \frac{1/2}{(z-w)^2}\psi(w) + \frac{1}{z-w}\partial\psi(w) . \qquad (1.103)$$

Hence, the ψ (or $\bar{\psi}$) is a primary field of conformal weight $(1/2, 0)$ (or $(0, 1/2)$).

On a cylinder, the MW fermions can have either periodic or anti-periodic boundary conditions. Since $h = 1/2$ for them, a periodic (anti-periodic) fermion on the cylinder becomes an anti-periodic (periodic) fermion on the plane:

$$\begin{aligned} \text{NS}: \quad \psi(e^{2\pi i}z) &= +\psi(z) \\ \text{R}: \quad \psi(e^{2\pi i}z) &= -\psi(z) . \end{aligned} \qquad (1.104)$$

A presence of the two types of fermionic boundary conditions (called the **Neveu-Schwarz**-type (NS) and **Ramond**-type (R), resp.) is ultimately related to the fact that *spinors* naturally live on the double covering of a punctured plane. [12]

The associated mode expansions for the fermions are

$$\begin{aligned} \text{NS}: \quad i\psi(z) &= \sum_{n \in \mathbb{Z}+1/2} \psi_n z^{-n-1/2} , \\ \text{R}: \quad i\psi(z) &= \sum_{n \in \mathbb{Z}} \psi_n z^{-n-1/2} , \end{aligned} \qquad (1.105)$$

so that the Ramond-type $\psi(z)$ has a square root cut from 0 to ∞. The OPE's given above can equally be considered as just another equivalent representation of the quantum

---

[12]The conformally invariant objects of the type $\psi(z)(dz)^{n+1/2}$ with integer n naturally define spinors. The consistent sign choice for the root implies that the space where they live has to be a *spin manifold*.

commutators between the operatorial modes and vice versa. In particular, for free bosons and fermions, one has

$$i\partial_z\phi(z) = \sum_n \alpha_n z^{-n-1} \; ,$$

$$i\psi(z) = \sum_n \psi_n z^{-n-1/2} \; . \tag{1.106}$$

Eq. (1.106) can be inverted to produce

$$\alpha_n = \oint \frac{dz}{2\pi i} \, z^n i\partial_z\phi(z) \; ,$$

$$\psi_n = \oint \frac{dz}{2\pi i} \, z^{n-1/2} i\psi(z) \; , \tag{1.107}$$

where all the contours are supposed to enclose the origin. The use of eqs. (1.92) and (1.102) reproduces the canonical commutation rules

$$[\alpha_n, \alpha_m] = n\delta_{n+m,0} \; ,$$

$$\{\psi_n, \psi_m\} = \delta_{n+m,0} \; . \tag{1.108}$$

One could equally start from eq. (1.108) and derive eqs. (1.92) and (1.102).

For the anti-periodic case, it is sometimes useful to introduce the **twist field** $\sigma(z)$ satisfying the OPE [33, 34]

$$\psi(z)\sigma(w) \sim (z - w)^{-1/2}\mu(w) \; , \tag{1.109}$$

where $\mu(z)$ is another twist field with the same conformal weight as σ. Because of the square-root branch in eq. (1.109), the twist field σ can be used to change the boundary conditions on ψ when ψ is transported around σ. In other words, the combination $\sigma(0)$ and $\sigma(\infty)$ creates a cut from the origin to infinity. The fermion ψ flips sign when passing through the cut. One can equally view the state $\sigma(0)|0\rangle$ as a new incoming vacuum, and only the Ramond-type fermions with anti-periodic boundary conditions (or integer modes) are allowed to be applied to that vacuum, which results in overall single-valued states. Hence, we can write [17]

$$\langle\psi(z)\psi(w)\rangle_{\mathrm{R}} \equiv \langle 0| \, \sigma(\infty)\psi(z)\psi(w)\sigma(o) \, |0\rangle \tag{1.110}$$

to represent the two-point function in the anti-periodic case.

The evaluation of eq. (1.110) goes in a straightforward way, when using eq. (1.108) and taking into account the presence of a zero mode ψ_0 ($\psi_0^2 = \frac{1}{2}$) :

$$- \langle\psi(z)\psi(w)\rangle_{\mathrm{R}} = \left\langle \sum_{n=0}^{\infty} \psi_n z^{-n-1/2} \sum_{m=0}^{\infty} \psi_m w^{-m-1/2} \right\rangle_{\mathrm{R}}$$

$$= \sum_{n=1}^{\infty} z^{-n-1/2} w^{n-1/2} + \frac{1}{2}\frac{1}{\sqrt{zw}} = \frac{1}{\sqrt{zw}}\left(\frac{w}{z-w}+\frac{1}{2}\right) = \frac{\frac{1}{2}\left(\sqrt{\frac{z}{w}}+\sqrt{\frac{w}{z}}\right)}{z-w} . \tag{1.111a}$$

The propagator (1.111) has the same short distance behaviour as the one in the periodic case, and it also changes sign when either z or w makes a loop around 0 or ∞. The r.h.s. of eq. (1.111a) is the unique function with such properties [17].

The conformal dimension h_σ of the twist field $\sigma(w)$ can be extracted from its OPE with the stress tensor, and it turns out to be (exercise # I-15) $h_\sigma = 1/16$ [33, 34, 35, 36].

The bosonic twist field could also be introduced by defining the two-point function as [17]

$$\langle \partial\phi(z)\partial\phi(w)\rangle_{\rm R} \equiv \langle 0|\, \sigma(\infty)\partial\phi(z)\partial\phi(w)\sigma(0)\,|0\rangle$$

$$= -\frac{1}{2}\frac{\left(\sqrt{\frac{z}{w}}+\sqrt{\frac{w}{z}}\right)}{(z-w)^2} . \tag{1.111b}$$

The twist field $\sigma(w)$ has dimension $h_\sigma = 1/16$ and satisfies the equation

$$\partial\phi(z)\sigma(w) \sim (z-w)^{-1/2}\tau(w) , \tag{1.112}$$

where the new twist field τ of dimension $h_\tau = h_\sigma + 1/2$ has been introduced [17].

A nice intuitive picture can be provided for treating the twist fields, where a cut along which fermions change sign is equivalent to an $SO(2)$ gauge field concentrated along the cut (the Dirac string singularity). The gauge field strength can then be adjusted to give a phase change π for a parallel transport around the endpoints of the cut. Given this picture, the twist field looks like a point magnetic vortex, whereas changing the position of the cut corresponds to a gauge transformation of the $SO(2)$ gauge field [17].

Exercises

I-12 ▷ Perform a detailed calculation leading to eq. (1.96).

I-13 ▷ Check eq. (1.98).

I-14 ▷ Derive eq. (1.108) from eqs. (1.92) and (1.102).

I-15 ▷ (*) Determine the conformal dimension of the twist field σ by evaluating the expectation value of the stress tensor (i.e. the leading term in the OPE of T with σ):

$$\langle T(z)\rangle = \frac{1}{2}\left(\partial_w \langle \psi(z)\psi(w)\rangle_{\rm R} - \frac{1}{(z-w)^2}\right)_{z=w} = \frac{1}{16}z^{-2} . \tag{1.113}$$

I.6   Conformal ghost systems

The next very simple but important example of 2d conformal free fields is provided by a pair of Grassmann-odd fields b and c of conformal dimensions $(\lambda, 0)$ and $(1 - \lambda, 0)$, resp., $\lambda \geq \frac{1}{2}$. Such conformal fields naturally appear as ghosts in the covariant quantization of strings (Ch. VIII). Their action takes the form [37, 38]

$$S_{b,c} = \frac{1}{2\pi} \int \left(b \bar{\partial} c \right) + \text{h.c.} \; , \tag{1.114}$$

and their infinitesimal conformal transformations are [32]

$$\begin{aligned}
\delta b &= \varepsilon \partial b + \lambda (\partial \varepsilon) b \; , \\
\delta c &= \varepsilon \partial c + (1 - \lambda)(\partial \varepsilon) c \; .
\end{aligned} \tag{1.115}$$

The variation of the action (1.114) w.r.t. the transformations (1.115),

$$\delta S_{b,c} = \frac{1}{2\pi} \int (-\bar{\partial} \varepsilon) T_{b,c} \; , \tag{1.116}$$

determines the associated stress tensor in the form

$$T_{b,c} = -\lambda b \partial c + (1 - \lambda)(\partial b) c \; . \tag{1.117}$$

The free equations of motion,

$$\bar{\partial} b = \bar{\partial} c = 0 \; , \tag{1.118}$$

imply the mode expansions

$$c(z) = \sum_{n=-\infty}^{+\infty} c_n z^{-n-(1-\lambda)} \; , \quad -ib(z) = \sum_{-\infty}^{+\infty} b_n z^{-n-\lambda} \; . \tag{1.119}$$

To quantize this theory, one uses the parametrization $z = e^{\tau + i\sigma}$, and τ as the evolution parameter. This yields the canonical equal-time anticommutator

$$\{c(\sigma_1), b(\sigma_2)\} = 2\pi i \delta(\sigma_1 - \sigma_2) \; , \tag{1.120}$$

or, equivalently, the Fock space anticommutation relations

$$\{c_n, b_m\} = \delta_{n+m,0} \; , \quad \{c_n, c_m\} = \{b_m, b_n\} = 0 \; . \tag{1.121}$$

The propagator of the (b, c) system follows from the equation

$$\bar{\partial} \langle c(z) b(w) \rangle = \pi \delta^2(z - w) \; , \tag{1.122}$$

which is easily solved,

$$\langle c(z)b(w) \rangle = \frac{1}{z - w} \ . \tag{1.123}$$

To be consistent with the mode expansions in eq. (1.119), the vacuum state $|0\rangle$ should satisfy the relations

$$b_n |0\rangle = 0 \ \text{ for } \ n > \lambda - 1 \ ,$$
$$c_n |0\rangle = 0 \ \text{ for } \ n > -\lambda \ , \tag{1.124}$$

which respect the $SL(2, \mathbf{C})$ global conformal invariance.

Eq. (1.123) implies the OPE

$$c(z)b(w) \sim \frac{1}{z - w} \ , \tag{1.125}$$

which could also be derived directly from eq. (1.121). In quantum theory, the expression (1.117) for the stress tensor has to be taken in the normally ordered form:

$$T_{b,c}(z) =: -\lambda b\partial c + (1 - \lambda)(\partial b)c : \ . \tag{1.126}$$

The OPE's between the stress tensor $T_{b,c}$ and the fields b, c reproduce the transformation properties of the latter, as they should:

$$T_{b,c}(z)b(w) \sim \frac{\lambda}{(z - w)^2}b(w) + \frac{1}{z - w}\partial_w b(w) \ ,$$
$$T_{b,c}(z)c(w) \sim \frac{1 - \lambda}{(z - w)^2}c(w) + \frac{1}{z - w}\partial_w c(w) \ . \tag{1.127}$$

Nothing beyond the Wick theorem is needed to check these relations.

We are now in a position to calculate the OPE of $T_{b,c}$ with itself:

$$T_{b,c}(z)T_{b,c}(w) \sim \frac{c/2}{(z - w)^4} + \frac{2T_{b,c}(w)}{(z - w)^2} + \frac{1}{z - w}\partial_w T_{b,c}(w) \ , \tag{1.128}$$

which yields the celebrated result [37]

$$c_{b,c} = 2(-1 + 6\lambda - 6\lambda^2) \ . \tag{1.129}$$

The (β, γ) conformal ghost system differs from the (b, c) conformal system only in statistics (the fields β and γ are *bosonic* or Grassmann-even). The obvious changes are

$$\beta(z)\gamma(w) \sim -\frac{1}{z - w} \ , \tag{1.130}$$

and

$$c_{\beta,\gamma} = (-2)\left(-1 + 6\lambda - 6\lambda^2\right) . \tag{1.131}$$

The stress tensor of the (β, γ) system is given by

$$T_{\beta,\gamma}(z) = : \lambda\beta\partial\gamma + (1-\lambda)\gamma\partial\beta : . \tag{1.132}$$

The important example of the $\beta - \gamma$ conformal system is provided by the superconformal ghosts of the superstring to be introduced in Ch. VIII.

The simple examples of conformal systems introduced above are particularly useful to introduce the concept of **quantum equivalence** in CFT. The action (1.114) admits a global symmetry $b \to e^{i\theta}b$, $c \to e^{-i\theta}c$, associated with the chiral fermion (ghost) number conservation in the classical theory. After quantization, this symmetry becomes anomalous, which results in the violation of the conservation law for the corresponding ghost number current $j_z = -bc$. It can be seen in the following way. Eq. (1.125) implies the OPE's

$$\begin{aligned} j_z(z)b(w) &\sim -\frac{1}{z-w}b(w) , \\ j_z(z)c(w) &\sim \frac{1}{z-w}c(w) . \end{aligned} \tag{1.133}$$

It follows

$$j_z(z)j_w(w) \sim \frac{1}{(z-w)^2} . \tag{1.134}$$

The conformal field j_z clearly has (conformal) dimension 1, and it can be used to define a dimension-2 operator

$$T_{zz} \equiv : \tfrac{1}{2}j_z^2 + \tfrac{1}{2}Q\partial_z j_z : , \tag{1.135}$$

where the constant Q has been introduced, and the normal ordering has been defined w.r.t. j-modes. The operator T_{zz} satisfies the OPE of eq. (1.128) defining the Virasoro algebra with the *same* central charge (1.129) provided

$$Q = (2\lambda - 1) . \tag{1.136}$$

Therefore, the operator T_{zz} can be identified with the stress tensor of the (b, c) ghost system. When inserted in correlation functions, the stress tensors defined by eqs. (1.126) and (1.135) give rise to the same VEV's. Therefore, they can be identified inside the CFT correlators, which just means quantum equivalence. The ghost number anomaly follows from the OPE

$$T_{zz}(z)j_w(w) \sim -\frac{Q}{(z-w)^3} + \frac{1}{(z-w)^2}j_z , \tag{1.137}$$

where the anomalous contribution is represented by the first term on the r.h.s. The use of a stress tensor in eq. (1.126) immediately leads to eqs. (1.136) and (1.137). More examples of quantum equivalence can be found in the next Chapters.

Exercises

I-16 ▷ (*) Consider the (b, c) ghost system defined on a genus-h Riemann surface Σ. It is the standard result of QFT [39] that the ghost number anomaly of the current $j_z = - : bc :$ is given by

$$\nabla^z j_z = -\tfrac{1}{2}(2\lambda - 1)\sqrt{g}R , \tag{1.138}$$

where the 2d scalar curvature R in terms of the metric g has been introduced. Equivalently, the index theorem [28, 40] implies for this case

$$\#(c \text{ zero modes}) - \#(b \text{ zero modes}) = \tfrac{1}{2}(2\lambda - 1)\chi(\Sigma) , \tag{1.139}$$

where the *Euler characteristic* $\chi(\Sigma)$ of the Riemann surface Σ has been introduced,

$$\chi(\Sigma) = \frac{1}{2\pi} \int_\Sigma d^2z \sqrt{g}R = 2 - 2h . \tag{1.140}$$

Verify that these results are consistent with the anomaly formula displayed by eq. (1.137), and re-derive eq. (1.136) for the case of the Riemann sphere with $h = 0$.

I-17 ▷ (*) Calculate the central charge in eq. (1.129) from the VEV of eq. (1.128) by using the stress tensor either of eq. (1.126) or of eq. (1.135).

I-18 ▷ Calculate the OPE's for products of the stress tensor (1.135) with the fields b and c, and derive eq. (1.136) from them.

Chapter II

Representations of the Virasoro Algebra

When dealing with representations (modules) of the Virasoro algebra or CFT correlation functions, the 'field-theoretical' approach discussed so far can be supplemented by the other approaches based on either 'group-theoretical' or 'BRST' structure of the modules. In this Chapter, the so-called (BPZ) 'minimal models' will be the focus of our discussion. They are the simplest non-trivial representations of the Virasoro algebra with central charge $c < 1$. 'Free field realizations' of the minimal models are very useful to determine their correlation functions. Among the free field constructions, the so-called 'Coulomb gas picture' is the special and most important example. We begin with a discussion of the remarkable correspondence between fields and states in CFT.

II.1   Fields and states

According to the previous Chapter, the operator spectrum of CFT consists of the primary fields $\{\phi_p(z, \bar{z})\}$ and their derivatives (descendants) $\phi_p^{\{-\vec{k}, -\vec{k}\}}(z, \bar{z})$ defined by eq. (1.60). The index p runs over all primaries, $p = \{h, \bar{h}\}$.

The in-*vacuum* $|0\rangle$ in CFT is always defined to be $SL(2, \mathbf{C})$ invariant. It can formally be introduced by an insertion of the unit operator I at the origin of z-plane, since this origin corresponds to the 'infinite past' (sect. I.2). The $\hat{L}_n$-operators for $n \geq -1$ have no poles at $z = 0$ while the unit operator is a scalar, so that one has

$$L_n |0\rangle = 0 , \quad n \geq -1 . \tag{2.1}$$

35

Eq. (2.1) implies a regularity of $T(z)|0\rangle$ at $z = 0$. The $L_{-n}|0\rangle$ for $n > 2$ are the non-trivial states.

The CFT *in-states* are naturally defined by applying the CFT operators to the vacuum,

$$|A_{\text{in}}\rangle \equiv \lim_{z,\bar{z}\to 0} A(z,\bar{z})|0\rangle \ . \tag{2.2}$$

To introduce 'out'-states, first we use eq. (2.1) to get

$$\langle 0| L_{-n} = 0 \ , \quad n \geq -1 \ . \tag{2.3}$$

In CFT's, conformal invariance relates a parametrization near ∞ on the Riemann sphere $\mathbb{C} \cup \infty$ to that of a neighbourhood about the origin via the map $z = 1/w$. Hence, if the $\tilde{A}(w,\bar{w})$ are the operators in the coordinates in which $w \to 0$ corresponds to the point at infitiny, it is quite natural to introduce the out-states as follows:

$$\langle A_{\text{out}}| \equiv \lim_{w,\bar{w}\to 0} \langle 0| \tilde{A}(w,\bar{w}) \ . \tag{2.4}$$

We now have to relate $\tilde{A}(w,\bar{w})$ to $A(z,\bar{z})$. For primary fields, their transformation law under $w \to z = 1/w$ yields

$$\tilde{A}(w,\bar{w}) = A(1/w, 1/\bar{w})(-w^{-2})^{h}(-\bar{w}^{-2})^{\bar{h}} \ . \tag{2.5}$$

Therefore, the CFT *out-states* are given by

$$\langle A_{\text{out}}| = \lim_{z,\bar{z}\to\infty} \langle 0| A(z,\bar{z})z^{2h}\bar{z}^{2\bar{h}} \ . \tag{2.6}$$

These definitions suggest introducing the notion of the *adjoint* in CFT, of the form

$$[A(z,\bar{z})]^{+} \equiv A\left(\frac{1}{z}, \frac{1}{\bar{z}}\right) \frac{1}{z^{2h}} \frac{1}{\bar{z}^{2\bar{h}}} \ , \tag{2.7}$$

because of the line of reasoning:

$$\langle A_{\text{out}}| = \lim_{w,\bar{w}\to 0} \tilde{A}(w,\bar{w}) = \lim_{z\to 0,\bar{z}\to 0} \langle 0| A\left(\frac{1}{z}, \frac{1}{\bar{z}}\right) \frac{1}{z^{2h}} \frac{1}{\bar{z}^{2\bar{h}}}$$

$$= \lim_{z,\bar{z}\to 0} \langle 0| [A(z,\bar{z})]^{+} = \left[\lim_{z,\bar{z}\to 0} A(z,\bar{z})|0\rangle\right]^{+} = |A_{\text{in}}\rangle^{+} \ . \tag{2.8}$$

The procedure outlined above defines asymptotic states as the eigenstates of the *exact* Hamiltonian, rather than that of some free (non-interacting) Hamiltonian, as it is usually the case in QFT. This is because 2d CFT is an example of *solvable* QFT.

The (physically motivated) hermiticity of the stress tensor w.r.t. the definition (2.7) implies $L_m^+ = L_{-m}$. We can immediately see that $c \geq 0$ in a positively-definite Hilbert space of states since

$$c/2 = \langle 0| [L_2, L_{-2}] |0\rangle = \langle 0| L_2 L_2^+ |0\rangle = \| L_2^+ |0\rangle \|^2 \geq 0 . \qquad (2.9)$$

The *primary* states are defined by

$$\left|h, \bar{h}\right\rangle = \lim_{z,\bar{z}\to 0} \Phi_{h,\bar{h}}(z,\bar{z}) |0\rangle , \qquad (2.10)$$

where $\Phi_{h,\bar{h}}(z,\bar{z})$ is a primary field. From the OPE between the stress tensor and the primary field Φ we get

$$[L_n, \Phi_{h,\bar{h}}(z,\bar{z})] = \oint \frac{dw}{2\pi i} w^{n+1} T(w) \Phi_{h,\bar{h}}(z,\bar{z})$$

$$= \left(z^{n+1} \frac{d}{dz} + (n+1)z^n h \right) \Phi_{h,\bar{h}}(z,\bar{z}) , \qquad (2.11)$$

so that

$$L_n \left|h, \bar{h}\right\rangle = 0, \quad n > 0, \qquad L_0 \left|h, \bar{h}\right\rangle = h \left|h, \bar{h}\right\rangle, \quad \bar{L}_0 \left|h, \bar{h}\right\rangle = \bar{h} \left|h, \bar{h}\right\rangle . \qquad (2.12)$$

Eq. (2.12) defines the so-called *highest-weight states*. The states of the associated highest-weight representation of the Virasoro algebra are created by acting with arbitrary polynomials in $\{L_{-n}, \bar{L}_{-m}; \ m, n \geq 1\}$ on the primary state $\left|h, \bar{h}\right\rangle$. They represent the *descendant* states,

$$\Phi_p^{\{-\vec{k}, -\vec{\bar{k}}\}} \equiv L_{-k_1} \cdots L_{-k_M} \bar{L}_{-\bar{k}_1} \cdots \bar{L}_{-\bar{k}_{M'}} \Phi_p(0,0) , \qquad (2.13)$$

which obviously correspond to the modes of the descendant fields expanded near the origin. A descendant state can be regarded as the result of the action of a descendant field on the vacuum: $L_{-n} |h\rangle = L_{-n}(\Phi(0) |0\rangle) = (\hat{L}_{-n}\Phi)(0) |0\rangle = \Phi^{(-n)}(0) |0\rangle$.

A highest-weight representation of the Virasoro algebra is also called a **Verma module**. Clearly, it is an infinite-dimensional representation, and it is completely characterized by its central charge and the dimension of the highest-weight state. In part, the CFT vacuum is a trivial highest-weight state which defines the trivial module corresponding to the identity operator. The Virasoro operators L_n in a Verma module act like the raising and lowering (ladder) operators, the L_0 being a grading operator measuring the conformal dimension of a state because of $[L_0, L_{-n}] = nL_{-n}$. The conformal symmetry implies that the CFT field operators, as well as the CFT states, can be organized into the conformal families, each generated by a primary field (or a highest-weight state). They comprise the irreducible representations (irreps) of the Virasoro algebra. All this can equally applied to the anti-holomorphic counterpart.

It is easy to see that we must have $c > 0$ and $h > 0$ for non-trivial *unitary* representations, if any. If the Hilbert space of CFT states has to have a positive norm, the evaluation of

$$0 < \langle h| L_{-n}^+ L_{-n} |h\rangle = \left(2nh + \frac{c}{12}(n^3 - n)\right) \langle h | h\rangle \qquad (2.14)$$

implies $c > 0$ and $h > 0$. The same conclusion could be reached [41] when taking the 2×2 matrix of scalar products in the basis of $L_{-2n} |h\rangle$ and $L_{-n}^2 |h\rangle$, and requiring it to be positive (exercise II-1). Clearly, one has $h = 0$ only if $L_{-1} |h\rangle = 0$, i.e. if $|h\rangle = |0\rangle$. Similarly, one can check that the only unitary representation of the Virasoro algebra with $c = 0$ is trivial. It follows from eqs. (2.11) and (2.14) that $[L_{-1}, \Phi] = \partial\Phi$, $[\bar{L}_{-1}, \Phi] = \bar{\partial}\Phi$, so that a field Φ of conformal dimension $(h, 0)$ is purely holomorphic.

The one-to-one correspondence between states and local operators in CFT is crucially dependent on (i) the conformal invariance, and (ii) the $SL(2, \mathbf{C})$ invariance of the CFT physical vacuum. In general, an insertion of a state is made by cutting a 'hole' on a manifold, inserting a wave function of the state on the boundary of the hole and then evaluating the functional integral with a weight given by the wave function. The conformal invariance makes it possible to shrink the hole to a 'point' of infinitesimally small size. The resulting *puncture* is equivalent to a local operator. However, under some circumstances (e.g., in the 2d Liouville theory, see sect. IX.2), the formally $SL(2, \mathbf{C})$ invariant vacuum $|0\rangle$ may not be a physical state, which may make the standard correspondence invalid, at least naively.

Exercise

II-1 ▷ Consider the 2×2 matrix of scalar products in the basis of $L_{-2n} |h\rangle$ and $L_{-n}^2 |h\rangle$. Requiring the matrix to be positively definite, find the associated restrictions on the central charge and the conformal dimension.

II.2 Correlation functions

Primaries are just the basic fields of CFT. As was shown in the previous Chapter, it is enough to calculate correlation functions of the primary fields. The complete operator product of two primary fields takes the form [12]

$$\Phi_n(z, \bar{z})\Phi_m(w, \bar{w}) = \sum_p \sum_{\{\vec{k}, \bar{\vec{k}}\}} C_{nm}^{p\{-\vec{k}, -\bar{\vec{k}}\}} z^{(h_p - h_n - h_m + |\vec{k}|)}$$

$$\times \bar{z}^{(\bar{h}_p - \bar{h}_n - \bar{h}_m + |\bar{\vec{k}}|)} \Phi_p^{\{-\vec{k}, -\bar{\vec{k}}\}}(w, \bar{w}) , \qquad (2.15a)$$

where some coefficients $C_{nm}^{p\{-\vec{k}, -\bar{\vec{k}}\}}$ have been introduced (*cf* eq. (1.65)). We always assume that the normalization (1.64) is used to fix the norms of the fields. Eq. (2.15a) means, in particular, that there are no other new operators (beyond descendants) produced from the OPE's of the primary fields, which is required by consistency of CFT.

In this Chapter and in what follows, we will normally use the *lower case* Greek letters to represent primary fields. As a rule, we explicitly indicate the holomorphic dependence only, in order to avoid the unnecessary complications in our formulae. In particular, eq. (2.15a) could equally be represented as [13]

$$\phi_2(z)\phi_1(0) = \sum_p \sum_{\{\vec{k}\}} C_{21}^{p\{-\vec{k}\}} z^{(h_p - h_1 - h_2 + |\vec{k}|)} \phi_p^{\{-\vec{k}\}}(0) . \qquad (2.15b)$$

Actually, the double expansion in powers of z and $\bar{z}$ is supposed to be in eq. (2.15b), but all the additional anti-holomorphic dependence can be trivially recovered, as it is in eq. (2.15a).

The conformal dimension of the descendant $\phi_h^{\{-\vec{k}\}}(z)$ is $h + |k|$. The number of descendants at level n (of dimension $h + n$) is clearly given by $P(n)$, the number of partitions of n into positive integers. The $P(n)$ has the generating function known in number theory [42],

$$\sum_{n=0}^{\infty} P(n)q^n = \frac{1}{\prod_{n=1}^{\infty}(1 - q^n)} , \qquad P(0) = 1 . \qquad (2.16)$$

The stress tensor $T(z)$ itself can be classified as a level-two descendant in the family of the identity operator $\phi_{h=0}(z) \equiv I = $ const, since

$$I^{(-2)}(w) = \int_{C_w} \frac{dz}{2\pi i}(z - w)^{-1} T(z) I = T(w) . \qquad (2.17)$$

The OPE coefficients for descendants can be expressed in terms of those of primaries, when using eq. (1.70) and the Virasoro algebra (1.78). Eq. (1.70) can be conveniently represented in the form

$$\left\langle \phi_1(w_1)\cdots\phi_{M-1}(w_{M-1})(\hat{L}_{-n}\phi_M)(w_M)\right\rangle$$

$$= \mathcal{L}_{-n}\left\langle \phi_1(w_1)\cdots\phi_{M-1}(w_{M-1})\phi_M(w_M)\right\rangle , \qquad (2.18)$$

where the differential operator $(n \geq 2)$

$$\mathcal{L}_{-n} = -\sum_{j=1}^{M-1}\left[\frac{(1-n)h_j}{(w_j-w_M)^n} + \frac{1}{(w_j-w_M)^{n-1}}\frac{\partial}{\partial w_j}\right] \qquad (2.19)$$

has been introduced $(n \geq 2)$. Eq. (2.18) can be further generalized to the correlation functions of arbitrary descendants.

We now want to see how the full conformal invariance in two dimensions restricts the M-point correlation functions in CFT. As to the 2-, 3- and 4-point functions, they were already considered in sect. I.2. In particular, eq. (1.45) can be generalized to the M-point correlation functions by using the conformal Ward identities as follows:

$$\left\langle (\hat{L}_{-n}\phi)(z)\phi_1(z_1)\cdots\phi_M(z_M)\right\rangle \equiv \left\langle \oint_{C_z}\frac{dw}{2\pi i}(w-z)^{-n+1}T(w)\phi(z)\phi_1(z_1)\cdots\phi_M(z_M)\right\rangle$$

$$= (-1)^{n-1}\sum_{j=1}^{M}\left[\frac{(1-n)h_j}{(z-z_j)^n} - \frac{1}{(z-z_j)^{n-1}}\frac{\partial}{\partial z_j}\right]\left\langle \phi(z)\phi_1(z_1)\cdots\phi_M(z_M)\right\rangle . \qquad (2.20)$$

Since $L_{\pm 1}\left|0\right\rangle = L_0\left|0\right\rangle = 0$, we have, in particular, for $\phi = 1$ and $z = 0$ (and similarly for $\bar{L}_{\pm 1}$ and $\bar{L}_0$)

$$\sum_{j=1}^{M}\frac{\partial}{\partial z_j}\left\langle \phi_1(z_1)\cdots\phi_M(z_M)\right\rangle = 0 ,$$

$$\sum_{j=1}^{M}(z_j\frac{\partial}{\partial z_j}+h_j)\left\langle \phi_1(z_1)\cdots\phi_M(z_M)\right\rangle = 0 , \qquad (2.21)$$

$$\sum_{j=1}^{M}(z_j^2\frac{\partial}{\partial z_j}+2h_jz_j)\left\langle \phi_1(z_1)\cdots\phi_M(z_M)\right\rangle = 0 ,$$

reflecting translational, dilatational and special conformal invariance, resp.

Using the known transformation properties of the correlators under the $SL(2,\mathbf{C})$ transformations $g(z) = (az+b)(cz+d)^{-1}$,

$$\left\langle \phi_1(g(z_1))\cdots\phi_M(g(z_M))\right\rangle = \left(\prod_{j=1}^{M}(cz_j+d)^{2h_j}\right)\left\langle \phi_1(z_1)\cdots\phi_M(z_M)\right\rangle , \qquad (2.22)$$

the correlation function $\langle \phi_1 \cdots \phi_M \rangle$ can be written as a function of $M - 3$ $SL(2, \mathbf{C})$-invariant ratios (like $\eta = w_{12}w_{34}/w_{14}w_{32}$, $w_{ij} = w_i - w_j$, for any four points w_i) multiplied by the pre-factor which ensures the transformation law (2.22):

$$\langle \phi_1(z_1) \cdots \phi_M(z_M) \rangle = \left(\prod_{i<j} z_{ij}^{-\gamma_{ij}} \right) f(\eta_a) ; \quad a = 1, \ldots, M - 3 . \qquad (2.23)$$

To respect eq. (2.22), the γ_{ij} should satisfy the relations

$$\gamma_{ij} = \gamma_{ji} , \qquad \sum_{j \neq i} \gamma_{ij} = 2h_i . \qquad (2.24)$$

In particular, as far as the 2-, 3- and 4-point functions are concerned, the previous results given by eqs. (1.46), (1.47) and (1.49), resp., are reproduced.

The OPE coefficients appearing in eq. (2.15) are symmetric, and they coincide with the numerical factors in the 3-point functions of primary or secondary fields,

$$\langle \Phi_1 | \Phi_2(z, \bar{z}) | \Phi_3 \rangle = \langle \Phi_1(\infty) \Phi_2(z, \bar{z}) \Phi_3(0) \rangle$$

$$= C_{123} z^{h_1 - h_2 - h_3} \bar{z}^{\bar{h}_1 - \bar{h}_2 - \bar{h}_3} . \qquad (2.25)$$

It is the BPZ theorem [12] that the OPE coefficients for descendants can be expressed in terms of those for primaries,

$$C_{mn}^{p\{-\vec{k}, -\vec{\bar{k}}\}} = C_{mnp} \beta_{mn}^{p\{-\vec{k}\}} \bar{\beta}_{mn}^{p\{-\vec{\bar{k}}\}} , \quad (\text{no sum over } m, n\,!) , \qquad (2.26)$$

where the C_{mnp}'s are the OPE coefficients for *primary* fields, and $\beta(\bar{\beta})$ are some functions of four parameters h_m, h_n, h_p and c ($\bar{h}_m, \bar{h}_n, \bar{h}_p$ and $\bar{c}$). The formal proof can be done by performing a conformal transformation of both sides of eq. (2.15) and comparing terms. The β-coefficients can also be computed this way. Moreover, the 3-point function of any three descendants can be calculated from that of their associated primaries. The primary C_{mnp}'s coefficients actually determine all the 3-point functions for any members of the families $[\phi_m]$, $[\phi_n]$ and $[\phi_p]$, where the $[\phi_p]$ denotes a collection of a primary field and all of its descendants (a conformal family). In CFT, the coefficients C_{mnp} play the role of Clebsch-Gordan coefficients: given a product of two conformal highest-weight representations, they determine which representations are contained in the decomposition of the product into the conformal irreps [21].

To this end, we outline the derivation of the crucial equation (2.26), following ref. [13]. Unfortunately, there is no explicit formula for a generic coefficient $C_{mn}^{p\{-\vec{k}, -\vec{\bar{k}}\}}$ in this

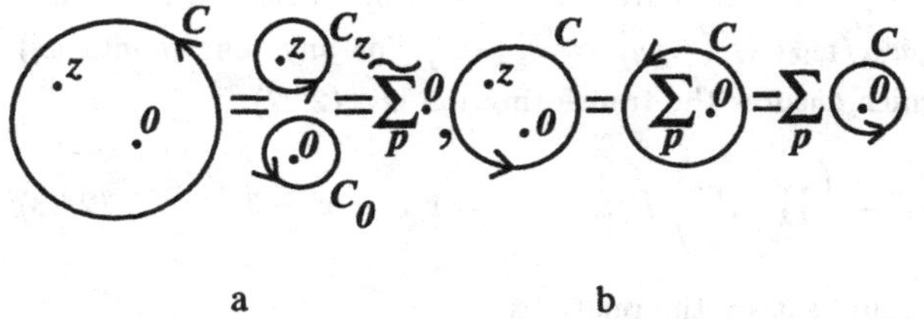

a   b

Fig. 5. For the calculation of the integral on the r.h.s. of eq. (2.27).

equation, and the actual calculations become very tedious at high levels (see refs. [12, 13] for more). Nevertheless, the general statement represented by eq. (2.26) can be made clear for understanding and practical calulations as well.

To simplify the notation even more, the dependence of the kinematical β-coefficients upon the indices m, n is suppressed. Consider the operator product on the l.h.s. of eq. (2.15b) and apply the operator L_n on it ($n \geq 1$):

$$L_n(\phi_2(z)\phi_1(0)) = \int_C \frac{d\zeta}{2\pi i}\zeta^{n+1}T(\zeta)\phi_2(z)\phi_1(0) \ . \tag{2.27}$$

Since the operator L_n is defined at the origin and it acts on both fields ϕ_1 and ϕ_2, the contour C should enclose both points z and 0 (see the l.h.s. of Fig. 5). There are now two different ways to proceed with eq. (2.27).

The first way is to deform the contour as it is shown in Fig. 5a. Then the operator L_n acts separately on ϕ_1 and ϕ_2. The contribution of C_0 vanishes, whereas the integral over C_z yields

$$L_n(\phi_2(z)\phi_1(0)) = (L_n\phi_2(z))\phi_1(0) + \phi_2(z)(L_n\phi_1(0))$$

$$= \left[(n+1)z^n h_2 + z^{n+1}\partial_z\right]\phi_2(z)\phi_1(0) \ . \tag{2.28}$$

Having expanded the product $\phi_2\phi_1$ according to eq. (2.15), we get

$$L_n(\phi_2(z)\phi_1(0)) = \left[(n+1)z^n h_2 + z^{n+1}\partial_z\right]\sum_p \frac{C_{21}^p}{z^{h_1+h_2-h_p}}$$

$$\times \left\{\sum_{\{\vec{k}\}} z^{|\vec{k}|}\beta_p^{\{-\vec{k}\}}\phi_p^{\{-\vec{k}\}}(0)\right\} \ . \tag{2.29}$$

Taking now, in part, the operators L_1 and L_2 results in

$$L_1(\phi_2(z)\phi_1(0)) = \sum_p \frac{C_{21}^p}{z^{h_1+h_2-h_p}}$$

$$\times \left\{ z(h_2 - h_1 + h_p)\phi_p(0) + z^2(h_2 - h_1 + 1)\beta_p^{(-1)}\phi_p^{(-1)}(0) + \ldots \right\} , \qquad (2.30)$$

and

$$L_2(\phi_2(z)\phi_1(0)) = \sum_p \frac{C_{21}^p}{z^{h_1+h_2-h_p}} \left\{ z^2(2h_2 - h_1 + h_p)\phi_p(0) + \ldots \right\} . \qquad (2.31)$$

The second way is to expand *first* the product $\phi_2\phi_1$ on the l.h.s. of eq. (2.28) and *then* apply the operator L_n to that expansion (Fig. 5b). This results in the expansion

$$L_n(\phi_2(z)\phi_1(0)) = L_n \sum_p \frac{C_{21}^p}{z^{h_1+h_2-h_p}} \left\{ \sum_{\{\vec{k}\}} \beta_p^{\{-\vec{k}\}} \phi_p^{\{-\vec{k}\}}(0) \right\}$$

$$= \sum_p \frac{C_{21}^p}{z^{h_1+h_2-h_p}} \left\{ \sum_{\{\vec{k}\}} \beta_p^{\{-\vec{k}\}} (L_n \phi_p^{\{-\vec{k}\}}(0)) \right\} . \qquad (2.32)$$

In particular, for the operators L_1 and L_2, this gives

$$L_1(\phi_2(z)\phi_1(0)) = \sum_p \frac{C_{21}^p}{z^{h_1+h_2-h_p}}$$

$$\times \left\{ 0 + z\beta_p^{(-1)}2h_p\phi_p(0) + z^2[\beta_p^{(-1,-1)}2(2h_p+1) + \beta_p^{(-2)}3]\phi_p^{(-1)}(0) + \ldots \right\} , \qquad (2.33)$$

and

$$L_2(\phi_2(z)\phi_1(0)) = \sum_p \frac{C_{21}^p}{z^{h_1+h_2-h_p}}$$

$$\left\{ 0 + 0 + z^2 \left[\beta_p^{(-1,-1)}6h_p + \beta_p^{(-2)} \left(4h_p + \tfrac{1}{2}c \right) \right] \phi_p(0) + \ldots \right\} , \qquad (2.34)$$

where the definition (1.60) of the descendants,

$$\phi_p^{\{-\vec{k}\}} = \phi^{\{-n_1,-n_2,\ldots,-n_k\}} = L_{-n_1}L_{-n_2} \cdots L_{-n_k}\phi_p , \qquad (2.35)$$

had been used, and the Virasoro operators L_1 and L_2 have been commuted through those of $\phi_p^{\{-\vec{k}\}}$.

Comparing eqs. (2.30) and (2.31) with eqs. (2.33) and (2.34), resp., yields the relations:

$$\beta_p^{(-1)}2h_p = h_2 - h_1 + h_p , \qquad (2.36a)$$

$$\beta_p^{(-1,-1)}2(2h_p + 1) + 3\beta_p^{(-2)} = (h_2 - h_1 + h_p + 1)\beta_p^{(-1)} , \qquad (2.36b)$$

$$\beta_p^{(-1,-1)}6h_p + \beta_p^{(-2)}\left(4h_p + \tfrac{1}{2}c\right) = (2h_2 - h_1 + h_p) \ . \tag{2.36c}$$

These equations determine the coefficients $\beta_p^{(-1)}$, $\beta_p^{(-1,-1)}$ and $\beta_p^{(-2)}$. When comparing coefficients at higher power terms in the above equations, all the β-coefficents can, in principle, be determined. It is sufficient to consider only the action of the operators L_1 and L_2 on a product of two primary fields, since the action of the other Virasoro operators follows, when using the Virasoro algebra.

We conclude that the CFT is completely specified by the central charge c, the conformal weights (anomalous dimensions) $\{h_i, \bar{h}_i\}$ of the Virasoro highest-weight states and the OPE coefficients (couplings) C_{mnp} between the primary fields. The latter are not fixed by conformal invariance, and they have to be determined from dynamics of the underlying physical system.

Exercises

II-2 ▷ (*) Consider the conformal Ward identity (1.69) in the limit $z \to w_M$, and derive eq. (2.18) from it. *Hint*: expand eq. (1.69) in powers of $(z - w_M)$, and use eqs. (1.56) and (1.57).

II-3 ▷ Check eq. (2.18) directly, by commuting $\hat{L}_{-n}$ to the left on the l.h.s. *Hint*: take $w_M = 0$ in eq. (2.18) and use eq. (2.11).

II-4 ▷ Take the section $\bar{z} = z^*$ and prove that this leads to the *spin quantization*, $h - \bar{h} \in \mathbf{Z}$, when requiring the correlation functions be single-valued. Is there a similar restriction on the scaling dimension $h + \bar{h}$?

II.3 Conformal bootstrap

So far we only considered the local constraints imposed by the conformal algebra. To find the allowed dimensions h's and couplings C_{ijk}'s in CFT, we need some *dynamical* principles or some global constraints on the correlation functions. These constraints arise from the *associativity* of the operator algebra in eq. (1.65), or in eq. (2.15). To see this, we study a 4-point function

$$\langle \Phi_i(z_1, \bar{z}_1)\Phi_j(z_2, \bar{z}_2)\Phi_l(z_3, \bar{z}_3)\Phi_m(z_4, \bar{z}_4)\rangle \tag{2.37}$$

in two ways. First, taking the limit $z_1 \to z_2$, $z_3 \to z_4$, we find the result schematically pictured on the l.h.s. of Fig. 6, where the sum (over p) goes over both primary and

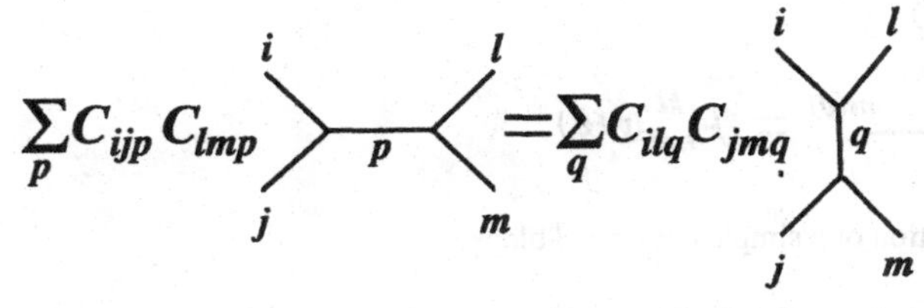

$$\sum_p C_{ijp} C_{lmp} \overset{i}{\underset{j}{\rangle}} p \overset{l}{\underset{m}{\langle}} = \sum_q C_{ilq} C_{jmq} \overset{i\ \ \ l}{\underset{j\ \ \ m}{\times}} q$$

Fig. 6. The crossing (duality) symmetry of a 4-point function.

secondary fields. Second, taking another limit $z_1 \to z_3$, $z_2 \to z_4$, we can alternatively evaluate eq. (2.37) with the result diagrammatically represented on the r.h.s. of Fig. 6. The associativity of the operator algebra means that these two methods should give rise to the same result (Fig. 6). This consistency condition is known as the **crossing** or **duality** symmetry of the 4-point functions. In QFT, an s-channel Feynman diagram [1] has to be supplemented by the similar t- and u-contributions in order to describe a full scattering amplitude. The duality means that the s-contribution alone represents the full amplitude.

The relation shown in Fig. 6 places the *infinite* number of restrictions. They become the algebraic constraints C_{ijk}'s must satisfy, after performing (in principle) the infinite sum over all descendant states. This procedure is known as the **conformal bootstrap**. In practice, it is very difficult to implement, even in two dimensions where the situation is considerably simpler since we only need to consider primary fields. Fortunately, there are the special values of c's and h's where things are dramatically simplified [12], as we are going to discuss in what follows.

To convert the diagrammatic relation of Fig. 6 to an analytic expression, we choose three reference points to be $z_1 = \infty$, $z_2 = 1$ and $z_4 = 0$, while $z_3 = z$. We are now interested in the Green's function

$$G^{kl}_{nm}(z, \bar{z}) = \langle k| \Phi_l(1,1)\Phi_n(z,\bar{z}) |m\rangle \ . \tag{2.38}$$

Substituting eqs. (2.15) and (2.26) into eq. (2.38) yields

$$\langle k| \Phi_l(1,1)\Phi_n(z,\bar{z}) |m\rangle = \langle k| \Phi_l(1,1) \sum_p C_{nmp} z^{h_p - h_n - h_m} \bar{z}^{\bar{h}_p - \bar{h}_n - \bar{h}_m}$$

---

[1] We refer to the standard (Mandelstam) kinematical variables s, t, u and the corresponding channels normally used to describe particle scattering [1, 43].

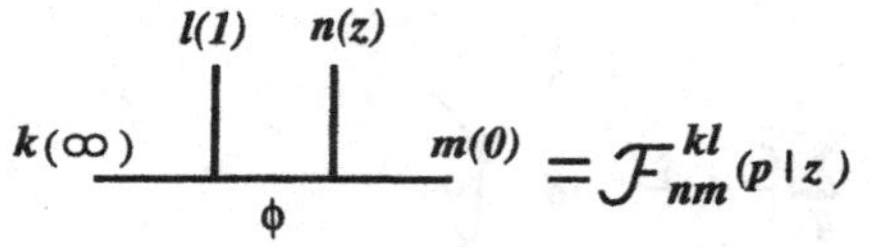

$$\begin{array}{cc} l(1) & n(z) \\ \Big| & \Big| \end{array}$$
$$k(\infty) \underbrace{\qquad\qquad}_{\phi}\; m(0) \;=\; \mathcal{F}^{kl}_{nm}(p\,|\,z)$$

Fig. 7. The graphical representation of a simple conformal block.

$$\times \sum_{\{\vec{k},\vec{\bar{k}}\}} \beta^{p\{-\vec{k}\}}_{nm}\beta^{p\{-\vec{\bar{k}}\}}_{nm} z^{|\vec{k}|}\bar{z}^{|\vec{\bar{k}}|}\Phi^{\{-\vec{k},-\vec{\bar{k}}\}}_{p}(0,0)\,|0\rangle \; . \tag{2.39}$$

Let us write down the contribution to the 4-point function from the 'intermediate states' belonging to *one* conformal family $[\phi_p]$ in the form $\mathcal{F}^{kl}_{nm}(p|z)\overline{\mathcal{F}^{kl}_{nm}(p|z)}$, where

$$\mathcal{F}^{kl}_{nm}(p|z) = z^{h_p-h_n-h_m}\sum_{\{\vec{k}\}}\beta^{p\{-\vec{k}\}}_{nm}\frac{\langle k|\,\Phi_l(1,1)L_{-k_1}\cdots L_{-k_M}\,|p\rangle}{\langle k|\,\Phi_l(1,1)\,|p\rangle}z^{|\vec{k}|} \; . \tag{2.40}$$

The amplitudes projected onto a single conformal family take a *factorized* form because the sums over descendants in the holomorphic and anti-holomorphic families $[\phi_p]$ and $[\bar{\phi}_p]$ (generated by T and $\bar{T}$, resp.) are independent. The Green's function becomes

$$G^{kl}_{nm}(z,\bar{z}) = \sum_p C_{nmp}C_{klp}\mathcal{F}^{kl}_{nm}(p|z)\overline{\mathcal{F}^{kl}_{nm}(p|z)} \; , \tag{2.41}$$

where we have used eq. (2.25) at $z = 1$.

The functions $\mathcal{F}^{kl}_{nm}(p|z)$ depend on the parameters h_n, h_m, h_l, h_k, h_p and c, and they are known as the (chiral) **conformal blocks**, [2] since any correlation function can be built from them. Diagrammatically, each function $\mathcal{F}$ can be represented as a skeleton diagram (Fig. 7), which indicates the order of taking the operator product. Each block is supposed to be normalized by the condition [12]

$$\mathcal{F}^{lk}_{nm}(p|z) \simeq z^{h_p-h_n-h_m}(1+\ldots) \; , \tag{2.42}$$

when $z \to 0$.

The associativity of OPE implies the desired equations on C_{ijk}'s in the form known as the **bootstrap**, **duality** or **crossing** relations:

$$\sum_p C_{klp}C_{nmp}\mathcal{F}^{kl}_{nm}(p|z)\overline{\mathcal{F}^{kl}_{nm}(p|z)}$$

---

[2]When the chiral (holomorphic) symmetry algebra of CFT is bigger than just the Virasoro algebra, we use another term, namely, *chiral blocks*.

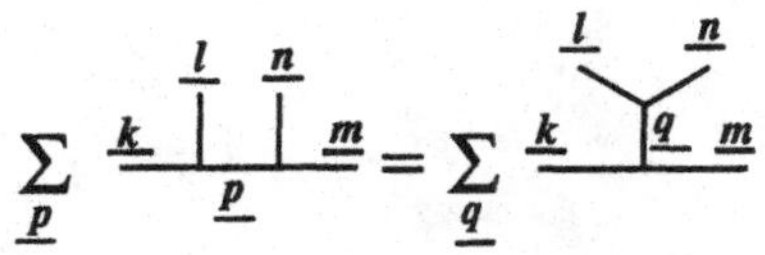

Fig. 8. The bootstrap, duality or crossing equation.

$$= \sum_q C_{kmq} C_{lnq} \mathcal{F}_{nl}^{km}(q|1-z) \overline{\mathcal{F}_{nl}^{km}(q|1-z)} \, , \tag{2.43}$$

where we have used the fact that

$$G_{nm}^{kl}(z, \bar{z}) = G_{nl}^{km}(1-z, 1-\bar{z}) \tag{2.44}$$

under the $SL(2, \mathbf{C})$ transformation $z \to 1 - z$. Similarly, when $z \to 1/z$, one has the transformation law

$$G_{nm}^{kl}(z, \bar{z}) = z^{-2h_n} \bar{z}^{-2\bar{h}_n} G_{nk}^{ml}\left(\frac{1}{z}, \frac{1}{\bar{z}}\right) \, . \tag{2.45}$$

A graphical representation of eq. (2.43) is given in Fig. 8. This equation is apparently very complicated: the conformal blocks themselves are first to be computed which is a non-trivial task, while the sum goes over all the 3-point couplings which are infinite in number. The conformal blocks are determined by the conformal Ward identities. Given all the conformal blocks $\mathcal{F}$, eq. (2.43) places the set of very non-trivial equations characterizing all possible OPE coefficients. The problem of classifying all 2d CFT's is therefore equivalent to solving eq. (2.43) for all possible values of the central charge c (and adding the modular invariance condition, see Ch. III). At the special values of c and h corresponding to the minimal models, the conformal blocks can be determined as solutions to the linear differential equations resulting from a presence of certain null states (see next sections). An investigation of general corollaries of eq. (2.43) is one of the central problems in CFT.

The two basic properties of CFT we have used so far were (i) locality, and (ii) associativity of OPE's. Although conformal blocks are generically *multi-valued*, physical correlators must always be single-valued. The multi-valuedness or the non-trivial **monodromy** of conformal blocks means that they are not just ordinary functions, but rather sections of some non-trivial vector bundle.

The notion of duality can be generalized to the M-point correlation functions (and, in fact, to arbitrary-genus Riemann surfaces as well). A general correlation function $G_{i_1 \cdots i_M}(z_1, \ldots, \bar{z}_M)$ can be expressed in terms of more general conformal blocks pictured in Fig. 9.

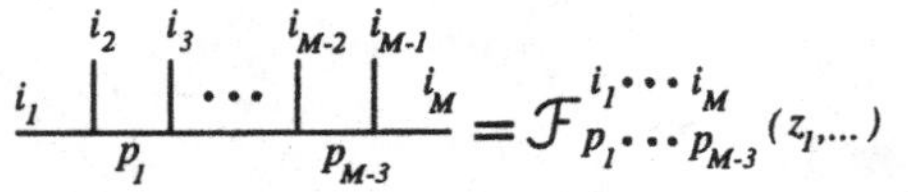

$$\frac{\overset{\displaystyle i_2 \quad i_3 \qquad i_{M-2} \; i_{M-1}}{\big| \quad \big| \quad \cdots \quad \big| \quad \big|}}{\underset{p_1 \qquad\qquad p_{M-3}}{\rule{0pt}{0pt}}} {}_{i_M} = \mathcal{F}^{i_1 \cdots i_M}_{p_1 \cdots p_{M-3}}(z_1, \dots)$$

Fig. 9. The graphical representation of a general
conformal block.

Any physical correlator has to be a *monodromy-invariant* combination of $\mathcal{F}_{\vec{p}_1}$ and $\overline{\mathcal{F}}_{\vec{p}_2}$, of the form

$$G_{i_1 \cdots i_M}(z_1, \bar{z}_1; \dots; z_M, \bar{z}_M) = \sum_{\vec{p}_1, \vec{p}_2} h_{\vec{p}_1, \vec{p}_2} \mathcal{F}^{i_1 \cdots i_M}_{\vec{p}_1}(z_1, \dots) \overline{\mathcal{F}^{i_1 \cdots i_M}_{\vec{p}_2}(z_1, \dots)}, \qquad (2.46)$$

where an invariant metric $h_{\vec{p}_1, \vec{p}_2}$ is to be constructed out of the structure constants (couplings) C_{ijk}'s. Quite generally, the Green's function G should be (i) a single-valued real analytic function of the coordinates z_i's and moduli of the Riemann surface, (ii) independent of the order of its arguments, and (iii) independent of the basis of conformal blocks used to compute it. As far as conformal blocks are concerned, this is *not* true: a function $\mathcal{F}$ is, in general, multi-valued. It is the monodromy (or braid) transformation laws, that are to be used to define $\mathcal{F}$ for some ordering of its arguments, and to analytically continue it to different domains (sect. XII.6).

II.4   Null states

'Solving' a CFT means finding its full spectrum and all correlation functions. Having the infinite-dimensional conformal symmetry, the general strategy towards formulating solvable CFT's is to impose as many constraints as possible, in a way to be consistent with this symmetry. This way to proceed is just one of the ways to realize the '*bootstrap*' idea: imposing all possible conformal constraints should result in not just very special CFT's but, presumably, the *all* of them which are internally consistent, with no additional symmetries used.

The closely related idea is to start to look for *unitary* representations of the Virasoro algebra. Each CFT state can be expressed as a linear conbination of primary and secondary states. The only relevant representations having a chance to contain unitary Virasoro irreps are just the highest-weight representations, since they have 'energy' levels bounded from below. [3] Starting from a highest-weight state $|h\rangle$, we can build up the associated Verma module. However, this module does not necessarily have a positively definite scalar product, i.e. it may not be a Hilbert space, in general. This depends on the structure of the Verma module for given values of c and h.

The descendant state $|\chi\rangle$ satisfying the equations

$$L_0 |\chi\rangle = (h + N) |\chi\rangle , \quad L_n |\chi\rangle = 0 \ \text{ for } \ n > 0 , \tag{2.47}$$

is known as a **null state** or a **singular vector**. The null state is simultaneously a primary and a descendant state, and it is a highest-weight state also. To obtain a non-degenerate representation, one has to eliminate all null states and all of their descendants, and consider the reduced theory. Later we will see how it helps to evaluate conformal blocks, which form a (fully reducible) representation of the duality transformations. The elimination of null states will result in non-trivial linear differential equations for conformal blocks.

We are now looking for degenerate representations of the Virasoro algebra which contain null states. Eq. (2.47) can be equally interpreted as the requirement of conformal invariance of the constraint $|\chi\rangle = 0$ we want to impose.

Consider a Verma module $\mathcal{V}(c, h)$ of the Virasoro algebra, with a highest-weight vector $|h\rangle$. If there is a null vector $|\chi\rangle$ satisfying eq. (2.47), this means that $\langle\psi\,|\,\chi\rangle = 0$ for any state $\langle\psi|$ of the theory. Setting $|\chi\rangle = 0$ is equivalent to taking the quotient by the null vectors and submodules they generate. At the level one, the only possibility is to take $|\chi\rangle = L_{-1} |h\rangle = 0$ or $\hat{L}_{-1}\phi_h(z) = \partial_z\phi_h(z) = 0$ (see the third line of eq. (1.59)). This obviously leads to the identity operator and $|h\rangle = |0\rangle$, which is a trivial representation. At the level two, it may happen that

$$|\chi\rangle = L_{-2} |h\rangle + a L_{-1}^2 |h\rangle = 0 \tag{2.48}$$

for some value of the parameter a. At least, eq. (2.48) respects the scaling symmetry because both terms on the l.h.s. have the same scaling dimension. Since all L_n operators of the Virasoro algebra at $n > 0$ are generated (in the algebraic sense) by L_1 and L_2, applying L_1 and L_2 to eq. (2.48) is enough to ensure the whole conditions for a null state in eq. (2.47).

---

[3] The 'energy' levels are always the eigenvalues of the Hamiltonian. In the case under consideration, they coincide with the 'levels' – the eigenvalues of the operator L_0.

Acting with L_1 on eq. (2.48) and using the Virasoro algebra yields $[3 + 2a(2h + 1)]L_{-1}|h\rangle = 0$, whereas acting with L_2 results in $[4h + c/2 + 6ah]|h\rangle = 0$. Hence, we find

$$a = -\frac{3}{2(2h+1)}\ , \qquad c = \frac{2h(5-8h)}{2h+1}\ , \tag{2.49}$$

and, therefore,

$$|\chi_h\rangle = \left(L_{-2} - \frac{3}{2(2h+1)}L_{-1}^2\right)|h\rangle\ , \tag{2.50}$$

where

$$h = \frac{1}{16}\left(5 - c \pm \sqrt{(c-1)(c-25)}\right)\ . \tag{2.51}$$

Eq. (2.50) can be rewritten to the form

$$\chi_h(z) \equiv \hat{L}_{-2}\phi_h(z) - \frac{3}{2(2h+1)}\hat{L}_{-1}^2\phi_h(z) = 0\ . \tag{2.52}$$

We are now going to show that eq. (2.52) leads to a second-order partial differential equation for the correlation functions containing the degenerate primary field $\phi_h(z)$. Eq. (2.52) implies that the correlation functions of ϕ_h with other fields are annihilated by the differential operator $\mathcal{L}_{-2} - \frac{3}{2(2h+1)}\mathcal{L}_{-1}^2$ because of eq. (2.18). Since

$$\hat{L}_{-1}\phi_h(z) = \frac{\partial}{\partial z}\phi_h(z)\ , \tag{2.53}$$

and the definition of $\mathcal{L}_{-n}$ in eq. (2.19), one gets

$$\langle\phi_1(z_1)\cdots\phi_M(z_M)\chi_h(z)\rangle = \left[\frac{3}{2(2h+1)}\frac{\partial^2}{\partial z^2} - \sum_{j=1}^{M}\frac{h_j}{(z-z_j)^2}\right.$$
$$\left. - \sum_{j=1}^{M}\frac{1}{z-z_j}\frac{\partial}{\partial z_j}\right]\langle\phi_1(z_1)\cdots\phi_M(z_M)\phi_h(z)\rangle = 0\ . \tag{2.54}$$

Eq. (2.54) is the second-order partial differential equation, whose solutions are expressible in terms of standard hypergeometric functions. The monodromy conditions pick up physically relevant particular solutions (sect. 7). As to the 4-point function, using the $SL(2,\mathbf{C})$ invariance of eq. (2.54) leads to the ordinary hypergeometric differential equation, whose solutions could be found, for instance, via Laplace method. This leads to hypergeometric functions as the solutions.

The procedure of selecting null states can, of course, be repeated level by level (see excercises II-5 and II-6, as to the level three). In general, the states of the Verma module $\mathcal{V}(c,h)$ at level N form a linear space spanned by linear combinations of

$$\left\{\left|h, \{-\vec{k}\}\right\rangle = L_{-n_1}\cdots L_{-n_k}|h\rangle\right\}\ , \qquad \sum n_i = N\ . \tag{2.55}$$

The level-N null state equation is given by

$$\left| \chi^{(N)} \right\rangle \equiv \sum b_{\{-\vec{k}\}} \left| h, \{-\vec{k}\} \right\rangle = 0 \ . \tag{2.56}$$

The scalar products of the states (2.55) form a matrix

$$\langle h | \, (L_{+n_k} \cdots L_{+n_1})(L_{-m_1} \cdots L_{-m_{k'}}) \, | h \rangle \equiv \hat{M}^{(N)}_{\{n\}\{m\}} \ , \qquad \sum n_i = \sum m_i = N \ , \tag{2.57}$$

which can be calculated by commuting $\{L_{+n_i}\}$ through $\{L_{-m_j}\}$ and applying them to the highest-weight state $|h\rangle$ eventually. The determinant of this matrix is known as the **Kač determinant** det $\hat{M}^{(N)}(c, h)$ [44, 45],

$$\det \hat{M}^{(N)}(c, h) = \prod_{k=1}^{N} \prod_{mn=k} \left[h - h(m, n) \right]^{P(N-k)} \ , \tag{2.58}$$

where m and n are positive integers, and

$$h(m, n) = \frac{1}{48} \Big[(13 - c)(m^2 + n^2) - 24mn$$

$$-2(1 - c) + \sqrt{(1 - c)(25 - c)(m^2 - n^2)} \Big] \ . \tag{2.59a}$$

Zeroes of the Kač determinant (i.e. zero eigenvectors of the matrix $\hat{M}^{(N)}(c, h)$) determine the values of (c, h) whose associated Verma modules contain the level-N null states of zero norm. These states must vanish in a positively definite Hilbert space. The zeroes of the Kač determinant det $\hat{M}_{(N)}(c, h)$ are quite clear from eq. (2.58). The Verma module $\mathcal{V}(c, h)$ of central charge c has a null vector at level $N = m \times n$ if and only if [44, 45]

$$h = h(m, n) = -\alpha_0^2 + \tfrac{1}{4}(n\alpha_+ + m\alpha_-)^2 \ , \tag{2.59b}$$

where

$$c = c(\alpha_+, \alpha_-) = 1 - 24\alpha_0^2 \ , \qquad 2\alpha_0 \equiv \alpha_+ + \alpha_- \ , \qquad \alpha_+ \alpha_- = -1 \ , \tag{2.60a}$$

or

$$\alpha_\pm = \alpha_0 \pm \sqrt{\alpha_0^2 + 1} = \frac{\sqrt{1 - c} \pm \sqrt{25 - c}}{\sqrt{24}} \ . \tag{2.60b}$$

Given a central charge value c, $0 < c < 1$, this solution represents the discrete set parametrized by two positive integers $n, m \geq 1$, which is known as the **Kač table**. The discrete set of conformal dimensions means the discrete set of primary fields $\{\phi_{(n,m)}; h_{(n,m)}\}$ in the corresponding CFT. The parametrization chosen in eqs. (2.60) is closely related with the 'Coulomb gas picture' to be discussed in sect. 6 (see also refs. [45, 46]). It is important to note that the corresponding CFT *closes* [12, 13], namely, there are no new operators beyond $\{\phi_{(n,m)}; h_{(n,m)}\}$ on the r.h.s. of OPE for any two of them,

$$\phi_1(z_1)\phi_2(z_2) \sim \sum_p \frac{C_{12}^p}{(z_1 - z_2)^{h_1+h_2-h_p}} \phi_p(z_2) \ . \tag{2.61}$$

This can be proved, for instance, by studying singularities of the 4-point functions and the characteristic equation of the differential equations they satisfy. One finds that, if h_1 and h_2 belong to the set $\{h_{(n,m)}\}$, the h_p belongs to the same set too [12]. This will be done in more detail in sect. 6.

When $c > 25$, eqs. (2.59) and (2.60) imply that both α_+ and α_- are purely imaginary. When the parameters n, m grow, some dimensions $h(n,m)$ become negative, so that the associated representation is definitely non-unitary. When $25 > c > 1$, the conformal dimensions are generically complex. The dimensions are *real* and positive only if $c \leq 1$.

Remarkably, when α_-/α_+ is a *rational* number, each module $\mathcal{V}(c,h)$ contains the *infinite* number of null states. After deleting all the sub-modules associated with the null highest-weight states, the operator algebra truncates to a *finite* set, as was first shown in ref. [12] (see sect. 8 also). In general, CFT's with a *finite* number of primary fields are called **rational** conformal field theories (RCFT's). The **minimal models** of Belavin, Polyakov and Zamolodchikov (BPZ) are the particular RCFT's of central charge $c < 1$ [12]. They are called *minimal* since (i) they are based on a finite number of scalar primary fields, (ii) they have no multiplicities in their spectra of conformal dimensions, and (iii) they contain no additional symmetries except the conformal symmetry [12, 13].
[4] The minimal models are characterized by

$$\alpha_-/\alpha_+ = -p/q \ , \quad c = 1 - \frac{6(p-q)^2}{pq} \ ,$$

$$h(n,m) = \frac{1}{4pq}\left[(nq - mp)^2 - (p-q)^2\right] \ , \tag{2.62}$$

where p and q are positive integers having no non-trivial common divisors. Because of the reflection property,

$$h(n,m) = h(p-n, q-m) \ , \tag{2.63}$$

we can restrict the values of n, m to the rectangle:

$$0 < n < p, \quad 0 < m < q, \quad p < q \ . \tag{2.64}$$

---

[4] Non-minimal RCFT's with additional symmetries and/or $c \geq 1$ are discussed in Chs. IV, V and VII. A CFT with irrational central charge is called an *irrational* CFT (ICFT) [47].

An investigation of unitary representations of the Virasoro algebra can also be based on a study of the Kač determinant. If this determinant is negative at some level, this means there are negative norm states and the representation is not unitary. If $\det \hat{M}^{(N)}(c, h) \geq 0$ for all N, it still requires further checks for unitarity.

As far as the region $c > 1$, $h \geq 0$, is concerned, it is easy to see that the Kač determinant has no zeroes at any level. The non-vanishing $\det \hat{M}^{(N)}$ implies that all the eigenvalues of the matrix $\hat{M}^{(N)}$ are positive. This is because for large h, this matrix becomes dominated by its diagonal elements (they are of the highest order in h). These matrix elements are all positive, so that the matrix $\hat{M}^{(N)}(h)$ has all positive eigenvalues for large h. Since the Kač determinant never vanishes for $c > 1, h \geq 0$, all of the eigenvalues must be positive in the entire region.

On the boundary $c = 1$, the determinant vanishes at the points $h = n^2/4$ [17], but does not become negative. Hence, we conclude that the Kač determinant poses no obstacles to having unitary representations of the Virasoro algebra for any $c \geq 1$, $h \geq 0$.

The region $0 < c < 1$, $h > 0$ needs a careful treatment. Most of studies in CFT were restricted to the case of the RCFT's (2.62), since a general analysis of irrational CFT's turned out to be quite involved [12, 47] and it has yet to find a practical use (see, however, ref. [47]). The idea is to draw the vanishing curves $h = h(m, n; c)$ in the (h, c) plane and consider the behaviour near $c = 1$. One finds that any point in the region $0 < c < 1$, $h > 0$, can be connected to the $c > 1$ region by a path that crosses a single vanishing curve of the Kač determinant at some level (Fig. 10). The vanishing of the determinant is due to a single eigenvalue crossing through zero, so the determinant reverses its sign when passing through the vanishing curve. This means that there must be a negative norm state, which excludes unitary representations of the Virasoro algebra at all points in this region except those *on* the vanishing curves where the determinant vanishes. The subsequent detailed analysis of the determinant shows [48, 49, 50] that there is an additional negative norm state everywhere on the vanishing curves except at the certain points where they intersect (Fig. 10).

This discrete set of points, where unitary representations are not excluded, has $q = p + 1$ and $p \geq 2$. According to eq. (2.62), the central charge is then given by

$$c = 1 - \frac{6}{p(p + 1)} \ . \tag{2.65}$$

One can associate $p(p - 1)/2$ allowed values of conformal dimension h with each such

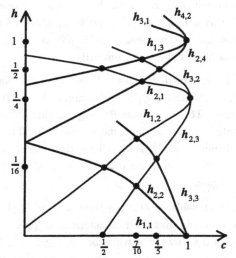

Fig. 10. The first few vanishing curves $h=h(m,n;c)$ in the h,c plane.

value of central charge c, in accordance with eq. (2.62),

$$h(n,m) = \frac{1}{4p(p+1)} \left[(n(p+1) - mp)^2 - 1 \right] .\qquad(2.66)$$

Therefore, the *necessary* conditions for the existence of unitary highest-weight representations of the Virasoro algebra are $c \geq 1$ and $h \geq 0$, *or* eqs. (2.65) and (2.66). More details about the unitarity proof can be found in ref. [50]. The *sufficiency* of the last two conditions, in eqs. (2.65) and (2.66), will be shown via the coset construction in Ch. VI.

Some of the CFT's introduced in this section are closely related to various statistical systems at their phase transition points in the continuous limit. This connection is illustrated on the particular examples in sect. 7.

Exercises

  # II-5 ▷ Consider a third-level constraint equation having the form

$$L_{-3} \left| \tilde{h} \right\rangle + aL_{-2}L_{-1} \left| \tilde{h} \right\rangle + bL_{-1}^3 \left| \tilde{h} \right\rangle = 0 ,\qquad(2.67)$$

where $\left| \tilde{h} \right\rangle$ is a primary state of dimension $\tilde{h}$, and a, b are some coefficients. The conformal invariance conditions are equivalent to the annihilation of the state (2.67) by the Virasoro operators L_1 and L_2. Prove that it results in

$$a = -\frac{2}{\tilde{h}} \ , \quad b = \frac{1}{\tilde{h}(1 + \tilde{h})} \ . \tag{2.68}$$

II-6 ▷ Derive the third-order partial differential equation for the correlation functions of a null operator $\phi_{\tilde{h}}$ with primary fields, provided the operator $\phi_{\tilde{h}}$ corresponds to the level-3 null state from the exercise # II-5.

II.5   Fusion rules

Since the OPE's for any descendants inside conformal families are determined by those of primaries, it is a good idea to introduce the so-called **fusion rules** for the conformal families, which formally read

$$[\phi_i] \times [\phi_j] \simeq \sum_k [\phi_k] \ . \tag{2.69}$$

These rules determine which conformal families $[\phi_k]$ have their primaries or descendants occuring in an OPE of any two members of the conformal families $[\phi_i]$ and $[\phi_j]$, as well as their multiplicities.

The linear differential equations like eq. (2.54) can be solved, in principle, for any set of degenerate operators. When being applied to the 3-point functions, this method does not determine the OPE coefficients C_{ijk} in full, but provides useful selection rules, namely, the fusion rules. To illustrate how the fusion rules can be obtained, let us take $h = h(1,2)$ or $h = h(2,1)$ as an example [12]. In this case, the null vector appears already at level 2, and the associated differential equation obtained by its decoupling is just eq. (2.54). The OPE between $\phi_h(z)$ and $\phi_{h_1}(z_1)$ takes the form

$$\phi_h(z)\phi_{h_1}(z_1) = \text{const.} (z - z_1)^k \left[\phi_{h'}(z_1) + b(z - z_1)\phi_{h'}^{(-1)}(z_1) + \dots \right] \ . \tag{2.70}$$

The consistency between eqs. (2.54) and (2.70) yields the relation

$$\frac{3}{2(2h+1)} k(k-1) - h_1 + k = 0 \ . \tag{2.71}$$

In addition, the scale invariance implies

$$k = h' - h_1 - h \ . \tag{2.72}$$

Parametrizing dimensions as in eq. (2.58),

$$h_1 = -\alpha_0^2 + \tfrac{1}{4}(n\alpha_+ + m\alpha_-)^2 \ , \quad h' = -\alpha_0^2 + \tfrac{1}{4}(n'\alpha_+ + m'\alpha_-)^2 \ , \tag{2.73}$$

we find

$$
\begin{aligned}
h &= h(1,2) \ : \ n' = n \ , \quad m' = m \pm 1 \ , \\
h &= h(2,1) \ : \ n' = n \pm 1 \ , \quad m' = m \ .
\end{aligned}
\tag{2.74}
$$

Hence, in this case, the fusion rules are

$$
\begin{aligned}
\left[\phi_{(1,2)}\right] \times \left[\phi_{(n,m)}\right] &= \left[\phi_{(n,m-1)}\right] + \left[\phi_{(n,m+1)}\right] \ , \\
\left[\phi_{(2,1)}\right] \times \left[\phi_{(n,m)}\right] &= \left[\phi_{(n-1,m)}\right] + \left[\phi_{(n+1,m)}\right] \ .
\end{aligned}
\tag{2.75}
$$

Since the normalization of ϕ's still remains to be determined, some of the families on the r.h.s. of eq. (2.75) may not really appear. Similar arguments can be used to derive fusion rules in a general CFT, once its null states are explicitly known.

Considering multiple insertions of ϕ_h (with $h = h(1,2)$ or $h = h(2,1)$) and using associativity of OPE in the minimal models, it is possible to derive the BPZ selection rule for the non-vanishing correlators $\left\langle \phi_{(m_1,n_1)} \phi_{(m_2,n_2)} \phi_{(m_3,n_3)} \right\rangle$ of degenerate fields $\phi_{(m,n)}$. They take the following form [12]:

$$\left[\phi_{(m_1,n_1)}\right] \times \left[\phi_{(m_2,n_2)}\right] = \sum_{k=(n_1-n_2)+1}^{n_1+n_2-1} \ \sum_{l=(m_1-m_2)+1}^{m_1+m_2-1} \left[\phi_{(k,l)}\right] \ . \tag{2.76}$$

If $|n_1 - n_2|$ is odd (even), then the sum over k runs over only even (odd) values, and similarly for $|m_1 - m_2|$ and l. The selection rules (2.76) resemble the $SU(2)$ branching rules, but they are not identical to them. Instead, they are the selection rules for what is known as the affine Lie (or Kač-Moody) algebra $\widehat{SU}(2)$ (Ch. IV).

Eq. (2.76) contains only the holomorphic parts of fields, satisfying a *commutative* and *associative* algebra. In a more specific form, the fusion rules read [51]

$$[\phi_i] \times [\phi_j] = \sum_k N_{ij}{}^k [\phi_k] \ , \tag{2.77}$$

where $[\phi_k]$'s denote primary fields, and $N_{ij}{}^k$'s are non-negative integers, which can be interpreted as the numbers of independent fusion paths from ϕ_i and ϕ_j to ϕ_k. Clearly, $N_{ij}{}^k = 0$ implies $C_{ij}{}^k = 0$. The inverse statement is also true [52]. This property of CFT is known as *naturality*.

The coefficients $N_{ij}{}^k$ are automatically symmetric in i and j, and satisfy a quadratic equation due to associativity of the multiplication in eq. (2.77). It can be shown [53] that for CFT with a *finite* number of primaries (i.e. for a RCFT), all the dimensions h_i are *rational* numbers. The minimal models with $0 < c < 1$ are the examples of RCFT. Another important class of RCFT's is provided by the WZNW theories (Ch. IV). The rationality condition means the indices of $N_{ij}{}^k$ are running over a finite set of values, and, hence, summations over them are well-defined. Using the matrix notation, $(N_i)_j{}^k \equiv N_{ij}{}^k$, the associativity of the OPE implies $[N_i, N_j] = 0$ or $N_i N_j = \sum_k N_{ij}{}^k N_k$. Therefore, the matrices N_i themselves form a commutative and associative matrix representation of the fusion rules (2.77).

In a RCFT, the Virasoro algebra may be extended to a more general symmetry algebra, normally represented by a W algebra (Ch. VII) and called a *chiral algebra* (see Chs. IV and VII for more). The problem of classifying all RCFT's is therefore reduced to finding all possible chiral algebras and automorphisms of the associated fusion rules [52]. Unfortunately, there are no general rules to determine the fusion rules in CFT.

II.6 Coulomb gas picture

As was shown by Dotsenko and Fateev [54, 55], all the essential features of the minimal models can be obtained by using a single scalar field interacting with a background charge. This approach is known as the **'Coulomb gas'** or **Dotsenko-Fateev** (DF) construction. Their method is very effective in computing correlations functions, as will be illustrated on a particular example in sect. 7.

Let us consider a reparametrization-invariant 2d action of a bosonic scalar field $\Phi(z, \bar{z})$ on the Riemann sphere $S^2 \simeq \mathbf{C} \cup \{\infty\}$,

$$S_{\text{DF}}(\Phi) = \frac{1}{8\pi} \int d^2 z \, (\partial_z \Phi \partial_{\bar{z}} \Phi + 2i\alpha_0 \sqrt{g} R \Phi) \ , \tag{2.78}$$

where the additional term proportional to the 2d (scalar) curvature R has been introduced. In eq. (2.78), α_0 is a real constant. The correlation functions are formally defined by eq. (1.3) with the action (2.78).

The curvature of the sphere can be concentrated at the north pole, representing the point at infinity, by the appropriate choice of metric. The Riemann sphere can then be described by the two coordinate charts, the nothern one in the vicinity of the north pole and the southern one everywhere outside it. For the latter, we can take the metric in the form (1.29), just as that for a plane. The curvature term in the action (2.78) results in the additional term in the stress tensor having the form

$$T_{\mathrm{DF}}(z) = -\tfrac{1}{4} : \partial\varphi(z)\partial\varphi(z) : +i\alpha_0\partial^2\varphi(z) , \tag{2.79}$$

where the solution to the equations of motion everywhere except at the infinity (the north pole), $\Phi(z, \bar{z}) = \varphi(z) + \bar{\varphi}(\bar{z})$, has been used. [5] As is usual, the normal ordering has been introduced in eq. (2.79), in order to subtract a singular term in the operator product. The main achievement is that the field φ can still be considered as a free field, with the constant $-2\alpha_0$ representing the *background charge* (see eq. (2.93) below).

Eq. (2.79) is exactly the modification needed for the scalar field stress tensor to obey the Virasoro algebra with the central charge $c_{\mathrm{DF}} = 1 - 24\alpha_0^2$ (*cf* eq. (2.60a)). The extra term in eq. (2.79) is the derivative of the well-defined conformal field $\partial\varphi$ and, hence, it does not affect the status of $T_{\mathrm{DF}}(z)$ as the generator of conformal transformations. Since the stress tensor $T_{\mathrm{DF}}(z)$ has the imaginary part, the theory it defines is not unitary for an *arbitrary* α_0. The modification of $T_\varphi(z)$ to $T_{\mathrm{DF}}(z)$ in eq. (2.79) can be interpreted as the one forced by a presence of the background charge $-2\alpha_0$ at infinity. The $U(1)$ current $j(z) = i\partial\varphi(z)$ is anomalous in this theory:

$$T_{\mathrm{DF}}(z)j(w) \sim \frac{1}{(z-w)^2}j(z) - \frac{2\alpha_0}{(z-w)^3} ,$$

$$j(z)j(w) \sim \frac{1}{(z-w)^2} , \tag{2.80}$$

and, therefore, it is not a primary field unless $\alpha_0 = 0$.

The DF primary fields are just the exponentials of the free field φ since

$$T_{\mathrm{DF}}(z) : e^{i\alpha\varphi(w)} : \sim \frac{\alpha(\alpha - 2\alpha_0)}{(z-w)^2} : e^{i\alpha\varphi(w)} : +\frac{1}{z-w}\partial_w e^{i\alpha\varphi(w)} . \tag{2.81}$$

These primary fields

$$V_\alpha(z) \equiv : e^{i\alpha\varphi(z)} : \tag{2.82}$$

are interpreted as the *chiral vertex operators* in string theory (Ch. VIII), or as the physical operators in statistical models (sect. 7). In what follows they will be often referred to as the vertex operators.

Compared to eq. (1.96), the background charge results in the shift of the conformal dimension of $: e^{i\alpha\varphi} :$ from α^2 to $\alpha(\alpha - 2\alpha_0)$. Eq. (1.97) implies

$$\left\langle : e^{i\Phi(z,\bar{z})} : : e^{-i\Phi(w,\bar{w})} : \right\rangle = |z - w|^{-2} = \left\langle b(z)c(w)\bar{b}(\bar{z})\bar{c}(\bar{w}) \right\rangle , \tag{2.83a}$$

---

[5] The normalization of the scalar field here differs from the normalization adopted in sect. I.5 by the factor of $\sqrt{2}$.

or just

$$\langle : e^{i\varphi(z)} : : e^{-i\varphi(w)} : \rangle = (z - w)^{-1} = \langle b(z)c(w) \rangle . \qquad (2.83b)$$

The bosonic operator $: e^{i\Phi(z,\bar{z})} :$ corresponds to the fermionic bilinear $(b\bar{b})(z, \bar{z})$. We are going to consider the chiral fields $: e^{\pm i\varphi(z)} :$ of the theory (2.78) as the *bosonized* version of the (b, c) ghost system. The anomaly of the bosonic $U(1)$ current j should then coincide with the anomaly of the (chiral) fermion number current, $-bc$, which leads to the identification $-2\alpha_0 = 2\lambda - 1$. This can be seen by comparing eqs. (1.136) and (1.137) with eq. (2.80), and using the quantum equivalence. Then, on the one hand, the Riemann-Roch theorem [28] implies the anomalous divergence equation, $\bar{\partial}(-bc) = -\frac{1}{2}(2\lambda - 1)\sqrt{g}R$ (see eq. (1.138)), while, on the other hand, one easlily finds the 'anomalous' equation of motion, $\bar{\partial}(i\partial\varphi) = \frac{1}{2}(-2\alpha_0)\sqrt{g}R$, from eq. (2.78). Hence, we should take $\alpha = 1$ for the $(b\bar{b})$ composite field and $\alpha = -1$ for the $(c\bar{c})$ composite field, in accordance with the conformal dimensions of the fields b and c.

Let us now turn to a discussion of the correlation functions of the vertex operators on the sphere, and consider the case when $\alpha_0 = 0$ first. Since the vacuum has to be neutral, the only non-vanishing correlation functions are those with the vanishing total charge defined w.r.t. the current $j(z) = i\partial\varphi(z)$,

$$Q = \oint \frac{d\zeta}{2\pi i} j(\zeta) . \qquad (2.84)$$

The OPE between the current $j(w)$ and the vertex operator $V_\alpha(z)$ takes the form

$$j(w)V_\alpha(z) \sim \frac{\alpha}{w - z} V_\alpha(z) , \qquad (2.85)$$

which implies

$$[Q, V_\alpha(z)] = \alpha V_\alpha(z) . \qquad (2.86)$$

Hence, the operator $V_\alpha(z)$ inserts at the point z the charge α, which is the eigenvalue of the charge operator Q.

The defining equation (1.3) can be used to directly calculate the two-point non-chiral correlation function as follows:

$$\langle V_\alpha(z_1, \bar{z}_1) V_{-\alpha}(z_2, \bar{z}_2) \rangle = \exp\left[-\frac{\alpha^2}{2} \left(2\langle \varphi^2 \rangle - 2\langle \varphi(z_1, \bar{z}_1)\varphi(z_2, \bar{z}_2) \rangle \right) \right]$$

$$= \exp\left[-4\alpha^2 \left(\ln\frac{\mu}{\varepsilon} - \ln\frac{\mu}{|z_1 - z_2|} \right) \right] = \left| \frac{\varepsilon}{z_1 - z_2} \right|^{4\alpha^2} \sim \frac{1}{|z_1 - z_2|^{4\alpha^2}} , \qquad (2.87)$$

where the propagator

$$\langle \varphi(z_1, \bar{z}_1) \varphi(z_2, \bar{z}_2) \rangle = 4 \ln \frac{\mu}{|z_1 - z_2|} , \tag{2.88}$$

has been used. The renormalization parameters μ and ε represent the *infra-red* (IR) and *ultra-violet* (UV) cut-off scales, resp., which are always needed in describing quantized massless scalar fields in two dimensions [1, 27]. Eq. (2.87) implies

$$\langle V_\alpha(z_1) V_{-\alpha}(z_2) \rangle \sim \frac{1}{(z_1 - z_2)^{2\alpha^2}} , \tag{2.89}$$

The 4-point function can also be found this way, *viz.*

$$\langle V_{\alpha_1}(z_1) V_{\alpha_2}(z_2) V_{-\alpha_3}(z_3) V_{\alpha_4}(z_4) \rangle \sim \prod_{i<j} (z_i - z_j)^{2\alpha_i \alpha_j} , \tag{2.90}$$

where the vertex operator charges $\{\alpha_j\}$ are subject to the condition

$$\sum_j \alpha_j = 0 . \tag{2.91}$$

This is just the condition that ensures cancellation of all the IR regulating factors μ in the correlation functions. These regulators explicitly enter the correlators via the propagator (2.88). According to sect. I.5, the CFT under consideration (without a background charge) has the central charge $c = 1$.

The extension to the case of $c < 1$ can be done by modifying the quantization prescription in a presence of background charge, and by taking into account the anomaly structure due to the curvature singularity. Quite generally, the anomaly leads to a dependence of a correlator not only upon the points where the vertex operators are located but also upon a point of metric's singularity. The latter has been chosen to be the infinity above. To ensure proper transformation properties of the correlation functions, one should either insert the additional vertex operator (the so-called '**vacuum charge**' [54]) at the infinity or, equivalently, assign this charge to the vacuum state. Because of the anomaly, we have $Q |0\rangle = 0$ but $\langle 0| Q \neq 0$, with our choice of the location of metric's singularity at $z = \infty$.

In other words, correlation functions of the vertex operators should vanish unless the vacuum (background) charge is screened by charges of the vertex operators:

$$\langle V_{\alpha_1}(z_1) \cdots V_{\alpha_M}(z_M) \rangle = \begin{cases} \prod_{i<j} (z_i - z_j)^{2\alpha_i \alpha_j} & \text{if } \sum \alpha_i = 2\alpha_0 , \\ 0 & \text{otherwise;} \end{cases} \tag{2.92}$$

One of the operators should now represent the vacuum charge which effectively reproduces the anomaly. The correlators can be renormalized in such a way that only their
dependence upon the vertex operator insertions has to be considered. The condition

$$\sum \alpha_i = 2\alpha_0 \tag{2.93}$$

just means the overall charge neutrality of correlation functions. Eq. (2.93) can also be
derived from the Ward identities resulting from the invariance of the theory (2.78) under
constant shifts of the field Φ, $\Phi \to \Phi + c$ (exercise # II-7).

The DF correlation functions are defined by

$$\langle V_{\alpha_1}(z_1) \cdots V_{\alpha_M}(z_M) \rangle = \lim_{\mu \to \infty} \left\langle \mu^{8\alpha_0^2} V_{\alpha_1}(z_1) \cdots V_{\alpha_M}(z_M) V_{-2\alpha_0}(\infty) \right\rangle_{(0)} , \tag{2.94}$$

where the correlator on the r.h.s. is supposed to be calculated in the usual (naive) way
indicated by the subscript (0).

In particular, the 2-point function takes the form

$$\langle V_\alpha(z_1) V_{2\alpha_0 - \alpha}(z_2) \rangle \sim \frac{1}{(z_1 - z_2)^{2\alpha(\alpha - 2\alpha_0)}} . \tag{2.95}$$

It is to be compared with the orthogonality condition in eq. (1.64), where the 2-point
correlators are only non-vanishing for the fields with the same conformal dimensions.
Therefore, there should be $h_\alpha = h_{2\alpha_0 - \alpha} = \alpha^2 - 2\alpha\alpha_0$, which is in exact agreement with
eq. (2.81).

The OPE between the field j and the vertex operator V_α still takes the form of
eq. (2.85), which implies eq. (2.86) as before, even though j is no longer a primary
field because of the anomaly in eq. (2.80). There exist fields of dimension one (called
currents), which can be used to define conformally invariant charges, namely

$$J_\pm =: e^{i\alpha_\pm \varphi} : , \tag{2.96}$$

where

$$\alpha_+ + \alpha_- = 2\alpha_0 , \quad \alpha_+ \alpha_- = -1 . \tag{2.97}$$

This gives back the definitions made in eq. (2.60) for the Kač formula! Clearly, both
operators, $V_\alpha(z)$ and $V_{2\alpha_0 - \alpha}(z)$, have the same conformal dimension, and we have a
choice to use either of them to represent a vertex operator in correlation functions. The
existence of these currents allows us to introduce the so-called **screening** [13, 54, 55] or
Feigin-Fuchs [45] operators (charges) $Q_\pm$,

$$Q_\pm = \oint_C dz \, J_\pm(z) , \tag{2.98}$$

which are conformally invariant. The integration in eq. (2.98) is going over a non-contractable contour C. The Feigin-Fuchs operator $Q_\pm$ cannot be local since, otherwise, it can only be the identity operator.

A non-vanishing correlator may have one or more insertions of $Q_\pm$, in addition to the vacuum charge, in order to meet the condition of the vanishing total charge, while keeping the conformal properties of the vertex operators. The natural appearance of the screening operators in the correlation functions can also be seen in the Lagrangian formulation of those CFT's which can be realized as WZNW models (Ch. IV).

Since the screening currents $J_\pm(z)$ have dimension one, the screening operators commute with the Virasoro generators,

$$[Q_\pm, L_n] = 0 .$$

(2.99)

This implies that the *screened* vertex operator defined by the equation

$$V^{(1)}_{2\alpha_0-\beta}(z) \equiv \oint \frac{d\zeta}{2\pi i} J_+(\zeta) V_{2\alpha_0-\beta-\alpha_+}(z)$$

(2.100)

is a primary field. It has conformal dimension

$$h^{(1)}_{2\alpha_0-\beta} = (\beta + \alpha_+)(\beta + \alpha_+ - 2\alpha_0)$$

(2.101)

and charge $2\alpha_0 - \beta$. Remarkably, there actually exist only specific values of the parameter β, at which the screened vertex operator can be consistently defined. The quantization condition on β comes since the integration contour in eq. (2.100) has to be closed and non-trivial [56, 57]. The OPE of J_+ with $V_{2\alpha_0-\beta-\alpha_+}$,

$$J_+(w)V_{2\alpha_0-\beta-\alpha_+}(z) = (w-z)^{2\alpha_+(2\alpha_0-\beta-\alpha_+)} : e^{i\alpha_+\varphi(w)+i(2\alpha_0-\beta-\alpha_+)\varphi(z)} : ,$$

(2.102)

implies that it is true if and only if

$$2\alpha_+(2\alpha_0 - \beta - \alpha_+) = -m - 1 , \quad m \geq 1 .$$

(2.103)

Hence, we get

$$\beta = \tfrac{1}{2}(1 - m)\alpha_- ,$$

(2.104)

where eq. (2.97) has been used. The ordinary (non-screened) vertex operator $V_{2\alpha_0-\beta}$ has the same charge $2\alpha_0 - \beta$ but the different dimension,

$$h_{2\alpha_0-\beta} = -\alpha_0^2 + \frac{1}{4}(\alpha_+ + m\alpha_-)^2 ,$$

(2.105)

where the β-value (2.104) has been substituted. Eq. (2.105) is to be compared with eq. (2.101) for the dimension of the screened vertex operator under the restriction (2.104):

$$h^{(1)}_{2\alpha_0-\beta} = -\alpha_0^2 + \frac{1}{4}(\alpha_+ + m\alpha_-)^2 + m , \qquad (2.106)$$

which exhibits the difference $h_{2\alpha_0-\beta} - h^{(1)}_{2\alpha_0-\beta} = -m$. Comparing it with the Kač formula (2.59) suggests to identify $V^{(1)}_{2\alpha_0-\beta}$ with the null field descended from $V_{2\alpha_0-\beta}$ at level m. A good check is provided by the case $n = 2$, when $\beta = -\alpha_-/2$ and $2\alpha_0 - \beta - \alpha_+ = 3\alpha_-/2$. One finds in this case [21]

$$V^{(1)}_{2\alpha_0-\beta}(z) = \oint \frac{d\zeta}{2\pi i} J_+(\zeta) V_{2\alpha_0-\beta-\alpha_+}(z)$$

$$= \oint \frac{d\zeta}{2\pi i} (z - \zeta)^{-3} : e^{i\alpha_+\varphi(\zeta)+i3\alpha_-\varphi(\zeta)/2} :$$

$$= -\tfrac{1}{2} \left[\alpha_+^2 \partial_z\varphi(z)\partial_z\varphi(z) - i\alpha_+ \partial_z^2\varphi(z) \right] V_{2\alpha_0-\beta}(z)$$

$$= \mathcal{N} \left[\hat{L}_{-2} + \frac{3}{2(2h_{2\alpha_0-\beta}+1)} \hat{L}_{-1}^2 \right] V_{2\alpha_0-\beta}(z) , \qquad (2.107)$$

where $\mathcal{N}$ is a non-vanishing normalization constant.

It turns out to be true in general [56, 57] that the screened vertex operators appear as the Virasoro descendants of the ordinary vertex operators of the same charge. The complete set of null fields follows when considering multiple (n times) applications of the screening operators to the vertex operators $V_{2\alpha_0-\beta-n\alpha_+}$. It results in the quantization condition for the β in the form [56, 57]

$$\beta = \tfrac{1}{2}[(1 - n)\alpha_+ + (1 - m)\alpha_-] . \qquad (1.108)$$

This construction simultaneously fixes the conformal dimensions of the operators V_β and $V_{2\alpha_0-\beta}$, and leads to the degenerate Virasoro representations with

$$h(n, m) = -\alpha_0^2 + \tfrac{1}{4}(n\alpha_+ + m\alpha_-)^2 . \qquad (2.109)$$

Their spectrum exactly coincides with the Kač formula in eq. (2.59) !

The alternative way to come to the same conclusion is to consider the correlation functions in the DF construction [13, 54, 55]. The 3-point functions still follow the pattern given in sect. I.2, which is the same for the 2-point functions. When trying to define a non-vanishing 4-point function of the operators $V_\alpha(z)$, we must meet the two requirements: $\sum_i \alpha_i = 2\alpha_0$, and $\alpha_i = \alpha$ or $2\alpha_0 - \alpha$. The correlators like $\langle V_\alpha V_\alpha V_{2\alpha_0-\alpha} V_{2\alpha_0-\alpha} \rangle$ or

$\langle V_\alpha V_\alpha V_\alpha V_{2\alpha_0-\alpha}\rangle$ naively vanish because of $\sum_i \alpha_i \neq 2\alpha_0$. In the second case, the restriction in question is actually given by the relation $3\alpha + 2\alpha_0 - \alpha = 2\alpha + 2\alpha_0$, and there exists a possibility to construct a non-vanishing correlator provided α is *quantized* as

$$2\alpha = -n\alpha_+ - m\alpha_- \,, \tag{2.110}$$

We can then insert n screening operators Q_+ and m screening operators Q_- into the correlator to obtain a non-vanishing result. The spectrum of the vertex operators with non-vanishing 4-point functions is therefore given by

$$\alpha = \alpha_{n,m} = \frac{1-n}{2}\alpha_+ + \frac{1-m}{2}\alpha_- \,, \quad n, m \geq 1 \,, \tag{2.111}$$

with the associated dimensions

$$h(n, m) = \alpha_{n,m}^2 - 2\alpha_{n,m}\alpha_0$$
$$= \frac{(n\alpha_+ + m\alpha_-)^2 + (\alpha_+ - \alpha_-)^2}{4}$$
$$= -\alpha_0^2 + \tfrac{1}{4}(n\alpha_+ + m\alpha_-)^2 \,, \tag{2.112}$$

once again reproducing the Kač spectrum (2.59) of the degenerate conformal families! Given eq. (2.112), the 4-point correlation functions are given by analytic integrals in the so-called *Feigin-Fuchs representation*:

$$\oint_{C_1} du_1 \cdots \oint_{C_{m-1}} du_{m-1} \oint_{S_1} dv_1 \cdots \oint_{S_{n-1}} dv_{n-1} \tag{2.113}$$

$$\left\langle V_{\alpha_{n,m}}(z_1)V_{\alpha_{n,m}}(z_2)V_{\alpha_{n,m}}(z_3)V_{2\alpha_0-\alpha_{n,m}}(z_4)J_-(u_1)\cdots J_-(u_{m-1})J_+(v_1)\cdots J_+(v_{n-1})\right\rangle \,.$$

An explicit calculation of the correlator can be completed by using integral representations of the hypergeometric functions [54] (see sect. 7 for an example).

As an application of this construction, let us derive the fusion rules (2.76). To find the fields $\phi_{(k,l)}$ appearing in the OPE for the product of $\phi_{(m,n)}$ and $\phi_{(r,s)}$, consider the 3-point function in the Coulomb gas picture, and use $SL(2, \mathbf{C})$ invariance to write down three equivalent ways to represent it:

$$\left\langle V_{\overline{(k,l)}}(\infty)V_{(m,n)}(1)V_{(r,s)}(0)Q_+^{\cdots}Q_-^{\cdots}\right\rangle \,, \tag{2.114a}$$

$$\left\langle V_{(k,l)}(\infty)V_{\overline{(m,n)}}(1)V_{(r,s)}(0)Q_+^{\cdots}Q_-^{\cdots}\right\rangle \,, \tag{2.114b}$$

$$\left\langle V_{(k,l)}(\infty)V_{(m,n)}(1)V_{\overline{(r,s)}}(0)Q_+^{\cdots}Q_-^{\cdots}\right\rangle \,, \tag{2.114c}$$

where $V_{(m,n)} \equiv V_{\alpha(m,n)}$ and $V_{\overline{(m,n)}} \equiv V_{2\alpha_0 - \alpha(m,n)}$. According to eqs. (2.111) and (2.114a), the correlator will not vanish if

$$k \leq m + r - 1 \quad \text{with} \quad k - m - r - 1 \quad \text{even},$$
$$l \leq n + s - 1 \quad \text{with} \quad l - n - s - 1 \quad \text{even}, \tag{2.115}$$

and similarly for eqs. (2.114b) and (2.114c). It results in the following selection rules

$$k + m + r \ \text{odd} \begin{cases} k \leq m + r - 1 \\ m \leq r + k - 1 \\ r \leq k + m - 1 \end{cases}, \ l + n + s \ \text{odd} \begin{cases} l \leq n + s - 1 \\ n \leq s + l - 1 \\ s \leq l + n - 1 \end{cases}, \tag{2.116}$$

which imply

$$\left[\phi_{(m,n)}\right] \times \left[\phi_{(r,s)}\right] = \sum_{\substack{k=|m-r|+1 \\ k+m+n=\text{odd}}}^{m+r-1} \sum_{\substack{l=|n-s|+1 \\ l+n+s=\text{odd}}}^{n+s-1} \left[\phi_{(k,l)}\right]. \tag{2.117}$$

As far as the minimal models are concerned, one has $h(n,m) = h(p - n, q - m)$ in eq. (2.63). Hence, one finds

$$\left[\phi_{(n,m)}\right] \times \left[\phi_{(r,s)}\right] = \sum_{a=|n-r|+1}^{n+r-1}{}' \sum_{b=|m-s|+1}^{m+s-1}{}' \left[\phi_{(a,b)}\right], \tag{2.118a}$$

and

$$\left[\phi_{(p-n,q-m)}\right] \times \left[\phi_{(p-r,q-s)}\right] = \sum_{a=|n-r|+1}^{2p-n-r-1}{}' \sum_{b=|m-s|+1}^{2q-m-s-1}{}' \left[\phi_{(a,b)}\right], \tag{2.118b}$$

where $\sum'$ means the sum $\sum$ with the constraints dictated by eq. (2.116) to be rewritten appropriately for each case. Eqs. (2.118a) and (2.118b) are compatible as long as

$$\left[\phi_{(n,m)}\right] \times \left[\phi_{(r,s)}\right] = \sum_{a=|n-r|+1}^{\min(n+r-1,2p-1-n-r)} \sum_{b=|m-s|+1}^{\min(m+s-1,2q-1-m-s)} \left[\phi_{(a,b)}\right]_{\substack{0<n<p \\ 0<m<q}}. \tag{2.119}$$

Eq. (2.119) gives the fusion rules for the minimal models. The fusion rules for the unitary series are obtained by setting $q = p + 1$ [48].

Exercises

\# II-6 ▷ Prove that all states in a Verma module have the same charge when $\alpha_0 = 0$, i.e.

$$[Q, L_n] = 0. \tag{2.120}$$

The charge Q has been defined in eq. (2.84). How does it get to be modified with a non-vanishing background charge? *Hint*: use the OPE between the stress tensor T_{DF} and the current j in eq. (2.80).

$\#$ II-7 ▷ Derive eq. (2.93) from the Ward identities for the DF correlation functions, resulting from the invariance of the theory (2.78) under constant shifts of the field Φ.

$\#$ II-8 ▷ (*) Check the relation

$$: e^{i\alpha\varphi(z)} : \sim e^{i\alpha\varphi(z)} + \frac{\alpha^2}{2}\left\langle \varphi^2 \right\rangle \sim \frac{1}{\varepsilon^{\alpha^2}} e^{i\alpha\varphi(z)} \tag{2.121}$$

at $\alpha_0 = 0$, where the normal ordering has been defined as

$$: A(z)B(z) := \lim_{z,w \to (z+w)/2} [A(z)B(w) - \langle A(z)B(w)\rangle] . \tag{2.122}$$

Eq. (2.121) represents just another way of identifying the anomalous dimension $h_\alpha = \alpha^2$ of the vertex operator $: V_\alpha(z) = e^{i\alpha\varphi(z)} :$. In general, the conformal dimension h of a composite operator $C[\varphi](z)$ can be determined from the following relation, if any,

$$: C[\varphi](z) : \sim \Lambda^h C[\varphi](z) , \quad \Lambda \sim \frac{1}{\varepsilon} . \tag{2.123}$$

$\#$ II-9 ▷ Given $\alpha_0 \neq 0$, what is the modification of the transformation law of the scalar field $\varphi(z)$ under the conformal transformations $z \to f(z)$, which is necessary to get the right transformation law of the vertex operator $V_\alpha(z)$ as a primary field? *Hint*: use the stress tensor (2.79), and take into account the transformation of the UV cut-off ε in eq. (2.121).

$\#$ II-10 ▷ (*) Verify that the screened vertex operator $V_{2\alpha_0-\beta}(z)$ defined by eq. (2.100) is a primary field having the conformal dimension $h_{2\alpha_0-\beta} = (\beta + \alpha_+)(\beta + \alpha_+ - 2\alpha_0)$.

II.7   Ising and other statistical models

Conformally invariant quantum field theories are well suitable to describe the critical behaviour of statistical systems at second order phase transitions. The two-dimensional **Ising model** [7, 9] is the canonical example, where the spins $\sigma_i = \pm 1$ are assigned to be on the sites of a square lattice. The Ising partition function $Z = \sum_{\{\sigma\}} \exp{(-E/T)}$ is defined in terms of the energy $E = -e\sum_{<ij>} \sigma_i\sigma_j$, where the summation $< ij >$

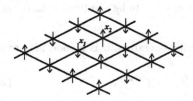

Fig. 11. The Ising lattice of spins pointing either up or down.

is going over the nearest neighbour sites on the lattice (Fig.11). [6] This model has a high temperature disordered phase (with the expectation value $\langle \sigma \rangle = 0$) and a low temperature ordered phase (with $\langle \sigma \rangle \neq 0$), and there is a second-order phase transition between them.

The Ising model is expected to be described *in the continuum limit* by the scalar field whose values give the average local magnetization of the system. At criticality, where typical configurations have fluctuations of all length scales, the field theory describing the Ising model should be conformally invariant.

Having obtained eqs. (2.65) and (2.66), we notice that the simplest non-trivial minimal model is characterized by $c = 1/2$, $p = 3$ and $q = 4$. There are only two non-trivial conformal families labeled by the primary fields $\phi_{2,2}$ (or $\phi_{1,2}$) and $\phi_{2,1}$ (or $\phi_{1,3}$) which have scaling dimensions 1/16 and 1/2, resp., since the grid of conformal dimensions according to eq. (2.66) is given by

$$h(1,1) = h(2,3) = 0 \, ,$$
$$h(1,2) = h(2,2) = 1/16 \, , \qquad (2.124)$$
$$h(2,1) = h(1,3) = 1/2 \, .$$

There is a 'doubling' of the primary fields in this example, which is the feature of the technique used, not of the physics [13, 58]. From the physical viewpoint, there should be just one primary field for each given conformal dimension, so that the primaries with the same conformal dimensions have to be identified. The doubling can be avoided by considering only the operators $\phi_{m,n}$ with $m + n$ even.

The Ising model is characterized by two *local* lattice quantities, the local magnetization (or spin) $\sigma(x)$ and the local energy density $\varepsilon(x)$, so that the simplest of the minimal

---

[6] Here T is temperature, $\beta \equiv e/T$ and e is a constant.

models has the right type and number of fields in order to be identified with the critical Ising model.

To describe a physical meaning of the scaling operators, it is a good idea to study the 2-point function:

$$\langle \phi_{p,q}(z_1)\phi_{p,q}(z_2)\rangle \sim |z_1 - z_2|^{-\Delta_{p,q}} , \tag{2.125}$$

where

$$2\Delta_{2,2} = 4h(2,2) = \frac{1}{4} , \quad 2\Delta_{2,1} = 4h(2,1) = 2 . \tag{2.126}$$

We now wish to compare this CFT data with the known critical behaviour of the Ising model in the case of the nearest neighbour (n.n.) interaction:

$$S = -\beta \sum_{\text{bonds}} \sigma(x_1)\sigma(x_2) , \tag{2.127}$$

where x_1 and x_2 are sites on the lattice, and $\sigma(x) = \pm 1$ are the values of the Ising spin at site x (Fig. 11). The order parameter is the value of the local magnetization at x, $\langle \sigma(x)\rangle$, which vanishes in the disordered (high temperature) phase, fluctuates wildy at the critical point, and freezes into a non-vanishing value in the ordered (low temperature) phase. The local energy density at position x_1 is given by:

$$\varepsilon(x_1) = \sum_{x_2 = n.n.} \sigma(x_1)\sigma(x_2) . \tag{2.128}$$

The two basic (microscopic) *critical exponents* η and ν determine the behaviour of the correlators of σ and ε as the functions of $x = |x_1 - x_2|$ near criticality:

$$\begin{aligned} \langle \sigma(x_1)\sigma(x_2)\rangle &\sim x^{-\eta} , \\ \langle \varepsilon(x_1)\varepsilon(x_2)\rangle &\sim x^{-2(2-\frac{1}{\nu})} . \end{aligned} \tag{2.129}$$

The *microscopic* exponents are related to the *macroscopic* exponents describing the behaviour of thermodynamical quantities near criticality through the *scaling laws* which are necessarily true if the thermodynamical functions are quasi-homogeneous in terms of the reduced variables [7, 9]:

$$\begin{aligned} \alpha + 2\beta + \gamma &= 2 \quad (Rushbrooke) , \\ \gamma &= \beta(\delta - 1) \quad (Widom) , \\ \gamma &= (2 - \eta)\nu \quad (Fisher) , \\ 2\nu &= 2 - \alpha \quad (Josephson) . \end{aligned} \tag{2.130}$$

In the case of Ising model, Onsager's exact solution or numerical simulations [6, 7, 9] imply that

$$\eta = \frac{1}{4} \ , \quad \nu = 1 \ . \tag{2.131}$$

Therefore, comparing eqs. (2.125) and (2.129), we find $\sigma = \phi_{2,2}$ and $\varepsilon = \phi_{2,1}$. The operator $\phi_{1,1}$ is just the identity operator. Having obtained this identification, we can completely solve the model in the sense that any correlation function (and, thereby, any observable quantity) can be computed. In particular, the fusion rules of the Ising model follow from eq. (2.76). One finds

$$[\varepsilon] \times [\varepsilon] = [1] \ , \quad [\varepsilon] \times [\sigma] = [\sigma] \ , \quad [\sigma] \times [\sigma] = [1] + [\varepsilon] \ . \tag{2.132}$$

The next representatives in the series of unitary minimal models have $p = 4$: $c = 7/10$, $p = 5$: $c = 4/5$ and $p = 6$: $c = 6/7$. They were identified with the *tricritical Ising model*, Z_3 *Potts model* and the *tricritical Z_3 Potts model*, resp., in the continuum limit [48, 59]. In the Z_3 (3-state) models, the spin variable σ takes just three values, $\sigma = \exp(2\pi i k/3)$, $k = 0, 1, 2$.

Another CFT series can be recognized when the operator $\phi_{1,2}$ plays the role of the energy operator [13, 58]. Some of the known statistical systems can be identified with certain minimal models, when comparing their spectra by using the known exact solutions of the lattice models (see, e.g., the exact solution of Baxter [60] for the Z_3 Potts model). The more general Q-*component Potts models* [61, 62], [7] as well as the solvable RSOS (*restricted solid-on-solid*) lattice models [63], [8] can also be introduced and studied along the similar lines [58, 62, 64]. They are in fact capable to realize all the minimal models.

The lattice approach is powerful enough to get exact analytic solutions for many non-trivial strongly interacting systems, like critical models in statistics (see the book [6] of Baxter for details). CFT provides us with the very different alternative approach, which has at least a comparable significance, and it nicely complements the lattice approach. The real power of the CFT methods can already be seen in the exact calculation of the correlation functions and OPE's of the minimal models [12]. To illustrate how CFT

---

[7]In the Q-component Potts model the spin σ takes Q different values.

[8]The RSOS model is defined in terms of the height variables l_i, $l_i = 1, \ldots, p$, living at the sites of a square lattice, where the nearest neibours are constrained to satisfy $l_i = l_j \pm 1$. The Boltsmann weights are given by the 4-height interactions around plaquettes of the lattice – the so-called 'interations round a face' [17]. The parameter p here is the same as that for the minimal model in eq. (2.65).

works for the Ising model – the simplest one among the minimal models – a sample calculation of its physical correlation functions in the Coulomb gas picture, when only one screening operator is required, is considered below.

The simplest non-trivial example is provided by the correlator

$$\left\langle \varphi_{\alpha_{m,n}} \varphi_{\alpha_{2,1}} \varphi_{\alpha_{2,1}} \varphi_{2\alpha_0 - \alpha_{m,n}} \right\rangle ,$$
(2.133)

which has the charge $(2\alpha_0 - \alpha_+)$ and, hence, only a single screening charge Q_+ is required. Using the two-point function $\langle \varphi(z)\varphi(w) \rangle = -2\ln(z - w)$ and Wick rules, one obtains [54, 55]

$$\left\langle \varphi_{\alpha_{m,n}} \varphi_{\alpha_{2,1}} \varphi_{\alpha_{2,1}} \varphi_{2\alpha_0 - \alpha_{m,n}} \right\rangle$$

$$\sim \oint_C d\zeta \left\langle V_{\alpha_{m,n}}(z_1) V_{\alpha_{2,1}}(z_2) V_{\alpha_{2,1}}(z_3) V_{2\alpha_0 - \alpha_{m,n}}(z_4) J_+(\zeta) \right\rangle$$

$$= \left(\prod_{i<j} z_{ij}^{2\alpha_i \alpha_j} \right) \oint_C d\zeta \prod_i (z_i - \zeta)^{2\alpha_i \alpha_+} ,$$
(2.134)

where the list of α_i's, $(i = 1, 2, 3, 4)$, is fixed by eq. (2.133), whereas the contour C is yet to be specified.

The projective $SL(2, \mathbf{C})$ invariance of the correlation functions (Ch. I) allows us to fix the three points as $z_1 \to \infty$, $z_2 \to 1$, $z_3 \to z$, $z_4 \to 0$, where

$$z = \frac{z_{12} z_{34}}{z_{13} z_{24}} .$$
(2.135)

It follows

$$\oint_C d\zeta \left\langle \prod_i V_{\alpha_i}(z_i) J_+(\zeta) \right\rangle = \left(\frac{z_{12} z_{14}}{z_{24}} \right)^{h_2 + h_3 + h_4 - h_1}$$

$$\times \frac{(1 - z)^{2\alpha_2 \alpha_3} z^{2\alpha_3 \alpha_4}}{z_{12}^{2h_2} z_{13}^{2h_3} z_{14}^{2h_4}} \oint_C d\zeta \, (1 - \zeta)^{2\alpha_2 \alpha_+} (z - \zeta)^{2\alpha_3 \alpha_+} \zeta^{2\alpha_4 \alpha_+} .$$
(2.136)

The really non-trivial piece to be calculated is given by the contour integral in eq. (2.136). The integrand has branch cuts at $\zeta = 0, 1, z, \infty$, and, hence, the contour C need not to be closed. Instead, it can have its end points among those with the coordinates $\zeta = 0, 1, z, \infty$, assuming either the convergence of the integral at the ends or the analytic continuation to the 'ends' from the region of convergence. There are actually two independent choices of the contour both shown in Fig. 12, and leading to the integral calculable in terms of the hypergeometric function [54, 55]:

$$I_1(a, b, c; z) = \int_1^\infty dt \, t^a (t - 1)^b (t - z)^c$$

Fig. 12. The integration contours for eq. (2.136). The points 0, η, 1 and ∞
(infinity) are the branch points of the integrand. These contours are
closed on the Riemann surface associated with the integrand.

$$= \frac{\Gamma(-a-b-c-1)\Gamma(b+1)}{\Gamma(-a-c)} F(-c, -a-b-c-1, -a-c; z) , \qquad (2.137a)$$

$$I_2(a, b, c; z) = (z)^{1+a+c} \int_0^1 dt\, t^a (1-t)^c (1-tz)^b$$

$$= (z)^{1+a+c} \frac{\Gamma(1+a)\Gamma(1+c)}{\Gamma(2+a+c)} F(-b, 1+a, 2+a+c; z) , \qquad (2.137b)$$

where $a = 2\alpha_+\alpha_{m,n}$, $b = c = 2\alpha_+\alpha_{1,2}$. To derive eq. (2.137), the integral representation
of the hypergeometric function has been used,

$$F(a, b, c; z) = \frac{\Gamma(c)}{\Gamma(b)\Gamma(c-b)} \oint_0^1 t^{b-1}(1-t)^{c-b-1}(1-tz)^{-a}\, dt . \qquad (2.138)$$

The two functions in eq. (1.137) are just the two independent solutions to the second-
order differential equation (2.54), each one corresponding to the exchange of one conformal block. The generalization to higher order correlation functions is now clear.

As to the Ising model, one has, in particular,

$$\varepsilon =: e^{i\alpha_{2,1}\varphi} : , \quad \sigma =: e^{i\alpha_{1,2}\varphi} : ,$$

$$h_{2,1} = \frac{1}{2} , \quad \alpha_{2,1} = -\alpha_+/2 . \qquad (2.139)$$

Eq. (2.137) now implies

$$\langle \varepsilon\varepsilon\varepsilon \rangle = \frac{1}{z_{13}z_{24}} [z(1-z)]^{2/3}$$

$$\times \begin{cases} [z(1-z)]^{-5/3} F(-2, -1/3, -2/3; z) & \text{for } C_1 , \\ F(4/3, 3, 8/3; z) & \text{for } C_2 . \end{cases} \qquad (2.140)$$

Given the contour C_2, the point $z = 0$ is not a branch point and, hence, the C_2 does *not*
contribute. This is equivalent to the fusion rule $[\varepsilon] \times [\varepsilon] = [1]$. Therefore, we get

$$\langle \varepsilon\varepsilon\varepsilon\varepsilon \rangle = \frac{1}{z_{13}z_{24}} \frac{1-z-z^2}{z(1-z)} . \qquad (2.141)$$

Similarly, all the other conformal blocks of the Ising model can be found [54, 55].

To compute the physical correlation functions and OPE coefficients, one has to combine the left and right blocks into monodromy-invariant combinations, because the conformal functions provided by the integrals above are not uniquely defined on the complex plane. As functions of z, they have $0, 1, \infty$ as the singular points. In the minimal CFT we describe, the primary fields $\phi_{h,\bar{h}}$ have no spin ($h = \bar{h}$), so that the full correlator should have the monodromy-invariant form

$$G(z, \bar{z}) \equiv \langle \phi_1(0,0)\phi_2(z,\bar{z})\phi_3(1,1)\phi_4(\infty,\infty) \rangle$$

$$\sim \sum_{ij} X_{ij} I_i(z)\overline{I_j(z)} \ . \tag{2.142}$$

The coefficients X_{ij} in this equation are dictated by the analyticity properties of the hypergeometric functions, when their argument is rotated around the singularities. In other words, the physical correlators should be single-valued when $\bar{z} = z^*$ [21]. In statistical models, the primary operators like operators of energy and spin should have non-trivial scaling dimensions $h + \bar{h}$, and their correlation functions should be uniquely defined on a plane. It is straightforward (but tedious) to get the result [13, 54, 55]

$$G(z, \bar{z}) = \frac{s(a)s(b)}{s(a+c)} |I_1(z)|^2 + \frac{s(b)s(a+b+c)}{s(a+c)} |I_2(z)|^2 \ , \tag{2.143}$$

where $s(y) \equiv \sin(\pi y)$. Eq. (2.143) gives the 4-point physical correlator in the minimal CFT. It has crossing symmetry, as it should. [9] More general correlation functions and conformal blocks can also be computed this way [13, 54, 55]. For an arbitrary correlation function with more than one insertion of the screening operators, there will be more choices for the integration contours used to define conformal blocks. Determining which of these contours contribute is equivalent to determining the fusion rules. There is a strong evidence (but no proof yet!) that any RCFT correlation function on the sphere can be expressed in terms of the hypergeometric functions.

As far as the Ising model is concerned, choosing the standard conventions for the propagators (the 2-point functions), in accordance with eq. (1.64), as

$$\langle \varepsilon(z, \bar{z})\varepsilon(w, \bar{w}) \rangle = \frac{1}{|z-w|^2} \ ,$$

$$\langle \sigma(z, \bar{z})\sigma(w, \bar{w}) \rangle = \frac{1}{|z-w|^{1/4}} \ , \tag{2.144}$$

---

[9]The crossing symmetry is a consequence of the associativity of the operator product algebra.

we learn that $C_{\varepsilon\varepsilon}^{\varepsilon} = C_{\sigma\sigma}^{\sigma} = 1$. The only non-trivial OPE coefficient $C_{\varepsilon\varepsilon}^{\sigma}$ follows from the computation of the correlator $\langle \sigma\sigma\sigma \rangle$, and it turns out to be

$$C_{\varepsilon\varepsilon}^{\sigma} = 1/2 \ . \tag{2.145}$$

Exercises

\# II-11 ▷ (*) Calculate the spectra of conformal dimensions for the tricritical Ising model, Z_3 Potts model and tricritical Z_3 Potts model.

\# II-12 ▷ Prove that the monodromy transformation matrix of the basic functions $\{I_i(z)\}$ defined by eq. (2.137) is diagonal. *Hint:* consider the monodromy transformation properties of the hypergeometric functions along the contours enclosing the points 0 and 1 on a complex plane.

II.8 Operator algebra and bootstrap in the minimal models

The purpose of this section is to show that the consistency requirements (bootstrap) for the correlation functions of CFT naturally lead to the miminal models as the only consistent RCFT's *without* additional (beyond conformal) symmetries [12, 13].

Eqs. (2.142) and (2.143) determine a 4-point correlator of the minimal models in the form [10]

$$\langle \phi_s(0)\phi_n(z, \bar{z})\phi_m(1)\phi_k(\infty) \rangle \sim \sum_p A_{snmk}^p \left| F_p(z) \right|^2 \ , \tag{2.146}$$

where the conformal blocks of eq. (2.41) have been written in terms of the hypergeometric functions. So, let them to be normalized by eq. (2.42). When $z \to 0$, we have

$$\langle \phi_s(0)\phi_n(z, \bar{z})\phi_m(1)\phi_k(\infty) \rangle \sim \sum_p \frac{A_{snmk}^p}{|z|^{\Delta_s + \Delta_n - \Delta_p}} \ , \tag{2.147}$$

where only the singular terms have been kept on the r.h.s. Δ's are the full scaling dimensions of the fields. We already know from sect. II.3 that the structure coefficients

---

[10] As in the previous section, we keep only the relevant factors in our correlators and use the $SL(2, \mathbf{C})$ projective symmetry. See eqs. (2.135) and (2.136) in order to restore the rest of correlators.

A^p_{snmk} of the 4-point functions are actually factorized into the product of the coefficients C_{ijk}, appearing in the operator algebra (1.65), as

$$A^p_{snmk} = C^p_{sn} C^p_{mk} .$$ (2.148)

In the DF construction, duality is automatic and it is, in fact, equivalent to the monodromy-invariance discussed in the previous section. Having obtained the conformal blocks and the 4-point function coefficients $\{A^p_{snmk}\}$, our next task is to determine the OPE coefficients $\{C^p_{sn}\}$. It is most convenient to get them from the *symmetric* 4-point functions [13]

$$\langle \phi_s \phi_n \phi_n \phi_s \rangle = \sum A^p_{snns} |F_p(z)|^2 ,$$ (2.149)

since eq. (2.148) implies for them the relation

$$A_{snns} = (C^p_{sn})^2 .$$ (2.150)

One comment is in order here. The structure constants of the 4-point functions (2.146) derived in the Coulomb gas picture generically have an asymmetric overall numerical factor, because their construction explicitly involves one vertex operator which is different from the rest (sect. 7). This can be cured by proper normalization of these functions in such a way that they become symmetric [65]. This normalization is fixed, in fact, by eq. (1.64), which implies $C^I_{nn} = 1$, where I is the identity operator.

We now want to illustrate the results [65] by focusing on the subset of the operators $\{\phi_{1,n}\}$ which decouple from the rest, as also follows from the calculation. Their operator product coefficients are given by [65]

$$(C^p_{sn})^2 = (D^p_{sn})^2 a_s a_n (a_p)^{-1} ,$$ (2.151)

where

$$a_n = \prod_{j=1}^{n-1} \frac{\Gamma(j\gamma)\Gamma(2 - \gamma(1+j))}{\Gamma(1 - j\gamma)\Gamma(-1 + \gamma(1+j))} ,$$ (2.152a)

$$D^p_{sn} = \prod_{j=1}^{k-1} \frac{\Gamma(j\gamma)}{\Gamma(1 - j\gamma)} \prod_{j=0}^{k-2} \frac{\Gamma(1 - \gamma(s - 1 - j))\Gamma(1 - \gamma(n - 1 - j))\Gamma(-1 + \gamma(p + 1 + j))}{\Gamma(\gamma(s - 1 - j))\Gamma(\gamma(n - 1 - j))\Gamma(2 - \gamma(p + 1 + j))} ,$$ (2.152b)

and

$$\gamma \equiv \alpha^2_+ , \quad k \equiv \frac{s + n - p + 1}{2} .$$ (2.152c)

Among the properties of these coefficients, just a few are important to explain the mechanism of decoupling in the minimal models. First, the lowest value of k in the

j-product of eq. (2.152b) should make the products to be equal to 1. This explains why $k_{min} = 1$. Given that, eq. (2.152c) yields

$$p_{max} = s + n - 1 .\qquad(2.153)$$

The value of k_{max} is fixed by the first appearance of the pole in the Γ-functions, with the relevant factor given by

$$\prod_{j=0}^{k-2} \frac{1}{\Gamma(\gamma(s-1-j))\Gamma(\gamma(n-1-j))} .\qquad(2.154)$$

When $k = k_{max} + 1$ and there is a pole in eq. (2.154), the coefficient C_{sn}^p must vanish. In addition, eq. (2.152c) implies

$$p_{min} = |s - n| + 1 .\qquad(2.155)$$

Hence, p has to vary between the values given by eqs. (2.153) and (2.155), with the step 2.

Second, the operator algebra is closed by a *finite* number of operators when γ is a *rational* number q/p [65]. Let us consider, for example, the simplest case of the Ising model again, which corresponds to $\gamma = 4/3$ [13]. The equations above yield in this case [11]

$$\phi_{2,1}\phi_{2,1} \sim C_{2,2}^1 \phi_{1,1} + C_{2,2}^3 \phi_{3,1} ,\qquad(2.156)$$

where $C_{2,2}^1 = 1$, as it should. What is important is that the $C_{2,2}^3$ vanishes when $\gamma = 4/3$, because of the pole in one of the Γ's appearing in eq. (2.152a) for the $a_{31} \equiv a_3$. Indeed, a_3 then becomes divergent and, hence, $C_{2,2}^3$ vanishes due to eq. (2.151). But this just means that the operator $\phi_{3,1}$ decouples. Similar things happen in the other minimal models (although, in a less obvious way as above) [13, 65].

The minimal models are in fact the only consistent CFT's which can be constructed this way by the conformal bootstrap! To see this, let us take a look at the Kač formula (2.112) again, and notice that some of the values of $h(n, m)$ become negative near the line in the (n, m) plane where

$$\alpha_+ n + \alpha_- m = \alpha_+ n - |\alpha_-| n = 0 .\qquad(2.157)$$

The 2-point functions of the corresponding primary operators grow with distance and, hence, they are to be considered as 'pathological' in the sense of the ordinary field

---

[11] For simplicity, the scaling factors are dropped in what follows, see eq. (1.65).

theory. Therefore, it makes sense to get rid of such operators in CFT. Having a set of 'good' primaries with positive dimensions, their mutual OPE's could generate some new operators which have to be included into the set in order to close the operator algebra. Unless the operator algebra closes on a finite set, among the new operators to be ultimately included there will be some with negative dimensions, thus making the theory inconsistent. In particular, only the family with $q = p + 1$ has all positive conformal dimensions (exercise II-14). All the other models fail to be unitary [48]. This explains why we have chosen to stick to the RCFT's or just to the minimal models, since they seem to be the only ones which are fully consistent as QFT's. The unitarity of CFT may not be so crucial in statistical physics, where correlation functions may grow with distance.

Exercises

II-13 ▷ (*) Consider the spectrum of the tricritical Ising model from the exercise II-11, and calculate the operator product of two $\phi_{3,1}$ operators. It has the generic form

$$\phi_{3,1}\phi_{3,1} \sim D_{3,3}^1\phi_{1,1} + D_{3,3}^3\phi_{3,1} + D_{3,3}^5\phi_{5,1} \ . \tag{2.158}$$

Verify that $D_{3,3}^3$ is finite, but $D_{3,3}^5$ is vanishing. *Hint*: consider the a_5-coefficient in eq. (2.152a) and prove that it diverges at $\gamma = 6/5$.

II-14 ▷ (*) Consider the rational values of $\gamma = \alpha_+^2 = q/p$ and prove that conformal dimensions inside a finite $(q - 1) \times (p - 1)$ table are all positive when $q = p + 1$, but have negative values when $q \neq p + 1$. *Hint*: use the Kač formula (2.112).

II.9   Felder's (BRST) approach to minimal models

The Fock space of states of a minimal model is the direct sum of some finite number of irreducible highest weight modules of the Virasoro algebra Vir $\oplus$ $\overline{\text{Vir}}$, [12]

$$\mathcal{H} = \underset{n,n'}{\oplus} \mathcal{H}_{n,n'} \otimes \mathcal{H}_{n,n'} \ . \tag{2.159}$$

---

[12] Our notation in this section is very close to that used in Felder's original paper [66], although some efforts have been made to make it compatible with ours. In particular, Vir $= \{L_k\} \equiv \{L_k \otimes 1\}$ and $\overline{\text{Vir}} = \{\bar{L}_k\} \equiv \{1 \otimes \bar{L}_k\}$. The two pairs of integers used to parametrize the central charge c and the highest weight spectrum $\{h_{n,n'}\}$ of the minimal models in eq. (2.62) are denoted in this section by $(p, q) \to (p, p')$ and $(n, m) \to (n, n')$.

Each module $\mathcal{H}_{n,n'}$ in eq. (2.159) is degenerate in the sense that it is given by a quotient of the Verma module $\mathcal{V}(h_{n,n'}, c)$ by its maximal submodule generated by the null states (2.50), and it is unitary if $p' = p + 1$ (sect. 4).

Because of the correspondence between fields and states in CFT (sect. 1), there exists a primary field $\Phi_{n,n'}(z, \bar{z})$ associated to a highest weight vector $v_{n,n'}$ of $\mathcal{H}_{n,n'}$. The decomposition (2.159) implies that the primary field can be represented in a factorized form as

$$\Phi_{n,n'}(z, \bar{z}) = \sum_{\{l,l'\},\{m,m'\}} C^{\{l,l'\}}_{\{n,n'\}\{m,m'\}} \phi^{\{l,l'\}}_{\{n,n'\}\{m,m'\}}(z) \otimes \phi^{\{l,l'\}}_{\{n,n'\}\{m,m'\}}(\bar{z}) \,, \tag{2.160}$$

where each block $\phi^{\{l,l'\}}_{\{n,n'\}\{m,m'\}}$ maps $\mathcal{H}_{\{m,m'\}}$ to $\mathcal{H}_{\{l,l'\}}$, and it is defined to be zero on $\mathcal{H}_{k,k'}$ when $(k, k') \neq (m, m')$. By definition, $\phi^{\{l,l'\}}_{\{n,n'\}\{m,m'\}}(z)$ is a holomorphic field of weight $h_{n,n'}$:

$$\left[L_k, \phi^{\{l,l'\}}_{\{n,n'\}\{m,m'\}}(z)\right] = \left(z^{k+1}\frac{d}{dz} + h_{n,n'}(k+1)z^k\right) \phi^{\{l,l'\}}_{\{n,n'\}\{m,m'\}}(z) \,. \tag{2.161}$$

The constants $C^{\{l,l'\}}_{\{n,n'\}\{m,m'\}}$ represent finitely many 3-point function coefficients (CFT structure constants). The normalization of the block fields $\phi^{\{l,l'\}}_{\{n,n'\}\{m,m'\}}(z)$ is conveniently fixed by the condition [66] [13]

$$\left(v_{l,l'}, \phi^{\{l,l'\}}_{\{n,n'\}\{m,m'\}}(1)v_{m,m'}\right) = 1 \,, \tag{2.162}$$

whereas the products of these fields can be recognized (sect. 3) as the conformal blocks $\mathcal{F}_{\{\vec{k}\}}(z_1, \ldots, z_k)$ labeled by the intermediate representations $\{\vec{k}\}$, [14]

$$\mathcal{F}_{\{\vec{k}\}}(z_1, \ldots, z_k) = \left(v_{11}, \prod_{j=1}^{k} \phi^{\{m_{j-1}, m'_{j-1}\}}_{\{n_{j-1}, n'_{j-1}\}\{m_{j-1}, m'_{j-1}\}}(z_j)v_{11}\right) \,. \tag{2.163}$$

The CFT is fixed by the structure constants $C^{\{l,l'\}}_{\{n,n'\}\{m,m'\}}$ which are to be determined, in principle, from the bootstrap equations. In practice, however, this is very difficult to implement unless there are singular vectors (null states), which lead to partial differential equations on the correlators (sect. 4). However, even this is not enough since explicit expressions for the null states may not be available, and boundary conditions for the

---

[13]The scalar product (,) known as the *Shapovalov form* on $\mathcal{H}_{n,n'} \times \mathcal{H}_{n,n'}$, is uniquely defined by the two conditions: $(v_{n,n'}, v_{n,n'}) = 1$, and $(L_n\xi, \eta) = (\xi, L_{-n}\eta)$ is a non-degenerate hermitian form. The Shapovalov form is positively definite in the unitary case [31].

[14]We have $\{\vec{k}\} \simeq \{\vec{k}, \vec{\vec{k}}\} = (m_1, m'_1; \ldots; m_{k-1}, m'_{k-1})$, $m_0 = m'_0 = m_k = m'_k = 1$.

partial differential equations in question are not obvious. It is the DF approach that gives the calculable ansatz for conformal blocks in the Feigin-Fuchs integral representation within the Coulomb gas picture. This approach allows us to express the CFT correlators in terms of the expectation values of the screened vertex operators and thus get the explicit results (sect. 7). The conformal blocks $\mathcal{F}_{\{\vec{k}\}}$ with different $\vec{k}$ correspond to the different choices of integration contours in the DF construction. To explain the underlying symmetry structure of this construction, Felder [66] introduced the *Becchi-Rouet-Stora-Tyutin (BRST) charge* [67, 68] in a series of Coulomb gas models and proved the Feigin-Fuchs representation for the minimal models. To this end, we follow the work [66]. In the 'Coulomb gas picture' (sect. 7), the 'Hilbert' space of states is a direct sum of 'charged' bosonic Fock spaces F_{α,α_0}, with charge α and Virasoro central charge $1 - 24\alpha_0^2$. Each space F_{α,α_0} gives a representation of the *Heisenberg algebra* in terms of the creation and annihilation operators (*cf* free string description in Ch. VIII)

$$[a_n, a_m] = 2n\delta_{n+m,0} , \quad n, m \in \mathbf{Z} , \tag{2.164}$$

built upon a vacuum state v_{α,α_0} with $a_n v_{\alpha,\alpha_0} = 0$ for $n > 0$ and $a_0 v_{\alpha,\alpha_0} = 2\alpha v_{\alpha,\alpha_0}$,

$$F_{\alpha,\alpha_0} = \bigoplus_{k=0}^{\infty} \bigoplus_{1 \le n_1 \le \cdots \le n_k} C a_{-n_1} \cdots a_{-n_k} v_{\alpha,\alpha_0} . \tag{2.165}$$

The action of the Virasoro generators in this space is given by [12, 54] (see Ch. VIII also)

$$L_n = \tfrac{1}{4} \sum_{k=-\infty}^{\infty} a_{n-k} a_k - \alpha_0(n+1)a_n , \quad n \neq 0 ,$$

$$L_0 = \tfrac{1}{2} \sum_{k=1}^{\infty} a_{-k} a_k + \tfrac{1}{4}\alpha_0^2 - \alpha_0 a_0 . \tag{2.166}$$

The Virasoro module F_{α,α_0} is graded by eigenspaces of the L_0,

$$F_{\alpha,\alpha_0} = \bigoplus_{n=0}^{\infty} (F_{\alpha,\alpha_0})_n , \tag{2.167}$$

where the vacuum state v_{α,α_0} of F_{α,α_0} is the highest weight state,

$$L_0 v_{\alpha,\alpha_0} = \left(\alpha^2 - 2\alpha\alpha_0\right) v_{\alpha,\alpha_0} . \tag{2.168}$$

The dimension of eigenspace $(F_{\alpha,\alpha_0})_n$ corresponding to the eigenvalue $\alpha^2 - 2\alpha\alpha_0 + n$ and the level n is $P(n)$ (see eq. (2.16)). The *dual* Virasoro module F_{α,α_0}^* can also be introduced as a direct sum of the duals to the homogeneous components in eq. (2.167). If $<\,,\,>$ denotes duality, let us define

$$< L_n\omega, \xi > = < \omega, L_{-n}\xi > , \quad \omega \in F_{\alpha,\alpha_0}^* , \quad \xi \in F_{\alpha,\alpha_0} . \tag{2.169}$$

The dual module F^*_{α,α_0} is also a Fock space, and it is isomorphic (as a Virasoro module) to the Fock space $F_{2\alpha_0-\alpha,\alpha_0}$ [66].

The vertex operator of charge α maps F_{α_1,α_0} to $F_{\alpha_1+\alpha,\alpha_0}$ for any α_1, and takes the form (*cf* string vertex operator in Ch. VIII)

$$V_\alpha(z) = z^{\alpha a_0} \exp\left(\alpha \sum_{n=1}^\infty \frac{a_{-n}}{n} z^n\right) \exp\left(-\alpha \sum_{n=1}^\infty \frac{a_n}{n} z^{-n}\right) . \qquad (2.170)$$

The vertex operator (2.170) is a primary field of (conformal) dimension $\alpha^2 - 2\alpha\alpha_0$ (sect. 7). The correlation function of the primary fields is thus given by the VEV of a product of the vertex operators (2.170), and it is, therefore, straightforwardly calculable by using, e.g., the coherent state techniques or eq. (2.92),

$$< \Omega^*, V_{\alpha_1}(z_1)\cdots V_{\alpha_M}(z_M)\Omega > = \prod_{i<j}(z_i - z_j)^{2\alpha_i\alpha_j} , \quad |z_1| > \ldots > |z_M| , \quad \sum \alpha_i = 0 , \qquad (2.171)$$

where the vacuum Ω of the zero-charge Fock space F_{0,α_0} has been introduced, and the background charge $-2\alpha_0$ has been included among α_i's. Since we are only interested in the minimal models, we fix α_0 to be equal to $\frac{(p-p')^2}{4pp'}$ and confine ourselves to the set of charges given by eqs. (2.111) and (2.112). [15]

The *screened* vertex operators depend on integer labels $n, n', r \geq 0, r' \geq 0$, and take the following form in Felder's approach [66]:

$$V^{r,r'}_{n,n'}(z) = \oint_{C_i,S_j} V_{\alpha_{n,n'}}(z) V_{\alpha_-}(u_1)\cdots V_{\alpha_-}(u_{r'}) V_{\alpha_+}(v_1)\cdots V_{\alpha_+}(v_r) \prod_{i=1}^{r'} du_i \prod_{j=1}^{r} dv_j . \quad (2.172)$$

The integration contours C_i of u_i and S_j of v_j (Fig. 13) are drawn in the region $|z| > |u_1| > \ldots > |v_{r'}|$, where the product inside the integrand is well-defined. The integrals of eq. (2.172) have to be regularized by either the point splitting (i.e. replacing the closed contours by the open ones and taking the end point coincidence limit), or the analytic continuation (from a region of the complex α_- where the integrals converge). This choice of the integration contours is clearly different from that used in the DF construction (sect. 7), but they are equivalent inside the correlation functions, where they can be deformed to each other.

---

[15]To simplify the notation even further, we write $F_{n,n'}$ and $v_{n,n'}$ instead of $F_{\alpha_{n,n'},\alpha_0}$ and $v_{\alpha_{n,n'},\alpha_0}$, resp. When comparing this with the equations of sect. 7, use $q \to p'$ and $m \to n'$. In particular, eqs. (2.63) and (2.64) imply $p < p'$, $\alpha_{n,n'} = \alpha_{n+p,n'+p'}$, $n \neq 0 \,(\mathrm{mod}\,p)$ and $n' \neq 0 \,(\mathrm{mod}\,p)'$.

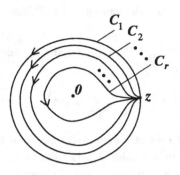

Fig. 13. The integration contours for eq. (2.172).

The screened vertex operator maps $F_{m,m'}$ to $F_{m+n-2r-1,m'+n'-2r'-1}$, and it is still a primary field of dimension $h_{n,n'}$,

$$[L_k, V_{n,n'}^{r,r'}(z)] = \left(z^{k+1}\frac{d}{dz} + h_{n,n'}(k+1)z^k\right)V_{n,n'}^{r,r'}(z) , \qquad (2.173)$$

since the screening charge operators $V_{\alpha_\pm}$ have dimension $h = 1$ and they do not contribute to the conformal dimension of $V_{n,n'}^{r,r'}$, although they do to its charge.

Felder's BRST operator Q_m maps $F_{m,m'}$ to $F_{-m,m'}$, and it is defined by [46]

$$Q_m = c \oint V_{\alpha_+}(v_0) \cdots V_{\alpha_+}(v_{m-1}) \prod_{j=0}^{m-1} dv_j , \qquad (2.174)$$

where the integration contours over $v_1, \ldots, v_{m-1}$ go from v_0 to v_0, like that for the screened vertex operators, whereas v_0 is integrated over the unit circle. The normalization constant in eq. (2.174) is equal to $c = m^{-1}(e^{2\pi i \alpha_+^2 m} - 1)/(e^{2\pi i \alpha_+^2} - 1)$ [66]. The corresponding BRST current acting on $F_{m,m'}$ is given by

$$J_m(z) = cV_{-1,1}^{m-1,0}(z) . \qquad (2.175)$$

These definitions are consistent since the BRST current $J_m(z)$ is *single-valued* as z goes around the origin, and it has dimension one indeed. This means that for all k we have

$$[L_k, Q_m] = 0 , \qquad (2.176)$$

where no boundary terms can appear because of closure of the integration contour. The meaning of the definitions above becomes clear after taking into account the following properties [66]:

- Q_m is the BRST charge, so that it can be used to construct the null states in the Fock space $F_{-m,m'}$ as $Q_m v_{m,m'}$, since $Q_m Q_{p-m} = 0$;

- the screened vertex operators are all BRST invariant ;

- the subspace of the physical states $B_{m,m'} \equiv \operatorname{Ker} Q_m / \operatorname{Im} Q_{p-m}$ is isomorphic to the irreducible highest weight Virasoro module $\mathcal{H}_{m,m'}$ of dimension $h_{m,m'}$ and central charge $c = 1 - 6(p - p')^2/pp'$.

The BRST invariance of the screened vertex operators means that they are all commuting with the BRST charge (up to a phase, which could be eliminated by their rescaling). The action of these operators is illustrated by the diagram [66]:

$$
\begin{array}{ccc}
F_{m,m'} & \xrightarrow{\;V_{n,n'}^{r,r'}(z)\;} & F_{m+n-2r-1,m'+n'-2r'-1} \\
\Big\downarrow{\scriptstyle Q_m} & & \Big\downarrow{\scriptstyle Q_{m+n-2r-1}} \\
F_{-m,m'} & \xrightarrow{\;V_{n,n'}^{n-r-1,r'}(z)\;} & F_{-m-n+2r+1,m'+n'-2r'-1}
\end{array}
\qquad (2.177)
$$

Because of the identities [16]

$$
Q_m V_{n,n'}^{r,r'}(z) = e^{2\pi i m \alpha_{n,n'} \alpha_+} V_{n,n'}^{r+m,r'}(z) \,,
$$

$$
V_{n,n'}^{r,r'}(z) Q_m = V_{n,n'}^{r+m,r'}(z) \,, \qquad\qquad (2.178)
$$

one gets [66]

$$
Q_{m+n-2r-1} V_{n,n'}^{r,r'}(z) = e^{2\pi i \alpha_{n,n'} \alpha_+ (m+n-2r-1)} V_{n,n'}^{n-r-1,r'}(z) Q_m \,. \qquad (2.179)
$$

The phases appearing in this equation are not essential since they can easily be eliminated by appropriate rescalings of the screened vertex operators. It is the BRST invariance of the screened vertex operators that makes them well-defined on the BRST cohomology, so that they actually map the physical states into the physical states. Being the primary fields, they can therefore be identified (up to a normalization constant) with the block fields $\phi_{\{n,n'\}\{m,m'\}}^{\{l,l'\}}(z)$. The normalization is given by

$$
N_{\{n,n'\}\{m,m'\}}^{\{l,l'\}} = \langle v_{l,l'}^*, V_{n,n'}^{r,r'}(1) v_{m,m'} \rangle \,, \qquad (2.180)
$$

where $l = n + m - 2r - 1$ and $l' = n' + m' - 2r' - 1$. It is straightforward to compute eq. (2.180) by reducing it to one of the standard DF integrals [55], like those considered

---

[16]The phase in the first equation (2.178) appears as the result of commuting $V_{\alpha_{n,n'}}$ with all the V_{a_+}'s comprising the BRST charge Q_m.

in sect. 7. The value of this constant turns out to be non-vanishing only if the minimal-models fusion rules (2.119) are satisfied [66] ! Simultaneously, eqs. (2.163) and (2.172) provide the integral representation for the conformal blocks,

$$\mathcal{F}_{\{k\}}(z_1, \ldots, z_k) = \text{const} < v_{11}^* , \prod_{j=1}^{k} V_{n_j, n_j'}^{r_j, r_j'}(z_j) v_{11} > , \qquad (2.181)$$

where $k_{j-1} = n_j + m_j - 2r_j - 1$ and $k_{j-1}' = n_j' + m_j' - 2r_j' - 1$. The contours drawn in Fig. 13 can be deformed in such a way that the conformal blocks (2.181) become linear conbinations of the DF-type integrals (sect. 7). This completes the Felder's proof of the Feigin-Fuchs integral representation on a plane. In the case of the WZNW theories (Ch. IV), this representation is proved in sect. IV.4. Felder's constriction of the minimal models can be straightforwardly generalized to Riemann surfaces of arbitrary genus. The case of torus (genus-1) will be considered in sect. III.5.

Since Felder's approach is based on the fundamental BRST structure [67, 68], one may ask whether his construction originates from the gauge fixing of an underlying gauge theory. This gauge theory, if any, is yet to be found.

To conclude this section, we note that although the minimal models are the simplest examples of CFT indeed, from the viewpoint of their BRST structure they are simultaneously the most complicated ones because of their highest level of degeneracy. Using free field realizations is very useful in analyzing CFT's, and it always implies three necessary ingredients [69]:

- realization of CFT chiral operator algebra $\mathcal{A}$ by free fields,

- 'null-state decoupling' by projecting the free field Fock spaces to the irreps of $\mathcal{A}$,

- realization of the vertex operators.

The BRST structure should also exist in any CFT formulated on an arbitrary Riemann surface, with the BRST symmetry playing a fundamental role.

Chapter III

Partition Functions and Bosonization

In this Chapter we consider CFT's on Riemann surfaces of genus $h \geq 1$. The case of torus, $h = 1$, will be most important for us. A consistent formulation of CFT on a torus, or on a Riemann surface of arbitrary genus, normally gives rise to further constraints on CFT operator contents. The string perturbation theory (Ch. VIII) gives the natural area for applications. Chiral bosonization on Riemann surfaces and Felder's (BRST) formulation of the minimal models on a torus are also considered in this Chapter.

III.1 Free fermions on a torus

The general strategy for constructing CFT's on a torus is to make use of the local properties of the CFT operators already constructed on a conformal plane, map this plane to a cylinder via the exponential map, and then make the torus via discrete identification. Though this procedure preserves all local properties of the operators, it does not necessarily preserve all of their *global* symmetries. A torus maps to an *annulus* on a plane, so that only L_0 and $\bar{L}_0$ survive as the global symmetry generators. The *global* symmetry group is therefore reduced to $U(1) \times U(1)$.

Boundary conditions on conformal fields are also affected by the passage from the plane to the cylinder (or to the torus). A primary field $\phi_{h,\bar{h}}(z,\bar{z})$ transforms under the map $w \to z = e^w$ as

$$\phi_{\text{cyl}}(w, \bar{w}) = \left(\frac{dz}{dw}\right)^h \left(\frac{d\bar{z}}{d\bar{w}}\right)^{\bar{h}} \phi(z, \bar{z}) = z^h \bar{z}^{\bar{h}} \phi(z, \bar{z}) \ . \tag{3.1}$$

Hence, under $z \to e^{2\pi i}z$, CFT primary fields with *integer* spin $s = h - \bar{h}$ have the *same* boundary conditions on the plane *and* the cylinder. On the contrary, CFT primary fields with *half-integer* spin *change* the periodic boundary conditions to the anti-periodic ones and vice-versa, when passing from the plane to the cylinder, as we have already noticed in eqs. (1.104) and (1.105). As far as the CFT stress tensor $T(z)$ is concerned, which is not even a tensor under the conformal transformations, the anomalous piece proportional to the Schwartzian derivative arises:

$$w \to z = e^w \ : \ S(e^w, w) = -1/2 \ :$$

$$T_{\text{cyl}}(w) = \left(\frac{\partial z}{\partial w}\right)^2 T(z) + \frac{c}{12} S(z, w) = z^2 T(z) - \frac{c}{24} \ . \tag{3.2}$$

Substituting the mode expansion $T(z) = \sum_n L_n z^{-n-2}$ into eq. (3.2) yields

$$T_{\text{cyl}}(w) = \sum_{n \in \mathbf{Z}} L_n z^{-n} - \frac{c}{24} = \sum_{n \in \mathbf{Z}} (L_n - \frac{c}{24} \delta_{n,0}) e^{-nw} \ , \tag{3.3}$$

and, hence,

$$(L_0)_{\text{cyl}} = L_0 - \frac{c}{24} \ . \tag{3.4}$$

The shift of the vacuum energy does make definite sense in CFT since the scale and rotational invariances on the plane naturally fix the L_0 and $\bar{L}_0$ eigenvalues to be zero for an $SL(2, \mathbf{C})$-invariant vacuum.

To introduce the CFT **partition function** on a torus, let us define first the real Hamiltonian (H) and momentum (P) operators as the generators of translations in the 'time' and 'space' directions, resp. It is convenient to redefine at this point the coordinate variable, $w \to iw$, so that $\text{Re}(w)$ and $\text{Im}(w)$ now become the space and time coordinates, resp. Since $(L_0 \pm \bar{L}_0)$ generate dilatations and rotations on the plane, one can identify H with $(L_0)_{\text{cyl}} + (\bar{L}_0)_{\text{cyl}}$, and P with the imaginary part of $(L_0)_{\text{cyl}} - (\bar{L}_0)_{\text{cyl}}$. For the torus with a modular parameter $\tau = \tau_1 + i\tau_2$, we identify two periods of w as $w \equiv w + 2\pi$ and $w \equiv w + 2\pi\tau$ (or, equivalently, $\text{Im}(w) \equiv \text{Im}(w) + 2\pi\tau_2$ and $\text{Re}(w) \equiv \text{Re}(w) + 2\pi\tau_1$). Therefore, the operator which generates translations in the Euclidean (imaginary) time is $e^{-2\pi\tau_2 H}$, whereas the accompanying translation operator in the Euclidean space is $e^{-2\pi\tau_1 P}$. This suggests a definition of the partition function in the form:

$$Z(\tau) = \text{Tr} \, e^{-2\pi\tau_2 H} \, e^{-2\pi\tau_1 P}$$

$$= \text{Tr } e^{2\pi i \tau_1 [(L_0)_{cyl} - (\bar{L}_0)_{cyl}]} e^{-2\pi \tau_2 [(L_0)_{cyl} + (\bar{L}_0)_{cyl}]}$$

$$= \text{Tr } e^{2\pi i \tau (L_0)_{cyl}} e^{-2\pi i \bar{\tau}(\bar{L}_0)_{cyl}} = \text{Tr } q^{(L_0)_{cyl}} \bar{q}^{(\bar{L}_0)_{cyl}}$$

$$= \text{Tr } q^{L_0 - c/24} \bar{q}^{\bar{L}_0 - \bar{c}/24} = q^{-c/24} \bar{q}^{-\bar{c}/24} \text{Tr } q^{L_0} \bar{q}^{\bar{L}_0} \, , \tag{3.5}$$

where $q \equiv e^{2\pi i \tau}$, and the hamiltonian trace runs over a CFT module. In particular, for the $c = \bar{c} = 1/2$ theory of a single holomorphic fermion $\psi(w)$ and a single anti-holomorphic fermion $\bar{\psi}(\bar{w})$ on a torus, we have

$$Z(\tau) = (q\bar{q})^{-1/48} \text{Tr } q^{L_0} \bar{q}^{\bar{L}_0} \, . \tag{3.6}$$

The full partition function is obtained by integrating $Z(\tau)$ over the modular space of the torus, with the modular-invariant measure known as the *Weyl-Peterson measure* [28]: $Z \equiv \int e^{-S} = \int \frac{d^2\tau}{(\text{Im}\,\tau)^2} Z(\tau)$. The CFT partition functions $Z(\tau)$ can be directly connected with the so-called τ-*functions* of certain 2d integrable systems (see Ch. XI).

Given a highest weight representation $\mathcal{V}$ of the Virasoro algebra with a highest weight vector h and central charge c, the **Virasoro character** of this representation is defined by

$$\chi_{\mathcal{V}}(q) = \text{Tr}_{\mathcal{V}} \, q^{L_0 - c/24} = q^{h - c/24} \sum_{n=0}^{\infty} a(n) q^n \, , \tag{3.7}$$

where n is the level, and $a(n)$ is the number of states at this level. If there are no singular vectors (null states), the convenient basis in the module $\mathcal{V}$ is given by

$$\mathcal{V} = \{ L_{-n_1}, \dots, L_{-n_k} | h \rangle \, , \quad n > 0 \} \, . \tag{3.8}$$

The basis is obtained by applying any polynomial in the L_{-n}'s ($n > 0$) to the highest weight state. Since $[L_0, L_{-n}] = n L_{-n}$, the character of $\mathcal{V}$ is

$$\chi_{\mathcal{V}}(q) = \frac{q^{h - c/24}}{\prod_{n=1}^{\infty}(1 - q^n)} \, , \tag{3.9}$$

where $\prod_{n=1}^{\infty}(1 - q^n)^{-1} = \sum_{n=0}^{\infty} p(n) q^n$ and $a(n) = p(n)$ is the number of partitions of n.

When there are singular vectors, one has to *subtract* the submodules generated by them. Therefore, in order to obtain the characters of the minimal models, eq. (3.9) has to be modified. The solution to this combinatorial problem can be obtained from the Kač formula (2.62), and it is known as the *Rocha-Caridi* formula for the Virasoro characters of the minimal models [70]. It could also be derived as the solution to the differential equations induced by inserting the null states into the correlation functions [57]. The Rocha-Caridi formula reads [71]

$$\chi_{n,m}(q) = \frac{q^{-c/24}}{\prod_{l=1}^{\infty}(1-q^l)} \sum_{k \in \mathbf{Z}} \left(q^{a(k)} - q^{b(k)} \right) , \qquad (3.10)$$

where

$$a(k) \equiv h(n+2pk, m) = \frac{(2pqk + qn - mp)^2 - (p-q)^2}{4pq} ,$$

$$b(k) \equiv h(-n-2pk, m) = \frac{(2pqk + qn + mp)^2 - (p-q)^2}{4pq} . \qquad (3.11)$$

In the case of $c = 1/2$, the same equation (2.62) implies that we should expect to obtain *free fermionic* correlators for the three unitary irreducible representations of the Virasoro algebra with $h = \{h_{1,1}, h_{2,1}, h_{2,2}\} = \{0, \frac{1}{2}, \frac{1}{16}\}$.

Let us first naively consider the states in the *anti-periodic* (A) sector of the fermionic field defined on a torus,

$$\psi_{-n_1} \dots \psi_{-n_k} |0\rangle , \qquad n_i \in \mathbf{Z} + 1/2 , \qquad (3.12)$$

and order them according to their eigenvalues w.r.t. $L_0 = \sum_{n>0} n\psi_{-n}\psi_n$. This allows us to write down the expression for the trace in the straightforward way,

$$\mathrm{Tr}_A \, q^{L_0} = 1 + q^{1/2} + q^{3/2} + q^2 + q^{5/2} + q^3 + q^{7/2} + 2q^4 + \dots . \qquad (3.13)$$

The states (3.12) form a reducible representation of the Virasoro algebra with $c = 1/2$. It is obvious from the inspection of the eigenvalues of L_0 that this representation is a direct sum $[0] \oplus [1/2]$ of the two highest weight representations with $h = 0$ and $h = 1/2$. Since there is only a single state with $h = 0$ and only a single state with $h = 1/2$, each of these two representations appears with unit multiplicity. All states in the representation $[0]$ clearly have an even fermion number F and, hence, $L_0 \in \mathbf{Z}$ for them, whereas all the states of $[1/2]$ are odd, and, hence, we should have $L_0 \in \mathbf{Z} + 1/2$ for them. The projection operators $\frac{1}{2}(1 \pm (-1)^F)$ are useful to disentangle these two representations,

$$q^{-1/48}\mathrm{Tr}_A \tfrac{1}{2}\left[1+(-1)^F\right] q^{L_0} = q^{-1/48}\left(1 + q^2 + q^3 + 2q^4 + \dots\right) \equiv \chi_0 ,$$

$$q^{-1/48}\mathrm{Tr}_A \tfrac{1}{2}\left[1-(-1)^F\right] q^{L_0} = q^{-1/48}\left(q^{1/2} + q^{3/2} + q^{5/2} + \dots\right) \equiv \chi_{1/2} , \qquad (3.14)$$

where $\chi_0 = q^{-1/48}\mathrm{Tr}_{h=0} \, q^{L_0}$ and $\chi_{1/2} = q^{-1/48}\mathrm{Tr}_{h=1/2} \, q^{L_0}$ are the characters of the $h = 0$ and $h = 1/2$ representations of the $c = 1/2$ Virasoro algebra, resp. In string theory

(Ch. VIII), the projection on the states of a given value of $(-1)^F$ is known as the **Gliozzi-Scherk-Olive** (GSO) projection [72].

In the *periodic* (P) sector of the fermion on the torus, one gets instead that $L_0 = \sum_{n>0} n\psi_{-n}\psi_n + 1/16$, $n \in \mathbf{Z}$. Going as before and noticing that in this case the fermion zero mode algebra requires the *two* ground states $|1/16\rangle_{\pm}$ with the eigenvalues ± 1 under $(-1)^F$: $\psi_0 |1/16\rangle_{\pm} = 1/\sqrt{2}\,|1/16\rangle_{\pm}$, [1] one finds the two irreducible representations of the $c = 1/2$ Virasoro algebra with the highest weight $h = 1/16$ and the Virasoro character

$$q^{-1/48}\mathrm{Tr}_P \tfrac{1}{2}\left[1 \pm (-1)^F\right] q^{L_0} = q^{1/24}\left(1 + q + q^2 + 2q^3 + \dots\right)$$

$$= q^{-1/48}\mathrm{Tr}_{h=1/16}\,q^{L_0} \equiv \chi_{h=1/16}\,. \qquad (3.15)$$

Actually, we have $\mathrm{Tr}_P (-1)^F q^{L_0} = 0$ due to the cancellation between equal numbers of states with the opposite values of $(-1)^F$ at each level.

To introduce the (Lagrangian) path integral formalism for the fermions living on a torus, first we need to specify the boundary conditions (of the P- or A-type) around the two non-trivial cycles of torus. By choosing the boundary conditions we assign the so-called fermionic **spin structure** [73] on the genus-one Riemann surface. In the real coordinates (σ^0, σ^1), this amounts to choosing signs in $\psi(\sigma^0 + 1, \sigma^1) = \pm\psi(\sigma^0, \sigma^1)$ and $\psi(\sigma^0, \sigma^1 + 1) = \pm\psi(\sigma^0, \sigma^1)$, where we have rescaled the coordinates for a later convenience in this section as $\sigma^0, \sigma^1 \in [0, 1)$, $w = \sigma^1 + \tau\sigma^0$, $\psi(w) \equiv \psi_{\mathrm{cyl}}(w)$. The sign ambiguity appears since spinors actually live on a double cover of the Riemann surface, as was already mentioned in Ch. I.

Let the symbol $X\,\overset{Y}{\square}$, where X, Y are either P or A, be the result of performing the path integral $\int \exp(-\int \psi\bar{\partial}\psi)$ over fermions with a fixed spin structure (X, Y) [17]. This functional integral can be regarded as the square root of the determinant of the operator $\bar{\partial}$ for various choices of boundary conditions. Due to the presence of the zero mode (a constant function) allowed by the PP boundary condition, we have, in particular

$$P\,\overset{P}{\square} = (\det_{PP} \bar{\partial})^{1/2} = 0\,. \qquad (3.16)$$

Any string theory (Ch. VIII) is usually required to be *modular*-invariant, which is crucial for its consistency. A breakdown of modular invariance may be thought of as a

---

[1] The states $|1/16\rangle_+$ and $|1/16\rangle_-$ can be identified with the $\sigma(0)|0\rangle$ and $\mu(0)|0\rangle$, resp., where the fields σ and μ have been introduced in eq. (1.109) [17].

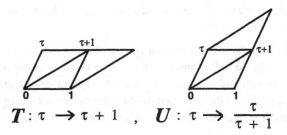

$$T : \tau \rightarrow \tau + 1 \quad , \quad U : \tau \rightarrow \frac{\tau}{\tau + 1}$$

Fig. 14. The generating modular transformations for a torus.

global anomaly in the fundamental reparametrizational invariance of the strings. It is expected that the same modular invariance should also be crucial for the consistency of CFT's to be defined on arbitrary Riemann surfaces [7, 17]. The group of modular transformations is the group of topologically non-trivial diffeomorphisms of a torus (in the genus-one case), generated by cutting along either of the non-trivial cycles and then regluing them after a twist by 2π. Cutting along a line of constant τ and regluing give one generating transformation, $T : \tau \rightarrow \tau + 1$, while cutting and then regluing along a line of constant σ give the other one, $U : \tau \rightarrow \tau/(\tau + 1)$ (Fig. 14).

These two transformations generate the group,

$$\tau \rightarrow \frac{a\tau + b}{c\tau + d} , \quad \begin{pmatrix} a & b \\ c & d \end{pmatrix} \in SL(2, \mathbf{Z}) , \tag{3.17}$$

which is called the **modular group** $PSL(2, \mathbf{Z}) \equiv SL(2, \mathbf{Z})/\mathbf{Z}_2$. Modding by the $\mathbf{Z}_2$ factor is necessary here because reversing the sign of all of a, b, c, d in eq. (3.17) leaves the action on τ unchanged. By a modular transformation, one can always take τ to lie in a *fundamental domain* (Fig. 15).

Usually one uses the transformations $T : \tau \rightarrow \tau + 1$ and $S = T^{-1}UT^{-1} : \tau \rightarrow -1/\tau$ to generate the whole modular group. Notably, $S^2 = (ST)^3 = 1$.

There exists the remarkable relation between the modular transformation matrix S and the fusion algebra (2.77) [51, 52, 74, 75]. The matrix S *diagonalizes* the fusion rules [51, 52], namely, $N_{ij}{}^k = \sum_n S_j{}^n \lambda_i^{(n)} (S^+)_n{}^k$, where $\lambda_i^{(n)}$'s are the eigenvalues of the matrix N_i. This relation can be solved for the integer $N_{ij}{}^k$'s in terms of the entries of the matrix S. Using $i = 0$ to specify the character for the identity family, one gets

$N_{0j}{}^k = \delta_j^k$. Hence, the N-eigenvalues satisfy the relation $\lambda_i^{(n)} = S_i{}^n/S_0{}^n$, which implies

$$N_{ij}{}^k = \sum_n \frac{S_j{}^n S_i{}^n (S^+)_n{}^k}{S_0{}^n} . \tag{3.18}$$

Eq. (3.18) is known as the *Verlinde formula* [51, 52, 74, 75, 76, 77].

It is the actual evaluation of the functional integrals or traces in eqs. (3.14) and (3.15) that reveals the modular transformation properties of various fermionic spin structures. This happens since the results of calculations can always be expressed in terms of the standard (Jacobi) *theta* functions $\vartheta_i = \vartheta_i(0, \tau)$ and the Dedekind *eta* function $\eta(q) = q^{1/24} \prod_{n=1}^{\infty}(1 - q^n)$, all having the nice transformation properties under the modular transformations (sect. XII.3). For example, the path integral $A\,\square^{A}$ for a single holomorphic fermion is given by $q^{-1/48}\mathrm{Tr}_A\, q^{L_0}$, where the factor $q^{-1/48}$ results from the vacuum energy. Similarly, $P\,\square^{A} = q^{-1/48}\mathrm{Tr}_A\,(-1)^F q^{L_0}$, since $(-1)^F \psi = -\psi(-1)^F$ and, hence, $(-1)^F$ flips the boundary condition in the time direction. For the P-sector, it is natural to define $A\,\square^{P} = \frac{1}{\sqrt{2}}q^{-1/48}\mathrm{Tr}_P\, q^{L_0}$ and $P\,\square^{P} = \frac{1}{\sqrt{2}}q^{-1/48}\mathrm{Tr}_P\,(-1)^F q^{L_0}\;(= 0)$, where the factor $\frac{1}{\sqrt{2}}$ has been added to simplify the behaviour under modular transformations.

It is straightforward (and even elementary) to calculate the traces. Taking the 2×2 basis $(|0\rangle, \psi_{-n}|0\rangle)$ for the n-th fermionic mode, one easily finds

$$q^{n\psi_{-n}\psi_n} = \begin{pmatrix} 1 & 0 \\ 0 & q^n \end{pmatrix} , \quad \mathrm{tr}\, q^{n\psi_{-n}\psi_n} = 1 + q^n . \tag{3.19}$$

Similarly, one gets

$$\mathrm{tr}\,(-1)^F q^{n\psi_{-n}\psi_n} = 1 - q^n . \tag{3.20}$$

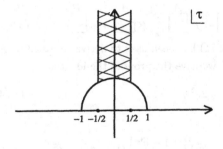

Fig. 15. A fundamental domain for genus 1.

The relevant operator reads

$$q^{L_0} = q^{\sum_{n>0} n\psi_{-n}\psi_n} = \prod_{n>0} q^{n\psi_{-n}\psi_n} = \prod_{n>0} \begin{pmatrix} 1 & 0 \\ 0 & q^n \end{pmatrix} . \tag{3.21}$$

Because the trace of the direct product of matrices $\otimes_i M_i$ obviously satisfies the relation $\text{tr} \otimes_i M_i = \prod_i \text{tr} M_i$, one finds the partition functions for a single $c = 1/2$ holomorphic fermion in the following form:

$$A\overset{A}{\boxed{}} = \chi_0 + \chi_{1/2} = q^{-1/48}\text{Tr}_A \, q^{L_0} = q^{-1/48} \prod_{n=0}^{\infty} \left(1 + q^{n+1/2}\right) \equiv \sqrt{\frac{\vartheta_3}{\eta}} , \tag{3.22a}$$

$$P\overset{A}{\boxed{}} = \chi_0 - \chi_{1/2} = q^{-1/48}\text{Tr}_A \, (-1)^F q^{L_0} = q^{-1/48} \prod_{n=0}^{\infty} \left(1 - q^{n+1/2}\right) \equiv \sqrt{\frac{\vartheta_4}{\eta}} , \tag{3.22b}$$

$$A\overset{P}{\boxed{}} = \sqrt{2}\chi_{1/16} = \frac{1}{\sqrt{2}}q^{-1/48}\text{Tr}_P \, q^{L_0} = \frac{1}{\sqrt{2}}q^{1/24} \prod_{n=0}^{\infty} \left(1 + q^n\right) \equiv \sqrt{\frac{\vartheta_2}{\eta}} , \tag{3.22c}$$

$$P\overset{P}{\boxed{}} = \frac{1}{\sqrt{2}}q^{-1/48}\text{Tr}_P \, (-1)^F q^{L_0} = \frac{1}{\sqrt{2}}q^{1/24} \prod_{n=0}^{\infty} \left(1 - q^n\right) \equiv \sqrt{\frac{\vartheta_1}{i\eta}} \, (= 0) . \tag{3.22d}$$

The transformation properties of the spin structures under the modular transformations are now simply deduced from the known properties (12.75) and (12.76) of the theta functions,

$$\tau \to \tau + 1 : \quad A\overset{A}{\boxed{}} \to e^{-i\pi/24} P\overset{A}{\boxed{}} , \quad P\overset{A}{\boxed{}} \to e^{-i\pi/24} A\overset{A}{\boxed{}} ,$$

$$A\overset{P}{\boxed{}} \to e^{i\pi/12} A\overset{P}{\boxed{}} ; \tag{3.23a}$$

$$\tau \to -1/\tau : \quad A\overset{A}{\boxed{}} \to A\overset{A}{\boxed{}} , \quad A\overset{P}{\boxed{}} \to P\overset{A}{\boxed{}} , \quad P\overset{A}{\boxed{}} \to A\overset{P}{\boxed{}} , \tag{3.23b}$$

where the Poisson resummation formula (12.71) has been used to derive eq. (3.23b). As a by-product of the derivation of eq. (3.22), we have the proof of the identity

$$\vartheta_2\vartheta_3\vartheta_4 = 2\eta^3 , \tag{3.24}$$

since

$$\sqrt{\frac{\vartheta_2\vartheta_3\vartheta_4}{\eta^3}} = \sqrt{2} \prod_{n=1}^{\infty} (1 - q^{2n-1})(1 + q^n) = \sqrt{2} \prod_{n=1}^{\infty} \left[\frac{1 - q^n}{1 - q^{2n}}\right](1 + q^n) = \sqrt{2} . \tag{3.25}$$

Eq. (3.22) can also be interpreted as $(\det \bar{\partial})^{1/2}$, where the determinant is supposed to be defined separately for each spin structure. This determinant could, of course, be calculated independently by using, e.g., the ζ-function regularization, which leads to the same results [28].

It is eq. (3.23) that implies the modular invariance of the following combination of the spin structures:

$$\frac{1}{2}\left(A\bar{A} \overset{A\bar{A}}{\Box} + P\bar{P} \overset{A\bar{A}}{\Box} + A\bar{A} \overset{P\bar{P}}{\Box} \right) = \chi_0\bar{\chi}_0 + \chi_{1/2}\bar{\chi}_{1/2} + \chi_{1/16}\bar{\chi}_{1/16} \ , \qquad (3.26)$$

Eq. (3.26) represents the simplest *diagonal* invariant in the A-D-E classification [78, 79, 80] of the modular-invariant partition functions (sect. IV.5). This invariant in fact gives the *Ising model partition function* on a torus.

Exercises

III-1 ▷ (*) Use the normal ordering prescription and the ζ-function regularization to calculate the shifts in the vacuum energy of a fermion on a cylinder for the both integer and half-integer modding,

$$(L_0)_{\text{cyl}} = \frac{1}{2}\sum_n n : \psi_{-n}\psi_n := \sum_{n>0} n\psi_{-n}\psi_n - \frac{1}{2}\sum_{n>0} n$$

$$= \sum_{n>0} n\psi_{-n}\psi_n + \begin{cases} -\frac{1}{2}\zeta(-1) & = +\frac{1}{24}, \quad \text{when} \ n \in \mathbf{Z} \\ -\frac{1}{2}\left[-\frac{1}{2}\zeta(-1)\right] & = -\frac{1}{48}, \quad \text{when} \ n \in \mathbf{Z}+1/2. \end{cases} \qquad (3.27)$$

and compare them with the corresponding shifts on a torus for the P- and A-type boundary conditions given above. The convenience of the ζ-function regularization technique here is due to its compatibility with the conformal *and* modular invariances [17].

III-2 ▷ Consider the critical Ising model on a torus. This model is equivalent to CFT of a free (M) fermion on the torus [7, 17]. Use eq. (3.26) to deduce the operator contents (i.e. primary fields) of this theory. The partition function (3.26) corresponds to certain boundary conditions on the Ising spins, which should be periodic along both homology cycles of the torus. Verify that the change of the boundary conditions on the Ising spins (e.g. 'twisting' them along the 'time' or 'space' directions) results in the breaking of the modular invariance for the partition function.

III-3 ▷ The Verlinde formula can be used to put some restrictions on the allowed values of central charge c and conformal dimensions h in CFT. Consider a CFT having

only two primaries $\mathbf{1}$ and ϕ, with the single non-trivial fusion rule $[\phi] \times [\phi] = \mathbf{1} + m[\phi]$, where m is a non-negative integer, and the matrix S has the form

$$S = \begin{pmatrix} \cos\theta & \sin\theta \\ \sin\theta & -\cos\theta \end{pmatrix} . \tag{3.28}$$

Show that requiring S to diagonalize the fusion rules implies $l_\pm = \tan\theta$, where $l_\pm$ are the roots of the quadratic equation $l^2 = 1 + ml$. Verify that the constraint $(ST)^3 = 1$ then implies [77]

$$\cos(2\pi h) = -\tfrac{1}{2}ml_\pm , \quad c = 12h + 6 \mod 8 . \tag{3.29}$$

The realizations of this CFT for $m = 0$ are given by the $A_1^{(1)}$ and $E_7^{(1)}$ WZNW theories at level one (Ch. IV). The case of $m = 1$ corresponds to the fusion rules in the non-unitary CFT describing the so-called *Lee-Yang edge singularity* of $c = -\frac{22}{5}$ and $h = -\frac{1}{5}$ (see sect. X.2 also).

III.2   Free bosons on a torus

The next simplest example of CFT defined on a torus is the theory of a free boson, with the action

$$S[\Phi] = \frac{1}{4\pi} \int \left(\partial\Phi \bar\partial\Phi \right) , \tag{3.30}$$

where the bosonic field Φ is assumed to be compactified on a circle of radius r: $\Phi \equiv \Phi + 2\pi r$. When calculating the functional integral $\int e^{-S}$, one now has to take into account all the *instanton* sectors $n_1 \begin{array}{c} n_2 \\ \boxed{} \end{array}$ specified by the boundary conditions

$$\Phi_0(z + 1, \bar z + 1) = \Phi_0(z, \bar z) + 2\pi r n_2 ,$$
$$\Phi_0(z + \tau, \bar z + \bar\tau) = \Phi_0(z, \bar z) + 2\pi r n_1 . \tag{3.31}$$

The classical equations of motion $\partial\bar\partial\Phi_0 = 0$, subject to the above boundary conditions, can be explicitly solved as

$$\Phi_0^{(n_1,n_2)}(z, \bar z) = \frac{2\pi r}{2i\tau_2} \left[n_1(z - \bar z) + n_2(\tau\bar z - \bar\tau z) \right] . \tag{3.32}$$

It is a good idea to separate a constant piece (zero mode) as $\Phi(z, \bar z) = \tilde\Phi + \Phi'(z, \bar z)$, where Φ' is orthogonal to the constant $\tilde\Phi$, so that the path integral measure $[d\Phi]$ takes the form $d\tilde\Phi [d\Phi']$. The gaussian functional integral $\int [d\Phi] \exp\left(-\frac{1}{4\pi} \int \Phi^2\right)$ is supposed to be normalized to unity (this convention is known in string theory as the *ultra-locality principle* [81]), which implies

$$\int [d\Phi'] \exp\left[-\frac{1}{4\pi}\int (\Phi')^2\right] = \left(\int d\tilde{\Phi} \exp\left[-\frac{1}{4\pi}\int \tilde{\Phi}^2\right]\right)^{-1}$$

$$= \left(\frac{\pi}{\frac{1}{4\pi}\int 1}\right)^{-1/2} = \frac{\sqrt{\tau_2}}{\pi} , \tag{3.33}$$

since the measure on the torus in the coordinates $z = \sigma^1 + \tau\sigma^0$ takes the form $2idz \wedge d\bar{z} = 4\tau_2 d\sigma^1 \wedge d\sigma^0$), so that $\int 1 = 4\tau_2$. The integration over the constant piece should clearly contribute $\int d\tilde{\Phi} = 2\pi r$, in order to avoid overcounting and because of $S[\tilde{\Phi}] = 0$.

Putting everything together allows us to rewrite the path integral representing the partition function to the form

$$\int e^{-S[\Phi]} = 2\pi r \frac{\sqrt{\tau_2}}{\pi} \frac{1}{(\det'\square)^{1/2}} \sum_{n_1,n_2=-\infty}^{+\infty} \exp\left\{-S[\Phi_0^{(n_1,n_2)}]\right\}$$

$$= 2r\sqrt{\tau_2}\frac{1}{(\det'\square)^{1/2}} \sum_{n_1,n_2=-\infty}^{+\infty} \exp\left\{4\tau_2\frac{1}{4\pi}\left(\frac{2\pi r}{2i\tau_2}\right)^2 (n_1 - \bar{\tau}n_2)(n_1 - \tau n_2)\right\}$$

$$= 2r\sqrt{\tau_2}\frac{1}{(\det'\square)^{1/2}} \sum_{n_1,n_2=-\infty}^{+\infty} \exp\left\{-\pi\left[\frac{1}{\tau_2}(n_1 r - \tau_1 n_2 r)^2 + \tau_2 n_2^2 r^2\right]\right\} . \tag{3.34}$$

The $\det'\square \equiv \det'(-\partial\bar{\partial})$ can be evaluated as a formal product of eigenvalues in the following basis of the eigenfunctions,

$$\omega_{nm}(z,\bar{z}) = \exp\left\{2\pi i\frac{1}{2i\tau_2}[n(z - \bar{z}) + m(\tau\bar{z} - \bar{\tau}z)]\right\} , \tag{3.35}$$

where the eigenfunction at $n = m = 0$ has to be omitted ('):

$$\det'\square \equiv \prod_{\{m,n\}\neq\{0,0\}} \frac{\pi^2}{\tau_2^2}(n - \tau m)(n - \bar{\tau}m) . \tag{3.36}$$

The ζ-function regularization is very efficient in evaluating the infinite products like that in eq. (3.36), see also ref. [82]. One has

$$\prod_{n=1}^{\infty} a = a^{\zeta(0)} = a^{-1/2} , \quad \prod_{n=-\infty}^{+\infty} a = a^{2\zeta(0)+1} = 1 , \tag{3.37}$$

where

$$\zeta(s) \equiv \sum_{n=1}^{\infty} n^{-s} : \quad \zeta(-1) = -\frac{1}{12} , \quad \zeta(0) = -\frac{1}{2} , \quad \zeta'(0) = -\ln\sqrt{2\pi} . \tag{3.38}$$

Eqs. (3.37) and (3.38) imply

$$\prod{}' \left(\frac{\pi^2}{\tau_2^2}\right) = \frac{\tau_2^2}{\pi^2} \,, \quad \prod_{n=1}^{\infty} n^\alpha = e^{-\alpha\zeta'(0)} = (2\pi)^{\alpha/2} \,, \tag{3.39}$$

and, therefore,

$$\det{}' \Box = \prod_{\{m,n\}\neq\{0,0\}} \frac{\pi^2}{\tau_2^2}(n-m\tau)(n-m\bar{\tau}) = \frac{\tau_2^2}{\pi^2}\left(\prod_{n\neq 0} n^2\right) \prod_{m\neq 0, n\in\mathbf{Z}} (n-m\tau)(n-m\bar{\tau})$$

$$= \frac{\tau_2^2}{\pi^2}(2\pi)^2 \prod_{m>0, n\in\mathbf{Z}} (n-m\tau)(n+m\tau)(n-m\bar{\tau})(n+m\bar{\tau})$$

$$= 4\tau_2^2 \prod_{m>0} \left(e^{-\pi im\tau} - e^{\pi im\tau}\right)^2 \left(e^{-\pi im\bar{\tau}} - e^{\pi im\bar{\tau}}\right)^2 = 4\tau_2^2 \prod_{m>0} (q\bar{q})^{-m}(1-q^m)^2(1-\bar{q}^m)^2$$

$$= 4\tau_2^2(q\bar{q})^{1/12} \prod_{m>0} (1-q^m)^2(1-\bar{q}^m)^2 = 4\tau_2^2\eta^2\bar{\eta}^2 \,, \tag{3.40}$$

where the identity

$$\prod_{n=-\infty}^{\infty} (n+a) = a \prod_{n=1}^{\infty} (-n^2)(1-a^2/n^2) = 2i\sin(\pi a) \tag{3.41}$$

has been used. Therefore, one finds

$$2r\sqrt{\tau_2}\frac{1}{(\det{}' \Box)^{1/2}} = \frac{r}{\sqrt{\tau_2}}\frac{1}{\eta\bar{\eta}} \,. \tag{3.42}$$

The transformation property (12.74) of the η-function under the modular transformation $\tau \to -1/\tau$ guarantees the modular invariance of eq. (3.42) because of $\tau_2 \to \tau_2/|\tau|^2$.

To determine the instanton contributions to eq. (3.34), the summation over the winding number n_1 in the 'time' direction has to be exchanged for a sum over the conjugate momentum, by using the Poisson resummation formula (12.71). In order to do so, we need the Fourier transform of the function $f(n_1 r) = \exp\left[-\frac{\pi}{\tau_2}(n_1 r - \tau_1 n_2 r)^2\right]$:

$$\tilde{f}(p) = \int_{-\infty}^{+\infty} dx \, e^{2\pi ixp} f(x) = \sqrt{\tau_2}e^{2\pi i\tau_1 n_2 rp - \pi\tau_2 p^2} \,. \tag{3.43}$$

Substituting eqs. (12.71) and (3.42) into eq. (3.34) yields

$$\int e^{-S} = \frac{1}{\eta\bar{\eta}} \sum_{n,m=-\infty}^{+\infty} \exp\left\{-\pi\tau_2 n^2 r^2 + 2\pi i\tau_1 nm - \pi\tau_2(m/r)^2\right\}$$

$$= \frac{1}{\eta\bar{\eta}} \sum_{n,m=-\infty}^{+\infty} q^{\frac{1}{4}\left(\frac{m}{r}+nr\right)^2} \bar{q}^{\frac{1}{4}\left(\frac{m}{r}-nr\right)^2} = \frac{1}{\eta\bar{\eta}} \sum_{n,m=-\infty}^{+\infty} q^{\frac{1}{4}(p+w)^2} \bar{q}^{\frac{1}{4}(p-w)^2}$$

$$= \frac{1}{\eta \bar{\eta}} \sum_{n,m=-\infty}^{+\infty} q^{\frac{1}{2}p_L^2} \bar{q}^{\frac{1}{2}p_R^2} \equiv Z_{\text{circ}}(\tau) , \qquad (3.44)$$

where the momentum $p = m/r$, the winding $w = nr$, and the 'chiral' momenta $p_{L,R} = (1/\sqrt{2})(p \pm w) = (1/\sqrt{2})(\frac{m}{r} \pm nr)$ have been introduced. The partition function $Z_{\text{circ}}(\tau)$ possesses the duality symmetry, $Z_{\text{circ}}(r) = Z_{\text{circ}}(\frac{1}{r})$, in which the roles of winding and momentum are interchanged.

The Z_{circ} in eq. (3.44) can be equally represented in the form $(q\bar{q})^{-c/24} Tr\, q^{L_0} \bar{q}^{\bar{L}_0}$, with the hamiltonian trace over Hilbert space sectors $|m, n\rangle$, for which

$$L_0 |m,n\rangle = \frac{1}{4} \left(\frac{m}{r} + nr \right)^2 |m,n\rangle , \quad L_0 = \sum_{m>0} \alpha_{-m}\alpha_m + \frac{1}{2}p_L^2 ,$$
$$\bar{L}_0 |m,n\rangle = \frac{1}{4} \left(\frac{m}{r} - nr \right)^2 |m,n\rangle , \quad \bar{L}_0 = \sum_{m>0} \bar{\alpha}_{-m}\bar{\alpha}_m + \frac{1}{2}p_R^2 . \qquad (3.45)$$

The factor of $(\eta\bar{\eta})^{-1}$ in eq. (3.44) now receives the obvious hamiltonian interpretation. We can calculate $\text{Tr}\, q^{L_0}$ along the lines of the fermionic case, but this time in the bosonic Fock space generated by α_{-n}, as follows:

$$\text{Tr}\, q^{L_0'} = \text{Tr}\, q^{\sum_{n=1}^{\infty} \alpha_{-n}\alpha_n} = \prod_{n=1}^{\infty} \left(1 + q^n + q^{2n} + \ldots \right) = \prod_{n=1}^{\infty} \frac{1}{1-q^n} , \qquad (3.46)$$

where $L_0' \equiv L_0 - \frac{1}{2}p_L^2$, and similarly for $\bar{L}_0'$. Hence, we find

$$(q\bar{q})^{-c/24} \text{Tr}\, q^{L_0'} \bar{q}^{\bar{L}_0'} = (q\bar{q})^{-1/24} \sum_{n,m=0}^{\infty} P(n)P(m)q^n\bar{q}^m = \frac{1}{\eta\bar{\eta}} . \qquad (3.47)$$

It is not difficult to check the modular invariance of the partition function (3.44). First, under $\tau \to \tau + 1$, the relevant phase factor $\exp\left[2\pi i \cdot \frac{1}{2}(p_L^2 - p_R^2) \right]$ acquired by each term in eq. (3.44) equals to unity: $\frac{1}{2}(p_L^2 - p_R^2) = pw = mn \in \mathbf{Z}$. Second, under $\tau \to -1/\tau$, the roles of 'space' and 'time' are interchanged, and the modular invariance can be seen after performing the Poisson resummations for both m- and n- summations in eq. (3.44), and using the η-transformation property (12.74).

The general statement is that any partition function of the form

$$Z_{\Gamma^{r,s}} = \frac{1}{\eta^r \bar{\eta}^s} \sum_{(p_L, p_R) \in \Gamma^{(r,s)}} q^{\frac{1}{2}p_L^2} \bar{q}^{\frac{1}{2}p_R^2} \qquad (3.48)$$

is modular invariant provided that $\Gamma^{(r,s)}$ is a $(r+s)$-dimensional *even* and *self-dual* Lorentzian lattice of signature (r,s) [299]. The even property, $p_L^2 - p_R^2 \in 2\mathbf{Z}$, guarantees the invariance under $\tau \rightarrow \tau + 1$ (up to a possible phase from $\eta^{-r} \bar{\eta}^{-s}$ when $r - s \neq 0 \mod 24$), while the self-duality property guarantees the invariance under $\tau \rightarrow -1/\tau$.

The partition function (3.44) could also be expressed in terms of the $c = 1$ Virasoro characters. However, unlike the previously considered case of free fermions, the analogous expression for eq. (3.44) may involve infinite number of the Virasoro characters [84].

The generalizations of this theory to arbitrary-genus Riemann surfaces and/or a non-vanishing background charge are given in sect. 4.

Exercises

III-4 ▷ (*) Derive the shift in the vacuum energy of a bosonic field defined on a torus from the vacuum normal ordering on a cylinder:

$$(L_0)_{\text{cyl}} = \tfrac{1}{2} \sum_n : \alpha_{-n} \alpha_n := \sum_{n>0} \alpha_{-n} \alpha_n + \tfrac{1}{2} \sum_{n>0} n$$

$$= \sum_{n>0} \alpha_{-n} \alpha_n + \begin{cases} \tfrac{1}{2} \zeta(-1) & = -\tfrac{1}{24}, \quad \text{when} \ \ n \in \mathbf{Z} \\ \tfrac{1}{2} \left[-\tfrac{1}{2} \zeta(-1) \right] & = \tfrac{1}{48}, \quad \text{when} \ \ n \in \mathbf{Z} + 1/2; \end{cases} \tag{3.49}$$

III-5 ▷ Verify the modular invariance of the partition function in eq. (3.48) for an even and self-dual Euclidean lattice.

III-6 ▷ Under what conditions can the partition function (3.44) be expressed in terms of a *finite* number of the $c = 1$ Virasoro characters?

III.3   Chiral bosonization

Given 2d bosonic fields taking their values in a *compact* region, it is possible to *fermionize* them, i.e. to replace one physical (RM+LM) boson by four real (two RM and two LM) fermions, or just one chiral (say, LM) boson by a pair of real chiral LM fermions of the same chirality. The reverse procedure is known as (chiral) **bosonization** [35]. The target space of the bosonic fields is usually a unit circle, the compactification being necessary to insure holomorphic factorization. The fermion-boson correspondence in two dimensions can actually be extended to the Riemann surfaces of arbitrary genus [85, 86, 87, 88, 89, 90, 91, 92]. We consider some examples below.

On the Riemann sphere $\mathbf{C} \cup \infty$, two chiral fermions of central charge $c = 1/2$ are equivalent to a chiral boson of central charge $c = 1$. [2] Indeed, let us compare the two conformal chiral free fields – the bosonic one $\phi(z)$ and the fermionic one $\psi(z)$ – which are characterized by the equations (sect. I.5)

$$\phi(z)\phi(w) \sim -\ln(z-w) , \quad T_\phi = -\tfrac{1}{2} : \partial\phi\partial\phi : , \quad c_\phi = 1 ,$$
$$\psi(z)\psi(w) \sim -\frac{1}{z-w} , \quad T_\psi = \tfrac{1}{2} : \psi\partial\psi : , \quad c_\phi = 1/2 , \tag{3.50}$$

and ask ourselves whether we could realize ψ's in terms of ϕ's only? The answer to this question is in affirmative, and it is based on the observation from sect. I.5 that the fields

$$\psi^+(z) =: e^{i\phi(z)} : \quad \text{and} \quad \psi^-(z) =: e^{-i\phi(z)} : \tag{3.51}$$

have the right conformal dimension $h = 1/2$ to represent spinors. Rewriting them to the form

$$\psi^\pm = \frac{i}{\sqrt{2}}(\psi^1 \pm \psi^2) , \tag{3.52}$$

we get the OPE's

$$\psi^i(z)\psi^j(w) \sim -\frac{\delta^{ij}}{z-w} , \tag{3.53}$$

as they should. Using the equivalence relations (3.51) and comparing order by order the other terms in the complete OPE's

$$\psi^+(z)\psi^+(w) = (z-w) : \partial\psi^+(w)\psi^+(w) : + \dots ,$$
$$\psi^+(z)\psi^-(w) = \frac{1}{z-w} + : \psi^+(w)\psi^-(w) : + (z-w) : \partial\psi^+(w)\psi^-(w) : + \dots , \tag{3.54}$$

and

$$: e^{i\phi(z)} :: e^{i\phi(w)} := (z-w) : e^{2i\phi(z)} : + \dots ,$$
$$: e^{i\phi(z)} :: e^{-i\phi(w)} := (z-w)^{-1} : e^{i\phi(z)-i\phi(w)} := (z-w)^{-1}$$
$$+i\partial\phi(w) + (z-w)[-\tfrac{1}{2} : (\partial\phi)^2 : +\tfrac{i}{2}\partial^2\phi] + \dots , \tag{3.55}$$

resp., we find the bosonization rules for the fermion number current,

$$: \psi^+(w)\psi^-(w) := i\partial\phi(w) \equiv j(w) , \tag{3.56}$$

and for the stress tensor (the quantum equivalence!),

$$T_\psi = \tfrac{1}{2} : \psi^1\partial\psi^1 : +\tfrac{1}{2} : \psi^2\partial\psi^2 := -\tfrac{1}{2}(: \psi^+\partial\psi^- : + : \psi^-\partial\psi^+ :)$$

---

[2] In two dimensions, bosons and fermions have integer and half-integer conformal dimensions, resp. Accordingly, we use commuting and anti-commuting fields for them.

$$= \tfrac{1}{2}\left(-\tfrac{1}{2} : (\partial\phi)^2 : +\tfrac{i}{2}\partial^2\phi - \tfrac{1}{2} : (\partial\phi)^2 : -\tfrac{i}{2}\partial^2\phi\right) = -\tfrac{1}{2} : (\partial\phi)^2 := T_\phi , \qquad (3.57)$$

as well as some other bosonization relations,

$$: e^{\pm in\phi} := \frac{1}{(n-1)!} : \partial^{n-1}\psi^\pm \partial^{n-2}\psi^\pm \cdots \partial\psi^\pm\psi^\pm : . \qquad (3.58)$$

The next natural question is how to bosonize the conformal ghost systems introduced in sect. I.6. The quantum equivalence between the stress tensors $T_{b,c}$ of the (b,c) system in eq. (1.126) and that defined by eq. (1.135) in terms of the current j_z of charge Q given by eq. (1.136) suggests us to use a free scalar field $\phi(z)$ of background charge Q, with the $U(1)$ current $j_z(z) = i\partial\phi(z)$, just like that used in the Coulomb gas picture of sect II.6. [3] The equations to be compared are

$$b(z)c(w) \sim \frac{1}{z-w} , \quad T_{b,c} =: (\lambda-1)c\partial b + \lambda(\partial c)b : ,$$

$$h[b] = \lambda , \quad h[c] = 1-\lambda , \quad c_{b,c} = -2(6\lambda^2 - 6\lambda + 1) . \qquad (3.59a)$$

for the (b,c) ghost system, and

$$\phi(z)\phi(w) \sim -\ln(z-w) , \quad T_{\rm DF} = -\tfrac{1}{2} : \partial\phi\partial\phi : +\frac{i}{2}Q\partial^2\phi ,$$

$$h[: e^{i\alpha\phi(z)} :] = \tfrac{1}{2}\alpha(\alpha - Q) , \quad c_{\rm DF}(Q) = 1 - 3Q^2 , \qquad (3.59b)$$

for the 'bosonized' version, resp. Therefore, we are in a position to identify

$$c(z) =: e^{i\phi(z)} : , \quad b(z) =: e^{-i\phi(z)} : ,$$

$$j_{b,c} \equiv - : bc := i\partial\phi \equiv j_\phi , \quad T_{b,c} = T_{\rm DF} , \qquad (3.60)$$

provided

$$Q = 2\lambda - 1 . \qquad (3.61)$$

To distinguish a DF scalar from an ordinary scalar and to get the real bosonized stress tensor $T_{\rm DF}$, we make the substitution

$$\sigma(z) = i\phi(z) . \qquad (3.62)$$

The field σ is purely imaginary, and it has the wrong sign at the kinetic term in its stress tensor

$$T_\sigma = +\tfrac{1}{2} : \partial\sigma\partial\sigma : +\tfrac{1}{2}Q\partial^2\sigma . \qquad (3.63)$$

---

[3]Compared to the notation of sect. II.6, here we use the scalar field $\phi(z) = \frac{1}{\sqrt{2}}\varphi(z)$ and the background charge $Q = 2\sqrt{2}\alpha_0$. This makes contact with the results of sect. I.6 too.

We have, in addition

$$\sigma(z)\sigma(w) \sim \ln(z-w) \ , \quad h[: e^{\alpha\sigma(z)} :] = \tfrac{1}{2}\alpha(\alpha - Q) \ , \quad c_\sigma = 1 - 3Q^2 \ ,$$

$$c(z) =: e^{\sigma(z)} : \ , \quad b(z) =: e^{-\sigma(z)} : \ , \quad j_{b,c} = \partial\sigma \equiv j_\sigma \ . \tag{3.64}$$

Another useful opportunity is to use the pure imaginary background charge, $Q \to -iQ$, which leads to the central charge $c = 1 + 3Q^2$. Let $\rho(z)$ be the corresponding DF field. It follows

$$\rho(z)\rho(w) \sim -\ln(z-w) \ , \quad h[: e^{\alpha\rho(z)} :] = -\tfrac{1}{2}\alpha(\alpha - Q) \ , \quad c_\rho = 1 + 3Q^2 \ . \tag{3.65}$$

The (β, γ) conformal system of the *commuting* ghost fields (sect. I.6) cannot be naively bosonized. This is because the defining relations

$$\gamma(z)\beta(w) \sim \frac{1}{z-w} \ , \quad \beta(z)\gamma(w) \sim -\frac{1}{z-w} \ ,$$

$$T_{\beta,\gamma} = -(\lambda - 1) : \gamma\partial\beta : -\lambda : (\partial\gamma)\beta : \ , \quad j_{\beta,\gamma} = - : \gamma\beta : \ ,$$

$$T_{\beta,\gamma}(z)j_{\beta-\gamma}(w) \sim -\frac{Q}{(z-w)^3} + \frac{1}{(z-w)^2}j_{\beta,\gamma}(z) \ , \tag{3.66}$$

imply

$$Q_{\beta,\gamma} = -(2\lambda - 1) \ , \quad c_{\beta,\gamma} = -(1 - 3Q^2) = 3Q^2 - 1 \ , \tag{3.67}$$

while the natural candidate for the 'bosonized' stress tensor

$$T_j \equiv - \left(\tfrac{1}{2} : j_{\beta-\gamma}^2 : + \frac{Q}{2}\partial j_{\beta,\gamma} \right) \ ,$$

$$T_j(z)j_{\beta,\gamma}(w) \sim -\frac{Q}{(z-w)^3} + \frac{1}{(z-w)^2}j_{\beta,\gamma}(z) \ , \tag{3.68}$$

yields $c = 1 + 3Q^2 \neq 3Q^2 - 1$ instead. It follows from comparing the conformal dimensions of what would be the 'naive' bosonization relations:

$$\gamma \overset{?}{=} : e^\rho : \ , \quad \beta \overset{?}{=} : e^{-\rho} : \ , \tag{3.69}$$

that

$$h[\gamma] = 1 - \lambda \neq h[: e^\rho :] = -\lambda \ , \quad h[\beta] = \lambda \neq h[: e^{-\rho} :] = \lambda - 1 \ , \tag{3.70}$$

according to eq. (3.65). Hence, even though we can identify

$$j_{\beta,\gamma} \equiv - : \gamma\beta := -\partial\rho \equiv j_\rho \ , \quad T_j = T_\rho \ , \tag{3.71}$$

when $Q = -(2\lambda - 1)$, we still has to add to the field ρ some new auxiliary anticommuting fields (say, a (b,c)-type (η,ξ) conformal system) of central charge $c_{\eta,\xi} = -2$, in order to bosonize the commuting conformal ghost system. Eq. (3.59a) then implies

$$\lambda_{\eta,\xi} = 1 \ , \quad Q_{\eta,\xi} = 1 \ , \quad T_{\eta,\xi} =: (\partial\xi)\eta : \ , \quad j_{\eta,\xi} =: \xi\eta : \ . \tag{3.72}$$

The correct bosonization rules are given by [35, 93, 94, 95]

$$\gamma(z) =: e^{\rho(z)} : \eta(z) =: e^{\rho(z)-\chi(z)} : \ , \quad \beta(z) =: e^{-\rho(z)} : \partial\xi(z) =: e^{-\rho(z)+\chi(z)} : \partial\chi \ , \tag{3.73}$$

where the (η,ξ) conformal system has been bosonized as

$$\xi =: e^{\chi} : \ , \quad \eta =: e^{-\chi} : \ , \quad : \xi\eta := \partial\chi \ . \tag{3.74}$$

The $\beta - \gamma$ fermionization of the DF scalar is described by the relations involving the Dirac delta-function (*cf* eq. (3.70)) [4]

$$: e^{\rho(z)} := \delta(\beta(z)) \ , \quad : e^{-\rho(z)} := \delta(\gamma(z)) \ . \tag{3.75}$$

The bosonization rules above lead to non-trivial identities among the bosonic and fermionic correlation functions. These rules and identities turned out to be the powerful computational tools both in string theory and statistics [93, 94, 95]. We begin with the case of sphere, the case of torus is briefly discussed at the end of the section. The bosonization rules for the correlators on arbitrary Riemann surfaces are addressed in the next section.

Let

$$V_q(z) = \mu^{-\frac{1}{2}q(q-Q)} : e^{iq\phi}(z) : \tag{3.76}$$

be the screened vertex operators defined in terms of the DF scalar field $\phi(z)$ with the propagator (sect. II.6)

$$\langle \phi(z)\phi(w) \rangle = -\frac{\ln(z-w)}{\mu} \ , \tag{3.77}$$

where the IR-regulator μ has been introduced. One finds (sect. II.6)

$$\langle V_1(z)V_{-1}(w) \rangle = \mu^{-1}e^{-\ln(z-w)/\mu} = \frac{1}{z-w} \ ,$$

---

[4]The delta-function is defined through its standard Fourier representation. As far as the anticommuting (b,c) ghost system is concerned, we still have $: e^{\sigma(z)} := \delta(c(z)) = c(z)$, $: e^{-\sigma(z)} := \delta(b(z)) = b(z)$.

$$\langle V_{q_1}(z) \cdots V_{q_M}(z_M) \rangle = \mu^{-\frac{1}{2}\sum_j q_j(q_j - Q)} \prod_{i<j} \left[\frac{z_i - z_j}{\mu} \right]^{q_i q_j}$$

$$= \mu^{-\frac{1}{2}(\sum_i q_i)(\sum_j q_j - Q)} \prod_{i<j}(z_i - z_j)^{q_i q_j} \equiv \mu^{-\frac{1}{2}(\sum_i q_i)(\sum_j q_j - Q)} Z\left(\sum_i q_i z_i \right) , \qquad (3.78)$$

where the argument of Z has to be understood in the divisor sense at the fixed order of the points. Eq. (3.78) vanishes when $\mu \to \infty$ unless $\sum_j q_j = 0$ or $\sum_j q_j = Q$, as it should (sect. II.6).

The bosonized correlators having the form (3.78) are to be compared with the correlators of the initial (b, c) conformal ghost system of charge $Q = 2\lambda - 1$. It is elementary, in particular, to get the correlator of the c-ghosts by using the fermionic state structure of their Fock space [5]

$$\langle c(z_1)c(z_2)c(z_3) \cdots c(z_Q) \rangle = \pm \begin{vmatrix} 1 & 1 & 1 & \dots & 1 \\ z_1 & z_2 & z_3 & \dots & z_Q \\ \vdots & \vdots & \vdots & \dots & \vdots \\ z_1^{Q-1} & z_2^{Q-1} & z_3^{Q-1} & \dots & z_Q^{Q-1} \end{vmatrix}$$

$$= \det_{j,r}(z_j)^{r-1} = \prod_{i<j}(z_i - z_j) \equiv Z\left(\sum_j z_j \right) , \quad 1 \le j, r \le Q . \qquad (3.79)$$

Similarly, in the general case, one finds by successive application of Wick's theorem that

$$\left\langle \prod_{i=1}^M b(x_i) \prod_{j=1}^{M+Q} c(y_j) \right\rangle = \sum_p \text{sign}(p)\, Z(\sum_j y_j')\, \det_{ij} Z(x_i - y_j'') , \qquad (3.80)$$

where the set of y_j's has been divided into two sets: $\{y_j\}_{j=1}^{M+Q} = \{y_j'\}_{j=1}^Q \cup \{y_j''\}_{j=1}^M$,

$$\det_{ij} Z(x_i - y_j'') = \sum_{\pi \in S_M} \text{sign}(\pi) \prod_{j=1}^M \frac{1}{x_j - y_{\pi(j)}''} , \qquad (3.81)$$

p goes over all the partitions, $p = (\{y_j'\}, \{y_j''\})$, and S_M is the permutation group. The corresponding correlator of the bosonic vertex operators takes the form

$$\left\langle \prod_{i=1}^M\, : e^{\sigma(x_i)} : \prod_{j=1}^{M+Q}\, : e^{-\sigma(y_j)} : \right\rangle$$

$$= \frac{\prod_{i<i'}(x_i - x_{i'}) \prod_{j<j'}(y_i - y_{i'})}{\prod_{i,j}(x_i - y_j)} = Z\left(\sum_i x_i - \sum_j y_j \right) . \qquad (3.82)$$

---

[5]There is non-trivial monodromy associated with chiral correlators (sect. II.6). In particular, there is the overall sign ambiguity in defining the (b, c) correlators.

In particular, we have

$$\left\langle : e^{\sigma(z_1)} : \cdots : e^{\sigma(z_Q)} : \right\rangle = \prod_{i<j}^{Q} e^{\ln(z_i - z_j)} = \prod_{i<j}^{Q} (z_i - z_j) \ . \tag{3.83}$$

The chiral bosonization implies the identity of the r.h.s. of eqs. (3.80) and (3.82),

$$\sum_p \text{sign}(p) \, Z\!\left(\sum_j y_j'\right) \det_{ij} Z(x_i - y_j'') = Z\!\left(\sum_i x_i - \sum_j y_j\right) \ , \tag{3.84}$$

which is just the simple particular case (at genus 0) of *Fay's trisecant identities* known in the theory of Riemann surfaces and theta functions [96, 97]. The corresponding bosonization formulae for the correlators of the commuting (β, γ) system were found in ref. [92]. In particular,

$$\left\langle \prod_{i=1}^{M} \beta(x_i) \prod_{j=1}^{M} \gamma(y_j) \prod_{k=1}^{Q} \delta(\gamma(z_k)) \right\rangle$$

$$= (-)^M \frac{Z\left(\sum x_i - \sum y_j\right)}{Z\left(\sum x_i - \sum y_j + \sum z_k\right)} \underset{ij}{\text{Sym}} \, Z(x_i - y_j) \ , \tag{3.85}$$

where the Sym-operation has been introduced,

$$\underset{ij}{\text{Sym}} \, A_{ij} \equiv \sum_{\pi \in S_M} \prod_i A_{i\pi(i)} \ . \tag{3.86}$$

As fas the torus (genus 1) case is concerned, let us consider two Dirac fermions $\psi_1(z)$, $\psi_2(z)$ and their conjugates, all having the same spin structure. On the one hand, the fermionic partition function is given by the modular invariant combination

$$Z_{\text{Dirac}} = \frac{1}{2} \left[A^2 \bar{A}^2 \overset{A^2 \bar{A}^2}{\Box} + P^2 \bar{P}^2 \overset{A^2 \bar{A}^2}{\Box} + A^2 \bar{A}^2 \overset{P^2 \bar{P}^2}{\Box} + P^2 \bar{P}^2 \overset{P^2 \bar{P}^2}{\Box} \right]$$

$$= \frac{1}{2} \left(\left| \frac{\vartheta_3}{\eta} \right|^2 + \left| \frac{\vartheta_4}{\eta} \right|^2 + \left| \frac{\vartheta_2}{\eta} \right|^2 + \left| \frac{\vartheta_1}{\eta} \right|^2 \right) \ , \tag{3.87}$$

where we have used eq. (3.22) to express Z_{Dirac} in terms of the ϑ-functions. On the other hand, we can use another representation for the ϑ-functions, via the Jacobi triple product identity (12.63). Substituting it into eq. (3.87), we obtain the bosonization formula for the torus in the form

$$Z_{\text{Dirac}} = \frac{1}{\eta \bar{\eta}} \sum_{n,m=-\infty}^{\infty} \left[q^{\frac{1}{2}n^2} \bar{q}^{\frac{1}{2}m^2} + q^{\frac{1}{2}(n+1/2)^2} \bar{q}^{\frac{1}{2}(m+1/2)^2} \right] \frac{1}{2} \left[1 + (-1)^{n+m} \right]$$

$$= \frac{1}{\eta\bar{\eta}} \sum_{n,m'=-\infty}^{+\infty} \left[q^{\frac{1}{2}(n+m')^2} \bar{q}^{\frac{1}{2}(n-m')^2} + q^{\frac{1}{2}(n+\frac{1}{2}+m')^2} \bar{q}^{\frac{1}{2}(n+\frac{1}{2}-m')^2} \right]$$

$$= \frac{1}{\eta\bar{\eta}} \sum_{n,m=-\infty}^{+\infty} q^{\frac{1}{2}\left(\frac{m}{2}+n\right)^2} \bar{q}^{\frac{1}{2}\left(\frac{m}{2}-n\right)^2} = Z_{\mathrm{circ}}(r=\sqrt{2}) \,. \tag{3.88}$$

Due to the appearance of the bosonic partition function at $r = \sqrt{2}$ on the r.h.s. of this equation, the vertex operator $: e^{\pm i\sqrt{2}\phi(z)} :\equiv: e^{\pm i\varphi(z)} :$ has (conformal) dimension one, and it is single-valued under the shift $\varphi \to \varphi + 2\pi$. The connection with the real fermions is given by: $: e^{\pm i\phi(z)} := \frac{i}{\sqrt{2}} \left[\psi_1(z) \pm i\psi_2(z) \right]$, $: e^{\pm i\bar{\phi}(\bar{z})} := \frac{i}{\sqrt{2}} \left[\bar{\psi}_1(\bar{z}) \pm i\bar{\psi}_2(\bar{z}) \right]$.

The similar considerations could be equally applied to the case of *orbifolds*, after the proper modding the initial theory admitting a discrete symmetry, in the modular invariant way. In particular, two Majorana fermions bosonize onto an $S^1/\mathbf{Z}_2$ orbifold at radius $r = \sqrt{2}$ [17].

Exercises

III-7 ▷ (*) How could the bosonization rules take care about the different statistics of bosons and fermions inside the correlation functions? *Hint*: use the property of log-function: $\ln(z - w) = i\pi + \ln(w - z)$.

III-8 ▷ Use eq. (3.61) to check that the bosonized b and c ghosts in eq. (3.60) have the right conformal dimensions λ and $1 - \lambda$, resp.

III-9 ▷ After the substitutions $Q \to -iQ$ and $\alpha \to -i\alpha$ in the bosonization rules of the (b,c) anticommuting ghost system, one gets the central charge $c = 1 + 3Q^2$ which is formally 'larger' than one, while the conformal dimensions of the fields $: e^{\alpha\phi(z)} :$ are still real. What will be the difference between the initial ghost system and the field system resulting from the fermionization of its bosonized version with the imaginary background charge?

III-10 ▷ Use the (β, γ) bosonization rules (3.71) to directly check the validity of the defining equations (3.66).

III-11 ▷ Derive the OPE's

$$\gamma(z)\delta(\gamma(w)) = (z - w)\partial\gamma\delta(\gamma)(w) + \dots ,$$

$$\beta(z)\delta(\gamma(w)) = -\frac{1}{z - w}\delta'(\gamma)(w) + \dots ,$$

$$\delta(\beta(z))\delta(\gamma(w)) = (z - w) + \dots . \tag{3.89}$$

III-12 ▷ Check explicitly the bosonization identity (3.84) for the simplest non-trivial case of the 4-point function with $Q = 0$.

III-13 ▷ (*) The (b, c) ghost contribution to the bosonic string BRST current (Ch. VIII) is given by $j_{\text{BRST}}^{b,c}(z) = \frac{1}{2} : c(z)T_{b,c}(z) :$. Prove that this contribution can be bosonized as follows:

$$\frac{1}{2} : cT_{b,c} := \frac{1}{2} : \left(-\frac{\lambda}{2} : \partial\sigma\partial\sigma + \frac{\lambda}{2}\partial^2\sigma \right) e^{\sigma} : . \tag{3.90}$$

Hint: use the splitting-point resolution:

$$\frac{1}{2} : c(z)T_{b,c}(z) := \frac{1}{2} \lim_{w \to z} \left[c(w) : T_{b,c}(z) + \frac{\lambda - 1}{(w - z)^2}c(z) + \frac{\lambda}{w - z}\partial c(z) \right] . \tag{3.91}$$

III-14 ▷ (*) Take $\lambda = 2$ in the exercise # III-13 and show that the ghost contribution to the BRST charge $Q_{\text{BRST}} \equiv \oint (d\zeta)/(2\pi i) \, j_{\text{BRST}}(\zeta)$ takes the form

$$Q_{\text{BRST}}^{b,c} = \oint (d\zeta)/(2\pi i) : e^{\sigma(\zeta)}T_{\sigma}(\zeta) : , \tag{3.92}$$

where T_{σ} is given by eq. (3.63) at $Q = 3$. *Hint*: use integration by parts.

III.4   Chiral bosonization on Riemann surfaces

In this section the theory of a single chiral (holomorphic) circle-valued bosonic field is solved, and its correlation functions are related with those of chiral fermions, all of them being defined on a Riemann surface [87]. This generalizes the results of the previous section to an arbitrary genus h. For simplicity, no background charge Q is introduced at the beginning. The basic facts about Riemann surfaces are collected in sect. XII.4.

Our starting point here is the scalar field action (3.30) defined on an arbitrary (closed) Riemann surface Σ of genus h. Eq. (3.30) can be rewritten as an integral of a $(1, 1)$ form as

$$S[\Phi] = \frac{i}{2\pi} \int_{\Sigma} \partial\Phi \wedge \bar{\partial}\Phi , \tag{3.93}$$

where the 1-form $d\Phi \equiv (\partial + \bar{\partial})\Phi$ is closed but not necessarily exact. Its winding numbers along cycles of the canonical homology basis label the solitonic sectors of the theory,

$$\oint_{a_i} d\Phi = 2\pi n_i , \qquad \oint_{b_j} d\Phi = 2\pi m_j , \tag{3.94}$$

so that $d\Phi$ can always be represented in the form

$$d\Phi = d\Phi' + \Phi_{m,n} ,\qquad (3.95)$$

where Φ' is a single-valued scalar with the zero-mode (a constant) removed, while $\Phi_{m,n}$ has the winding numbers (m, n). Hence, the action and the partition function take the form

$$S[\Phi] = S[\Phi_{m,n}] + S[\Phi'] ,$$

$$Z_B = \sum_{m,n} e^{-S[\Phi_{m,n}]} \int [d\Phi'] e^{-S[\Phi']} . \qquad (3.96)$$

The form $\Phi_{m,n}$ can be decomposed w.r.t. a basis of Abelian differentials $\{\omega_i\}_{i=1}^h$ on Σ,

$$\Phi_{m,n} = -i\pi (m + \bar{\Omega} n)_i (\mathrm{Im}\Omega)_{ij}^{-1} \omega_j + c.c. , \qquad (3.97)$$

which implies

$$S[\Phi_{m,n}] = \frac{\pi}{2} (m + \bar{\Omega} n)(\mathrm{Im}\Omega)^{-1}(m + \Omega n) . \qquad (3.98)$$

The contribution of Φ' to the partition function is given by the Gaussian integral

$$\int [d\Phi'] e^{-S[\Phi']} = \left[\frac{8\pi^2 \det' \Delta_g}{\int d^2 z \sqrt{g}} \right]^{-1/2} , \qquad (3.99)$$

whereas the solitonic sectors produce the factor

$$\sum_{m,n} \exp\left[-\frac{\pi}{2} (m + \bar{\Omega} n)(\mathrm{Im}\Omega)^{-1}(m + \Omega n) \right] . \qquad (3.100)$$

Rewriting the index m as $2(k + \delta'')$ with half-integer valued δ'' and applying the Poisson resummation formula (12.71) yields

$$(\det \mathrm{Im}\Omega)^{1/2} \sum_{\delta''} \sum_{n,k} e^{-\pi n (\mathrm{Im}\Omega) n/2} e^{-\pi k (\mathrm{Im}\Omega) k/2} e^{i(2\pi\delta'' + \pi n \mathrm{Re}\Omega)k} \qquad (3.101)$$

instead eq. (3.100). Introducing $p + q + 2\delta' = n$ and $p - q = k$ allows us to rewrite the summation over (n, k) as the summation over integers (p, q) and half-integers δ'. Eq. (3.101) then appears to be a sum over all spin structures of the Jacobi theta functions with characteristics and the vanishing argument (see sect. XII.4 for their definition),

$$\sum_{\delta'} |\vartheta[\delta](0, \Omega)|^2 . \qquad (3.102)$$

Putting all together, one finds [87]

$$Z_B = \sum_{\delta} Z_B^{\delta} , \qquad (3.103)$$

where

$$Z_{\mathrm{B}}^{\delta} = \left[\frac{8\pi^2 \mathrm{det}' \Delta_g}{\int d^2z \sqrt{g} \, \mathrm{det} \, \mathrm{Im}\Omega} \right]^{-1/2} |\vartheta[\delta](0, \Omega)|^2 \ . \tag{3.104}$$

Eq. (3.103) is just the partition function of a spin-$\frac{1}{2}$ fermionic field, summed over all its spin structures, since [6]

$$Z_{\mathrm{F}}^{\delta} = \int e^{S_{b-c}} = \mathrm{det} \, \Delta_{1/2}^{(-)} \tag{3.105}$$

also leads to the r.h.s. of eq. (3.104). Hence, $Z_{\mathrm{B}}^{\delta} = Z_{\mathrm{F}}^{\delta}$.

As far as the correlation functions in both theories are concerned, their derivation still requires to calculate only the Gaussian integrals with the propagators written in terms of the *prime form* (sect.XII.4). In the bosonic case, the result is (in the conformal gauge)

$$Z_{\mathrm{B}}(z_1, \ldots, w_M) = \int [d\Phi] e^{-S[\Phi]} \prod_{i=1}^{M} \rho^{1/2}(z_i) e^{i\Phi(z_i)} \prod_{i=1}^{M} \rho^{1/2}(w_i) e^{-i\Phi(w_i)}$$

$$= \sum_{m,n} e^{-S[\Phi_{m,n}]} \exp\left[-i\pi(m + \bar{\Omega}n)(\mathrm{Im}\Omega)^{-1} I(\sum z_i - \sum w_i) + \mathrm{c.c.} \right]$$

$$\times \int [d\Phi'] e^{-S[\Phi']} \prod_{i=1}^{M} \rho^{1/2}(z_i) e^{i\Phi(z_i)} \prod_{i=1}^{M} \rho^{1/2}(w_i) e^{-i\Phi(w_i)}$$

$$= \left[\frac{8\pi^2 \mathrm{det}' \Delta_g}{\int d^2z \sqrt{g} \, \mathrm{det} \, \mathrm{Im}\Omega} \right]^{-1/2} \sum_{\delta} \left| \vartheta[\delta]\left(\sum z_i - \sum w_i, \Omega \right) \right|^2$$

$$\times \left| \frac{\prod_{i<j} E(z_i, z_j) \prod_{i<j} E(w_i, w_j)}{\prod_{i,j} E(z_i, w_j)} \right|^2 , \tag{3.106}$$

where in each solitonic sector (m, n) the difference $\Phi(z_i) - \Phi(w_i)$ has been replaced by $\Phi(z_i) - \Phi(w_i) = \int_{w_i}^{z_i} \Phi_{m,n} + \Phi'(z_i) - \Phi'(w_i)$, and the *Abel map* $I(\sum z_i - \sum w_i)$ has been introduced (sect. XII.4). [7]

Given an arbitrary even spin structure δ, its contribution to eq. (3.106) takes the form

$$Z_{\mathrm{B}}^{\delta} \left| A_{\mathrm{B}}^{\delta}(z_1, \ldots, w_M) \right|^2 , \tag{3.107}$$

where the chiral amplitude has been introduced,

$$A_{\mathrm{B}}^{\delta}(z_1, \ldots, w_M) = \frac{\vartheta[\delta]\left(\sum z_i - \sum w_i, \Omega \right)}{\vartheta[\delta](0, \Omega)} \cdot \frac{\prod_{i<j} E(z_i, z_j) \prod_{i<j} E(w_i, w_j)}{\prod_{i,j} E(z_i, w_j)} . \tag{3.108}$$

---

[6]Given an odd spin structure and Dirac's zero mode, both Z_{B}^{δ} and Z_{F}^{δ} obviously vanish.

[7]It is the common notational convention used in eq. (3.106) that an explicit reference to the Abel map is omitted in the arguments of theta functions.

The chiral bosonic correlator given above has the same zeroes and poles, and it transforms in the same way as the correlator of the (b,c) conformal ghost system with spin structure δ. Hence, there should be the identity

$$A_B^\delta(z_1,\ldots,w_M) = \left\langle \prod_{i=1}^{M} b(z_i) \prod_{i=1}^{M} c(w_i) \right\rangle . \qquad (3.109)$$

As a rule, bosonization produces non-trivial identities when correlators of bosons and fermions can be computed *independently*. For instance, the (b,c) correlators in eq. (3.109) can also be expressed in terms of the *Szegö kernel* (sect. XII.4) as

$$\det S_\delta(z_i, w_j) , \qquad (3.110)$$

for an even spin structure. Taking, e.g., $M = 2$ and using the bosonized form (3.108) for the correlator result in the particular case of Fay's trisecant identities [96, 97], namely

$$\vartheta[\delta](z_1 + z_2 - w_1 - w_2, \Omega)\vartheta[\delta](0, \Omega)E(z_1, z_2)E(w_1, w_2)$$

$$= \vartheta[\delta](z_1 - w_1, \Omega)\vartheta[\delta](z_2 - w_2, \Omega)E(z_2, w_1)E(z_1, w_2) - (z_1 \to z_2) , \qquad (3.111)$$

which does not hold when Ω is not a period matrix of a Riemann surface!

As to the odd spin structures, the similar identities arise when the projections necessary to get rid of the zero mode are appropriately taken into account [28]. The generalization of the chiral bosonization formulae to the case of a non-vanishing background charge Q can be found in ref. [87] (see refs. [28, 98] also).

As far as the bosonization rules for the (b,c) and (β,γ) conformal ghost systems on arbitrary Riemann surfaces are concerned, the algebraic relations (3.79)–(3.86) can be generalized by imposing appropriate monodromy conditions for each homology cycle without spoiling single-valuedness, modular transformation properties and local singularity structure [92]. Therefore, we still need a generalization of the Z-function introduced for the genus-0 in eq. (3.78) for the bosonized correlator. The unique combination generalizing this function to arbitrary genus takes the form [87, 92]

$$Z_{\bar\delta,\lambda}\left(\sum_i q_i z_i\right) \equiv Z_1^{-1/2}\vartheta[\bar\delta]\left(\sum_i q_i z_i + Q\Delta, \Omega\right) \prod_{i<j} E(z_i, z_j)^{q_i q_j} \prod_i \varrho(z_i)^{-Qq_i} , \qquad (3.112)$$

where the holomorphic non-vanishing $h/2$-differential $\varrho(z)$ has been introduced. [8] This

---

[8] As far as the definitions of the Jacobi theta function with characteristics $\vartheta[\bar\delta](z, \Omega)$, the prime form $E(z_i, z_j)$, and the *Riemann vector of constants* Δ are concerned, see sect. XII.4. The normalization factor $Z_1(\Omega)$ has been computed in ref. [87].

actually assumes the validity of the 'zero-charge' constraint

$$\sum_i q_i + Q(h-1) = 0 \ .$$

(3.113)

The appearance of the Riemann vector of constants Δ in the argument of the theta function in eq. (3.112) is needed to reduce to zero a degree of the divisor $D = \sum_k z_k$. Hence, we have

$$\left\langle \prod_i \ : e^{q_i \sigma(z_i)} : \right\rangle_{\vec{\delta}} = Z_{\vec{\delta},\lambda}\left(\sum_i q_i z_i\right) \ ,$$

(3.114)

where

$$\langle \sigma(z)\sigma(w) \rangle = \ln E(z,w) \ .$$

(3.115)

The fermionic representation of the same correlator (3.114) can be immediately constructed for the case $Q = 0$ and even spin structure $\vec{\delta}$. The (b,c) system 2-point function

$$\langle b(z)c(w) \rangle = \frac{Z_{\vec{\delta},1/2}(z-w)}{Z_{\vec{\delta},1/2}(0)}$$

(3.116)

and Wick theorem imply in this case

$$\left\langle \prod_{i=1}^M b(x_i) \prod_{j=1}^M c(y_j) \right\rangle = (-)^{M(M-1)/2} Z_{\vec{\delta},1/2}(0)^{-M+1} \det_{ij} Z_{\vec{\delta},1/2}(x_i - y_j) \ .$$

(3.117)

The equivalence of eqs. (3.114) and (3.117) at $Q = 0$ gives the bosonization formula:

$$(-)^{M(M-1)/2} \det_{ij} Z_{\vec{\delta},1/2}(x_i - y_j) = Z_{\vec{\delta},1/2}(0)^{M-1} Z_{\vec{\delta},1/2}\left(\sum_i x_i - \sum_j y_j\right) \ .$$

(3.118)

In particular, when $M = 2$, this is exactly the Fay's trisecant identity (3.111). The identity (3.118) can be generalized to arbitrary divisors D of degree $Q(1-h)$ [92]:

$$(-)^{M(M-1)/2} \det_{ij} Z_{\vec{\delta},\lambda}(x_i - y_j + D) = Z_{\vec{\delta},\lambda}(D)^{M-1} Z_{\vec{\delta},\lambda}\left(\sum_i x_i - \sum_j y_j + D\right) \ .$$

(3.119)

Similarly, the generalization of eq. (3.85) for the commuting (β,γ) ghost system to higher-genus Riemann surfaces takes the form [92]

$$\left\langle \prod_{i=1}^M \beta(x_i) \prod_{j=1}^M \gamma(y_j) \prod_{k=1}^{Q(1-g)} \delta(\gamma(z_k)) \right\rangle$$

$$= (-)^{M(M+1)/2} Z_{\vec{\delta},\lambda}(D)^{-M-1} \underset{ij}{\text{Sym}} \ Z_{\vec{\delta},\lambda}(x_i - y_j + D) \ .$$

(3.120)

Eq. (3.120) has the right structure of singularities, monodromy and modular properties, and it reduces to eq. (3.85) on the Riemann sphere.

III.5   BRST approach to minimal models on a torus

In the previous sections, the bosonic and fermionic free fields, as well as the corresponding CFT's, have been extended to higher genus. The minimal models introduced in Ch. II can also be generalized this way, when using the Felder's BRST approach (sect. II.9). In this section the case of torus is considered along the lines of the original paper of Felder [66]. See sect. II.9 for the notation used in this section.

On a sphere (sect. II.9), the screened vertex operators map BRST physical states into BRST physical states, and no spurious (unphysical) states arise. On higher genus Riemann surfaces, when taking a trace over the Fock space in order to form a correlation function, one has to project out the propagating spurious states, which results in a more subtle BRST cohomology structure of the *BRST complex* $C_{m,m'}$ defined by

$$\ldots \xrightarrow{Q_m} F_{-m+2p,m'} \xrightarrow{Q_{p-m}} F_{m,m'} \xrightarrow{Q_m} F_{-m,m'} \xrightarrow{Q_{p-m}} F_{m-2p,m'} \xrightarrow{Q_m} \ldots . \tag{3.121}$$

The BRST cohomology groups are

$$H^j(C_{m,m'}) = \mathrm{Ker}Q^{(j)}/\mathrm{Im}Q^{(j-1)} , \tag{3.122}$$

where

$$Q^{(2j)} \equiv Q_m \; : \; F_{m-2jp,m'} \xrightarrow{Q_m} F_{-m-2jp,m'} , $$
$$Q^{(2j+1)} \equiv Q_{p-m} \; : \; F_{-m-2jp,m'} \xrightarrow{Q_{p-m}} F_{m-2(j+1)p,m'} , \tag{3.123}$$

It is Felder's theorem [66] that

$$H^0(C_{m,m'}) = \mathcal{H}_{m,m'} , \qquad H^j(C_{m,m'}) = 0 , \quad j \neq 0 . \tag{3.124}$$

just like that for a plane or a sphere. The proof is based on the detailed analysis of the BRST structure of the modules $F_{m,m'}$ and the map Q itself in terms of the highest weight states, and it can be found in ref. [66].

For the torus with the Teichmüller (moduli) parameter $q = \exp(2\pi i\tau)$, $|q| < 1$, the conformal blocks are given by

$$\mathcal{F}_{\{k\}}^{\mathrm{torus}}(z_1, \ldots, z_k) = \mathrm{Tr}_{\mathcal{H}_{m,m'}} \prod_{j=1}^{k} \phi_{\{n_{j-1},n'_{j-1}\}\{m_{j-1},m'_{j-1}\}}^{\{m_{j-1},m'_{j-1}\}}(z_j)q^{L_0} , \tag{3.125}$$

$m = m_0$, $m' = m'_0$, instead of eq. (2.163), whereas the (unrenormalized) Green's functions still have the form (2.46), now with $1 > |z_1| > \ldots > |z_k| > |q|$. Since eq. (3.124), the conformal block (3.125) is actually given by

$$\mathcal{F}_{\{\vec{k}\}}^{\text{torus}}(z_1,\ldots,z_k|q) = \text{const.}\ \text{Tr}_{H^0(C_{m,m'})}\left(\prod_{j=1}^{k} V_{n_j,n_j'}^{r_j,r_j'}(z_j)q^{L_0}\right)\ . \tag{3.126}$$

Eq. (3.126) can be rewritten in terms of traces over the Fock spaces. First, eq. (2.179) yields

$$Q_m \prod_{j=1}^{k} V_{n_j,n_j'}^{r_j,r_j'}(z_j)q^{L_0} = e^{i\theta}\prod_{j=1}^{k} V_{n_j,n_j'}^{n_j-r_j-1,r_j'}(z_j)q^{L_0}Q_m\ ,$$

$$Q_{p-m} \prod_{j=1}^{k} V_{n_j,n_j'}^{n_j-r_j-1,r_j'}(z_j)q^{L_0} = e^{-i\theta}\prod_{j=1}^{k} V_{n_j,n_j'}^{r_j,r_j'}(z_j)q^{L_0}Q_{p-m}\ , \tag{3.127}$$

where

$$\theta = 2\pi \sum_{j=1}^{k} \alpha_{n_j,n_j'}\alpha_+\left[\sum_{i=1}^{j}(n_i-2r_i-1)+m\right]\ . \tag{3.128}$$

Second, after introducing the new 'effective' operators,

$$X^{(2j)} = \prod_{j=1}^{k} V_{n_j,n_j'}^{r_j,r_j'}(z_j)q^{L_0}\ ,\quad X^{(2j+1)} = e^{i\theta}\prod_{j=1}^{k} V_{n_j,n_j'}^{n_j-r_j-1,r_j'}(z_j)q^{L_0}\ , \tag{3.129}$$

acting on $F_{m-2jp,m'}$ and $F_{-m-2jp,m'}$, resp., we can draw the commutative diagram [66]:

$$\begin{array}{ccccccccc}
\longrightarrow & F_{m+2p,m'} & \xrightarrow{Q^{(-2)}} & F_{-m+2p,m'} & \xrightarrow{Q^{(-1)}} & F_{m,m'} & \xrightarrow{Q^{(0)}} & F_{-m,m'} & \xrightarrow{Q^{(1)}} \\
& \downarrow{\scriptstyle X^{(-2)}} & & \downarrow{\scriptstyle X^{(-1)}} & & \downarrow{\scriptstyle X^{(0)}} & & \downarrow{\scriptstyle X^{(1)}} & \\
\longrightarrow & F_{m+2p,m'} & \xrightarrow{Q^{(-2)}} & F_{-m+2p,m'} & \xrightarrow{Q^{(-1)}} & F_{m,m'} & \xrightarrow{Q^{(0)}} & F_{-m,m'} & \xrightarrow{Q^{(1)}}
\end{array} \tag{3.130}$$

in terms of which the trace $X = \oplus_{i\in\mathbb{Z}}X^{(i)}$ we want to compute over the physical Fock space can be represented as the alternating sum of the traces of $X^{(i)}$,

$$\text{Tr}_{H^0(C_{m,m'})}(X^{(0)}) = \text{Tr}_{H^{\bullet}(C_{m,m'})}(X)$$

$$= \sum_{j=-\infty}^{\infty} \text{Tr}_{F_{m-2jp,m'}}(X^{(2j)}) - \sum_{j=-\infty}^{\infty} \text{Tr}_{F_{-m-2jp,m'}}(X^{(2j+1)})\ . \tag{3.131}$$

Therefore, we arrive at the Feigin-Fuchs integral representation for the conformal blocks on the torus, in the form [66]

$$\mathcal{F}_{\{\vec{k}\}}^{\text{torus}}(z_1,\ldots,z_k|q) = \text{const}\left[\sum_{j=-\infty}^{\infty} \text{Tr}_{F_{m-2jp,m'}}\left(\prod_{i=1}^{k} V_{n_i,n_i'}^{r_i,r_i'}(z_i)q^{L_0}\right)\right.$$

$$\left. -e^{i\theta}\sum_{j=-\infty}^{\infty} \text{Tr}_{F_{-m-2jp,m'}}\left(\prod_{i=1}^{k} V_{n_i,n_i'}^{n_i-r_i-1,r_i'}(z_i)q^{L_0}\right)\right]\ . \tag{3.132}$$

The techniques described in sects. 1 and 2 can now be applied for computing the remaining traces of products of the vertex operators over the Fock spaces and expressing the results in terms of the conventional Jacobi theta functions defined on the torus (sect. XII.3). In particular, the formula replacing eq. (3.132) reads [66]

$$\sum_{l=-\infty}^{\infty} \text{Tr}_{F_{\pm m - 2lp, m'}} \left(V_{\alpha_1}(z_1) \cdots V_{\alpha_k}(z_k) q^{L_0} \right)$$

$$= \frac{1}{\eta(\tau)} q^{c/24} \sum_{l=-\infty}^{\infty} e^{\pi i \tau (2p'pl + pm' \mp p'm)^2 / 2pp')} e^{2\pi i l w}$$

$$\times \prod_{i=1}^{k} z_i^{\alpha_i(2\alpha_{\pm m, m'} - \alpha_i)} \prod_{i<j} \left(2\pi i \frac{\vartheta_1(w_i - w_j | \tau)}{\vartheta_1'(0|\tau)} \right)^{2\alpha_i \alpha_j} , \qquad (3.133)$$

where the following notation has been introduced:

$$z_j = e^{2\pi i w_j} , \quad w = \sum_{i=1}^{k} [(n_i' - 1)p - (n_i - 1)p'] w_i , \quad \alpha_i = \alpha_{n_i, n_i'} . \qquad (3.134)$$

As to the partition (the 0-point) function, eq. (3.134) yields

$$\text{Tr}_{\mathcal{H}_{m,m'}} q^{L_0 - c/24} = \frac{1}{\eta(\tau)} \sum_{j=-\infty}^{\infty} \left(e^{\pi i \tau (2p'pj + pm' - p'm)^2 / 2pp'} - e^{\pi i \tau (2p'pj + pm' + p'm)^2 / 2pp'} \right) , \qquad (3.135)$$

which is nothing but the Rocha-Caridi character formula (sect. 1), as it should.

The generalizations of the Feigin-Fuchs construction to higher-genus Riemann surfaces, various examples of evaluation of conformal blocks, the inclusion of supersymmetry (see also Ch. V) and much more can be found in refs. [99, 100, 101, 102, 103, 104].

Chapter IV

AKM Algebras and WZNW Theories

Holomorphic symmetry algebra of 2d CFT is called **chiral algebra** $\mathcal{A}$. A CFT chiral algebra always contains the Virasoro algebra as a subalgebra. It is therefore quite natural to classify CFT's w.r.t. their chiral algebras encoding all of their relevant properties. The classification of CFT's is clearly related to the notion of quantum equivalence. Two CFT's are (quantum) equivalent if they have the same operator algebras. This implies that the chiral algebras of the quantum equivalent CFT's are the same, since the chiral algebra is always a subalgebra of the CFT operator algebra. Since the OPE coefficients for CFT *primary* fields are actually the only ones to be considered, the coincidence of the chiral algebras together with the same OPE coefficients for primary fields (this implies a one-to-one correspondence between the primary fields of the CFT's in question) are enough for the quantum equivalence.

In the previous Chapters, the BPZ minimal models have been introduced. They all have finite operator algebras, the BPZ correlators satisfy certain partial differential equations so that they can, in principle, be explicitly computed. The minimal models can be formulated on arbitrary Riemann surfaces when using, e.g., the BRST approach. The range of their central charge values is discrete and quite restrictive: $0 < c < 1$. It can be shown [84] that the minimal models are, in fact, the only consistent CFT's with a finite number of primaries at $c < 1$. Their chiral algebra is just the Virasoro algebra, which justifies their identification as the 'minimal' models.

The next question is about a contruction of other CFT's with $c > 1$. In the absence of

113

a satisfactory classification of CFT's at $c > 1$, it is worthy to look for such CFT's which share at least some of the properties of the minimal models. The RCFT's represent a more general class of CFT's having only a finite number of $\mathcal{A}$ representations [105]. Like the minimal models, the (generalized) RCFT Ward identities determine the correlation functions of descendants in terms of those of primaries, whereas the differential equations they satisfy follow from decoupling the null states w.r.t. the RCFT chiral algebra $\mathcal{A}$. The most known RCFT's arise after introducing a Lie group structure in 2d CFT. The associated conserved holomorphic (anti-holomorphic) currents of dimension $h = 1$ ($\bar{h} = 1$) define an **affine Kač-Moody** (AKM) algebra. [1] The AKM algebras have a lot of applications in modern mathematical physics, and they are relevant, in particular, for the classification program of 2d CFT's with arbitrary values of central charge c [106, 107]. The AKM algebras also play the important role in string theory (Ch. VIII), where they give world-sheet realizations of space-time gauge symmetries, help to analyze string interactions and to classify possible superstring vacua. These algebras also naturally arise in exactly solvable 2d models as the underlying (infinite-dimensional) symmetry structures (Ch. IX).

It this Chapter, the class of RCFT's with the additional symmetries to be represented by affine Kač-Moody algebras are considered. Compared to the minimal models, they represent the CFT's with enhanced chiral algebras. The simplest 2d RCFT with an AKM symetry and a finite number of the highest-weight modules associated with primary fields is given by the non-linear sigma-model on a compact Lie group manifold with the parallelizing torsion – the so-called *Wess-Zumino-Novikov-Witten* (WZNW) theory. The WZNW theory as a particular RCFT can also be formally defined as the 2d field theory whose fields belong to irreducible highest-weight AKM modules, and whose stress tensor is of the Sugawara-Sommerfeld form (sect. 2).

IV.1   AKM algebras and their representations

In the previous Chapters we considered only those CFT's, whose maximal symmetry algebra is the direct sum of two Virasoro algebras, $Vir \oplus \overline{Vir}$. In general, the symmetry of CFT is given by an algebra $\mathcal{A} = \mathcal{A}_L \oplus \mathcal{A}_R$ generated by holomorphic and anti-

---

[1]We use the more general term 'AKM algebras' for the *affine Lie (or current) algebras*, forming only a subclass of the Kač-Moody algebras, that we are actually going to consider in this Chapter (see sect. XII.2 for more).

holomorphic tensors of the type $(p, 0)$ and $(0, \bar{p})$, resp., and containing the Virasoro algebra $Vir \oplus \overline{Vir}$. Any $(p, 0)$ tensor or p-differential $S(z)$ obeys the conservation law

$$\bar{\partial} S(z) = 0 , \tag{4.1}$$

which obviously implies the infinite number of conserved quantities,

$$\bar{\partial} \left(z^n S(z) \right) = 0 . \tag{4.2}$$

The conserved quantities can be identified with the coefficients in the Laurent series for $S(z)$:

$$S(z) = \sum_{n \in \mathbf{Z}} S_n z^{-n-p} . \tag{4.3}$$

The Virasoro algebra Vir is the particular case of eq. (4.3) with $p = 2$ and $S(z) = T(z)$. Of special importance in CFT is the case when the chiral algebra $\mathcal{A}$ [2] is generated by the $(1, 0)$ conformal fields, called conformal *currents*. A free scalar field theory again provides us with the simplest case to introduce the $U(1)$ current $j(z) = i\partial\phi(z)$ with the OPE

$$\partial j(z) \partial j(w) \sim \frac{1}{(z-w)^2} , \tag{4.4}$$

where the mode expansion in terms of the bosonic oscillators $\{\alpha_n\}$,

$$j(z) = i\partial\phi(z) = \sum_{n \in \mathbf{Z}} \alpha_n z^{-n-1} , \tag{4.5}$$

and their commutation relations,

$$[\alpha_n, \alpha_m] = n\delta_{n+m,0} , \tag{4.6}$$

have been used.

Dimensional analysis constrains the OPE's of $(1, 0)$ conformal currents $J^a(z)$ to take the form generalizing eq. (4.4) as

$$J^a(z) J^b(w) \sim \frac{\tilde{k}^{ab}}{(z-w)^2} + \frac{if^{abc}}{z-w} J^c(w) , \tag{4.7}$$

where the 'structure constants' f^{abc} are antisymmetric in a and b. The associativity of the operator products implies that f^{abc}'s satisfy the Jacobi identity. This means that they are the structure constants of a Lie algebra $\mathcal{G}$. In what follows in this Chapter, the

---

[2] As a rule, we omit the subscript L in $\mathcal{A}_L$ since we are only going to discuss the chiral operator algebra in what follows.

Lie group G is assumed to be simple and *compact*. This is needed to have a positively definite Cartan metric. In addition, this allows us to choose a basis in which the central extension takes the diagonal form, $\tilde{k}^{ab} = \tilde{k}\delta^{ab}$. Expanding $J^a(z)$ in the way analogous to that in eq. (4.5),

$$J^a(z) = \sum_{n\in\mathbf{Z}} J^a_n z^{-n-1} \,, \qquad (4.8)$$

we find from eq. (4.7) the commutation relations

$$[J^a_m, J^b_n] = if^{abc}J^c_{m+n} + \tilde{k}m\delta^{ab}\delta_{m+n,0} \,, \qquad (4.9)$$

which generalize eq. (4.6), and have $m, n \in \mathbf{Z}$ and $a, b = 1, \ldots, \dim G$.

Eq. (4.7) or (4.9) defines the commutation relations of an *affine (untwisted) Kač-Moody* (AKM) algebra $\hat{\mathcal{G}}$ with central extension. The subalgebra of zero modes J^a_0 comprises the ordinary Lie algebra $\mathcal{G}$ called the *horizontal* Lie subalgebra of the AKM algebra:

$$[J^a_0, J^b_0] = if^{abc}J^c_0 \,. \qquad (4.10)$$

The procedure $\mathcal{G} \to \hat{\mathcal{G}}$ is known as the (infinite-dimensional) *affinization* of a finite-dimensional Lie algebra $\mathcal{G}$. See sect. XII.2 for more about Lie and AKM algebras.

When pulling $J(z)$ back to a cylinder,

$$J^a_{\text{cyl}}(w) = \sum_n J^a_n e^{-nw} \,, \qquad (4.11)$$

with real w, the modes J^a_n can be recognized as the infinitesimal generators of the group of gauge transformations $g(\sigma)\colon S^1 \to G$ on a circle S^1.

In the same way the Virasoro generators give a projective representation of the conformal algebra in eq. (1.16). The mode operators J^a_n are then associated with the *loop algebra* of G: $\mathcal{L}(z) = \sum_{a,n} \mathcal{L}^a_n t^a_{(r)} z^n$, where $\mathcal{L}^a_n$ are the loop algebra generators and $\{t^a_{(r)}\}$ are generators of a matrix representation (r) of $\mathcal{G}$.

The representation theory of AKM algebras has many features similar to that of the Virasoro algebra. The AKM highest-weight representations are characterized by the highest weights $\hat{\Lambda}$ (sect. XII.2). When $\hat{\Lambda}$ is an integral dominant weight, such representations are called *integrable*. They are the only ones we consider since, according to the *Gepner-Witten* theorem [108], only the integrable highest-weight modules can appear in the WZNW spectrum.

The regularity of $J(z)|0\rangle$ at $z = 0$ implies

$$J^a_n |0\rangle = 0 \quad \text{for} \quad n \geq 0 \,. \qquad (4.12)$$

Under the condition (4.12), CFT fields organize into the families, each providing an irreducible representation of AKM algebra. As in the Virasoro case, there will be primary and descendant fields. An AKM-primary field $\phi_{(r)}^k(z)$ has the simplest transformation law given by the OPE:

$$J^a(z)\phi_{(r)}(w) \sim \frac{t_{(r)}^a}{z-w}\phi_{(r)}(w) , \qquad (4.13a)$$

or, equivalently,

$$[J_n^a, \phi_{(r)}^j(w)] = \left(t_{(r)}^a\right)^{jk} \phi_{(r)}^k(w)w^n . \qquad (4.13b)$$

The module generated by the AKM-primary field has the AKM highest-weight state

$$|(r)\rangle = \phi_{(r)}(0)|(0)\rangle , \qquad (4.14)$$

which transforms in a representation (r) of the horizontal Lie algebra,

$$J_0^a|(r)\rangle = t_{(r)}^a|(r)\rangle ,$$
$$J_n^a|(r)\rangle = 0 \quad (n > 0) . \qquad (4.15)$$

The descendants are obtained by acting on the AKM highest-weight state $|(r)\rangle$ with the raising operators J_{-n}^a $(n > 0)$:

$$V_{\phi_{(r)}} = \{ J_{-n_1}^{a_1} \ldots J_{-n_N}^{a_n} |(r)\rangle ; \quad n_1, n_2, \ldots, n_N > 0 \} . \qquad (4.16)$$

The OPE's between the stress tensor, primaries or currents take the standard form:

$$T(z)\phi_{(r)}(z) \sim \frac{h}{(z-w)^2}\phi_{(r)}(w) + \frac{1}{z-w}\partial_w\phi_{(r)}(w) ,$$
$$T(z)J^a(w) \sim \frac{1}{(z-w)^2}J^a(w) + \frac{1}{z-w}\partial_w J^a(w) . \qquad (4.17)$$

Of course, in the complete OPE's, the secondary fields associated with the secondary states (4.16), as well as the Virasoro generators, will appear on the r.h.s. of eq. (4.17).

The dimension-1 currents J^a and $\bar{J}^a$ are to be considered on equal footing with the Virasoro dimension-2 generators. Together they form the full symmetry algebra which is the semi-direct sum of the AKM current algebra $\hat{\mathcal{G}} \times \hat{\bar{\mathcal{G}}}$ with the Virasoro algebra $Vir \oplus \overline{Vir}$.

The *Ward identities* associated with the AKM symmetry are derived in exactly the same way as those for the Virasoro symmetry. They take the form [108]

$$\left\langle J^a(z)\phi_{(r_1)}(w_1, \bar{w}_1) \cdots \phi_{(r_M)}(w_M, \bar{w}_M) \right\rangle$$

$$= \sum_{j=1}^{M} \frac{t_{(r_j)}^{a}}{z - w_j} \left\langle \phi_{(r_1)}(w_1, \bar{w}_1) \cdots \phi_{(r_M)}(w_M, \bar{w}_M) \right\rangle . \tag{4.18}$$

These equations alone can be used to derive the *first-order* differential equations for the Green's functions of the primary fields $\phi_{(r_j)}$ (see below). Similarly to the minimal models, the chiral Ward identities fix the correlators of descendants in terms of those of primaries [109].

The Virasoro symmetry implies, in particular, the $SL(2, \mathbf{C})$ Ward identities (2.21). Quite similarly, the AKM symmetry of the correlators implies the identity

$$\sum_{j=1}^{M} t_{(r_j)}^{a} \left\langle \phi_{(r_1)}(w_1, \bar{w}_1) \cdots \phi_{(r_M)}(w_M, \bar{w}_M) \right\rangle = 0 . \tag{4.19}$$

Eq. (4.19) simply means that the correlation functions have the global symmetry G.

The 2-, 3- and 4-point correlation functions of the AKM currents on a plane are computable from eq. (4.7), when using the $SL(2, \mathbf{C})$ and Bose symmetries as the input:

$$\left\langle J^a(z) J^b(w) \right\rangle = \frac{k \delta^{ab}}{(z - w)^2} ,$$

$$\left\langle J^a(z) J^b(w) J^c(y) \right\rangle = \frac{i k f^{abc}}{(z - w)(z - y)(w - y)} ,$$

$$\left\langle J^a(z) J^b(w) J^c(x) J^d(y) \right\rangle = k^2 \left[\frac{\delta^{ab} \delta^{cd}}{(z - w)^2 (x - y)^2} \right.$$

$$+ \frac{\delta^{ac} \delta^{bd}}{(z - x)^2 (w - y)^2} + \frac{\delta^{ad} \delta^{bc}}{(z - y)^2 (w - x)^2} \right] - \frac{k}{3} \left[\frac{f^{abe} f^{cde} + f^{ace} f^{bde}}{(z - w)(z - x)(w - y)(x - y)} \right.$$

$$+ \frac{f^{abe} f^{cde} + f^{ade} f^{cbe}}{(z - w)(z - y)(w - x)(x - y)} + \left. \frac{f^{ace} f^{bde} + f^{ade} f^{bce}}{(z - x)(z - y)(x - w)(w - y)} \right] . \tag{4.20}$$

The generators $J^a(z)$ can be decomposed to $H^i(z)$ and $E^{\pm \alpha}(z)$ in a Cartan-Weyl basis, where the index i, $i = 1, \ldots, \operatorname{rank} G$, labels the mutually commuting generators of the Cartan subalgebra, and the positive roots α label the raising and lowering operators (sect. XII.2). The highest weight state $|\lambda\rangle$ transforming in a vacuum representation (r) obeys the relations:

$$H_n^i |\lambda\rangle = E_n^{\pm \alpha} |\lambda\rangle = 0 \qquad n > 0 ,$$

$$H_0^i |\lambda\rangle = \lambda^i |\lambda\rangle , \qquad E_0^\alpha |\lambda\rangle = 0 \qquad \alpha > 0 . \tag{4.21}$$

The new states are created by acting on the state $|\lambda\rangle$ with products of $E_0^{-\alpha}$'s and J_{-n}^a's for $n > 0$.

Exercises

\# IV-1 ▷ Derive the 2- and 3-point correlation functions of the AKM-currents from eq. (4.7).

\# IV-2 ▷ (*) Verify that the 4-point function in eq. (4.20) is completely Bose symmetric, obeys the Virasoro-Ward identity, and is regular at infinity for each variable.

IV.2 Sugawara-Sommerfeld construction

In the previous section, the AKM currents and stress tensor were completely independent. However, given an AKM algebra, one can always build the Virasoro algebra. Indeed, AKM currents have dimension 1 while stress tensor is of dimension 2, which suggests us to use a bilinear in the AKM currents to form the stress tensor. This is known as the **Sugawara-Sommerfeld** (SS) construction [110, 111]. More insights are again provided by the theory of a single boson ϕ, where we know the $U(1)$ current, $J(z) = i\partial\phi$, and the stress tensor, $T(z) = -\frac{1}{2} : \partial\phi(z)\partial\phi(z) : = \frac{1}{2} : J(z)J(z) :$. It is now natural to try to generalize this prescription to the non-abelian case. The appropriate group-invariant generalization is given by

$$T(z) = \frac{1}{\beta} \sum_{a=1}^{|G|} : J^a(z)J^a(z) : , \tag{4.22}$$

where the normal ordering between the two currents is given by

$$\sum_a : J^a(z)J^a(z) : \equiv \lim_{z \to w} \left(\sum_a J^a(z)J^a(w) - \frac{\bar{k}|G|}{(z-w)^2} \right) , \tag{4.23}$$

while $|G| = \dim G$. The constant β above is fixed by requiring either $T(z)$ to satisfy the canonical OPE of eq. (1.75), or $J^a(z)$ to transform as the $(1,0)$ primary fields. Taking the latter, we finds that for

$$T(z)J^a(w) \sim \frac{J^a(w)}{(z-w)^2} + \frac{\partial J^a(w)}{z-w} \tag{4.24}$$

to hold, we need to have the commutation relations

$$[L_m, J_n^a] = -nJ_{m+n}^a , \tag{4.25}$$

where the L_n's are to be computed from eq. (4.22) and take the form

$$L_n = \frac{1}{\beta} \sum_{m=-\infty}^{+\infty} : J^a_{m+n} J^a_{-m} : . \tag{4.26}$$

Applying L_{-1} to a highest weight state and using eq. (4.15), one gets

$$L_{-1} |(r)\rangle = \frac{2}{\beta} J^a_{-1} t^a_{(r)} |(r)\rangle . \tag{4.27}$$

Applying now J^b_1 to both sides of this equation and using eqs. (4.9) and (4.25) yield

$$t^b_{(r)} |(r)\rangle = \frac{2}{\beta} \left(if^{bac} J^c_0 + \tilde{k}\delta^{ab} \right) t^a_{(r)} |(r)\rangle$$

$$= \frac{2}{\beta} \left(if^{bac} \tfrac{1}{2} if^{dca} t^d_{(r)} + \tilde{k} t^b_{(r)} \right) |(r)\rangle = \frac{2}{\beta} \left(\tfrac{1}{2} C_A + \tilde{k} \right) t^b_{(r)} |(r)\rangle , \tag{4.28}$$

where we have introduced the quadratic Casimir eigenvalue C_A of the adjoint representation: $f^{acd} f^{bcd} = C_A \delta^{ab}$. The consistency requires

$$\beta = 2\tilde{k} + C_A . \tag{4.29}$$

It is a good exercise (# IV-4) to check at this point that the SS stress tensor

$$T(z) = \frac{1/2}{\tilde{k} + C_A/2} \sum_{a=1}^{|G|} : J^a(z) J^a(z) : \tag{4.30}$$

does satisfy the correct OPE (1.75) indeed, with the central charge

$$c = \frac{\tilde{k}|G|}{\tilde{k} + C_A/2} . \tag{4.31}$$

Eq. (4.26) implies, in addition, that the Virasoro generators are all contained in the AKM enveloping algebra. This means, in particular, that all Virasoro descendants can be represented in terms of descendants of the AKM currents.

It is worthy to mention here that both numbers $\tilde{k}$ and $C_A/2$ introduced above depend on the normalization used for the structure constants f^{abc}. Having introduced the trace

$$\text{tr} \left(t^a_{(r)} t^b_{(r)} \right) = l_r \delta^{ab} \tag{4.32}$$

for an arbitrary $\mathcal{G}$-representation (r) of dimension d_r, the diagonal sum over $a, b = 1, \ldots, |G|$ gives

$$C_r d_r = l_r |G| , \tag{4.33}$$

where the quadratic Casimir eigenvalue C_r in the representation (r) has been introduced, $C_r \delta^{ij} = \sum_a \left(t^a_{(r)} t^a_{(r)} \right)^{ij}$. If the sum were restricted to the Cartan subalgebra of $\mathcal{G}$ ($a, b = 1, \ldots, r_G$), we would get instead

$$\sum_{k=1}^{d_r} \mu^2_{(k)} = l_r r_G , \qquad (4.34)$$

where r_G is the rank of the group G, and μ are the weights of the representation (r) of dimension d_r. For the adjoint representation, one has $d_A = |G|$ and $C_A = l_A = r_G^{-1} \sum_{a=1}^{|G|} \alpha^2_{(a)}$, where α's are the roots of $\mathcal{G}$. Denoting ψ the highest root, one can introduce the normalization-*independent* quantity $\tilde{h}_G \equiv C_A / \psi^2$, known as the *dual Coxeter number*:

$$\tilde{h}_G = \frac{C_A}{\psi^2} = \frac{1}{r_G} \left(n_L + \left(\frac{S}{L} \right)^2 n_S \right) , \qquad (4.35)$$

where n_S and n_L are the number of short (S) and long (L) roots of the algebra, resp. The $(S/L)^2$ is just the ratio of their lengths squared (the roots of simple Lie algebras come at most in two lengths, see sect. XII.2). The Lie algebras associated to Dynkin diagrams with only *single* lines have roots all of the same length, and they are known as the A, D, E series of Lie algebras, i.e. $SU(n)$, $SO(2n)$ and $E_{6,7,8}$. These algebras are usually referred to as the *simply-laced* Lie algebras.

Any positive root $\alpha \in \Delta_+$ of a Lie algebra $\mathcal{G}$ can be decomposed as a linear combination of simple roots $\beta \in \Pi$, $\alpha = \sum_{\beta \in \Pi} n^\beta_\alpha \beta$, with non-negative integer coefficients n^β_α. Among the positive roots, we can always distinguish the *maximal root* θ, such that $\theta + \alpha \notin \Delta \equiv \Delta_+ \cup (\Delta_- = -\Delta_+)$ for any $\alpha \in \Delta_+$. The *Coxeter number* h_G can then be defined as $h_G = \sum_{\beta \in \Pi} n^\beta_\theta + 1 \equiv ht_\theta + 1$, where the ht_θ is called the *height* of a root θ. The lattice *dual* to the root lattice is generated by *weights*, and the dual Coxeter number may be introduced in exactly the same way as the Coxeter number above, but now w.r.t. to the dual lattice and the associated *fundamental weights*. The dual Coxeter number coincides with the Coxeter number only for the simply-laced Lie algebras. The remaining algebras have roots of two lengths, their ratio (L/S) being either $\sqrt{2}$ for $SO(2n + 1)$, $Sp(2n)$ and F_4, or $\sqrt{3}$ for G_2 (see sect. XII.2 also).

The dual Coxeter numbers for all compact simple Lie algebras are given by eq. (4.35). It gives for the particular cases:

$$\mathbf{SU(n)} \ (n \geq 2) \ : \ \tilde{h}_{SU(n)} = n , \quad l_{(n)} = \tfrac{1}{2} \psi^2 ;$$

$$\mathbf{SO(n)} \ (n \geq 4) \ : \ \tilde{h}_{SO(n)} = n - 2 , \quad l_{(n)} = \psi^2 ;$$

$$\mathbf{E_6} \ : \ \tilde{h}_{E_6} = 12 \ , \ \ l_{(27)} = 3\psi^2 \ ; \quad \mathbf{E_7} \ : \ \tilde{h}_{E_7} = 18 \ , \ \ l_{(57)} = 6\psi^2 \ ;$$

$$\mathbf{E_8} \ : \ \tilde{h}_{E_8} = 30 \ , \ \ l_{(248)} = 30\psi^2 \ ;$$

$$\mathbf{Sp(2n)} \ (n \geq 1) \ : \ \tilde{h}_{Sp(2n)} = n + 1 \ , \ \ l_{(2n)} = \tfrac{1}{2}\psi^2 \ ;$$

$$\mathbf{G_2} \ : \ \tilde{h}_{G_2} = 4 \ , \ \ l_{(7)} = \psi^2 \ ; \quad \mathbf{F_4} \ : \ \tilde{h}_{F_4} = 9 \ , \ \ l_{(26)} = 3\psi^2 \ , \tag{4.36}$$

where the index l_r has been tabulated for some representations of low dimensions. The dual Coxeter number is always an integer.

The normalization-independent quantity

$$k = \frac{2\tilde{k}}{\psi^2} \tag{4.37}$$

is known as the **level** of AKM algebra. It will be shown in a moment that k is quantized to be an integer for the *unitary* highest weight representations. In terms of the integer quantities k and $\tilde{h}_G$, eq. (4.31) takes the form

$$c = \frac{k|G|}{k + \tilde{h}_G} \ . \tag{4.38}$$

To derive the quantization rule for the level k, consider the case $G = SU(2)$ first. The normalization $\psi^2 = 2$ corresponds to the $SU(2)$ structure constants $f^{ijk} = \sqrt{2}\varepsilon^{ijk}$. Because of $\sqrt{2}$ in the $SU(2)$ commutation relations, we need to take

$$I^{\pm} = \frac{1}{\sqrt{2}} \left(J_0^1 \pm i J_0^2 \right) \ , \quad I^3 = \frac{1}{\sqrt{2}} J_0^3 \ , \tag{4.39}$$

for the conventionally normalized $su(2)$ algebra $[I^+, I^-] = 2I^3$, $[I^3, I^{\pm}] = \pm I^{\pm}$, in which the $2I^3$ has *integer* eigenvalues in any finite-dimensional representation. Simultaneously, eq. (4.9) implies that the operators

$$\tilde{I}^+ = \frac{1}{\sqrt{2}} \left(J_{-1}^1 + i J_{-1}^2 \right), \ \tilde{I}^- = \frac{1}{\sqrt{2}} \left(J_{+1}^1 - i J_{+1}^2 \right), \ \tilde{I}^3 = \frac{1}{\sqrt{2}} J_0^3 - \frac{1}{2} k \ , \tag{4.40}$$

satisfy the $su(2)$ algebra as well: $[\tilde{I}^+, \tilde{I}^-] = 2\tilde{I}^3$, $[\tilde{I}^3, \tilde{I}^{\pm}] = \pm\tilde{I}^{\pm}$. Hence, the $2\tilde{I}^3 = 2I^3 - k$ also has integer eigenvalues. It follows that $k \in \mathbf{Z}$ for the unitary highest weight representations.

To generalize this argument to any Lie algebra $\mathcal{G}$, it is sufficient to use the canonical $su(2)$ subalgebra (in the Chevalley basis) defined by

$$I^{\pm} = E_0^{\pm\psi} \ , \quad I^3 = \psi \cdot H_0/\psi^2 \ , \tag{4.41}$$

which is generated by the highest root ψ of $\mathcal{G}$. Eq. (4.9) implies that

$$\tilde{I}^{\pm} = E_{\mp 1}^{\pm \psi} , \quad \tilde{I}^3 = \left(\psi \cdot H_0 - \tilde{k} \right) / \psi^2 \tag{4.42}$$

also form an $su(2)$ subalgebra. This means that the level $k = 2\tilde{k}/\psi^2 = 2I^3 - 2\tilde{I}^3$ is to be quantized as an integer. The different proof of the quantization condition on k will be given in sect. 4, in the Lagrangian approach.

Let us now discuss null states (or singular vectors) in CFT with AKM symmetry. In general, we may expect the three types of them: (i) purely AKM, (ii) purely Vir, and (iii) mixed. The issues relevant to (ii) were already discussed. The constraints imposed by (i) follow from the Ward identity (4.18). Decoupling the mixed singular vectors leads to the first-order differential equations for the chiral blocks of the theory. Being expanded in powers of z, the SS construction (4.30) tells us that

$$L_{-1} = \frac{1}{\tilde{k} + C_A/2} \left(J_{-1}^a J_0^a + J_{-2}^a J_1^a + \ldots \right) . \tag{4.43}$$

Acting on any AKM-primary state, we find the null state:

$$\left(L_{-1} - \frac{1}{\tilde{k} + C_A/2} J_{-1}^a t_{(r)}^a \right) \varphi_{(r)} = 0 . \tag{4.44}$$

Since

$$\left(\hat{J}_{-1}^a \varphi_{(r)} \right) (z) = \oint_z \frac{dw}{2\pi i} \frac{1}{w - z} J^a(w) \varphi_{(r)}(z) , \tag{4.45}$$

decoupling the singular vector (4.44) from any correlation function,

$$\left\langle \varphi_{(r_1)}(z_1) \cdots \left(\hat{L}_{-1} - \frac{1}{\tilde{k} + C_A/2} \hat{J}_{-1}^a \hat{J}_0^a \right) \varphi_{(r_i)}(z_i) \cdots \varphi_{(r_M)}(z_M) \right\rangle = 0 , \tag{4.46}$$

implies the first-order differential equation

$$\left[\left(\tilde{k} + C_A/2 \right) \frac{\partial}{\partial z_i} + \sum_{j \neq i} \frac{t_{(r_j)}^a t_{(r_i)}^a}{z_j - z_i} \right] \left\langle \varphi_{(r_1)}(z_1) \cdots \varphi_{(r_M)}(z_M) \right\rangle = 0 , \tag{4.47}$$

where eq. (4.18) has been used. Eq. (4.47) is known as the **Knizhnik-Zamolodchikov** equation [109] for the chiral blocks.

The conformal weight h_r of the primary multiplet $\varphi_{(r)}(z)$ can be determined by evaluating the L_0 eigenvalue of the vacuum state as

$$L_0 |(r)\rangle = \frac{1/2}{\tilde{k} + C_A/2} \sum_{a,m} : J_m^a J_{-m}^a : |(r)\rangle$$

$$= \frac{1/2}{\tilde{k} + C_A/2} \sum_a t^a_{(r)} t^a_{(r)} \,|(r)\rangle = \frac{C_r/2}{\tilde{k} + C_A/2} \,|(r)\rangle \ . \tag{4.48}$$

Thereby, one gets

$$h_r = \frac{C_r/2}{\tilde{k} + C_A/2} = \frac{C_r/\psi^2}{k + \tilde{h}_G} \ . \tag{4.49}$$

In the particular case of $SU(2)$, with the ground state transforming in the spin-j representation of the horizontal Lie subalgebra $su(2)$, one finds

$$h_j = \frac{j(j+1)}{k+2} \ . \tag{4.50}$$

It is not difficult to generalize this formula to any simply-laced Lie group G. Normalizing the roots to have length 2, the quadratic Casimir operator takes the form

$$C_r = \sum_i H_i^2 + \sum_\alpha \frac{\alpha^2}{2} E_\alpha E_{-\alpha} \ . \tag{4.51}$$

It is always possible to express the quadratic Casimir eigenvalue in terms of the highest weight Λ as $C_r |\Lambda\rangle = (\Lambda, \Lambda + 2\rho)|\Lambda\rangle$, where ρ is a half the sum of positive roots, $\rho = \frac{1}{2} \sum_{\alpha>0} \alpha$ (sect. XII.2 also). Hence, in general, we have

$$h_\Lambda = \frac{(\Lambda, \Lambda + 2\rho)}{2(k + \tilde{h}_G)} \ . \tag{4.52}$$

There exists a simple constraint on the possible vacuum representations (r) allowed to appear in a *unitary* AKM highest-weight representation at given level k [108, 107]. In the $SU(2)$ case with the spin-j representation for the vacuum, we have $2j + 1$ states labeled by their I^3 eigenvalue: $I^3 |m\rangle = m |m\rangle$. Using the other $su(2)$ generators of eq. (4.40), one obtains

$$0 \le \langle j|\, \tilde{I}^- \tilde{I}^+ \,|j\rangle = \langle j|\, [\tilde{I}^-, \tilde{I}^+]\, |j\rangle = \langle j|-2I^3 + k\,|j\rangle = -2j + k \ . \tag{4.53}$$

Therefore, only the ground state AKM representations with $2j \le k$ are allowed. They all have a null state which appears at level $k + 1 - 2j$. For an arbitrary Lie algebra, one should consider $|\lambda\rangle$ instead of $|j\rangle$, where λ is the highest weight of the vacuum representation, to obtain

$$2\psi \cdot \lambda/\psi^2 \le k \ . \tag{4.54}$$

As far as the Lie group $SU(n)$ is concerned, this just means that the width of the Young tableau, describing its irreducible representation, should be less then the level k.

Eq. (4.54) is, in fact, the general condition for unitarity [108, 107]. One always gets only a finite number of unitary representations this way.

Exercises

IV-3 ▷ Compute the coefficient β in eq. (4.22) from the known 4-point function in eq. (4.20).

IV-4 ▷ (*) Verify directly that the stress tensor defined by eq. (4.30) satisfies the OPE (1.75) with the central charge (4.31).

IV-5 ▷ Consider the $\widehat{SU(2)}_k$ highest-weight representation of spin $j = k/2$, and prove that it has a zero-norm state at level one.

IV.3   WZNW theories

In the previous section we discussed CFT with AKM symmetry and the dynamics determined by the SS construction. No Lagrangian local QFT was used to derive all of those results. The simple RCFT action having an AKM symmetry is [112]

$$kI_{\text{WZNW}}(g) = \frac{k}{8\pi} \int_{S^2} d^2x \, \text{tr} \left(\partial_m g \partial^m g^{-1} \right)$$

$$+\frac{k}{12\pi} \int_{B, \partial B = S^2} d^3y \, \varepsilon^{mnl} \, \text{tr} \left(g^{-1} \partial_m g g^{-1} \partial_n g g^{-1} \partial_l g \right) . \qquad (4.55)$$

This action represents the so-called **Wess-Zumino-Novikov-Witten** (WZNW) theory [113, 114, 115, 112]. The scalar field g is valued in a Lie group G and has the boundary condition: $g(0, \tau) = g(2\pi, \tau)$. The second term in eq. (4.55) is given by the integral over the three-dimensional ball B. The integrand of the second term is a total derivative, and, hence, it actually represents the *Wess-Zumino (WZ) topological term*, which can be rewritten as an ordinary action over the two-dimensional sphere S^2 by using Gauss's theorem.

The action (4.55) is classically invariant not only under the rather obvious scale transformations, but also under the much larger infinite-dimensional symmetry associated with the AKM current algebra and having the form

$$g(\zeta^+, \zeta^-) \to \bar{\Omega}(\zeta^-) g(\zeta^+, \zeta^-) \Omega(\zeta^+) , \qquad (4.56)$$

where $\Omega(\bar{\Omega})$ is an arbitrary group-valued function, and $\zeta^{\pm} = x_1 \pm x_2$. The symmetry (4.56) is just the 2d analogue of the chiral symmetry well-known in 4d field theories [1]. After introducing the complex variables z and $\bar{z}$ as $z = e^{ix^+}$ and $\bar{z} = e^{ix^-}$ (*cf* sect. I.1), the symmetry (4.56) becomes

$$g(z, \bar{z}) \rightarrow \bar{\Omega}(\bar{z}) g(z, \bar{z}) \Omega(z) \ . \tag{4.57}$$

Due to the boundary condition $g(0) = g(2\pi)$, the coordinates z and $\bar{z}$ are well-defined (no cuts in the complex plane). The transformation law (4.57) is nothing but the AKM-symmetry $\hat{G} \times \hat{\bar{G}}$, which persists in the quantized WZNW theory too. This symmetry is characterized by the currents

$$J(z) = J^a(z) t^a = kg^{-1}\partial_z g = \sum_{n=-\infty}^{+\infty} J_n z^{-n-1} \ ,$$

$$\bar{J}(\bar{z}) = \bar{J}^a(\bar{z}) t^a = -k(\partial_{\bar{z}} g) g^{-1} = \sum_{n=-\infty}^{+\infty} \bar{J}_n \bar{z}^{-n-1} \ , \tag{4.58}$$

satisfying the conservation law

$$\bar{\partial} J = \partial \bar{J} = 0 \ , \tag{4.59}$$

which is a consequence of the WZNW theory equations of motion.

The currents J and $\bar{J}$ generate two commuting AKM algebras (4.7) or (4.9) with the central extension (level) k. The associated stress tensor is given by the SS construction (4.30), with central charge (4.38).

The action (4.55) is the particular case of a more general theory described the action

$$kI(g) = \frac{1}{2\lambda^2} \int \text{tr} \, (g^{-1}\partial g)^2 + k\Gamma \ , \tag{4.60}$$

where the WZ term Γ is just $\frac{1}{12\pi} \int_B tr \, (g^{-1}\partial g)^3 \equiv \int_B H$. The theory (4.60) becomes conformally invariant when the coupling constant λ takes its critical value, $\lambda^2 = 4\pi/k$, which corresponds to an exact fixed point of the renormalization group β-function of the theory (4.60) [112]. From an even more general point of view, eq. (4.60) is the particular case of a 2d *non-linear σ-model* with a generalized WZ term (or torsion) [116]. The non-linear σ-model is a field theory whose fields take their values in a manifold (which is a group manifold in the case (4.60)).

To exhibit a topological nature of the WZ term in eq. (4.60) and to describe the associated *topological* quantization of the WZ coefficient k, we need to introduce the notions of *integer homology* and *DeRham cohomology* [40, 117]. The main point is that the kinetic term in eq. (4.60) is defined by the use of the field g mapping the sphere S^2 into a group manifold G, while the WZ term is defined as an integral of the (2+1)-form H over the (2+1)-dimensional ball B, with the boundary $\partial B = S^2$. The form H must be *closed*, but not necessarily *exact*. A closure of H guarantees a *local* existence of the 2-form b such that $H = db$, according to the Poincaré lemma, but this does *not* guarantee that the 2-dimensional action defined as $\int_{S^2} b$ will be globally well-defined because of possible topological obstructions. This situation is only possible when the group manifold G has a non-trivial third homology class, and it is implicit in the definition of the WZ term. To avoid confusion, one should clearly distinguish between the forms defined on the group manifold G and their pull-backs to B or S^2.

To explain precisely the relation between the integer homology and the WZ terms, let us define the k-cycles [3] C of a given *generically* d-dimensional manifold $\mathcal{W}$ as its k-dimensional submanifolds with no boundary, $\partial C = 0$. Linear combinations of cycles with *integer* coefficients are also defined to be cycles. The boundary of a boundary is always zero: $\partial^2 = 0$.

The *integer homology classes* $H_k(\mathcal{W}) \equiv H_k(\mathcal{W}, \mathbf{Z})$ are the equivalence classes of the cycles: two cycles C_1 and C_2 are said to be in the same homology class if $C_1 - C_2 = \partial B$, where B is a $(k+1)$-dimensional submanifold (then C_1 and C_2 are just the opposite boundaries of B). The homology classes obviously form an abelian group under the addition of cycles:

$$H_k(\mathcal{W}) = \underbrace{\mathbf{Z} \oplus \ldots \oplus \mathbf{Z}}_{b_k(\mathcal{W}) \text{ factors}} \oplus \mathbf{Z}_{p_1} \oplus \ldots \oplus \mathbf{Z}_{p_l} , \tag{4.61}$$

where we have decomposed $H_k(\mathcal{W})$ into a sum of the independent $\mathbf{Z}$-factors, each one being generated by a fundamental cycle C_i. The simplest example is provided by the torus with two independent non-trivial cycles and $H_1(\mathcal{W}) = \mathbf{Z} \oplus \mathbf{Z}$. The number of $\mathbf{Z}$-factors in eq. (4.61) is known as the k-th *Betti number* (for the torus, $b_1 = 2$). In general, there may be the so-called *torsion subgroups* in $H_k(\mathcal{W})$, which are also explicitly indicated in eq. (4.61). They are generated by some fundamental cycles $\mathcal{E}_i$ but, in contrast to the C_i-cycles, the $p_i \mathcal{E}_i$ (no sum!), with a positive integer p_i, is in the trivial class, i.e. although $\mathcal{E}_i$ cannot be expressed as a boundary of a $(k+1)$-dimensional submanifold, the $p_i \mathcal{E}_i$ can (see the example in Fig. 16) [117].

---

[3] We use the same letter k for both the dimension of a cycle and the coefficient at the WZ term since it is hard to confuse them.

Fig. 16. The two-sphere with antipodal points identified (RP^3). The one-cycle C given by the right half of the equator represents a nontrivial homology class, while the cycle $2C$ given by the boundary of the upper hemisphere D is in the trivial homology class (the topological torsion!).

A general k-cycle C can be expanded in terms of the fundamental cycles C_i and $\mathcal{E}_i$ as

$$C = \sum_{i=1}^{b_k(\mathcal{W})} n_i C_i + \sum_{i=1}^{l} m_i \mathcal{E}_i + \partial \mathcal{B}_0 \;, \tag{4.62}$$

with integers n_i and m_i, $0 \leq m_i \leq p_i - 1$. Since $p_i \mathcal{E}_i$ is trivial, $p_i(m_i \mathcal{E}_i) = \partial \mathcal{B}_i$ for some $(k+1)$-dimensional submanifolds $\mathcal{B}_i$, we can rewrite eq. (4.62) to the form

$$C = \sum_{i=1}^{b_k(\mathcal{W})} n_i C_i + \partial \mathcal{B}, \quad \mathcal{B} = \mathcal{B}_0 + \sum_{i=1}^{l} \mathcal{B}_i/p_i \;. \tag{4.63}$$

Clearly, a submanifold $\mathcal{B}$ is not unique but defined modulo a cycle multiplied by the fraction $1/p$, where p is the least common multiplier of $\{1, p_1, \ldots, p_l\}$.

The *DeRham cohomology classes* $H^k(\mathcal{W}, \mathbf{R})$ are defined as the equivalence classes of closed k-forms: two k-forms Ω_1 and Ω_2 are said to be equivalent (i.e. in the same cohomology class) if $\Omega_1 - \Omega_2 = d\Lambda$ for some $(k-1)$-form Λ. The non-trivial cohomology classes are therefore generated by the forms which are closed but not exact. It is the Stokes theorem, $\int_C d\Omega = \int_{\partial C} \Omega$, that provides the natural duality relation between the homology and cohomology classes. The precise statement is known as the *DeRham theorem*: there exists the basis for $H^k(\mathcal{W}, \mathbf{R})$ which consists of the k-forms ω_i satisfying

$$\int_{C_i} \omega_j = \delta^{ij} \;, \qquad i, j = 1, \ldots, b_k(\mathcal{W}) \;, \tag{4.64}$$

where C_i's are the fundamental cycles of $H_k(\mathcal{W})$. In particular, the dimension of $H^k(\mathcal{W}, \mathbf{R})$ is the Betti number $b_k(\mathcal{W})$. The ω_i's are called the *fundamental k-forms*. Using the decomposition (4.63), we find for a general k-cycle C that the integral of the fundamental form ω_j over C is always *integer*-valued.

We are now in a position to apply the decomposition (4.63) to the WZNW model. The image of the world-sheet $g(S^2)$ in the manifold G is a submanifold without boundary, i.e. a 2-cycle. Applying eq. (4.63) to $g(S^2)$:

$$g(S^2) = \sum_{i=1}^{b_2(G)} n_i \mathcal{C}_i + \partial B , \quad n_i \in \mathbf{Z} , \tag{4.65}$$

we can represent the WZ functional $k\Gamma$ in the form

$$k\Gamma[g] = k \sum_{i=1}^{b_2(G)} n_i \Gamma[\mathcal{C}_i] + k\Gamma[\partial B] , \tag{4.66}$$

where $\Gamma[\mathcal{C}_i]$ are arbitrary real numbers, and $\Gamma[\partial B] \equiv \int_B \check{H}$. Since the 3-form $\check{H}$ is not exact, the $\Gamma[\partial B]$ is *ambiguous* because of the ambiguity in the definition of B which is only defined up to a cycle multiplied by a fraction $1/p$. In Euclidean QFT, this ambiguity will nevertheless be harmless if it only affects the phase of the path integral,

$$\exp\{-k \int_{(1/p)\tilde{B}} \check{H}\} = 1 , \tag{4.67}$$

for every 3-cycle $\tilde{B}$. It is easy to find a solution to eq. (4.67) by using the fact that the integrals of the fundamental 3-forms $\omega_j \in H^3(G, \mathbf{R})$ over any cycle $\mathcal{C}$ are integer-valued. Therefore, we can choose $\check{H} = 2\pi i \omega_j$ if k/p is an *integer*, in order to satisfy eq. (4.67). The number of the independent WZ terms is therefore given by the Betti number $b_3(G)$. Thereby, the quantization condition for the coefficient k at the WZ term precisely means that k should be an integer multiple of p.

In the case of a 2d manifold $\mathcal{W}$ with the topology of a sphere S^2, the situation is even simpler because there are no torsion subgroups ($p = 1$). The coefficient k at the WZ term can therefore be *any* integer. The action (4.60) is quantum-mechanically well-defined in the sense that the theory it represents does not depend on the way it is parametrized. This completes our discussion about the topological nature of the WZ term and the associated topological quantization of the coefficient k. The topological method is complementary to the algebraic approach of the previous sections.

Exercise

IV-6 ▷ Consider the global (coordinate-independent) symmetry transformations (4.56) for the action (4.55) and compute the associated Noether currents. What is the relation between them and the AKM currents of eq. (4.58)? *Hint:* use eqs. (4.1)–(4.3).

IV.4   Free field AKM representations

In this section we discuss *free field representations* of RCFT's. The basic idea of this approach is to represent a RCFT in terms of some *free* fields. This is going to be very important for any universal treatment of RCFT's, since all formulae for chiral blocks and correlation functions of RCFT on Riemann surfaces could then be deduced as certain combinations of the known results about the free fields. In case of the Virasoro algebra, there are some useful conformal representations afforded by free bosons and fermions (Ch. I). Free fields can also be used to construct particular representations of AKM algebras. We already considered in this Chapter the abelian example provided by a single free boson.

Another very useful example is provided by a set of MW fermions $\psi^i(z)$, with the OPE

$$\psi^i(z)\psi^j(w) \sim -\frac{\delta^{ij}}{z-w} \ . \tag{4.68}$$

Given a Lie group G and its *real* representation $(t^a)_{ij}$, we can introduce the currents

$$J^a(z) = \frac{1}{2} : \psi(z)t^a\psi(z) : \ , \tag{4.69}$$

which define the AKM algebra with $\tilde{k}^{ab} = \frac{1}{2}\text{tr}\,(t^a t^b)$, as can be easily verified. For $G = SO(N)$ at level $k = 1$ and $N > 2$, eq. (4.38) gives

$$c = \frac{1 \cdot \frac{1}{2}N(N-1)}{1 + (N-2)} = \tfrac{1}{2}N \ , \tag{4.70}$$

as it should for N free fermions.

Similarly, for the case of N *complex* chiral fermions transforming in the vector representation of $SU(N)$, the currents

$$J^a(z) = \psi^*(z)t^a\psi(z) \tag{4.71}$$

realize the AKM algebra $\widehat{SU(N)}_1 \times \widehat{U(1)}$ of $\tilde{k}^{ab} = \text{tr}(t^a t^b)$, with the central charge

$$c_{U(1)} + c_{SU(N),k=1} = 1 + \frac{1 \cdot (N^2 - 1)}{1 + N} = N \ , \tag{4.72}$$

again in consistency with the central charge for N free complex fermions. Clearly, a proper basis for the Lie algebra, in which $\text{tr}(t^a t^b) = k\delta^{ab}$, does exist.

The third important example is the Frenkel-Kač **vertex operator construction** [118, 119] for the simply-laced Lie algebras $SU(n)$, $SO(2n)$ and $E_{6,7,8}$ at level one.

Taking r_G free bosons $\phi^i(z)$, $i = 1, \ldots, r_G$, where r_G is the rank of a simply-laced Lie algebra $\mathcal{G}$, let us choose $H^i(z) = i\partial\phi^i(z)$ to represent the Cartan subalgebra and $J^{\pm\alpha}(z) = c_\alpha : e^{\pm i\alpha\phi(z)} :$ to represent the remaining currents. Here α's are the positive roots normalized to $\alpha^2 = 2$, and c_α is a *cocycle* operator or *Klein factor* needed to give correct signs in the commutation relations [107]. The dual Coxeter number and the central charge are given by eqs. (4.35) and (4.38), resp., which in this case yield

$$\tilde{h}_G = \frac{|G|}{r_G} - 1, \quad c_G = r_G. \tag{4.73}$$

In order to generalize this construction to any level, one adds the new fields $\chi_\alpha(z)$ and one introduces the new currents as

$$J^{\pm\alpha}(z) = c_\alpha : e^{\pm i\alpha\phi(z)/\sqrt{k}} : \chi_\alpha(z) :, \tag{4.74}$$

in terms of the same r_G free bosons $\phi^i(z)$. $H^i(z) = i\sqrt{k}\partial\phi^i(z)$ can still be taken to represent the Cartan currents. The field $: \exp\left(\pm i\alpha \cdot \phi(z)/\sqrt{k}\right) :$ has the correct OPE with the Cartan currents, but its dimension, $h = \alpha^2/2k$, does not generically equal to one. It is the role of the new 'fermionic' fields χ_α in eq. (4.74) that they keep the right OPE (4.7). Hence, they should have dimension $h_\alpha = 1 - \alpha^2/2k$. These fields are known as **parafermions**, since they have fractional dimensions and obey neither commutation nor anticommutation relations [120, 121, 122]. The parafermions depend on the choice of G and k, and their central charge is given by $c_\chi = c_G(k) - r_G$. Clearly, the parafermionic representation does *not* involve only free fields.

The next important issue we are going to discuss is the free field representation for the WZNW theory [123]. To this end, we closely follow ref. [123].

As the representatives of free field conformal systems, let us choose (i) free scalar fields taking their values in some tori or orbifolds with the stress tensor (2.79), (ii) a Grassmannian (b, c) system with b of spin λ, and c of spin $1 - \lambda$, and the stress tensor (1.117), and (iii) a bosonic (β, γ) system with β of spin λ and γ of spin $1 - \lambda$, and the stress tensor (1.132).

The simplest case to consider is the $\widehat{sl(2)}_k$ current algebra having the following (*Wakimoto*) free field representation for its currents [124] :

$$J_+(z) = \frac{i}{\sqrt{2}}W(z), \quad H(z) = \frac{iq}{\sqrt{2}}\partial\phi(z) - W(z)\chi(z),$$

$$J_-(z) = \frac{i}{2}\left[W(z)\chi^2(z) - i\sqrt{2}q\chi(z)\partial\phi(z) + (2 - q^2)\partial\chi(z)\right], \tag{4.75}$$

where (W, χ) is a bosonic system ($\lambda = 1$), ϕ is a scalar field taking its values on a circle, and $q^2 = k + C_A/2 = k + 2$. It is straightforward to verify that the currents (4.75) indeed satisfy the commutation relations of the AKM algebra $\widehat{sl(2)}_k$ or, equivalently, the OPE's

$$H(z)J_\pm(w) \sim \pm \frac{1}{z - w} J_\pm(w) \, ,$$

$$J_+(z)J_-(w) \sim \frac{k/2}{(z - w)^2} + \frac{1}{z - w} H(w) \, . \tag{4.76}$$

The associated SS-constructed stress tensor

$$T = \frac{1}{q^2} : \left(J_+ J_- + J_- J_+ + H^2 \right) : \tag{4.77}$$

results in the *quadratic* (the free field system!) expression in terms of the fields (4.75):

$$T(z) =: W\partial\chi - \tfrac{1}{2}(\partial\phi)^2 - \frac{i}{\sqrt{2}q}\partial^2\phi : \ . \tag{4.78}$$

Having bosonized the (β, γ)-system (W, χ) as (sect. III.3)

$$W = -\partial\xi : e^{-u} := -i\partial\nu : e^{-u+i\nu} : \, , \quad \chi = \eta : e^u :=: e^{u-i\nu}; \, , \tag{4.79}$$

where the Grassmann-odd fields ξ and η, and the free scalar fields u and ν have been introduced with the standard OPE's: $\xi(z)\eta(w) \sim (z - w)^{-1}$, $u(z)u(w) \sim -\log(z - w)$, $\nu(z)\nu(w) \sim -\log(z - w)$, the stress-energy tensor (4.78) can be rewritten as a sum of the Dotsenko-Fateev tensors: $T = T_u + T_\nu + T_\phi$, where

$$T_\varphi = -\tfrac{1}{2}(\partial\varphi)^2 + i\sqrt{2}\alpha_{0,\varphi}\partial^2\varphi \, ,$$

$$\varphi = (u, \nu \ \phi) \, ,$$

$$\alpha_{0,u} = \frac{1}{2\sqrt{2}} \, , \quad \alpha_{0,\nu} = \frac{1}{2\sqrt{2}} \, , \quad \alpha_{0,\phi} = -\frac{1}{2q} \, . \tag{4.80}$$

All the powerful DF machinery in the Coulomb gas picture can now be used to calculate the correlation functions of the initial WZNW system in terms of that of the vertex operators $V_\beta = \exp(i\beta\varphi)$ for the three scalar fields.

Since the central charge of the WZNW theory,

$$c = c_\phi + c_u + c_\nu$$

$$= (1 - 24\alpha_{0,\phi}^2) + (1 - 24\alpha_{0,u}^2) + (1 - 24\alpha_{0,\nu}^2) \, , \tag{4.81}$$

$$= 3 - 6/q^2$$

differs from that of three free scalar fields, the 'vacuum charge' operator of conformal dimension $2/q^2 = (3 - c)/3$ has to be inserted inside the correlation functions, in order to ensure their proper transformation properties (sect. II.6).

The origin of the 'vacuum charge' operator can be understood from the Lagrangian approach to the WZNW theory described by the action (4.55), with the help of the *Gauss decomposition* [5, 125]:

$$g = \begin{pmatrix} 1 & \psi \\ 0 & 1 \end{pmatrix} \begin{pmatrix} e^\varphi & 0 \\ 0 & e^{-\varphi} \end{pmatrix} \begin{pmatrix} 1 & 0 \\ \chi & 1 \end{pmatrix} . \tag{4.82}$$

The invariant norm of the field g takes the form

$$\|\delta g\|^2 = \frac{1}{2} \int \gamma \, tr \, (g^{-1} \delta g)^2 = \int \gamma \left[(\delta\varphi)^2 + e^{-2\varphi} \delta\varphi \delta\chi \right] , \tag{4.83}$$

where γ formally represents a 2d metric in the conformal gauge. It is easy to verify that the matrix of currents takes the form

$$kg^{-1}dg = \begin{pmatrix} W\chi + kd\varphi & W \\ -W\chi^2 - 2k\chi d\varphi + kd\chi & -W\chi - kd\varphi \end{pmatrix} \equiv J_+ \sigma_+ + J_- \sigma_- + H\sigma_3 , \tag{4.84}$$

while for the kinetic and WZ terms in the Lagrangian (4.55) one gets

$$\frac{k}{2} tr \, (g^{-1}\partial_\mu g)^2 = k(\partial_\mu \varphi)^2 + W_\mu \partial_\mu \chi,$$

$$k \, tr \, (g^{-1}dg)^3 = d(W d\chi) , \quad \frac{k}{3} d^{-1}(g^{-1}dg)^3 = \varepsilon_{\mu\nu} W_\mu \partial_\nu \chi , \tag{4.85}$$

where d^{-1} stands for one extra coordinate integration, and

$$W \equiv ke^{-2\varphi}d\psi . \tag{4.86}$$

The (W, χ) fields are a particular example of the bosonic (β, γ) system with spins 1 and 0, resp. The scalar field φ takes its values on the circle, $\varphi \sim \varphi + 2\pi i$.

The anomalous functional determinant $\mathrm{Det}\,(e^{-2\varphi}\partial)$ associated with the change of variables (4.86) gives the additional piece to the Lagrangian due to the local anomaly

$$\frac{1}{4\pi} \left[2\partial\varphi\bar\partial\varphi + \varphi\partial\bar\partial \log \gamma \right] . \tag{4.87}$$

The anomaly (4.87) can be computed from the general formula [285]:

$$\lim_{M \to 0} \log \mathrm{Det} \left[F\bar\partial H\partial + M^2 \right]$$

$$\sim \frac{1}{48\pi} \int d^2 z \left[|\partial \log F|^2 + |\partial \log H|^2 - 4\partial \log F \bar{\partial} \log H \right] , \tag{4.88}$$

which can be proved, e.g., by the use of the proper-time technique [39] (eq. (4.87) follows, in particular, from the results of sect. IX.1).

The quantum Lagrangian of the $\widehat{sl(2)}_k$ WZNW model in the free field representation can therefore be represented in the form

$$4\pi L_q = - \left[W \bar{\partial} \chi + (k+2) \partial \varphi \bar{\partial} \varphi + R^{(2)} \varphi \right] , \tag{4.89}$$

where $R^{(2)} = \partial \bar{\partial} \log \gamma$, or, equivalently,

$$4\pi L_q = -W \bar{\partial} \chi + \tfrac{1}{2} \partial \phi \bar{\partial} \phi + \frac{i\sqrt{2}}{\sqrt{k+2}} R^{(2)} \phi , \tag{4.90}$$

after the appropriate rescaling of the scalar field, $\varphi = \frac{-i/\sqrt{2}}{\sqrt{k+2}} \phi$.

The Noether stress tensor for the Lagrangian (4.90) reads

$$T = W \partial \chi - \tfrac{1}{2} (\partial \phi)^2 - \frac{i}{\sqrt{2(k+2)}} \partial^2 \phi , \tag{4.91}$$

and it exactly coincides with that in eq. (4.78), as it should.

The change of variables (4.86), rewritten to the form

$$d\psi = \frac{1}{k} W \exp \left(\frac{-i\sqrt{2}}{q} \phi \right) \equiv \frac{1}{k} W e^{\phi/\alpha} , \tag{4.92}$$

also implies some additional global restrictions on the functional integral of the WZNW theory. These restrictions simply reflect the fact that the original variables ψ, χ and e^φ are the *single-valued* functions on a Riemann surface. Therefore, for any non-contractable cycle on the surface we have

$$\oint W e^{2\varphi} = \oint W e^{\phi/\alpha} = k \oint d\psi = 0 . \tag{4.93}$$

This means that the functional integral for the $\widehat{sl(2)}_k$-based WZNW theory should be defined with the δ-function insertions $\prod_C \delta \left(\oint_C W e^{\phi/\alpha} \right)$ for all non-contractable cycles C. The δ-function can be rewritten as $\int dx \, \exp \left[ix \oint W e^{\phi/\alpha} \right]$, and the exponent can then be expanded in series. Hence, the contour integral insertions $\oint W e^{\phi/\alpha}$ should be taken into account in the free field correlators representing the WZNW chiral blocks in the Coulomb gas picture. This ultimately explains why the screening (Feigin-Fuchs) operators should be included and what is their specific form. A number of their insertions is dictated by balance of charge.

Eq. (4.82) also suggests the way to proceed in the case of a general AKM algebra, in order to find its representation in terms of free fields. One should associate (β, γ) systems (W_α, χ_α) with each positive root $\alpha \in \Delta_+$, introduce scalar fields $\vec{\varphi}$ associated with the Cartan subalgebra H and taking their values in the Cartan torus, and then use the Gauss decomposition $g = g(\psi_\alpha, \vec{\varphi}, \chi_\alpha)$ and an appropriate change of variables: $\psi_\alpha, \vec{\varphi} \to W_\alpha, \vec{\phi}$, in order to rewrite the SS-constructed stress tensor to the form which would be quadratic in these fields:

$$T = \frac{1}{2(k + C_A/2)} : Tr\, J^2 :=: \sum_{\alpha \in \Delta_+} W_\alpha \partial \chi_\alpha - \tfrac{1}{2}(\partial \vec{\phi})^2 - \frac{i}{q} \vec{\rho} \partial^2 \vec{\phi} :\, , \qquad (4.94)$$

where

$$q^2 = k + C_A/2\,, \quad \vec{\rho} = \tfrac{1}{2} \sum_{\alpha \in \Delta_+} \vec{\alpha}\,. \qquad (4.95)$$

The corresponding free field Lagrangian (in the conformal gauge) is given by

$$4\pi L_q = - \sum_{\alpha \in \Delta_+} W_\alpha \bar{\partial} \chi_\alpha + \tfrac{1}{2} \partial \vec{\phi} \bar{\partial} \vec{\phi} + \frac{i}{q} R^{(2)} \vec{\rho} \vec{\phi}\,. \qquad (4.96)$$

Quite generally, the AKM free field representations can be divided into the three categories: (i) those in terms of free fermions (*free fermion construction*), (ii) those in terms of free bosons (*vertex operator construction*), and (iii) those in terms of free bosons and ghosts (generically called *Wakimoto constructions* [124]). The latter are the most general, in fact. Eq. (4.75) gives a particular case of the Wakimoto construction for the level-1 AKM algebra $A_1^{(1)}$. Free field representations considerably simplify calculations of correlators on a sphere for WZNW theories. Moreover, they allow us to naturally introduce the WZNW multi-loop chiral blocks on arbitrary Riemann surfaces. The explicit examples can be found in refs. [85, 86, 87, 123, 127].

Exercises

\# IV-7 ▷ (*) Prove that the currents in eq. (4.75) form the $\widehat{sl(2)}_k$ AKM algebra. What is the central charge associated with the SS-constructed stress tensor?

\# IV-8 ▷ (*) Take $F = \gamma^{j-1}$ and $H = \gamma^{-j}$ as the particular case in eq. (4.88), and calculate the corresponding local conformal anomaly. The result will be used in Chs. VIII and IX, when considering the conformal anomaly of the Polyakov string action.

IV.5   AKM characters, and the A-D-E classification of modular invariant partition functions

One of the most important and yet unsolved problems in CFT is to classify all modular invariant partition functions. As was explained in Ch. III, the modular invariant partition functions for the minimal models can be built from the characters of the Virasoro algebra. For the WZNW theories, a similar construction of the partition functions can be done in terms of the AKM characters (see sect. XII.2 for their definition). The AKM characters can also be expressed in terms of the Jacobi theta functions with known transformation properties under the modular group.

Since the Hilbert space of states in the WZNW theory decomposes into different representations $(R_s, \bar{R}_s)$ of the current algebra, where s runs over different representations R_s of the holomorphic (left) AKM algebra and that $(\bar{R}_s)$ of the anti-holomorphic (right) AKM algebra, the partition function is evidently a sum over the partition functions for each representation. Therefore, after taking into account the vacuum eigenvalues (4.49) of the L_0 operator, in the normalization $\psi^2 = 2$ we have

$$Z = \sum_s Tr_{(R_s, \bar{R}_s)} \exp\left[2\pi i \tau \left(N_s + \frac{C_{R_s}}{C_A + 2k}\right)\right]$$

$$\times \exp\left[-2\pi i \bar{\tau} \left(\bar{N}_s + \frac{C_{R_s}}{C_A + 2k}\right)\right], \tag{4.97}$$

where N_s and $\bar{N}_s$ are the number operators (levels of states) in a representation $(R_s, \bar{R}_s)$: $N = \sum_a n J^a_{-n} J^a_n$ ($\bar{N} = \sum_a n \bar{J}^a_{-n} \bar{J}^a_n$), and the trace goes over a representation space (module). Eq. (12.41) gives the connection between the AKM characters and the partition function (4.97),

$$Z = \sum_s A_{R_s} \bar{A}_{\bar{R}_s} \exp\left[2\pi i \tau C_{R_s}/(C_A + 2k)\right] \exp\left[-2\pi i \bar{\tau} C_{R_s}/(C_A + 2k)\right]. \tag{4.98}$$

The AKM characters can be expressed in terms of theta functions via eq. (12.50) which suggests us to replace $A_R(\tau)$ in favour of [4]

$$\chi_R(\tau) = \exp\left[2\pi i \tau C_R/(C_A + 2k)\right] \exp\left[-2\pi i \tau k|G|/24(k + C_A/2)\right] A_R. \tag{4.98}$$

This expression should have nice transformation properties under the modular group

---

[4]Eq. (12.53) has to be used here.

since it can be expressed in terms of the theta functions as follows: [5]

$$\chi_R(\tau) = \frac{\sum_{w \in W} \epsilon(w) \vartheta_{w(\hat{\Lambda}+\hat{\rho})}}{\sum_{w \in W} \epsilon(w) \vartheta_{w(\hat{\rho})}} (0, \tau, 0) . \tag{4.100}$$

The factor $k|G|/(k + C_A/2)$ contributes to the central charge of the system. For example, consider the Hilbert space of states for the closed bosonic string (Ch. VIII) moving in the space which is the product of space-time and a semi-simple group manifold, $M \times \prod_s G_s$, where M stands for the d'-dimensional Minkowski space and G_s are simple Lie groups. As far as the central charge is concerned, the flat space M contributes the number of its light-cone dimensions $d = d' - 2$ to the total central charge, while the G_s are going to contribute $k_s|G_s|/(k_s + C_A^s/2)$ each, resp. Their sum must be equal to 24, as in the totally flat space (sect. VIII.1). Thereby, we get the formula for the critical dimension of the bosonic closed string theory defined on $M \times \prod_s G_s$ (i.e. for the bosonic string string 'compactified' on a group manifold):

$$d' = 26 - \sum_s \frac{k_s|G_s|}{(k_s + C_A^s/2)} . \tag{4.101}$$

Combining eq. (4.101) with the standard result for the string partition function in the flat case (see excercise # IV-9 or ref. [128]), one finds the partition function of a string propagating on $M \times G$ in the form [108]

$$Z(\tau) = \left| (\text{Im } \tau)^{-d/2} \eta(\tau)^{-2d} \exp\left[-\frac{1}{24} 2\pi i \tau \left(24 - d - \sum_s \frac{k_s|G_s|}{k_s + C_A^s/2} \right) \right] \right|$$

$$\times \prod_s \sum_{R,\bar{R}} \chi_R \bar{\chi}_{\bar{R}} . \tag{4.102}$$

If we were interested in the modular invariant partition functions of the WZNW theories, we could ignore the exp-factor in eq. (4.102) when working in the *critical* dimension. Since the cancellation of all the local anomalies is a pre-requisite for considering any global anomaly cancellations, it is clearly meaningless to discuss the modular invariance when the total trace anomaly is non-vanishing. It should be also mentioned here that the factor $\left| (\text{Im } \tau)^{-d/2} \eta(\tau)^{-2d} \right|$ in eq. (4.102) is modular invariant by itself. We can therefore conclude that, in the critical dimension for the string we consider, the necessary and sufficient condition for the modular invariance is the invariance of

$$\prod_s \sum_{(R,\bar{R})} \chi_R \bar{\chi}_{\bar{R}} . \tag{4.103}$$

---

[5]See sect. XII.2 for our notation.

Assuming for simplicity that G is a simple group, one may ask for the most general modular invariant combinations of characters, which constitute the very convenient building blocks for constructing modular invariant partition functions and classifying them. These 'basic' partition functions generically read in terms of the AKM characters as follows:

$$Z = \sum_{(R,\bar{R})} N_{R,\bar{R}}\, \chi_R(\tau)\overline{\chi_{\bar{R}}(\tau)} \,, \tag{4.104}$$

with some coefficients $N_{R,\bar{R}}$ which are not necessarily of diagonal form.

Since under the T transformation, $\tau \to \tau + 1$, a character $\chi_R(\tau)$ changes by a phase (sect. XII.2), the first requirement on the $N_{R,\bar{R}}$ is given by

$$N_{R,\bar{R}} = 0 \quad \text{unless} \quad h_R - h_{\bar{R}} \in \mathbf{Z} \,. \tag{4.105}$$

The invariance of Z under the S transformation, $\tau \to -1/\tau$, is more complicated. First, we need to know the modular transformation properties of the characters. Since the complete classification of the modular invariant partition functions for an arbitrary group G is yet to be found (see, however, refs. [51, 129] for some recent attempts towards the resolution of this problem), let us restrict ourselves to the tractable case of $G = SU(2)$. In this case, we can use eqs. (12.57)–(12.62) to simplify the analysis considerably. In particular, eqs. (12.60), (12.61) and (12.62) give the characters of the spin j-representation of $\widehat{SU(2)}_k$ in the form [6]

$$\chi_j(\tau) = \frac{C_{2j+1,k+2}}{C_{1,2}}(0,\tau,0) = \frac{\vartheta_{2j+1,k+2} - \vartheta_{-2j-1,k+2}}{\vartheta_{1,2} - \vartheta_{-1,2}}(0,\tau,0) \,. \tag{4.106}$$

This formula is intended as a limit since both numerator and denominator vanish. The transformation property of the character (4.106) can now be directly deduced from the known transformation laws (12.58) and (12.59) of the ϑ-functions. The result reads

$$\chi_j(-1/\tau) = S^{(k)}_{jj'}\chi_{j'}(\tau) \equiv \sqrt{\frac{2}{k+2}}\sum_{j'}\sin\left[\frac{\pi(2j+1)(2j'+1)}{k+2}\right]\chi_{j'}(\tau) \,,$$

$$\chi_j(\tau+1) = T^{(k)}_{jj'}\chi_{j'}(\tau) \equiv \exp\left\{i\pi\left[\frac{(2j+1)^2}{2(k+2)} - \frac{1}{4}\right]\right\}\chi_{j'}(\tau) \,. \tag{4.107}$$

It now becomes clear from this explicit equation that the matrices S and T are unitary. This proves the modular invariance of the diagonal partition function known as the *A-invariant*:

$$Z_{\text{diag}} = \sum_{j=0}^{k/2}|\chi_j(\tau)|^2 \,. \tag{4.108}$$

---

[6]See sect. XII.2 for our notation.

It is straightforward to verify the defining relations of the modular group, which are given by $S^2 = (ST)^3 = I$.

The characters of the minimal models are given by the Rocha-Caridi formulae (3.10) and (3.11). After the change of notation,

$$c = 1 - \frac{6(r-s)^2}{rs} , \quad k = rs , \quad n_\pm = rp \pm sq , \tag{4.109}$$

they can also be expressed in terms of the theta functions as follows [71, 76],

$$\chi_{p,q}(\tau) = \frac{1}{\eta(\tau)} \left[\vartheta_{n_-,k}(0,\tau,0) - \vartheta_{n_+,k}(0,\tau,0) \right] . \tag{4.110}$$

The matrix S for the transformation

$$\chi_{p,q}(-1/\tau) = \sum_{p',q'} S_{pq}^{p'q'} \chi_{p'q'}(\tau) \tag{4.111}$$

can be determined along the similar lines. The result is [82, 84]

$$\chi_{p,q}(-1/\tau) = \sqrt{\frac{8}{(k+2)((k+3)}} \sum_{p',q'} (-1)^{(p+q)(p'+q')}$$

$$\times \sin\left(\frac{\pi p p'}{k+2}\right) \sin\left(\frac{\pi q q'}{k+3}\right) \chi_{p'q'}(\tau) . \tag{4.112}$$

A calculation of the T-transform yields the result

$$\chi_{p,q}(1+\tau) = e^{i\pi n_-^2/2k} \chi_{p,q}(\tau) . \tag{4.113}$$

Being applied to eq. (4.107), the Verlinde formula (3.18) leads to the fusion rules for the $\widehat{SU(2)}_k$ WZNW CFT in the form

$$[\phi_{j_1}] \times [\phi_{j_2}] = \sum_{j_3=|j_1-j_2|}^{\min(j_1+j_2,k-j_1-j_2)} [\phi_{j_3}] . \tag{4.114}$$

It is now straightforward to deduce the fusion rules (2.119) for the minimal models, directly from eq. (4.112).

Since the fusion rules for the minimal models follow those of $\widehat{SU(2)}_p \otimes \widehat{SU(2)}_{p'}$ CFT with $r = p+2$ and $s = p'+2$, it seems quite reasonable to expect that once the $SU(2)$ modular invariants are known, one can also construct all the modular invariants for the minimal models. This is indeed the case, as was explained by Capelli, Itzykson and Zuber in refs. [78, 79].

The $SU(2)$ modular invariants are classified by the same A-D-E series that classifies the simply-laced Lie algebras. To exhibit this relationship more explicitly, it is convenient to label the characters $\chi_j(\tau)$ as $\chi_{2j+1}(\tau)$. The results of refs. [78, 79] then take the form

$$k \geq 1: \qquad \sum_{\nu=1}^{k+1} |\chi_\nu|^2 \qquad (A_{k+1})$$

$$k = 4\rho, \ \rho \geq 1: \qquad \sum_{\substack{\nu \text{ odd} =1 \\ l \neq 2\rho+1}}^{4\rho+1} |\chi_\nu|^2 + |\chi_{2\rho}|^2$$

$$+ \sum_{\nu \text{ odd} =1}^{2\rho-2} \left(\chi_\nu \chi^*_{4\rho+2-\nu} + c.c. \right) \qquad (D_{2\rho+2})$$

$$k = 4\rho - 2, \ \rho \geq 2: \qquad \sum_{\nu \text{ odd} =1}^{4\rho-1} |\chi_\nu|^2 + |\chi_{2\rho}|^2$$

$$+ \sum_{\nu \text{ even} =2}^{2\rho-2} \left(\chi_\nu \chi^*_{4\rho-\nu} + c.c. \right) \qquad (D_{2\rho+1})$$

$$k + 2 = 12: \qquad |\chi_1 + \chi_7|^2 + |\chi_4 + \chi_8|^2 + |\chi_5 + \chi_{11}|^2 \qquad (E_6)$$

$$k + 2 = 18: \qquad |\chi_1 + \chi_{17}|^2 + |\chi_5 + \chi_{13}|^2 + |\chi_7 + \chi_{11}|^2 + |\chi_9|^2 \qquad (E_7)$$

$$k + 2 = 30: \quad |\chi_1 + \chi_{11} + \chi_{19} + \chi_{29}|^2 + |\chi_7 + \chi_{13} + \chi_{17} + \chi_{23}|^2 \ . \qquad (E_8) \quad (4.115)$$

The modular invariant associated with a given simply-laced group $G = A, \ D, \ E$ occurs for the affine $\widehat{SU(2)}_k$ theory at level $k = \tilde{h}_G - 2$. The A-series corresponds to the diagonal invariants. A connection, if any, between these affine $\widehat{SU(2)}_k$ invariants and the A-D-E classification of simply-laced Lie algebras still yet to be seen.

The non-diagonal invariants in eq. (4.115) can be interpreted in a variety of ways. For example, the D-invariants $D_{2\rho+2}$ can be thought of as the ones associated with the WZNW theory defined on $SO(3) = SU(2)/\mathbf{Z}_2$. Some of the partition functions in eq. (4.115), which look off-diagonal, may actually be diagonal in terms of a larger underlying symmetry algebra, which is usually given by a *W-algebra* (see Ch. VII about the W algebras). The simplest W_3 algebra of Zamolodchikov [130] is generated by the stress tensor T together with the operator $\phi_{4,1}$ of dimension $h_{4,1} = 3$. It has been noticed by Dijkgraaf and Verlinde in ref. [52] that the modular invariant partition functions of RCFT's, when expressed in terms of the characters $\hat{\chi}_i$ of the largest chiral algebra, are either diagonal, $\sum \hat{\chi}_i \hat{\bar{\chi}}_i$, or of the from $\sum \hat{\chi}_i P_{ij} \hat{\bar{\chi}}_j$, where P is a permutation of the chiral characters that preserves the fusion rules. Indeed, the RCFT's are characterized by the property that the matrix N appearing in eq. (4.104) is of *finite* rank [75, 535]. This means that a RCFT partition function has the generic form

$$Z(\tau) = \sum_{(h,\bar{h})} (\chi_h + \ldots)(\bar{\chi}_{\bar{h}} + \ldots) , \qquad (4.116)$$

where the dots stand for some other Virasoro characters, and the sum goes over a finite number of primaries labeled by $(h, \bar{h})$ and representing a chiral algebra of RCFT. The modular invariants above, which were non-diagonal in terms of χ, may become diagonal in terms of the characters $\hat{\chi}$ of the *extended* chiral algebra [131]. As an example, consider the exceptional modular invariant corresponding to the minimal unitary CFT of central charge $c = \frac{4}{5}$ [131],

$$Z(\tau) = |\chi_0 + \chi_3|^2 + \left|\chi_{2/5} + \chi_{7/5}\right|^2 + 2\left|\chi_{1/15}\right|^2 + 2\left|\chi_{2/3}\right|^2 , \qquad (4.117)$$

where subscripts denote conformal dimensions. The chiral algebra of this particular CFT is just the W_3 algebra (Ch. VII). In terms of the associated W_3 characters $\hat{\chi}$, the partition function (4.117) turns out to be diagonal, namely [131]

$$Z(\tau) \equiv \hat{Z}(\tau) = |\hat{\chi}_0|^2 + \left|\hat{\chi}_{2/5}\right|^2 + \left|\hat{\chi}_{1/15,+}\right|^2 + \left|\hat{\chi}_{1/15,-}\right|^2 + \left|\hat{\chi}_{2/3,+}\right|^2 + \left|\hat{\chi}_{2/3,-}\right|^2 , \quad (4.118)$$

where $\pm$ refer to the eigenvalues of a spin-3 field. This observation gives a good reason to believe that *any* modular invariant might be interpretable either as a diagonal invariant w.r.t. some maximally extended chiral algebra, or the one which is obtained from a diagonal invariant by an automorphism of the fusion rules. See ref. [77] for a review of the other approaches to find modular invariants, and for numerous examples of them as well, all based on the WZNW theories.

It is possible to construct modular invariant combinations for the minimal models by using the invariants (4.115) as the building blocks [78, 79]. The procedure is straightforward, and the results are

$$\frac{1}{2}\sum_{r=1}^{p'-1}\sum_{s=1}^{p-1} |\chi_{rs}|^2 \qquad (A_{p'-1}, A_{p-1})$$

$$
\begin{aligned}
p' = 4\rho + 1 \qquad & \frac{1}{2}\sum_{s=1}^{p-1}\left\{ \sum_{\substack{r \text{ odd} =1 \\ r\neq 2\rho+1}}^{4\rho+1} |\chi_{rs}|^2 + 2\left|\chi_{2\rho+1,s}\right|^2 \right. \\
p \geq 1 \qquad & \\
& + \left. \sum_{r \text{ odd} =1}^{2\rho-1} \left(\chi_{rs}\chi^*_{r,p-s} + c.c.\right)\right\} \qquad (D_{2\rho+2}, A_{p-1})
\end{aligned}
$$

$$p' = 4\rho \quad \frac{1}{2}\sum_{s=1}^{p-1}\left\{\sum_{r \text{ odd } =1}^{4\rho-1}|\chi_{rs}|^2 + |\chi_{2\rho,s}|^2\right.$$
$$p \geq 2$$

$$+ \sum_{r \text{ even } =1}^{2\rho-2}\left(\chi_{rs}\chi^*_{p'-r,s} + c.c.\right)\Bigg\} \qquad (D_{2\rho+1}, A_{p-1})$$

$$p' = 12 \quad \frac{1}{2}\sum_{s=1}^{p-1}\left\{|\chi_{1s} + \chi_{7s}|^2 + |\chi_{4s} + \chi_{8s}|^2\right.$$
$$\left. + |\chi_{5s} + \chi_{11s}|^2\right\} \qquad (E_6, A_{p-1})$$

$$p' = 18 \quad \frac{1}{2}\sum_{s=1}^{p-1}\left\{|\chi_{1s} + \chi_{17s}|^2 + |\chi_{5s} + \chi_{13s}|^2 + |\chi_{7s} + \chi_{11s}|^2\right.$$
$$\left. + |\chi_{9s}|^2 + [(\chi_{3s} + \chi_{15s})\chi_{9s} + c.c.]\right\} \qquad (E_7, A_{p-1})$$

$$p' = 30 \quad \frac{1}{2}\sum_{s=1}^{p-1}\left\{|\chi_{1s} + \chi_{11s} + \chi_{19s} + \chi_{29s}|^2\right.$$
$$\left. + |\chi_{7s} + \chi_{13s} + \chi_{17s} + \chi_{23s}|^2\right\} \qquad (E_8, A_{p-1}) \quad (4.119)$$

where

$$c = 1 - \frac{6(p-p')^2}{pp'}\,,$$
$$k = p - 2\,, \quad k' = p' - 2\,. \qquad (4.120)$$

It is now not difficult to show, by using the results of sect. XII.2, that the diagonal combination of left and right characters is always modular invariant for any AKM algebra $\hat{\mathcal{G}}$. Unfortunately, as was already mentioned above, it is *not* known how to classify all possible modular invariant combinations, and whether they have something to do with finite crystallographic subgroups of G, as was suggested in ref. [129].

Exercise

IV-9 (*) Derive the partition function of a free closed bosonic string propagating in the flat 26-dimensional space-time. *Hint*: consult sect. VIII.1 first.

Chapter V

Superconformal and Super-AKM Symmetries

Most of the algebraic structures discussed in the previous Chapters can be generalized to *supersymmetric* extensions of the Virasoro and affine Kač-Moody (current) algebras. Their definitions and the associated representation theory can be developed along the similar lines. So, in this Chapter, we normally skip the proofs. The superconformal and super-AKM symmetries are also relevant for the superstring theory (Ch. VIII). We introduce the N-extended superconformal algebras, where $1 \leq N \leq 4$ is the number of the 2d (extended) supersymmetry charges. The $N = 2$ extended superconformal symmetry is of particular interest for the superstring compactification, since it implies the $N = 1$ space-time supersymmetry of the compactified four-dimensional superstrings [128]. We are going to pay a special attention to the $N = 2$ *superconformal field theory* (SCFT) in this Chapter.

V.1 Superconformal algebras and their unitary representations

The supersymmetric extensions of the Virasoro algebra are usually obtained by generalizing the conformal transformations to the superconformal ones. The latter can be defined as the special superdiffeomorphisms acting in formal *superspace* $\mathbf{z} = (z, \theta)$ and preserving the supercovariant derivative $\mathbf{D} = \partial/\partial\theta + \theta\partial_z$ [132, 133]. This results in the $N = 1$ **superconformal algebra** (SCA) [128]. This algebra (with central extension) is

143

the symmetry algebra of the fermionic strings (Ch. VIII). The general $N = 1$ SCA can be defined for an arbitrary value of central charge, and it reads [59, 134, 135]

$$T(z)T(w) \sim \frac{3\hat{c}/4}{(z-w)^4} + \frac{2T(w)}{(z-w)^2} + \frac{\partial T(w)}{z-w} ,$$

$$T(z)G(w) \sim \frac{\frac{3}{2}G(w)}{(z-w)^2} + \frac{\partial G(w)}{z-w} ,$$

$$G(z)G(w) \sim \frac{\hat{c}}{(z-w)^3} + \frac{2T(w)}{z-w} , \tag{5.1}$$

where an anticommuting (Grassmannian) field $G(z)$ has been introduced, and $\hat{c} \equiv \frac{2}{3}c$. The conventional normalization is such that a single free chiral scalar *superfield* $\phi(\mathbf{z}) = \phi(z) + \theta\psi(z)$ has the central charge $\hat{c} = 1$, which implies $c = 1 + \frac{1}{2} = \frac{3}{2}$, similarly to a single bosonic chiral scalar field $\phi(z)$ having the central charge $c = 1$. The second line of eq. (5.1) is equivalent to the statement that the *supercharge* $G(z)$ is a primary field of dimension $h = 3/2$.

The OPE's (5.1) are equivalent to the (anti)commutation relations

$$[L_m, L_n] = (m - n)L_{m+n} + \frac{\hat{c}}{8}(m^3 - m)\delta_{m+n,0} ,$$

$$[L_m, G_n] = \left(\frac{m}{2} - n\right)G_{m+n} ,$$

$$\{G_m, G_n\} = 2L_{m+n} + \frac{\hat{c}}{2}\left(m^2 - \frac{1}{4}\right)\delta_{m+n} , \tag{5.2}$$

in terms of the moments (modes) L_n of $T(z)$ and that of $G(z)$,

$$G_n = \oint_0 \frac{dz}{2\pi i} z^{n+1/2} G(z) . \tag{5.3}$$

For the *integer* modding ($n \in \mathbf{Z}$) of G_n, eq. (5.2) is termed the **Ramond** (R) algebra, whereas for the *half-integer* modding ($n \in \mathbf{Z} + 1/2$), it is termed the **Neveu-Schwarz** (NS) algebra.

In the radial quantization (sect. I.2), the dilatation operator L_0 plays the role of a Hamiltonian. It is the corollary of eq. (5.2) that $G_0^2 = L_0 - \hat{c}/16$. Therefore, the global supersymmetry is not broken when the ground states are of dimension $h = \hat{c}/16$.

The two holomorphic fields $T(z) \equiv T_B(z)$ of $h_B = 2$ and $G(z) \equiv 2T_F(z)$ of $h_F = 3/2$ can also be formally represented by one superfield $T(z, \theta) = T_F(z) + \theta T_B(z)$, with the OPE [136, 137, 287]

$$T(\mathbf{z}_1)T(\mathbf{z}_2) \sim \frac{\hat{c}}{4z_{12}^3} + \frac{3\theta_{12}}{2z_{12}^2}T(\mathbf{z}_2) + \frac{\mathbf{D}_2 T(\mathbf{z}_2)}{2z_{12}} + \frac{\theta_{12}}{z_{12}}\partial_2 T(\mathbf{z}_2) , \tag{5.4}$$

where the following definitions have been used:

$$z_{12} = z_1 - z_2 - \theta_1 \theta_2 , \quad \theta_{12} = \theta_1 - \theta_2 ,$$

$$\mathbf{D}_2 = \partial/\partial\theta_2 + \theta_2 \partial_2 . \tag{5.5}$$

The superanalogue $\phi(\mathbf{z})$ of a primary field $\phi_B(z)$ of dimension $h + 1/2$,

$$\phi(\mathbf{z}) \equiv \phi(z, \theta) = \phi_F(z) + \theta \phi_B(z) , \tag{5.6}$$

has the components $\phi_F(z)$ and $\phi_B(z)$ of dimensions h and $h + 1/2$, resp., and its OPE with $T(\mathbf{z})$ takes the form [136, 137, 287]:

$$T(\mathbf{z}_1)\phi(\mathbf{z}_2) \sim h \frac{\theta_{12}}{z_{12}^2}\phi(\mathbf{z}_2) + \frac{1}{2z_{12}}\mathbf{D}_2\phi(\mathbf{z}_2) + \frac{\theta_{12}}{z_{12}}\partial_2\phi(\mathbf{z}_2) . \tag{5.7}$$

Such a superfield $\phi(\mathbf{z})$ is called an $N = 1$ **superprimary** field (see refs. [135, 139, 140, 141, 142, 143] for more about the SCFT primary fields).

The representation theory of the $N = 1$ SCA follows the lines of the bosonic case (Ch. II). In particular, the superprimary fields are associated with the highest weight states $|h\rangle$ which satisfy

$$L_0 |h\rangle = h |h\rangle , \quad L_n |h\rangle = G_n |h\rangle = 0 , \quad n > 0 . \tag{5.8}$$

Eq. (5.2) implies in the R-sector that a highest weight state has eigenvalue $h - \hat{c}/16$ under G_0^2. For $\hat{c} > 1$, the only restrictions imposed by unitarity are $h \geq 0$ (NS) and $h \geq \hat{c}/16$ (R). Except when the latter inequalities are saturated, the $N = 1$ superconformal Verma modules are irreducible (no null states) for $\hat{c} > 1$. On the other hand, for $\hat{c} < 1$ ($c < 3/2$), the *unitary* representations (no negative-norm states) of the $N = 1$ SCA (5.1) can only occur at the *discrete* values of central charge, which correspond to the $N = 1$ unitary *minimal models* [59, 134, 135, 144],

$$\hat{c} = 1 - \frac{8}{(k+2)(k+4)} , \quad k = 0, 1, 2, \ldots . \tag{5.9}$$

The spectrum of allowed conformal dimensions for the superprimary fields in the $N = 1$ unitary minimal models follows from calculating the Kač determinant in the NS- and R-sectors, and it is given by [59]

$$h_{m,n} = \frac{[(k+4)m - (k+2)n]^2 - 4}{8(k+2)(k+4)} + \frac{1}{32}\left[1 - (-1)^{m-n}\right] , \tag{5.10}$$

where m, n are positive integers, $1 \leq m < k+2$, $1 \leq n < k+4$. The NS (R) representations correspond to even (odd) values of $(m-n)$. When k is even, $h_{(k+2)/2,(k+4)/2} = \hat{c}/16$, so that the supersymmetry is unbroken.

Some of these models having $c < 1$ can be found in the list of the $N = 0$ minimal models when the appropriate supercurrent G of dimension $3/2$ exists. For instance, the first value of c in eq. (5.9), $c = 7/10$, coincides with the second non-trivial member of the $N = 0$ discrete series in eq. (2.65). This particular model is just the tricritical Ising model [59, 145], which is, therefore, supersymmetric. Since the tricritical Ising model can be experimentally realized (e.g., by absorbing helium-4 on krypton planted graphite [146]), this provides us with the real example of the superconformal symmetry in Nature!

SCFT has, in fact, more structure beyond that given by superfields alone [59], since it also contains fields intertwining between the NS- and R-sectors. These conformal fields, which are double-valued w.r.t. the fermionic parts of superfields, are called *spin fields*. The irreps they correspond to are the irreps of the Ramond algebra. In particular, the NS-vacuum $|0\rangle_{NS}$ is a ground state of the lowest energy, $h = 0$. The superprimary fields defined above are in one-to-one correspondence to the NS ground states. The Ramond ground states $|h^{\pm}\rangle_R$, where $|h^{-}\rangle_R = G_0 |h^{+}\rangle_R$, are created from the NS vacuum by the spin fields $S^{\pm}(z)$ of dimension h: $|h^{\pm}\rangle_R = S^{\pm}(0) |0\rangle_{NS}$. The (conserved) fermion parity operator $\Gamma = (-1)^F$ anticommuting with the fermionic parts of the superfields and commuting with the bosonic parts is called the *chirality* operator [59]. Since G_0 anticommutes with Γ, the paired ground states in the R-sector have opposite chirality. In a generic SCFT, spin fields of opposite chirality are non-local w.r.t. each other because of the appearance of fermionic fields in their OPE. A projection onto the $\Gamma = 1$ sector in SCFT gives the *local* SCFT called the *spin model* [59], whereas the projection itself is known as the *Gliozzi-Scherk-Olive* (GSO) projection [72]. [1]

The presence of null states in the $N = 1$ minimal models results in linear differential equations satisfied by the SCFT correlation functions, which can be used for an explicit calculation of the latter [59]. The alternative approach is provided by an appropriate supersymmetric extension of the DF (Coulomb gas) construction and of the Feigin-Fuchs integral representation [134]. The $N = 1$ supersymmetric AKM algebra and the supersymmetric SS construction are considered in sect. 2.

By definition, the N-extended SCA is an algebra which contains (i) the Virasoro algebra with central extension (T, c), (ii) N fermionic generators $G^{\alpha}(z)$ of dimension

---

[1] See Ch. VIII for more about the spin fields and space-time supersymmetry in superstrings.

3/2 and (iii) $2T(w)\delta^{\alpha\beta}/(z-w)$ on the r.h.s. of the OPE for $G^{\alpha}(z)G^{\beta}(w)$. The $N=2$ SCA [147, 148, 150] contains, in particular, the $U(1)$ current $J(z)$, in addition to the anticipated $T(z)$ and $G^{\alpha}(z)$, $\alpha = 1, 2$. The OPE's $T(z)T(w)$ and $T(w)G^{\alpha}(w)$ are as above, whereas the rest of the $N=2$ SCA reads as follows:

$$G^{\alpha}(z)G^{\beta}(w) \sim \left[\frac{2\tilde{c}}{(z-w)^3} + \frac{2T(w)}{z-w}\right]\delta^{\alpha\beta} + i\left[\frac{2J(w)}{(z-w)^2} + \frac{\partial J(w)}{z-w}\right]\varepsilon^{\alpha\beta},$$

$$T(z)J(w) \sim \frac{J(w)}{(z-w)^2} + \frac{\partial J(w)}{z-w},$$

$$J(z)G^{\alpha}(w) \sim i\varepsilon^{\alpha\beta}\frac{G^{\beta}(w)}{z-w},$$

$$J(z)J(w) \sim \frac{\tilde{c}}{(z-w)^2}, \tag{5.11}$$

where $\tilde{c} \equiv c/3 = \hat{c}/2$. The normalization of central charge $\tilde{c}$ is fixed by the condition $\tilde{c} = 1$ for a free chiral scalar $N=2$ superfield containing two scalars (each of $c=1$) and two MW fermions (each of $c=1/2$). It is sometimes convenient to deal with

$$G^{\pm}(z) = \frac{1}{\sqrt{2}}\left(G^1(z) \pm iG^2(z)\right) \tag{5.12}$$

instead of $G^{\alpha}(z)$. Then the OPE $G^{\pm}(z)G^{\pm}(w)$ becomes non-singular, whereas the only singular OPE is $G^+(z)G^-(w)$ (or $G^-(z)G^+(w)$).

In the case of $N=2$ SCA, we distinguish the three sectors: the *NS-sector* (integer modes for T and J, half-integer modes for G^1 and G^2), the *R-sector* (all operator modes are integers) and the *T-sector* (integer modes for T and G^2 (or G^1), half-integer modes for G^1 (or G^2) and J).

The *unitary* highest weight representations of the $N=2$ SCA having the central charge $c < 3$ form a discrete series. They are called the $N=2$ (unitary) *minimal models* [120, 151, 152, 153, 154]. The central charge values of the $N=2$ minimal models follow from the $N=2$ generalization of the Kač determinant formula given in refs. [144, 153]. They are

$$c = \frac{3k}{k+2}, \qquad k = 1, 2, \ldots . \tag{5.13}$$

The restricted (by unitarity) values of central charge c, when $c \geq 3$, form a continuous spectrum [153]. The boundary value, $\tilde{c} = 1$, is just realized in terms of a single free (complex) chiral scalar $N=2$ superfield. We restrict ourselves to the unitary representations of SCA's since they are the ones used in the most of applications.

When considering an arbitrary unitary representation of the $N = 2$ SCA, the NS *left-chiral* states are defined by the condition

$$G^+_{-\frac{1}{2}} |\phi\rangle = 0 , \tag{5.14}$$

The *left anti-chiral* states are similarly defined by replacing G^+ with G^-, whereas the *right* chiral and anti-chiral states follow the same pattern when replacing G by $\bar{G}$. [2] The chiral *primary* NS states satisfy

$$G^-_{n+\frac{1}{2}} |\phi\rangle = G^+_{n+\frac{1}{2}} |\phi\rangle = 0 , \qquad n \in \mathbf{Z} , \quad n \geq 0 , \tag{5.15}$$

in addition to eq. (5.14). The $N = 2$ SCA implies for such states the relation

$$\{G^-_{\frac{1}{2}}, G^+_{-\frac{1}{2}}\} |\phi\rangle = (2L_0 - J_0) |\phi\rangle = 0 . \tag{5.16}$$

Hence, the dimension h of a chiral primary state is a half of its charge q, $h = q/2$. Similarly, $h = -q/2$ for an anti-chiral primary state. In a unitary SCFT, the operator $\{G^-_{\pm\frac{1}{2}}, G^+_{\mp\frac{1}{2}}\}$ is positive, which implies the inequality $h \geq |q/2|$ for *any* state in the SCFT Hilbert space, since $G^{-\dagger}_{\frac{1}{2}} = G^+_{-\frac{1}{2}}$. The reversed argument leads to the statement that the chiral primary states are the only ones, which actually saturate the inequality (exercise # V-2).

Any NS state $|\phi\rangle$ of dimension h and charge q can be decomposed as [155]

$$|\phi\rangle = |\phi_0\rangle + G^+_{-\frac{1}{2}} |\phi_1\rangle + G^-_{\frac{1}{2}} |\phi_2\rangle , \tag{5.17}$$

where $|\phi_0\rangle$ is a chiral primary state. To prove eq. (5.17), one considers [155] the states $|\phi_1\rangle$ and $|\phi_2\rangle$ minimizing the form

$$\left\| \left(|\phi\rangle - G^+_{-\frac{1}{2}} |\phi_1\rangle - G^-_{\frac{1}{2}} |\phi_2\rangle \right) \right\|^2 . \tag{5.18}$$

Let $|\phi_0\rangle$ be their difference, as in eq. (5.17). The definition means in part that the norm of $|\phi_0\rangle$ does not change (to leading order) under the arbitrary infinitesimal transformations

$$|\phi_0\rangle \rightarrow |\phi_0\rangle + G^+_{-\frac{1}{2}} |\varepsilon_1\rangle + G^-_{\frac{1}{2}} |\varepsilon_2\rangle , \tag{5.19}$$

---

[2] For simplicity, we only consider the LM modes below.

which imply eq. (5.15) (exercise # V-2) and, hence, eq. (5.17). It is now easy to see from eq. (5.17) that, given a chiral state $|\phi\rangle$, we can take $|\phi_2\rangle = 0$ (exercise # V-3).

The $N = 2$ SCA implies, in addition, the inequality

$$h \leq \frac{c}{6} , \tag{5.20}$$

valid for any chiral primary field of $N = 2$ SCFT of central charge $\tilde{c} \equiv c/3$. To justify eq. (5.20), one uses the $N = 2$ SCA anticommutator for the particular case of a positive operator

$$\{G^-_{3/2}, G^+_{-3/2}\} = 2L_0 - 3J_0 + \frac{2c}{3} . \tag{5.21}$$

Taking the expectation value for any chiral primary state and substituting $h = q/2$ yields eq. (5.20). It will be shown in sect. 4 (see also ref. [155]) that there always exists a unique $N = 2$ chiral primary state saturating the inequality in eq. (5.20). In addition, eq. (5.20) implies that there exists only a *finite* number of chiral primary fields in the non-degenerate $N = 2$ SCFT's, in which the spectrum of the Hamiltonian L_0 is discrete [155].

As far as the operator algebra of chiral primary fields is concerned, there are no leading singularities to subtract, contrary to the case of ordinary CFT (Ch. I). This is because the naive product

$$(\phi\chi)(z) = \lim_{z' \to z} \phi(z')\chi(z) \tag{5.22}$$

is always *non-singular*. Indeed, the $U(1)$ charge of the fields is an additive quantity, and one always has

$$h_{\phi\chi} \geq \tfrac{1}{2}(q_\phi + q_\chi) = h_\phi + h_\chi . \tag{5.23}$$

When $\phi\chi$ is a primary field, the inequality in eq. (5.23) is saturated, while the leading singularity in eq. (5.22) proportional to $(z - z')^{h_{\phi\chi} - h_\phi - h_\chi}$ actually disappears. [3] If $\phi\chi$ is not a primary field, then the definition (5.22) sets $\phi\chi$ to zero when $z \to z'$ because of eq. (5.23). Therefore, given a finite number of chiral primary fields, eq. (5.22) allows us to introduce a (commutative, nilpotent and *finite*) **chiral ring** $\mathcal{K}$ of chiral primary fields, with the ring $\mathcal{K}$ being formally defined as the operator algebra *modulo* descendants [155]. There are, in fact, four pairwise charge-conjugated chiral rings depending on the chirality and the L(R)-choice.

---

[3]The product of two chiral primary fields generically gives a chiral field which is not necessarily a primary field.

As far as the Ramond sector of an $N = 2$ SCA unitary module is concerned, we have $h \geq c/24$, similarly to the $N = 1$ case considered above. The fermion number operator $(-1)^F$ can be defined in terms of the $U(1)$ current as

$$(-1)^F = \exp\left[i\pi(J_0 - \bar{J}_0)\right] . \tag{5.24}$$

It is usually assumed that the LM and RM $U(1)$ charges (q_L, q_R) of all NS and R states satisfy $q_L - q_R \in \mathbf{Z}$, which makes the eigenvalues of the fermion number operator to be ± 1. In particular, *Witten's index* [156] takes the form

$$\text{Tr}(-1)^F = \sum_{\mathcal{K}} \exp\left[i\pi(q_L - q_R)\right] . \tag{5.25}$$

The investigation of the properties of the $N = 2$ SCFT's will be continued in sect. 4.

The $N = 3$ and $N = 4$ SCA's [147, 149] are usually considered along the similar lines [154, 157, 158]. In the so-called '*small*' $N = 4$ SCA, the $SU(2)$ current algebra appears instead of the $U(1)$ current algebra in the $N = 2$ case. The 'small' $N = 4$ supercharges are the $SU(2)$ complex doublets. The level of this $SU(2)$-extended AKM algebra is fixed by central charge c, which is the only parameter of the 'small' $N = 4$ SCA.

The unitary representations of the N-extended SCA for $N \geq 3$ exist at the discrete values of central charge [159],

$$N = 3: \quad c = \frac{3k}{2} , \quad k = 1, 2, \ldots ,$$

$$N = 4: \quad c = 6k , \quad k = 1, 2, \ldots , \tag{5.26}$$

and no continuum values of central charge are allowed by unitarity.

The highest weight states of the 'small' $N = 4$ SCA are labeled by the two integers, dimension h and isospin l. The unitarity implies the inequalities: $h \geq k/4$ in the R-sector, and $h \geq l$ in the NS sector [154]. For the *massless* irreps, one finds

$$h = k/4 , \quad l = 0, 1/2, \ldots, k/2 \quad (R) ,$$

$$h = l , \quad l = 0, 1/2, \ldots, k/2 \quad (NS) , \tag{5.27a}$$

whereas

$$h > k/4 , \quad l = 1/2, 1, \ldots, k/2 \quad (R) ,$$

$$h > l , \quad l = 0, 1/2, \ldots, k/2 - 1/2 \quad (NS) , \tag{5.27b}$$

for the *massive* ones [154, 157, 158]. In the case (b) Witten's index is zero, which implies an equal number of bosons and fermions. The characters of the $N = 4$ unitary representations can be found in ref. [158].

Free field representations of the *gauged* (i.e. locally realized) N-extended 2d CSA's, $N \le 4$, lead to the N-extended fermionic string models (sect. VIII.3).

Exercises

V-1 ▷ Determine the value of the coefficient a at which the state

$$\left[G_{-3/2} + aL_{-1}G_{-1/2}\right]\left|h = \tfrac{1}{10}\right\rangle \tag{5.28}$$

becomes the null state in the tricritical Ising model.

V-2 ▷ (*) Let a state $|\phi\rangle$ in Hilbert space of a unitary $N = 2$ SCA module obey the relation $h = q/2$, where h is its (conformal) dimension and q is its $U(1)$ charge. Prove that the state $|\phi\rangle$ is a chiral and primary state. *Hint*: use the constraint $J_n |\phi\rangle = 0$ for $n > 0$, which is valid for any $N = 2$ SCA representation state (why?).

V-3 ▷ (*) Consider the decomposition (5.17) for a left chiral state $|\phi\rangle$, and show that one can take $|\phi_2\rangle = 0$ in this case indeed.

V.2 Super-AKM algebra and its representations

The **super-AKM** algebra is defined by adding to an AKM-current $J^a(z)$ its fermionic superpartner $j^a(z)$ in the *adjoint* representation and of dimension $h = 1/2$. The corresponding OPE's take the form

$$J^a(z)J^b(w) \sim \frac{k\delta^{ab}}{(z-w)^2} + \frac{if^{abc}J^c(w)}{z-w} \, ,$$

$$J^a(z)j^b(w) \sim \frac{if^{abc}j^c(w)}{z-w} \, ,$$

$$j^a(z)j^b(w) \sim \frac{k\delta^{ab}}{z-w} \, , \tag{5.29}$$

where we maintain to use the normalization of structure constants in which the highest root squared is equal to two, $\psi^2 = 2$.

The super-AKM currents can be unified into a single superfield as

$$J^a(\mathbf{z}) \equiv J_F^a(z) + \theta J_B^a(z) = j^a(z) + \theta J^a(z) \ . \tag{5.30}$$

The superspace form of eq. (5.29) is then given by

$$J^a(\mathbf{z}_1) J^b(\mathbf{z}_2) \sim \frac{k\delta^{ab}}{z_{12}} + i\frac{\theta_{12}}{z_{12}} f^{abc} J^c(\mathbf{z}_2) \ . \tag{5.31}$$

The simplest representation of the super-AKM algebra (5.29) with the lowest possible level is given by the MW *free* fermions ψ^a transforming in the adjoint representation of a Lie group G. This is known as the '*quark model*' [160].

First, we note that the currents

$$J_f^a(z) = \frac{i}{2} f^{abc} \psi^b(z) \psi^c(z) \tag{5.32}$$

define a representation of $\hat{\mathcal{G}}$ at level $k = \bar{h}_G = C_A/2$. The associated central charge comes from either the OPE satisfied by the free fermionic stress tensor $T_f = \frac{1}{2} : \psi^a \partial \psi^a :$, or from the SS constuction in terms of the currents (5.32). Its value is given by

$$c = \frac{1}{2}|G| = \frac{k|G|}{k + \bar{h}_G} = \frac{\bar{h}_G|G|}{\bar{h}_G + \bar{h}_G} \ . \tag{5.33}$$

Second, we choose the fermionic superpartner j_f of J_f in the form [4]

$$j_f^a(z) = i\sqrt{k}\psi^a(z) \ . \tag{5.34}$$

This implies that the bosonic currents $J_f^a(z)$ are equal to the fermion bilinears,

$$J_f^a(z) = -\frac{i}{2k} f^{abc} j^b(z) j^c(z) \ . \tag{5.35}$$

The set of $|G|$ free MW fermions can be used to realize the super-AKM algebra with the enveloping super-Virasoro algebra [161, 162, 163]. The spin-3/2 superpartner $G_f(z)$ of $T_f(z)$ is given by

$$G_f(z) = -\frac{1}{6\sqrt{C_A/2}} f^{abc} \psi^a \psi^b \psi^c \ . \tag{5.36}$$

---

[4] j_f^a should clearly be proportional to ψ^a, since this is the only way to satisfy the dimension $h = 1/2$ of j_f^a. The normalization of j_f^a is then determined by the algebra (5.29). The similar arguments are applied to the derivation of eqs. (5.35) and (5.36) below.

It is now easy to complete the list of OPE's by using the Wick rules and the Jacobi identity for structure constants. One gets

$$G_f(z)J_f^a(w) \sim \frac{1}{(z-w)^2}j_f^a(w) ,$$

$$G_f(z)j_f^a(w) \sim \frac{1}{z-w}J_f^a(w) . \tag{5.37}$$

We conclude that both super-AKM and superconformal symmetries in two dimensions can be non-linearly realized in terms of the free fermions transforming in the adjoint representation of a semi-simple Lie group G (without the $U(1)$ factors) [160, 161, 162, 163, 164].

As was pointed out by Kač and Todorov [165], a general representation of the super-AKM algebra can be obtained by forming a linear combination of the currents,

$$J^a(z) = J_f^a(z) + \hat{J}^a(z) , \tag{5.38}$$

where the currents $\hat{J}^a(z)$ define a level-$\hat{k}$ representation of the AKM-algebra $\mathcal{G}$, $a = 1, \ldots, |G|$. The currents $\hat{J}$'s may be thought of as that of the WZNW theory, but we don't need here the explicit representation for them. It is enough to know that $\hat{J}^a$ define an AKM-representation which is independent on the fermionic fields: $\hat{J}^a(z)j_f^b(w) \sim O(1)$, $\hat{J}^a(z)J_f^b(w) \sim O(1)$. We also assume that $\mathcal{G}$ is a *simply-laced* Lie algebra.

The currents J^a in eq. (5.38) define the new AKM-representation of level

$$k = \hat{k} + C_A/2 = \hat{k} + \bar{h}_G , \tag{5.39}$$

and of central charge

$$c = \frac{\hat{k}|G|}{\hat{k} + \bar{h}_G} + \tfrac{1}{2}|G| = \frac{3}{2}|G| - \frac{C_A}{2k}|G| . \tag{5.40}$$

The superconformal anomaly is given by

$$\hat{c} = \frac{2}{3}c = \left(1 - \frac{C_A}{3k}\right)|G| . \tag{5.41}$$

The AKM-representation $J^a(z)$ can be extended to a representation of the super-AKM algebra by adding the superpartner current in the form

$$j^a(z) = i\sqrt{k}\psi^a , \tag{5.42}$$

where the subscript (f) at $j^a(z)$ has been omitted.

We can associate the $N = 1$ (super-Virasoro) SCA with this construction by defining [165]

$$T(z) = \frac{1}{2k} \left[: \hat{J}^a(z)\hat{J}^a(z) : - : j^a(z)\partial j^a(z) : \right] , \tag{5.43a}$$

$$G(z) = \frac{1}{k} \left[j^a(z)\hat{J}^a(z) - \frac{i}{6k} f^{abc} j^a(z)j^b(z)j^c(z) \right] . \tag{5.43b}$$

Eq. (5.43) gives the $N = 1$ supersymmetric extension of the SS construction (4.22).

Exercises

\# V-4 ▷ (*) Verify that the central charge of the 'quark model' is given by eq. (5.33), using the SS construction.

\# V-5 ▷ (*) Calculate the normalizations in eqs. (5.34) and (5.36).

\# V-6 ▷ Check the super-Virasoro (SCA) OPE's for the super-SS construction (5.43).

V.3   Supersymmetric WZNW theories

In two dimensions, the fundamental spinor representation is one-dimensional, and it just represents a MW spinor. Therefore, a general two-dimensional supersymmetry algebra may have p real left-handed spinor supersymmetry generators and q real right-handed generators, where p, q are positive integers [166]. The basic case is given by the $(1, 0)$ supersymmetry algebra, which has only one MW supersymmetry charge [167]. The N-extended Majorana supersymmetry corresponds to the case $p = q = N$.

The standard pattern of a 2d field theory with $N = 1$ super-AKM and superconformal symmetry is given by the *supersymmetric* WZNW theory [139]. The corresponding action is just the $(1, 1)$-supersymmetric generalization of the bosonic action in eq. (4.55). This action can be constructed most easily in flat $(1, 1)$ superspace with the coordinates $(\sigma^0, \sigma^1, \theta^\alpha)$, where the two-component M-spinor θ^α represents anticommuting (Grassmannian) superspace coordinates. [5] The *non-linear sigma-model* (NLSM) action on a group manifold reads

$$S^{(1,1)}_{\text{sWZNW}} = -\frac{1}{4\lambda^2} \int d^2\sigma d^2\theta \, tr \left(\bar{D}^\alpha G D_\alpha G^{-1} \right)$$

---

[5] Our conventions for the $(1, 1)$ superspace are: (light-cone) $\sigma^{\pm\pm} = \frac{1}{\sqrt{2}} \left(\sigma^0 \pm \sigma^1\right)$, $\sigma^{+} \equiv \sigma^{++}$, $\sigma^{=} \equiv \sigma^{--}$; ($\gamma$-matrices) $\gamma^0 = i\sigma_2$, $\gamma^1 = \sigma_1$, $\gamma_3 = \sigma_3$; (M-spinors) $\theta = C_2\bar{\theta}^T$, $C_2 = \gamma^0$; (covariant derivatives) $D_\alpha = \partial/\partial\bar{\theta}^\alpha + i(\gamma \cdot \partial\theta)_\alpha = (-D^+, D^-)$, $D^\pm = \partial/\partial\theta^\mp + i\theta^\mp \partial/\partial\sigma^{\mp\mp}$; $\theta^\pm = (1 \pm \gamma_3)\theta$. We use the two-dimensional Lorentz eigenvalues (quantum helicity numbers) for both spinors and vectors.

$$-\frac{k}{16\pi}\int_0^1 dy \int d^2\sigma d^2\theta \, tr\left[\tilde{G}^{-1}\frac{\partial\tilde{G}}{\partial y}\bar{D}^\alpha\tilde{G}(\gamma_3)_{\alpha\beta}D^\beta\tilde{G}^{-1}\right] \,, \tag{5.44}$$

where $G = \exp\left[it^a\phi^a(\sigma,\theta)\right]$ is a Lie group-valued superfield

$$G(\sigma,\theta) = g(\sigma) + i\bar{\theta}\psi(\sigma) + \tfrac{1}{2}i\bar{\theta}\theta F(\sigma) \,, \tag{5.45}$$

$\tilde{G}(\sigma,\theta;y)$ is the extension of $G(\sigma,\theta)$, restricted by the boundary conditions

$$\tilde{G}\mid_{y=1} = G \,, \quad \tilde{G}\mid_{y=0} = 0 \,. \tag{5.46}$$

Quite similarly to its bosonic counterpart (4.55), the QFT defined by the action (5.44) has a *renormalization group* (RG) fixed point at

$$\lambda^2 = 4\pi/k \,, \tag{5.47}$$

where it is conformally invariant. Moreover, after rewriting the action (5.44) in terms of the light-cone variables $(\sigma^{\pm\pm}, \theta^\pm, D^\pm)$, it becomes clear that this action is also invariant under the $(1,1)$ super-AKM transformations

$$G \to \Omega_L G \Omega_R^{-1} \,, \tag{5.48}$$

where the Lie group-valued parameter superfields $\Omega_{L,R}$ are subject to the constraints

$$D^+\Omega_L = D^-\Omega_R = 0 \,. \tag{5.49}$$

In components, when using the algebraic equations of motion for the auxiliary field F and parametrizing the fermionic field ψ in terms of a new field χ as

$$\chi = g^{-1}\psi_+ + \psi_- g^{-1} \,, \tag{5.50}$$

the super-WZNW action (5.44) *at* $\lambda^2 = 4\pi/k$ becomes the sum of the bosonic WZNW action (4.55) and that of *free* fermions, namely

$$S_{\text{sWZNW}}^{(1,1)} = S_{\text{WZNW}} + \frac{ik}{16\pi}\int d^2\sigma \, tr\left(\bar{\chi}\gamma\cdot\partial\chi\right) \,. \tag{5.51}$$

Hence, the supersymmetric fermions completely decouple in the super-WZNW action, as they should (exercise # V-7).

The super-WZNW classical equations of motion have a general solution where all the fields take a factorized form, *viz.*

$$g = g_L(\sigma^=)g_R(\sigma^{\pm}) \,, \quad \chi_+ = \chi_+(\sigma^{\pm}) \,, \quad \chi_- = \chi_-(\sigma^=) \,. \tag{5.52}$$

The AKM transformations in components are

$$g \to \Omega_L g \Omega_R^{-1} , \quad \chi^- \to \Omega_L \chi^- \Omega_L^{-1} , \quad \chi^+ \to \Omega_R \chi^+ \Omega_R^{-1} , \tag{5.53}$$

whereas the accompanying super-AKM trasformations, in accordance with eq. (5.48), read as

$$\delta g = 0 , \quad \delta \chi^\pm = \beta^\pm , \tag{5.54}$$

where $\beta^\pm$ are two Grassmannian parameters, and

$$\beta^+ = \beta^+(\sigma^+) , \quad \beta^- = \beta^-(\sigma^=) . \tag{5.55}$$

In the super-WZNW theory, we clearly have *two* separate super-AKM invariances, the one in the L-sector and another one in the R-sector. After the Wick rotation, they are related to the holomorphic and anti-holomorphic parts of the theory, resp.

It is not difficult to construct the $(1,0)$ supersymmetric WZNW action, when using the superspace techniques [167, 166, 168]. In the flat $(1,0)$ superspace parameterized by the coordinates $(\sigma^+, \sigma^=, \theta^+)$, with the flat $(1,0)$ super-covariant derivatives $(\partial_+, \partial_=, D_+)$, $D_+ = \partial/\partial\theta^+ + i\theta^+\partial_+$, the $(1,0)$ NLSM action on a group manifold is given by [169]

$$S_{\text{sWZNW}}^{(1,0)} = -\frac{i}{2\lambda^2} \int d^2\sigma (d\theta)^- \, \text{tr} \left[(G^{-1} D_+ G)(G^{-1} \partial_= G) \right]$$

$$-\frac{ik}{8\pi} \int_0^1 dy \, \text{tr} \left[\left(\tilde{G}^{-1} \frac{\partial \tilde{G}}{\partial y} \right) \left\{ D_+ \tilde{G}^{-1} \partial_= \tilde{G} - \partial_= \tilde{G}^{-1} D_+ \tilde{G} \right\} \right] . \tag{5.56}$$

The action (5.56) can be constructed, e.g., by truncating the $(1,1)$ supersymmetric WZNW action (exercise # V-8). Here G and $\tilde{G}$ are the Lie group-valued $(1,0)$ superfields, related with each other by eq. (5.46).

The $(1,0)$ super-WZNW action is just the $(1,0)$ supersymmetric NLSM action (5.56) at the RG fixed point $\lambda^2 = 4\pi/k$. This theory has the superconformal and super-Kač-Moody invariances in its L-sector, but the conformal and AKM invariances in its R-sector. The super-AKM transformation law in superspace is given by [169]

$$G \to \Omega G \check{\Omega}^{-1} , \tag{5.57}$$

where the Lie group-valued $(1,0)$ superfield parameters Ω and $\check{\Omega}$ are restricted by the conditions:

$$D_+ \Omega = \partial_= \check{\Omega} = 0 . \tag{5.58}$$

The symmetry (5.58) is just the $(1,0)$ super-AKM invariance.

The simplest way to check the super-AKM invariance of the super-WZNW theory at its RG fixed point is to use the *Polyakov-Wiegmann* identity [170, 171]. In the case of the $(1,0)$ supersymmetric WZNW theory, this identity takes the form [169]

$$S_{\text{sWZNW}}^{(1,0)}(GH) = S_{\text{sWZNW}}^{(1,0)}(G) + S_{\text{sWZNW}}^{(1,0)}(H)$$

$$-\frac{i}{4\pi} \int d^2\sigma (d\theta)^- \, \text{tr}\left[(G^{-1}D_+G)(\partial_= H \cdot H^{-1})\right] . \qquad (5.59)$$

In general, the Polyakov-Wiegmann identities in the bosonic and supersymmetric WZNW theories are just the integrated variations of the corresponding classical actions (exercise # V-9).

The super-AKM invariant WZNW actions can also be constructed in the Hamiltonian approach by the method of *coadjoint orbits* [172], where a (super-)WZNW Lagrangian takes the form of an integral of the *Kirillov-Kostant form* to be defined on the orbit of a (super)AKM algebra [173]. To illustrate this approach, we are now going to consider the construction of the $(2,0)$ supersymmetric WZNW theory [174, 175].

The flat $(2,0)$ superspace is parametrized by the coordinates

$$z^M = (\sigma^m, \zeta^+, \bar{\zeta}^+) , \quad \sigma^m = (\tau, \sigma) , \quad \zeta^+ = \zeta_1^+ + i\zeta_2^+ , \qquad (5.60)$$

where each Grassmannian coordinate $\zeta_{1,2}^+$ is a MW spinor. The spinor supersymmetric derivative is $D_+ = \partial_+ + i\bar{\zeta}^+ \partial_{\#}$. Together with the other derivatives, they from a (graded) $(2,0)$ supersymmetry algebra:

$$\{D_+, D_+\} = 0 , \quad \{D_+, \bar{D}_+\} = 2i\partial_{\#} , \quad [D_+, \partial_m] = 0 , \quad [\partial_m, \partial_n] = 0 , \qquad (5.61)$$

where $\partial_m \equiv (\partial_{\#}, \partial_=)$. The $(2,0)$ superfields are complex superfunctions $U(z)$, with their components to be defined via the standard superspace expansion:

$$U(z) = \phi(\sigma) + \zeta^+ \beta_+(\sigma) + \bar{\zeta}^+ \psi_+(\sigma) + i\bar{\zeta}^+ \zeta^+ \omega_{\#}(\sigma) . \qquad (5.62)$$

The action of the most *general* $(2,0)$ supersymmetric NLSM in $(2,0)$ superspace takes the form [168]

$$S = \frac{i}{2} \int d^3 z^= \left[K_{\bar{m}}(\Phi, \bar{\Phi}) \partial_= \bar{\Phi}^{\bar{m}} - K_m(\Phi, \bar{\Phi}) \partial_= \Phi^m \right] , \qquad (5.63)$$

where we have introduced the real superspace measure, $d^3 z^= = d^2\sigma d\bar{\zeta}^+ d\zeta^+$, and the vector potential function $K_{\bar{m}} = (K_m)^*$, $m = 1, \ldots, m'$.

In order to get the standard Klein-Gordon and Dirac actions (in components) from the superspace action (5.63), in the free case of $K_{\underline{m}} = (\Phi_{\underline{m}})^*$, we must impose the analyticity (or chirality) conditions on the superfields:

$$\overline{D}_+ \Phi^m = D_+ \overline{\Phi}^{\hat{m}} = 0 , \quad \overline{\Phi}^{\hat{m}} = (\Phi^m)^* . \tag{5.64}$$

This automatically implies that the complex structure on the NLSM target space is canonical. At the same time, the constraints (5.64) ensure that our action (5.63) is a 'physical' action in the sense that it reproduces the standard supersymmetric NLSM action in components or $(1, 0)$ superfields [175].

The vector potential $K_{\underline{M}} \equiv (K_{\underline{m}}, K_{\hat{\underline{m}}})$ determines the geometry of the hermitian NLSM target manifold, i.e. the *metric* $g_{\underline{MN}}$ and the *torsion* $H_{\underline{MNP}}$, as follows

$$g_{\underline{m\hat{n}}} = K_{\underline{m},\hat{\underline{n}}} + K_{\hat{\underline{n}},\underline{m}} , \quad H_{\underline{mn\hat{p}}} = \frac{1}{2} \left(K_{\underline{m},\underline{n}\hat{\underline{p}}} - K_{\underline{n},\underline{m}\hat{\underline{p}}} \right) ,$$

$$g_{\underline{mn}} = g_{\hat{\underline{m}}\hat{\underline{n}}} = H_{\underline{mnp}} = H_{\hat{\underline{m}}\hat{\underline{n}}\hat{\underline{p}}} = 0 . \tag{5.65}$$

Varying the action (5.63),

$$\delta S = \frac{i}{2} \int d^3 z^= \delta \Phi^m \left[g_{\underline{m\hat{n}}} \partial_= \overline{\Phi}^{\hat{n}} + h_{\underline{mn}} \partial_= \Phi^n \right] + \text{h.c.} , \tag{5.66}$$

yields the equations of motion,

$$\overline{D}_+ J_{\underline{m}}^{\ddagger} \equiv \overline{D}_+ \left[g_{\underline{m\hat{n}}} \partial_= \overline{\Phi}^{\hat{n}} + h_{\underline{mn}} \partial_= \Phi^n \right] = 0 , \tag{5.67}$$

where we have introduced the torsion potential $h_{\underline{mn}}$ as $h_{\underline{mn}} = K_{\underline{m},\underline{n}} - K_{\underline{n},\underline{m}}$. Eq. (5.67) defines the conserved $(2, 0)$ supercurrent $J_{\underline{m}}^{\ddagger}(z)$.

The variation (5.66) can be integrated, and this leads to the equivalent action which, in the three-dimensional form, reads as

$$S = \frac{i}{2} \int d^3 z^= \int_0^1 dy \left[g_{\underline{m\hat{n}}} \partial_y \tilde{\Phi}^m \partial_= \overline{\tilde{\Phi}}^{\hat{n}} + h_{\underline{mn}} \partial_y \tilde{\Phi}^m \partial_= \tilde{\Phi}^n \right] + \text{h.c.} , \tag{5.68}$$

where the interpolating superfields $\tilde{\Phi}^m(z, y)$ have been introduced,

$$\tilde{\Phi}^m(z, 0) = 0 , \quad \tilde{\Phi}^m(z, 1) = \Phi^m(z) , \quad \overline{D}_+ \tilde{\Phi}^m(z, y) = 0 . \tag{5.69}$$

The torsion potential is, in general, only locally defined. To avoid this, we want to rewrite the torsion term in eq. (5.68) to the chiral form by integrating over one complex Grassmannian coordinate. This results in the following action [174]:

$$S = \frac{i}{2} \int_0^1 dy \int d^2 \sigma \left\{ \int d\zeta^+ d\bar{\zeta}^+ \, g_{\underline{m\hat{n}}} \partial_y \tilde{\Phi}^m \partial_= \overline{\tilde{\Phi}}^{\hat{n}} \right.$$

$$+ 2 \int d\zeta^+ \, H_{\underline{mnp}} \partial_y \tilde{\Phi}^{\underline{m}} \partial_= \tilde{\Phi}^{\underline{n}} \overline{D}_+ \tilde{\bar{\Phi}}^{\underline{p}} \Big\} + \text{h.c.} \tag{5.70}$$

The integrand of the second WZ-type term in eq. (5.70) is chiral since the torsion 3-form H is closed. This WZ-type term is related by extended supersymmetry to the kinetic term. Both actions (5.68) and (5.70) cannot be considered independently of the constraints (5.65). It is the restrictions (5.65) that make the physics of the $(2 + 1)$-dimensional actions (5.68) and (5.70) to be two-dimensional. In other words, it is the special feature of the $(2, 0)$ extended supersymmetry, unlike the $(1, 0)$ or $(1, 1)$ super-symmetry, that the field dynamics is not entirely determined by the form of the action but the kinematical conditions imposed on the superfields (they are either chiral or anti-chiral in our case) too.

Any supersymmetric NLSM action can be coupled to proper 2d supergravity. It is usually enough to replace a flat superspace by a curved superspace, and the flat superspace covariant derivatives by the curved superspace covariant derivatives subject to the supergravity constraints on their algebra.

Given a finite-dimensional semi-simple Lie algebra $\mathcal{G}$ with structure constants $f_{\underline{AB}}{}^{\underline{C}}$ and generators $t_{\underline{A}}$, $\underline{A} = 1, \ldots, \dim \mathcal{G}$, the group elements (connected to the identity) are the exponentials of the algebra elements: $G(\phi) = \exp\left(\phi^{\underline{A}} t_{\underline{A}}\right)$. The generators satisfy

$$[t_{\underline{A}}, t_{\underline{B}}] = f_{\underline{AB}}{}^{\underline{C}} t_{\underline{C}}, \quad \text{tr}\left(t_{\underline{A}} t_{\underline{B}}\right) = \gamma_{\underline{AB}}, \tag{5.71}$$

where the matrix $\gamma_{\underline{AB}}$ is supposed to be invertible, and, hence, it may be used to raise and lower indices.

Let $\mathcal{T}$ be a *differentiation* of the Lie algebra $\mathcal{G}$, i.e. a linear map $\mathcal{T} : \mathcal{G} \to \mathcal{G}$ satisfying the conditions

$$\mathcal{T}([\phi, \psi]) = [\mathcal{T}(\phi), \psi] + [\phi, \mathcal{T}(\psi)],$$

$$\text{tr}[\mathcal{T}(\phi)\psi] = -\text{tr}[\phi \mathcal{T}(\psi)], \tag{5.72}$$

for any two Lie algebra elements, $\phi \in \mathcal{G}$ and $\psi \in \mathcal{G}$. We can then associate with this map the matrix $T_{\underline{AB}}$ via the equation

$$\mathcal{T}(\phi) = T_{\underline{AB}} \phi^{\underline{A}} t_{\underline{C}} \gamma^{\underline{BC}}, \tag{5.73}$$

and rewrite eq. (5.72) to the equivalent form:

$$T_{\underline{AB}} = -T_{\underline{BA}}, \quad T_{\underline{A}|\underline{B}} f_{\underline{CD}|}{}^{\underline{A}} = 0. \tag{5.74}$$

Given the differentiation $\mathcal{T}$ defined on the algebra elements, we can always extend its definition to the universal enveloping algebra of $\mathcal{G}$, by using the rules

$$\mathcal{T}(1) = 0 \, ,$$

$$\mathcal{T}(\phi + \psi) = \mathcal{T}(\phi) + \mathcal{T}(\psi) \, ,$$

$$\mathcal{T}(\phi\psi) = \mathcal{T}(\phi)\psi + \phi\mathcal{T}(\psi) \, . \tag{5.75}$$

In the context of $(2,0)$ supersymmetry, the fields ϕ^A above have to be replaced by the scalar $(2,0)$ superfields $\Phi^A(z)$ This allows us to define the $(2,0)$ supercurrents as

$$\tilde{J}_=(z) = \tilde{G}^{-1}\partial_=\tilde{G} \, , \quad \tilde{J}_y(z) = \tilde{G}^{-1}\partial_y\tilde{G} \, . \tag{5.76}$$

The natural candidate for the super-WZNW action in the $(2,0)$ superspace is given by [176] (*cf* eq. (5.68))

$$S = \frac{k}{2\pi} \int d^3z^= \int_0^1 dy \, \text{tr} \left[\mathcal{T}\left(\tilde{J}_=\right) \tilde{J}_y \right] \, , \tag{5.77}$$

with a (z,y)-*independent* differentiation *ct*. The remarkable identity [175]

$$\delta \, \text{tr}[\mathcal{T}(\tilde{J}_=)\tilde{J}_y] = \partial_y \text{tr}[\tilde{G}^{-1}\delta\tilde{G}\mathcal{T}(\tilde{J}_=)] - \partial_= \text{tr}[\tilde{G}^{-1}\delta\tilde{G}\mathcal{T}(\tilde{J}_y)] \tag{5.78}$$

then implies that the integrand in eq. (5.77) is a *closed* 2-form w.r.t. the integration variables $(\sigma^=, y)$. This form is known as the *Kirillov-Kostant form* [172].

It is eq. (5.78) that makes it possible to rewrite the action (5.77) to the two-dimensional form as

$$S = \frac{k}{2\pi} \int d^3z^= W_A(\Phi)\partial_=\Phi^A \, , \tag{5.79}$$

where the interpolation-independent 1-form $W \equiv W_A d\Phi^A$ represents a cohomological class (modulo exact differentials).

The action (5.77) is trivially invariant under the $(2,0)$ loop group transformations

$$\tilde{G}(z,y) \to H\left(\sigma^{\ddagger}, \zeta^+, \bar{\zeta}^+\right) \tilde{G}(z,y) \, . \tag{5.80}$$

The associated algebra is nothing but the $(2,0)$ super-Kač–Moody algebra. In addition, the action (5.77) is $(2,0)$ superconformally invariant, and it can be easily coupled to the $(2,0)$ supergravity (sect. VIII.3). In the superconformal gauge, the supergravity fields decouple, and the surviving $(2,0)$ supercoordinate transformations are just those which represent the $(2,0)$ superconformal symmetry of the action (5.77).

However, this is not yet the end of the story. The action (5.77) is still *not* a physical action, if $\Phi^A(z)$ were chosen to be *general* or *unconstrained* $(2,0)$ superfields which is implicit in eq. (5.80). In order to check this, it is enough to consider the abelian case, where one gets $\partial_=\Phi^A = 0$ as the free equations of motion. To make contact with the general results about the $(2,0)$ NLSM introduced above, the $(2,0)$ superfields $\Phi^A(z)$ must be *chiral* or *anti-chiral*, $\Phi^A(z) = (\Phi^{\dagger}, \overline{\Phi}^{\bar{a}})$, in the canonical basis for the complex structure (see eq. (5.83) below), but this assignment is *not* compatible with the transformations (5.80) ! This is related to the fact that a general $(2,0)$ real scalar superfield $U(z)$ is decomposed into the three irreducible pieces,

$$U(z) = \Phi(z) + \overline{\Phi}(z) + U'(z) , \qquad (5.81)$$

where $\Phi(z)$ is chiral, $\overline{\Phi}(z)$ is anti-chiral, and $U'(z)$ is the non-chiral irreducible superfield, containing a vector $\omega'_{+}(\sigma)$ from the component decomposition (5.62) of the U superfield, and satisfying the superspace constraint $i[D_+, \overline{D}_+]U'(z) = 0$. The Lie group-valued superfield $G(z)$ cannot be chosen to be chiral while maintaining the physical meaning of the $(2,0)$ WZNW action. Therefore, we conclude that the $(2,0)$ WZNW action, written in terms of the (anti)chiral $(2,0)$ superfields, is *not* invariant under the $(2,0)$ super-AKM transformations since they do *not* preserve the complex structure. Being written in a general coordinate system defined on the NLSM group manifold, the chirality condition takes the form [176, 177]

$$G^{-1}D_{2+}G = \mathcal{J}\left(G^{-1}D_{1+}G\right) , \qquad (5.82)$$

where the $(2,0)$ superspace covariant derivatives D_{i+} associated with the anticommuting coordinates ζ_i^+ and the complex structure $\mathcal{J}$ have been introduced. In the complex coordinate system $\Phi^M = (\Phi^m, \overline{\Phi}^{\bar{m}})$, in which the complex structure $\mathcal{J}$ is canonical and the Maurer-Cartan 1-form $G^{-1}dG \equiv L^A{}_M t_A d\Phi^M$ is diagonal,

$$\mathcal{J}^A{}_B = \begin{pmatrix} i\delta^a{}_b & 0 \\ 0 & -i\delta^{\bar{a}}{}_{\bar{b}} \end{pmatrix} , \quad L^A{}_M = \begin{pmatrix} L^a{}_m & 0 \\ 0 & L^{\bar{a}}{}_{\bar{m}} \end{pmatrix} , \qquad (5.83)$$

eq. (5.82) takes the form

$$G(z) = \exp\left(\Phi^a(z)t_a + \overline{\Phi}^{\bar{a}}(z)t_{\bar{a}}\right) , \quad \overline{D}_+\Phi^a = D_+\overline{\Phi}^{\bar{a}} = 0 . \qquad (5.84)$$

Therefore, one can think of $\mathcal{J}$ as the linear (field-independent) map (or matrix) defined on the Lie algebra $\mathcal{G}$, with the properties

$$\mathcal{J}^2 = -1 , \quad \mathcal{J}^T = \mathcal{J} ,$$

$$\mathcal{N}\left(\phi,\psi\right) \equiv \left[\phi,\psi\right] - \left[\mathcal{J}(\phi),\mathcal{J}(\psi)\right] + \mathcal{J}\left(\left[\mathcal{J}(\phi),\psi\right]\right) + \mathcal{J}\left(\left[\phi,\mathcal{J}(\psi)\right]\right) = 0 \ . \qquad (5.85)$$

The 2-form $\mathcal{N}$ is called the *Nijenhuis* tensor [177], whose vanishing is necessary, in genaral, for an almost complex structure to be globally well-defined.

Being constrained by eq. (5.82), the action (5.77) is no longer $(2,0)$ super-AKM invariant, but it remains to be $(1,0)$ super-AKM invariant and it still has the $(2,0)$ superconformal symmetry. [6] Indeed, the action (5.77) in terms of the constrained superfields (5.82) can be represented in the form (5.63), while it is straightforward to integrate over one real Grassmannian coordinate ζ_2^+ there, and thus get the $(1,0)$ superspace form of this action.

Expanding G and G^{-1} in ζ_2^+ as

$$G = g + \zeta_2^+ g \mathcal{J}\left(g^{-1}D_+ g\right) \ , \quad G^{-1} = g^{-1} - \zeta_2^+ \mathcal{J}\left(g^{-1}D_+ g\right) g^{-1} \ , \qquad (5.86)$$

where the $(1,0)$ group-valued real superfield $g = G|$ and the $(1,0)$ superspace covariant derivative $D_+ \equiv D_{1+}$ have been introduced, an integration over ζ_2^+ in the action (5.77) yields

$$S = \frac{k}{2\pi} \int d^2\sigma \, (d\zeta)^- \, \mathrm{tr}\left[\left(g^{-1}D_+ g\right) \mathcal{J}\mathcal{T}\left(g^{-1}\partial_= g\right)\right.$$
$$\left. + \int_0^1 dy \, \left(\tilde{g}^{-1}D_+\tilde{g}\right) \mathcal{J}\mathcal{T}\left(\partial_=\tilde{g}^{-1}\partial_y\tilde{g} - \partial_y\tilde{g}^{-1}\partial_=\tilde{g}\right)\right] \ . \qquad (5.87)$$

The action (5.87) *differs* from the standard super-WZNW action in $(1,0)$ superspace by the factor $\mathcal{J}\mathcal{T}$, and it is *not* possible to ignore this factor by identifying $\mathcal{J}$ and $\mathcal{T}$ (then $\mathcal{J}\mathcal{T}$ would be proportional to the unit matrix), since the defining conditions (5.75) and (5.85) on the $\mathcal{T}$ and $\mathcal{J}$, resp., are *incompatible*. Nevertheless, it is still possible to prove the $(1,0)$ super-KM invariance of the action (5.87) when using, e.g., the Polyakov–Wiegmann identity.

The failure to incorporate the $(2,0)$ super-AKM symmetry within the $(2,0)$ WZNW theory does not seem to be surprising from the viewpoint of CFT: an investigation of the $N = 1$ super-AKM currents shows that they are *not* $N = 2$ primary fields w.r.t. the $N = 2$ superconformal algebra [174]. [7]

---

[6]The $(1,0)$ supersymmetric NLSM model is invariant under the $(2,0)$ extended supersymmetry if its target space admits a complex structure. A complex structure can be defined for any even-dimensional Lie algebra.

[7]The $(1,0)$ super-AKM invariance is sufficient to guarantee the integrability and finiteness of the super-WZNW theory.

Having rewritten the Lie group-valued $(2,0)$ superfield $G(z)$ to the form (5.84), we get the $(2,0)$ supersymmetric loop group transformations in the form

$$G(z) \to \exp\left[\varepsilon^{\underline{a}}(z)t_{\underline{a}} + \bar{\varepsilon}^{\underline{\bar{a}}}t_{\underline{\bar{a}}}\right]G(z) , \tag{5.88}$$

with the $\sigma^=$-independent and chiral (anti-chiral) $(2,0)$ superfield parameters $\varepsilon^{\underline{a}}$ $(\bar{\varepsilon}^{\underline{\bar{a}}})$. This is still consistent with the given parameterization *provided*

$$[t_{\underline{a}}, t_{\underline{\bar{b}}}] = 0 . \tag{5.89}$$

Eq. (5.89) holds e.g., when the group G is a product of two copies of an arbitrary Lie group with the naturally defined complex structure. The component structure of the associated $(2,0)$ supercurrent then resembles the $(1,0)$ counterpart.

The problem of constructing the most general WZNW action in the $(2,2)$ superspace is yet to be solved.

Exercises

V-7 ▷ (*) Calculate the component form of the superspace action (5.44), and make the substitution (5.50) to prove decoupling of the fermionic fields in the $(1,1)$ super-WZNW action.

V-8 ▷ Derive the $(1,0)$ supersymmetric NLSM action (5.56) by the truncation of the $(1,1)$ supersymmetric NLSM action in eq. (5.44).

V-9 ▷ (*) Compute variations of the bosonic and supersymmetric WZNW actions, and use their integrated forms to prove the Polyakov-Wiegmann identity for the (supersymmetric) WZNW theories.

V.4 Chiral rings and Landau-Ginzburg models

The $N = 2$ SCA has an important property known as the **spectral flow** [159]. To introduce it, let us define a one-parameter twisting of the $N = 2$ SCA by the boundary condition as

$$G^{\pm}(z) = e^{\pm 2\pi i \eta} G^{\pm}\left(e^{2\pi i}z\right) . \tag{5.90}$$

Eq. (5.90) implies that $T(z)$ and $J(z)$ are periodic, and they have integer modes. In particular, we have $\eta = 0 \bmod \mathbf{Z}$ for the NS-sector and $\eta = \frac{1}{2} \bmod \mathbf{Z}$ for the R-sector.

The $N = 2$ SCA spectral flow is a continuous transformation having the form

$$G_\eta^\pm(z) = z^{\pm\eta}G^\pm(z) ,$$

$$T_\eta(z) = T(z) + \eta J(z) + \frac{c\eta^2}{6z^2} ,$$

$$J_\eta(z) = J(z) + \frac{c\eta}{3z} . \tag{5.91}$$

The point is the η-deformed operators $T_\eta(z)$, $G_\eta^\pm(z)$ and $J_\eta(z)$ still satisfy the *isomorphic* $N = 2$ SCA for an arbitrary value of the parameter η. In particular, the zero-mode eigenvalues h of T and q of J are changed by the spectral flow as ($\tilde{c} \equiv c/3$)

$$h_\eta = h + \eta q + \tfrac{1}{2}\eta^2\tilde{c} , \quad q_\eta = q + \tilde{c}\eta . \tag{5.92}$$

One usually finds it useful to bosonize the $U(1)$ current as (exercise # V-10)

$$J(z) = i\sqrt{c/3}\partial\phi(z) , \tag{5.93}$$

and represent a state of charge (q_L, q_R) in the form

$$|q_L, q_R\rangle = \exp[i\sqrt{3/c}(q_L\phi_L - q_R\phi_R)]|\chi\rangle , \tag{5.94}$$

where χ is a neutral operator. In terms of the field ϕ, the spectral flow operator $\mathcal{U}_\eta$ takes the form [155]

$$\mathcal{U}_\eta = \exp[-i\eta\sqrt{c/3}(\phi_L - \phi_R)] , \tag{5.95}$$

which follows from eq. (5.92).

Allowing the spectral flow parameter η to vary from zero to $1/2$, we can establish a one-to-one correspondence between those NS primary states which are simultaneously left- and right-chiral, and the ground states of the R-sector. When η varies between integer values, the spectral flow moves NS to NS, and R to R. Furthermore, since Witten's index does not change under the spectral flow, one finds [155]

$$\mathrm{Tr}(-1)^F = \mathrm{Tr}_R\left[(-1)^{J_0-\bar{J}_0}q^{L_0-c/24}\bar{q}^{\bar{L}_0-c/24}\right]$$

$$= \mathrm{Tr}_{NS}\left[(-1)^{J_0-\bar{J}_0}q^{L_0-\frac{1}{2}J_0}\bar{q}^{\bar{L}_0-\frac{1}{2}\bar{J}_0}\right] = \sum_\kappa \exp[i\pi(q_L - q_R)] . \tag{5.96}$$

The index receives the non-vanishing contributions from either the ground states in the R-sector or the chiral primary states in the NS-sector, which is the reason for eq. (5.96), after taking into account the spectral flow. Since the difference between the charges of the NS chiral primaries and that of the R ground states is just $c/6$, the corresponding characters are related as [154, 155]

$$\text{Tr}_{\text{R}}\left[t^{J_0}\bar{t}^{\bar{J}_0}\right]\Big|_{G_0^\pm = \bar{G}_0^\pm = 0} = (t\bar{t})^{-c/6}\text{Tr}_{NS}\left[t^{J_0}\bar{t}^{\bar{J}_0}\right]\Big|_{\mathcal{K}} = (t\bar{t})^{-c/6}P(t,\bar{t}) , \tag{5.97}$$

where t and $\bar{t}$ can be regarded to be independent. Since the Ramond ground states are invariant under the charge conjugation, the polynomial $P(t,\bar{t}) = \text{Tr}_{\text{R}}[t^{J_0}\bar{t}^{\bar{J}_0}]$ has the duality property:

$$P(t,\bar{t}) = (t\bar{t})^{c/3}P(1/t,1/\bar{t}) . \tag{5.98}$$

Given an $N = 2$ SCFT, we can therefore associate with it the polynomial

$$P(t,\bar{t}) = \sum_{p,q=0}^{d} b_{p,q}t^p\bar{t}^q , \tag{5.99}$$

where $b_{p,q}$ denotes the number of chiral primary fields of charges (p,q), and $d = \bar{c}$. If all the $N = 2$ SCFT charges are integers, the polynomial (5.99) has the geometrical significance, since it can be identified [155] with the *Poincaré polynomial* to be defined by eq. (5.99) for some *complex* manifold $\mathcal{Y}$, provided the coefficients $b_{p,q}$ are dimensions of the Dolbeault cohomology groups (*Hodge numbers*) $H^{(p,q)}(\mathcal{Y})$ [40]. In particular, the well-known mathematical theorem [40] claiming that $\dim H^{(d,d)} = \dim H^{(0,0)} = 1$ now implies the existence of a unique $N = 2$ SCFT state ρ of the highest charge $(d,d) = (c/3, c/3)$ because of the uniqueness of the vacuum.

In terms of topology, the duality (5.98) is just the well-known *Poincaré duality* of the cohomology groups, $b_{p,q} = b_{\bar{c}-p,\bar{c}-q}$, known to be true for any complex manifold [40]. A complex manifold $\mathcal{Y}$ endowed with the Kählerian and Ricci-flat structure is known as a *Calabi-Yau manifold*. The Calabi-Yau manifolds are often used as the internal space manifolds for the superstring compactification. The geometrical interpretation of chiral rings implies a one-to-one correspondence between the elements of ring $\mathcal{K}$ and the harmonic (p,q) forms on $\mathcal{Y}$ representing the Dolbeault cohomology.

The chiral rings are closely related with the $N = 2$ **Landau-Ginzburg** (LG) models and *catastrophe theory*. The basic idea [178] is to use the $N = 2$ *chiral* scalar superfields Φ_i, $i = 1,\ldots,n$, satisfying the conditions $\bar{D}^+\Phi_i = \bar{D}^-\Phi_i = 0$, and a holomorphic *quasi-homogeneous* function $W(\Phi_i)$, having an isolated singularity at $\Phi_i = 0$, [8]

$$W(\lambda^{w_i}\Phi_i) = \lambda^d W(\Phi_i) , \tag{5.100}$$

---

[8]A singularity at the origin is called to be *isolated* when the n equations $\partial_i W(\Phi) = 0$ have $\Phi_j = 0$ as the only solution. The parameters w_i and d in eq. (5.100) are supposed to be integers.

in order to introduce an $N = 2$ NLSM action in the form

$$S_{\rm LG} = \int d^2z d^4\theta \, K(\Phi_i, \bar{\Phi}_i) + \left(\int d^2z d^2\theta \, W(\Phi_i) + \text{h.c.} \right) \, , \qquad (5.101)$$

and to make it conformally invariant by adjusting the *Kähler potential* $K(\Phi_i, \bar{\Phi}_i)$. It is the general (physically motivated) assumption of the whole LG approach [179] that (i) there exists a RG fixed point for the theory (5.101), and (ii) only the Kähler potential K changes under the RG flow towards a RG fixed point, whereas the polynomial *superpotential* W remains unchanged. Because of the (perturbative) *non-renormalization* theorem related with the $N = 1$ supersymmetry in four dimensions, and its $N = 2$ dimensionally-reduced counterpart in two dimensions [180], one actually needs to assume about a *non-perturbative* character of the non-renormalization theorem in supersymmetry. The theory (5.101) at its RG fixed point is called an $N = 2$ *Landau-Ginzburg* (LG) model. The $N = 2$ superconformal LG models can be identified with the $N = 2$ Toda-like CSFT's (Ch. XI), which shows a way of deriving their free field representations [181].

The ring $\mathcal{K}$ of chiral primary fields in the LG-corresponding $N = 2$ SCFT is isomorphic to the local ring of $W(\Phi)$, which is defined as the space $\mathcal{S}$ of all monomials of Φ_i modulo $\partial_j W(\Phi_i)$. According to the equations of motion in the LG theory, $\bar{D}^- \bar{D}^+ \bar{\Phi}_j \sim \partial_j W(\Phi_i)$, the polynomials $\partial_j W(\Phi_i)$ correspond to the descendant fields, so that one has [155]

$$\mathcal{K} = \frac{\mathcal{S}[\Phi_i]}{[\partial_j W(\Phi_i)]} \, . \qquad (5.102)$$

The dimension of the ring, $\dim \mathcal{K}$, is called the *multiplicity* of the LG potential W. Because of the left-right symmetry of charges of the LG chiral primary fields (exercise # V-14), a Poincaré polynomial is actually dependent on $t\bar{t}$ only, and its explicit form is given by the Poincaré polynomial of the corresponding singularity, known from the catastrophe theory [182, 183] and having the form (after replacing $t\bar{t} \to t^d$)

$$P(t) = \text{Tr}_{\mathcal{K}} \left[t^{dJ_0} \right] = \prod_{i=1}^{n} \frac{(1 - t^{d-w_i})}{(1 - t^{w_i})} \, . \qquad (5.103)$$

The Poincaré polynomial (5.103) corresponding to an $N = 2$ LG SCFT has to be finite, and with positive coefficients, which places the nessesary (but not sufficient [182, 183]) restrictions on the numbers (w_i, d).

Eqs. (5.97) and (5.103) imply for the Witten index of an $N = 2$ LG model the following expression [155]:

$$\text{Tr}(-1)^F = P(t=1) = \dim \mathcal{K} = \prod_{i=1}^{n} \frac{d-w_i}{w_i} = \prod_{i=1}^{n} \left(\frac{1}{q_i} - 1 \right) . \tag{5.104}$$

The chiral primary state ρ of the highest charge corresponds to the leading term in the Poincaré polynomial at large t. Eq. (5.103) gives the $U(1)$ charge and the conformal dimension of this state [155, 182, 183],

$$q_\rho = \frac{1}{d} \sum_{i=1}^{n} (d-2w_i) , \quad h_\rho = \sum_{i=1}^{n} (\tfrac{1}{2} - q_i) . \tag{5.105}$$

Since the state ρ also saturates the inequality in eq. (5.20), one finds [178, 155]

$$c = 6h_\rho = 6 \sum_{i=1}^{n} (\tfrac{1}{2} - q_i) , \tag{5.106}$$

The $N = 2$ LG models obviously describe only a part of all $N = 2$ SCFT's, namely the simplest ones having only one conjugate pair of chiral rings generated by products of finitely-many chiral primary fields with *equal* left and right $U(1)$ charges $q_L = q_R$ [155].

The LG superpotentials W relevant for the $N = 2$ unitary minimal models of central charge $d \equiv c/3 = k/(k+2)$, where $k = 0, 1, \ldots$, are labeled by a Lie group G of the A-D-E type, with the Coxeter number $h_G = k + 2$. They read [182, 183]

$$
\begin{aligned}
A_n &: \ W = X^{n+1} , \\
D_n &: \ W = X^{n-1} + XY^2 , \\
E_6 &: \ W = X^3 + Y^4 , \\
E_7 &: \ W = X^3 + XY^3 , \\
E_8 &: \ W = X^3 + Y^5 .
\end{aligned}
\tag{5.107}
$$

These LG potentials are the only ones which have no marginal deformations [182, 183]. The rings associated with eq. (5.107) are easily determined. For instance, the A_n-ring is given by polynomials in X modulo $X^n = 0$, so that its basic independent elements are $(1, X, X^2, \ldots, X^{n-1})$. Similarly, the E_6-ring comprises the polynomials in X and Y modulo $\partial_X W = X^2 = 0$ and $\partial_Y W = Y^3 = 0$, and its basic independent elements are $(1, X, Y, XY, Y^2, XY^2)$.

The $U(1)$ charges of the monomials follow from scaling behaviour of the fundamental fields, $X_i \to \lambda^{q_i} X_i$, under the condition that the superpotential W has charge one.

Hence, the fields Φ_i have charges $q_i = i/(k+2)$, where $i \in \mathbf{Z}$ and $0 \le i \le k$. Quite generally, given a LG model labeled by a simply-laced Lie group G, one finds that $(i+1)$ is just the exponent of G, related to the order $(i+2)$ of the corresponding Casimir operator of G. The data about the Coxeter numbers and the exponents of the simply-laced Lie groups is gathered in Table 4 (sect. XII.2).

Exercises

V-10 ▷ (*) Calculate the normalization of the bosonized current in eq. (5.93) from the $N = 2$ SCA.

V-11 ▷ What are the bosonized expressions for all the $N = 2$ SCA generators in terms of a single circle-valued scalar bosonic field of central charge $c = 1$?

V-12 ▷ (*) Consider the scalar bosonic field from the exercise # V-11 with the allowed winding modes having the form $\exp[i(n_L\phi_L - n_R\phi_R)/\sqrt{12}]$, $n_L - n_R = 0 \bmod 6$, and determine the chiral ring structure, Witten's index and the Poincaré polynomial.

V-13 ▷ (*) Let the $N = 2$ SCA be realized by one complex boson $x(z)$ and one complex fermion $\psi(z)$, both defined on a two-dimensional torus,

$$G^+(z) = \psi^*\partial x , \quad G^-(z) = \psi\partial x^* , \quad J(z) = \psi^*\psi , \tag{5.108}$$

and similarly for the right-movers. What are the corresponding chiral primary states, the chiral ring and the Poincaré polynomial ?

V-14 ▷ Use eqs. (5.100) and (5.101) to show that the LG chiral and anti-chiral primary fields have the same $U(1)$ charges $q_i = w_i/d$.

V-15 ▷ (*) Calculate a basis for the E_8-ring in the space of polynomials of two variables. *Hint*: use eq. (5.107).

Chapter VI

Coset Models

The enveloping Virasoro algebra associated with an AKM algebra has the central charge $c = c_G$ given by eq. (4.38). The central charge values are therefore restricted to the interval

$$\operatorname{rank} G \leq c_G \leq |G| \tag{6.1}$$

for any Lie group G. In particular, for $G = SU(2)$, one has

$$c_{SU(2)} = \frac{3k}{k+2}, \tag{6.2}$$

so that $1 \leq c_{SU(2)} \leq 3$. Therefore, the SS group construction is unable to describe the minimal models having $c < 1$. For the superstring compactification program (Ch. VIII), compact manifolds with a non-vanishing Euler characteristic χ are needed, but all the group manifolds have $\chi = 0$. Hence, we want to break somehow the SS-constructed stress tensor (4.22) into pieces, each with a smaller central charge. The general idea is to generalize the SS construction to the *cosets* G/H, where H is a subgroup of the compact Lie group G. The corresponding **coset models** give a variety of RCFT's (maybe, all of them) and, in particular, the minimal models in the very explicit form.

VI.1 Goddard-Kent-Olive construction

The **Goddard-Kent-Olive** (GKO) method [163] of constructing the unitary representations of the Virasoro algebra on cosets is capable to produce all the minimal models, and many other RCFT's as well. This method always gives rational values of central charge, i.e. the RCFT's, in fact.

169

To introduce the method, let $J_G^a(z)$ be the G-currents, $a = 1, \ldots, |G|$, and $J_H^i(z)$ the H-currents, $i = 1, \ldots, |H|$. We assume that the group indices run over the adjoint representation and the first $|H|$ currents in $\{J_G^a\}$ just represent the H-currents $\{J_H^i\}$. A normalization of structure constants is supposed to respect the highest root norm $\psi^2 = 2$, so that we have

$$J_G^a(z)J_G^b(w) \sim \frac{k_G \delta^{ab}}{(z-w)^2} + \frac{if^{abc}}{z-w} J_G^c(w) \, ,$$

$$J_H^i(z)J_H^j(w) \sim \frac{k_H \delta^{ij}}{(z-w)^2} + \frac{if^{ijk}}{z-w} J_H^k(w) \, . \tag{6.3}$$

The level k_H is generically determined by an embedding of H into G. For instance, if the simple roots of H form a subset of the simple roots of G, then $k_H = k_G$, or if $G = G_1 \otimes G_2$ and H is the diagonal subgroup, then $k_H = k_1 + k_2$.

We can now exploit the SS construction (sect. IV-2) to construct the G and H stress tensors, T_G and T_H resp., in the form

$$T_G(z) = \frac{1/2}{k_G + \bar{h}_G} \sum_{a=1}^{|G|} : J_G^a(z)J_G^a(z) : \, ,$$

$$T_H(z) = \frac{1/2}{k_H + \bar{h}_H} \sum_{i=1}^{|H|} : J_H^i(z)J_H^i(z) : \, , \tag{6.4}$$

with the Virasoro central charges

$$c_G = \frac{k_G|G|}{k_G + \bar{h}_G} \, , \quad c_H = \frac{k_H|H|}{k_H + \bar{h}_H} \, . \tag{6.5}$$

Since the currents J_H^i of dimension $h = 1$ are all the primary fields w.r.t. both stress tensors T_G and T_H of eq.(6.4), we have

$$T_G(z)J_H^i(w) \sim \frac{J_H^i(w)}{(z-w)^2} + \frac{\partial J_H^i(w)}{z-w} \, ,$$

$$T_H(z)J_H^i(w) \sim \frac{J_H^i(w)}{(z-w)^2} + \frac{\partial J_H^i(w)}{z-w} \, . \tag{6.6}$$

Thereby, the OPE of $T_{G/H} \equiv T_G - T_H$ with J_H^i is non-singular. The OPE of $T_{G/H}$ with T_H is also non-singular since the T_H was constructed in eq. (6.4) in terms of the H-currents J_H^i alone. Hence, we get the orthogonal decomposition of the Virasoro algebra generated by T_G into the two mutually commuting Virasoro subalgebras:

$$T_G = (T_G - T_H) + T_H = T_{G/H} + T_H \, , \quad [T_{G/H}, T_H] = 0 \, . \tag{6.7}$$

In terms of the modes, this means

$$\left[L_m^{G/H}, J_{H,n}^i\right] = \left[L_m^{G/H}, L_n^H\right] = \left[L_m^G - L_m^H, L_n^H\right] = 0 \; . \tag{6.8}$$

Therefore, we find

$$\left[L_m^{G/H}, L_n^{G/H}\right] = \left[L_m^G - L_m^H, L_n^G - L_n^H\right] = \left[L_m^G, L_n^G\right] - \left[L_m^H, L_n^H\right]$$

$$= (m - n)L_{m+n}^G + \frac{c_G}{12}(m^3 - m)\delta_{m+n,0} - (m - n)L_{m+n}^H - \frac{c_H}{12}(m^3 - m)\delta_{m+n,0}$$

$$= (m - n)L_{m+n}^{G/H} + \frac{c_{G/H}}{12}(m^3 - m)\delta_{m+n,0} \; , \tag{6.9}$$

i.e. $L_m^{G/H}$ satisfy the Virasoro algebra of central charge [163]

$$c_{G/H} = c_G - c_H = \frac{k_G|G|}{k_G + \tilde{h}_G} - \frac{k_H|H|}{k_H + \tilde{h}_H} \; . \tag{6.10}$$

The stress tensor $T_{G/H}(z)$ defines a representation of conformal symmetry which is obviously *unitary* since it is realized on a subspace of the unitary representation $T_G(z)$.

A particularly simple example is provided by the coset $\widehat{SU}(2)_k/\widehat{U(1)}$. In this case, eqs. (6.2) and (6.10) give the central charge

$$c_{SU(2)/U(1)} = \frac{3k}{k + 2} - 1 = \frac{2(k - 1)}{k + 2} \; . \tag{6.11}$$

Choosing $k = 2$ corresponds to the Ising model with $c = 1/2$. The choice of $k = 1$ yields zero on the r.h.s. of eq. (6.11) which means the quantum equivalence, $T_G = T_H$, between the two stress tensors: the SS-constructed stress tensor $T_{SU(2)}$ and the stress tensor of a single free (chiral) boson [107]. This simple observation can be easily generalized to the vertex operator construction of sect. IV-4. Taking above any simply-laced Lie group G divided by its Cartan subgroup, $G/U(1)^{r_G}$, results in $c_G - r_G = 0$ at level $k = 1$ because of eqs. (4.38) and (4.73). If G were not simply-laced or $k > 1$, $T_{G/U(1)^{r_G}}$ would be the parafermionic stress tensor of ref. [120]. The classification of the so-called *conformal embeddings* generating $c_{G/H} = 0$, can be found in refs. [184, 185] (see ref. [77] also).

The interesting examples of GKO construction are given by the cosets with a *diagonal* embedding, e.g.

$$\widehat{G}/\widehat{H} = \widehat{SU}(2)_k \otimes \widehat{SU}(2)_l / \widehat{SU}(2)_{k+l} \; , \tag{6.12}$$

of central charge

$$c_{G/H} = \frac{3k}{k + 2} + \frac{3l}{l + 2} - \frac{3(k + l)}{k + l + 2}$$

$$= 1 - \frac{6l}{(k+2)(k+l+2)} + \frac{2(l-1)}{l+2} \ . \tag{6.13}$$

Taking $l = 1$ in eq. (6.13) gives back the discrete series of central charge values in eq. (2.65) which correspond to the unitary minimal models ($p = k + 2$) ! It is actually the GKO approach that proves the existence of unitary Virasoro representations for all the discrete values of $c < 1$ and h resulting from an analysis of the Kač determinant, i.e. the unitary minimal models [163].

The $N = 1$ superconformal central charge series (5.9) is similarly obtained by taking $l = 2$ in eq. (6.13):

$$c_{G/H} = \frac{3k}{k+2} + \frac{3}{2} - \frac{3(k+2)}{(k+2)+2} = \frac{3}{2}\left(1 - \frac{8}{(k+2)(k+4)}\right) \ . \tag{6.14}$$

Again, this observation can be used for the unitarity proof of the $N = 1$ minimal models [163]. The appearance of the superconformal symmetry here is not surprising since we already know from sect. V-2 the way to construct the associated supercurrent (exercise # VI-1). [1] It is not difficult to show that the GKO construction based on the cosets

$$\frac{\widehat{G}_k \otimes \widehat{SO(|G|)}_1}{G_{k+\tilde{h}_G}} \ , \tag{6.15}$$

where $\tilde{h}_G$ is the dual Coxeter number, gives rise to the $N = 1$ SCFT's having super-AKM symmetry. Similarly, using the cosets

$$\frac{\widehat{SO(|G|)}_1 \otimes \widehat{SO(|G|)}_1}{G_{2\tilde{h}_G}} \tag{6.16}$$

for a GKO construction results in the unitary minimal $N = 2$ SCFT's considered in refs. [150, 151, 152, 153, 154].

The coset models (6.12) at $l > 2$ are also unitary, and they give rise to the new RCFT's with more complicated chiral operator algebras discussed in refs. [57, 186, 187, 188]. The coset models are, actually, very suitable for constructing more general RCFT's than the minimal models, since the G/H operators are always H-singlets and, when acting in a given H module, they always represent the conserved currents. This provides the reason to identify the RCFT Virasoro primaries with the full set of chiral currents [21]. When $c > 1$, there are the infinitely many chiral currents which are H-singlets [187].

---

[1] The AKM algbera $\widehat{SU(2)}_2$ of central charge $c = 3/2$ can be realized in terms of three free fermions in the adjoint representation of $SU(2)$ (exercise # VI-2).

It is worthy to mention here that GKO constructions are *not* unique. For example, there are different choices of cosets for the following $N = 1$ minimal models:

$$
\begin{aligned}
(k = 3, \quad c = 81/70) &: \quad (\widehat{E_8})_2/\widehat{SU}(3)_2 \otimes (\widehat{E_6})_2 , \\
(k = 7, \quad c = 91/66) &: \quad (\widehat{E_8})_2/(\widehat{G_2})_2 \otimes (\widehat{F_4})_2 , \\
(k = 8, \quad c = 7/5) &: \quad \widehat{Sp(6)}_1 \otimes \widehat{Sp(6)}_1/\widehat{Sp(6)}_2 .
\end{aligned}
\tag{6.17}
$$

Under the orthogonal decomposition of the Virasoro algebra $T_G = T_{G/H} + T_H$, any representation space (module) of $\widehat{G}_{k_G}$ of highest weight λ_G^a decomposes into a direct sum of irreps of $\widehat{H}_{k_H}$ as

$$
|c_G, \lambda_G\rangle = \oplus_j \left| c_{G/H}, h_{G/H}^j \right\rangle \otimes \left| c_H, \lambda_H^j \right\rangle ,
\tag{6.18}
$$

where $\left| c_{G/H}, h_{G/H}^j \right\rangle$ denotes the highest-weight irrep of $T_{G/H}$ of the L_0-eigenvalue $h_{G/H}$. The value of $h_{G/H}$ is clearly dependent on λ_G^a and λ_H^j. The states in a Virasoro representation of $T_{G/H}$ belong to the highest-weight representations w.r.t. H. Therefore, the associated affine characters should satisfy the relation

$$
\chi_G \equiv \chi_{\lambda_G^a}^{k_G}(\theta^i, \tau) = \sum_j \chi_{h_{G/H}(\lambda_G^a, \lambda_H^j)}^{c_{G/H}}(\tau)\chi_{\lambda_H^j}^{k_H}(\theta^i, \tau) \equiv \chi_{G/H} \cdot \chi_H ,
\tag{6.19}
$$

which can be used to determine the characters $\chi_{G/H}$. [2] Since we already know from sect. IV-5 about the modular transformation laws of the G and H characters, we can now use eq. (6.19) to deduce modular transformation properties of the G/H characters and construct modular invariants in terms of them, by using the known invariants in terms of the group characters. Equations of the type (6.19), expressing the characters of G in terms of those of H, are known as **branching rules**. It is generally believed that *all* RCFT's can be obtained by the coset constructions, since the branching 'coefficients' (they are, in fact, functions) in equations like eq. (6.19) behave as the characters of some CFT. Unfortunately, not all of the characters of the coset models are equal to the branching functions [77].

The $\widehat{SU}(2)_k/\widehat{U(1)}$ level-k parafermionic theory has the alternative description as the coset model

$$
\widehat{SU(k)}_1 \otimes \widehat{SU(k)}_1/\widehat{SU(k)}_2 ,
\tag{6.20}
$$

where $\widehat{SU(k)}_2$ is the diagonal subgroup. It is a simple exercise (exercise # VI-3) to verify that the central charge associated with the GKO construction (6.18) is the same

---

[2]The arguments θ^i in eq. (6.19) are supposed to be restricted to a Cartan subalgebra of H.

as that of the $\widehat{SU(2)}_k/\widehat{U(1)}$ parafermions. It is an advantage of the representation (6.20) that the parafermionic currents can be constructed in terms of primaries of the AKM algebra $\widehat{SU(k)}_1$ [21]. Indeed, the parafermionic algebra is generated by two currents χ_1 and $\chi_1^\dagger$, which can now be represented as bilinears in terms of the primary fields ϕ^a and ϕ_a transforming in the fundamental representation of $SU(k)$,

$$\chi_1(z) = \frac{1}{\sqrt{k}}\tilde{\phi}_a(z)\phi^a(z) \ ,$$

$$\chi_1^\dagger(z) = \frac{1}{\sqrt{k}}\tilde{\phi}^a(z)\phi_a(z) \ ,$$

(6.21)

More parafermionic fields can be introduced this way, when using other $SU(k)$ representations [21],

$$\chi_i(z) = \sqrt{\frac{(k-i)!}{i!k!}}\tilde{\phi}_{[a_1\cdots a_i]}(z)\phi^{[a_1\cdots a_i]}(z) \ ,$$

$$\chi_i^\dagger(z) = \sqrt{\frac{(k-i)!}{i!k!}}\tilde{\phi}^{[a_1\cdots a_i]}(z)\phi_{[a_1\cdots a_i]}(z) \ ,$$

(6.22)

for $i = 1,\ldots,k-1$. The (fractional) dimensions of the χ-fields above are given by

$$h_i = \frac{i(k-i)}{k} \ ,$$

(6.23)

which exactly correspond to the conformal dimensions of the parafermionic fields introduced in eq. (4.74). The operator algebra of $SU(k)$ primary fields implies the following parafermionic OPE's:

$$\chi_i(z)\chi_j(w) \sim c_{ij}(z-w)^{-2ij/k}\left[\chi_{i+j}(w) + \ldots\right] \ ,$$

$$\chi_i(z)\chi_i^\dagger(w) \sim (z-w)^{-2i(k-i)/k}\left[I + \frac{i(k-i)(k+2)}{k(k-1)}(z-w)^2T(w) + \ldots\right] \ ,$$

(6.24)

which give the algebra of the $\widehat{SU(2)}_k/\widehat{U(1)}$ parafermionic theory, which was originally postulated in ref. [120].

Eq. (6.20) could be further generalized by replacing $SU(k)$ for another Lie group G. When $G = SO(n)$, this leads to the $c = 1$ model which is equivalent to the single chiral boson compactified on a circle of radius $r = \sqrt{n}/2$. When $G = E_6$, one gets the tricritical Z_3 Potts model of central charge $c = 6/7$, whereas the coset construction (6.20) with $G = E_7$ leads to the tricritical Ising model of central charge $c = 7/10$. Other interesting models can be constructed by taking tensor products of parefermionic theories [21, 189].

Exercises

VI-1 ▷ (*) Derive the supercurrent of the $N = 1$ minimal models in the GKO approach (6.14). *Hint:* use the ansatz

$$G(z) =: \left[J^a(z) + \beta \bar{J}^a(z) \right] \phi^a(z) : ,\tag{6.25}$$

where J^a is the current of the first $SU(2)$ (see eq. (6.12)), $\bar{J}^a$ is the current of the second $SU(2)$, ϕ^a is the primary spin-1 field of the first $SU(2)$, and β is the constant to be determined.

VI-2 ▷ (*) Realize the AKM algbera $\widehat{SU}(2)_2$ of central charge $c = 3/2$ in terms of three free fermions transforming in the adjoint representation of $SU(2)$.

VI-3 ▷ Verify that the central charge of the GKO construction (6.20) is the same as that of the $SU(2)_k/U(1)$ parafermionic theory.

VI.2 Kazama-Suzuki construction

The $N = 2$ superconformal models are particularly needed to describe the internal part of superstring compactification (sect. VIII.2). The solvable unitary SCFT's of this type are the $N = 2$ minimal models of central charge (5.13). One of the ways to construct them is to take the $SU(2)$ parafermionic theory of Fateev and Zamolodchikov[120, 151] that can be obtained from the two distinct GKO constructions,

$$\frac{\widehat{SU}(2)_k}{\widehat{U}(1)} \quad \text{or} \quad \frac{\widehat{SU}(k)_1 \otimes \widehat{SU}(k)_1}{\widehat{SU}(k)_2} ,\tag{6.26}$$

and combine it with a free scalar CFT. The total central charge results in the desired value given by eq. (5.13) in both cases, *viz.*

$$c = \frac{2(k-1)}{k+2} + 1 = \frac{3k}{k+2} ,\tag{6.27a}$$

and

$$c = 2(k-1) - \frac{2(k^2-1)}{k+2} + 1 = \frac{3k}{k+2} ,\tag{6.27b}$$

respectively.

However, this is not the whole story. We still need to construct generators of the $N = 2$ SCA. A part of this problem can be solved by using the $N = 1$ supersymmetric generalization of the SS construction already described at the end of sect. V.2. The next problem is to find the conditions which would tell us when the so-constructed $N = 1$ SCFT's actually have the $N = 2$ extended superconformal symmetry. It has been shown in refs. [190, 191] that this distinguishes the subclass of cosets known as *hermitian symmetric spaces*. [192]. The **Kazama-Suzuki** (KS) construction generalizes that of the $N = 2$ minimal models. In the rest of this section, we are going to describe the KS construction in some detail.

Let G/H be a coset, where H is a semi-simple subgroup of G. Let f^{ijk} be structure constants of H. We continue to use here the notation for group indices adopted in the previous section: *early* lower case Latin letters are used for G-indices, while *middle* lower case Latin letters are used for H-indices. In addition, we introduce lower case Latin letters *with bars* to denote G/H-indices.

The first task is to find an orthogonal decomposition of the $N = 1$ SCA associated with the group G,

$$
\begin{aligned}
T_G(z) &= T_H(z) + T_{G/H}(z) \ , \\
G_G(z) &= G_H(z) + G_{G/H}(z) \ ,
\end{aligned}
\tag{6.28}
$$

where the H- and G/H- currents are mutually commuting. To achieve this, let us define the H-currents $J^i(z)$ according to eq. (5.38),

$$
J^i(z) = \tilde{J}^i(z) - \frac{i}{2k} f^{imn} j^m(z) j^n(z) \ ,
\tag{6.29}
$$

where the current $\tilde{J}^i(z)$ has been introduced as

$$
\tilde{J}^i(z) \equiv \hat{J}^i(z) - \frac{i}{2k} f^{i\bar{b}\bar{c}} j^{\bar{b}}(z) j^{\bar{c}}(z) \ .
\tag{6.30}
$$

Following eq. (5.29a), we now define the stress tensor $T_H(z)$ in the form

$$
T_H(z) = \frac{1}{2k} \left[\tilde{J}^i(z) \tilde{J}^i(z) - j^i(z) \partial j^i(z) \right] \ .
\tag{6.31}
$$

Similarly, in accordance with eq. (5.29b), we can introduce the supercurrent $G_H(z)$ as

$$
G_H(z) = \frac{1}{k} \left[j^i(z) \tilde{J}^i(z) - \frac{i}{6k} f^{imn} j^i(z) j^m(z) j^n(z) \right] \ .
\tag{6.32}
$$

Defining now $T_{G/H}(z)$ and $G_{G/H}(z)$ by eq. (6.28), it is not difficult to check that this really gives us the desired *orthogonal* decomposition. The point is that $T_{G/H}(z)$ and

$G_{G/H}(z)$ defined this way commute with the $J^i(w)$ and $j^i(w)$. The stress tensor $T_{G/H}$ can be explicitly written down, in the form [191]

$$T_{G/H} = \frac{1}{2k} \left[: \hat{J}^{\bar{a}} \hat{J}^{\bar{a}} : -\frac{\hat{k}}{k} : j^{\bar{a}} \partial j^{\bar{a}} : +\frac{i}{k} : \hat{J}^i f^{i\bar{b}\bar{c}} j^{\bar{b}} j^{\bar{c}} : \right.$$

$$\left. -\frac{1}{k} f_{\bar{a}\bar{c}\bar{d}} f_{\bar{b}\bar{c}\bar{d}} : j^{\bar{a}} \partial j^{\bar{b}} : -\frac{1}{4k^2} f_{\bar{a}\bar{b}\bar{c}} f_{\bar{a}\bar{d}\bar{e}} : j^{\bar{b}} j^{\bar{c}} j^{\bar{d}} j^{\bar{e}} : \right] , \tag{6.33}$$

where the four-fermion normally-ordered operator in the current product $\tilde{J}^i(z) \tilde{J}^i(z)$ has been replaced by the two-fermion normally-ordered product, because of their quantum equivalence [161]. The central charge of the resulting G/H Virasoro algebra is

$$c_{G/H} = c_G - c_H = \frac{3}{2} \dim(G/H) + \frac{1}{2k} [C_A(H) \dim(H) - C_A(G) \dim(G)]$$

$$= \frac{3\hat{k}}{2k} \dim(G/H) + \frac{1}{4k} f_{\bar{a}\bar{b}\bar{c}} f_{\bar{a}\bar{b}\bar{c}} . \tag{6.34}$$

For a *symmetric* space G/H we find, in particular

$$T_{G/H}(z) = \frac{1}{2k} \left[: \hat{J}^{\bar{a}}(z) \hat{J}^{\bar{a}}(z) : -\frac{\hat{k}}{k} : j^{\bar{a}}(z) \partial j^{\bar{a}}(z) : +\frac{i}{k} f_{\bar{a}\bar{b}\bar{c}} \hat{J}^i(z) j^{\bar{b}}(z) j^{\bar{c}}(z) \right] ,$$

$$G_{G/H} = \frac{1}{k} j^{\bar{a}}(z) \hat{J}^{\bar{a}}(z) . \tag{6.35}$$

The simplest non-trivial example is provided by the case of $\widehat{SU(2)}_{\hat{k}}/\widehat{U(1)}$, with

$$T = \frac{1/2}{\hat{k}+2} \left[: (\hat{J}^1)^2 : + : (\hat{J}^2)^2 : \right] + \frac{i}{(\hat{k}+2)^2} \hat{J}^3 j^1 j^2 - \frac{\hat{k}/2}{(\hat{k}+2)^2} \left[: j^1 \partial j^1 + j^2 \partial j^2 : \right] ,$$

$$G \equiv G^1 = \frac{1}{\hat{k}+2} \left[j^1 \hat{J}^1 + j^2 \hat{J}^2 \right] , \tag{6.36}$$

and the central charge

$$c = \frac{3}{2} \cdot 2 + \frac{1}{\hat{k}+2} (0 - 2 \cdot 3) = \frac{3\hat{k}}{\hat{k}+2} . \tag{6.37}$$

We are now in a position to identify those coset spaces for which the GKO-constructed SCFT's actually possess the $N = 2$ extended supersymmetry. Since the $N = 2$ SCA has the second supercurrent $G^2(z)$ and the $U(1)$ current $J(z)$ beyond the content of the $N = 1$ SCA, all we need is to construct new currents $G^2(z)$ and $J(z)$ in terms of the given currents $\hat{J}(z)$ and $j(z)$.

Table 1. The hermitian symmetric spaces, and the (Virasoro) central charges of the GKO-constructed $N = 2$ SCFT's, $\hat{k} = 1, 2, 3, \ldots$

G/H	$c_{G/H}$
$SU(n+m)/SU(m) \times SU(n) \times U(1)$	$3\hat{k}mn/(\hat{k}+m+n)$
$SO(n+2)/SO(n) \times SO(2)$	$3\hat{k}n/(\hat{k}+n) \quad n \geq 2$
$SO(3)/SO(2)$	$3\hat{k}/(\hat{k}+2)$
$SO(2n)/SU(n) \times U(1)$	$3\hat{k}n(n-1)/2(\hat{k}+2n-2)$
$Sp(2n)/SU(n) \times U(1)$	$3\hat{k}(n+1)/2(\hat{k}+n+1)$
$E_6/SO(10) \times U(1)$	$48\hat{k}/(\hat{k}+12)$
$E_7/E_6 \times U(1)$	$81\hat{k}/(\hat{k}+18)$

In their original papers, Kazama and Suzuki [190, 191] used the general ansatz for the currents in question, based on dimensional and analytic grounds. The equations they obtained by imposing the (anti)commutation relations of the $N = 2$ SCA imply that the coset space must be a special Kähler manifold, namely, the *hermitian symmetric space* [5, 192]. The list of such spaces and central charges of the associated $N = 2$ SCFT's are given in Table 1.

When rank G = rank H, the space G/H is a Kähler manifold. Generally speaking, a *Kähler* manifold is simply a Riemannian complex manifold whose (Christoffel) connection is consistent with its complex structure. In other words, a $2n$-dimensional Kähler manifold is a Riemannian manifold with $U(n)$ holonomy.

Instead of reproducing the full proof [191], we want to illustrate it on the example of $\widehat{SU(2)}_{\hat{k}}/\widehat{U(1)}$ coset, where the natural ansatz for the $U(1)$-current takes the form

$$J(z) = \alpha \hat{J}^3(z) + i\beta j^1(z)j^2(z) , \tag{6.38}$$

with some coefficients α and β to be determined. Requiring the OPE (5.11) between $T(z)$ of eq. (6.36) and $J(z)$ of eq. (6.38) already gives *two* homogeneous equations on α and β, since

$$T(z)J(w) \sim \frac{1}{(z-w)^2}\left[\left(\frac{2\alpha}{k}+\beta\right)\hat{J}^3 + i\left(\frac{2\alpha\hat{k}}{k^2}+\frac{\beta\hat{k}}{k}\right)j^1 j^2\right] + \ldots \tag{6.39}$$

Fortunately, they are consistent and yield $\alpha\hat{k} = \beta k$. Next, we examine the OPE (5.11) between $J(z)$ and $G^1(z)$, in order to guess first a generic form of $G^2(z)$, and to check then the OPE between $G^2(z)$ and $J(z)$. This results in

$$G^2(z) = \frac{1}{k}\left[j^1 \hat{J}^2 - j^2 \hat{J}^1\right] , \quad 2\alpha + \beta k = 1 . \tag{6.40}$$

Thereby, we conclude that $\alpha = 1/k$ and $\beta = \hat{k}/k^2$. The OPE of $J(z)$ with itself provides now the final check for the presence of the $N = 2$ SCA in the theory under consideration.

There is even more general class of solutions for the cosets G/H admitting the $N = 2$ SCA, among *non-symmetric* spaces [193]. In particular, it is always the case when H is chosen to be a Cartan subgroup of G.

As was already noticed in the previous section, the character relations like eq. (6.19) could also be used in the KS construction. They allow us to deduce modular transformation properties of the KS characters in terms of those of the AKM characters. In particular, the simple left-right symmetric diagonal combination of the coset characters is always modular invariant. A Lagrangian formulation of the $N = 2$ KS models is based on the gauged $(2,2)$ supersymmetric WZNW theories [194].

Exercise

\# VI-4 ▷ Complete the calculation of eqs. (6.39) and (6.40).

VI.3   KS construction and LG models

Some of the KS coset constructions introduced in the previous section can be represented by the LG models (sect. V.4), while their chiral primary rings can be identified with the cohomology rings of coset manifolds [155]. First of all, they need to satisfy the charge condition $q_L - q_R \in \mathbf{Z}$, which is certainly the case when G/H is a Kähler manifold. For the Kähler manifolds G/H, the Lie algebra $\mathcal{G}$ of G can be conveniently decomposed as

$$g = h \oplus t_+ \oplus t_- , \tag{6.41}$$

with t_+ and t_- forming closed subalgebras. As far as the hermitian symmetric Kählerian spaces are concerned, the Lie algeras t_+ and t_- are abelian. In addition, the group G should be a simply-laced group.

The G/H superconformal model can be represented by the coset [155]

$$\frac{\widehat{G}_k \otimes \widehat{SO(2d)}_1}{\widehat{H}_{k+\hbar_G-\hbar_H}} , \tag{6.42}$$

where $d = \frac{1}{2}\dim(G/H)$. The AKM algebra of $\widehat{SO(2d)}_1$ is usually chosen to be realized by d complex fermions ψ^α, where α runs over all roots of t_+.

Because of the known correspondence between the chiral primary states and the ground states $|\phi\rangle$ of Ramond sector in $N = 2$ SCFT's (sect. V.4), the former can be found among solutions to the equation

$$G_0^{\pm} |\phi\rangle = 0 \,, \tag{6.43}$$

where $G^{\pm}$ are the $N = 2$ supersymmetry generators (see the previous section). The space where the operators $G_0^{\pm}$ act is the Hilbert space of the superconformal G model. The non-trivial and non-equivalent solutions to eq. (6.43) can be identified with the non-trivial 'cohomology elements' defined *modulo* action of H currents. The ground state of the superconformal G model transforms in a finite-dimensional $\mathcal{G}$-representation of highest weight Λ tensored with two fundamental (spinor) representations of $so(2d)$. The spinor representation can be obtained by acting with components of ψ on the ground state spinor of the form $\left|-\frac{1}{2}, \ldots, -\frac{1}{2}\right\rangle$, where the $so(2d)$ spinor weights have been indicated [155]. The Ramond ground states then correspond to the exterior algebra $\wedge t_+$ of ψ's. Denoting V_Λ a finite-dimensional Λ-module of $\mathcal{G}$, the corresponding finite-dimensional Hilbert space is given by

$$\mathcal{H} = \wedge t_+ \otimes V_\Lambda \,. \tag{6.44}$$

One can think of ψ's as forming a basis for holomorphic differential forms. Finding the solutions to eq. (6.44) can then be treated as the cohomology problem in the *Lie algebra cohomology* of t_+, whose coefficients are in the representation Λ of $\mathcal{G}$ [192, 195]. Therefore, the non-trivial and non-equivalent solutions to eq. (6.43) are indeed encoded by the corresponding cohomology groups $H^*(t_+, V_\Lambda)$ which are representations of H. To identify the states of the G/H theory, one has to decompose $H^*(t_+, V_\Lambda)$ into the H-irreps and pick up just one representative from each irrep. A number of such H-irreps in $H^*(t_+, V_\Lambda)$ is known to be independent on the $\mathcal{G}$-representation Λ, and it is in fact equal to the ratio of dimensions of the *Weyl groups* $W(G)$ and $W(H)$ [196, 197],

$$\mu = \frac{\dim W(G)}{\dim W(H)} \,. \tag{6.45}$$

This approach could be generalized, when considering the whole AKM algebra $\hat{t}_+$ instead of t_+ and looking for the cohomology elements of $H^*(\hat{t}_+, V_\Lambda)$ instead of that of $H^*(t_+, V_\Lambda)$. The *affine* Weyl group (sect. XII.2), which is the semi-direct product of the Weyl group with the root lattice [44], now has to be considered. This may result in the extra factor on the r.h.s. of eq. (6.45), given by dimension $|\Gamma^r(G)/\Gamma^r(H)|$ of the finite group $\Gamma^r(G)/\Gamma^r(H)$, where Γ^r denote the corresponding root lattices. In the simplest cases where a detailed analysis is possible, this does not however happen, since the extra

Table 2. The weights of LG fields for the KS coset models defined on the hermitian symmetric spaces at level one [155].

G/H	weights
$SU(n+m)/SU(m) \times SU(n) \times U(1)$	$1, 2, \ldots, \min(n, m)$
$SO(n+2)/SO(n) \times SO(2)$	$1, n/2$ (n even)
$SO(2n)/SU(n) \times U(1)$	$4i - 2, \begin{cases} i = 1, \ldots, n/2, & \text{if } n \text{ even} \\ i = 1, \ldots, (n-1)/2, & \text{if } n \text{ odd} \end{cases}$
$E_6/SO(10) \times U(1)$	$1, 4$
$E_7/E_6 \times U(1)$	$1, 5, 9$

factor appears to lead to multiple counting of the same physical states of the coset model, at least for the states at level one [155].

The dimension of the chiral ring (i.e. the number of Ramond ground states) is given by eq. (6.45) in this case. In general, it takes the form [155]

$$\mu = N_G^k \cdot \frac{\dim W(G)}{\dim W(H)} |\Gamma^r(G)/\Gamma^r(H)| , \qquad (6.46)$$

where N_G^k is the total number of the highest-weight irreps of AKM algebra $\widehat{\mathcal{G}}_k$ of level k. As far as the level-one coset models defined on the hermitian symmetric spaces are concerned (Table 1), the corresponding charges of the LG primary fields are tabulated in Table 2.

It is really non-trivial that the chiral rings of all LG models are actually isomorphic to the cohomology rings of the corresponding cosets G/H. This was already checked on a case-by-case basis for the simply-laced hermitian symmetric spaces [155]. In particular, it implies the representation (5.102) for the latter, which is not certainly the case for general Kählerian cosets. For instance, the $N = 2$ (minimal) superconformal model based on $\mathbf{CP}^n = SU(n+1)/U(n)$ has the chiral ring

$$\mathcal{K} = \{1, x, x^2, \ldots, \text{ subject to } x^{n+1} = 0\} , \qquad (6.47)$$

which is just the definition of the $\mathbf{CP}^1$ cohomology ring [155].

As to an arbitrary coset manifold G/H of rank $G = $ rank $H = l$, the corresponding Poincaré polynomial $P(t)$ is known, and it takes the form [198]

$$P(t) = \prod_{j=1}^{l} \frac{(1 - t^{d_j(G)})}{(1 - t^{d_j(H)})} , \qquad (6.48)$$

Table 3. The hermitian symmetric spaces and the associated Poincaré polynomials of level $\hat{k}$, $\hat{k} = 1, 2, 3, \ldots$

G/H	Poincaré polynomial
$SU(n+m)/SU(m) \times SU(n) \times U(1)$	$\prod_{i=1}^{m} \prod_{j=1}^{n} \frac{(1-t^{d-(i+j-1)})}{(1-t^{(i+j-1)})}$
$SO(n+2)/SO(n) \times SO(2)$	$\frac{(1-t^{d-n/2})}{(1-t^{n/2})} \prod_{i=1}^{n-1} \frac{(1-t^{d-i})}{(1-t^{i})}$
$SO(2n)/SU(n) \times U(1)$	$\prod_{i=1}^{n-1} \prod_{j=1}^{n-1} \frac{(1-t^{d-2(i+j-1)})}{(1-t^{2(i+j-1)})}$
$Sp(2n)/SU(n) \times U(1)$	$\prod_{i=1}^{n} \prod_{j=i}^{n} \frac{(1-t^{d-(i+j)})}{(1-t^{(i+j)})}$
$E_6/SO(10) \times U(1)$	$\prod_{i=1}^{11} \frac{(1-t^{d-i})}{(1-t^{i})} \prod_{j=4}^{8} \frac{(1-t^{d-j})}{(1-t^{j})}$
$E_7/E_6 \times U(1)$	$\frac{(1-t^{d-9})}{(1-t^{9})} \prod_{i=1}^{17} \frac{(1-t^{d-i})}{(1-t^{i})} \prod_{j=5}^{13} \frac{(1-t^{d-j})}{(1-t^{j})}$

where the degrees $d_j(G)$ and $d_j(H)$ of the Casimir operators for G and H, resp., have been introduced. [3] The list of the Poincaré polynomials for the KS cosets is given in Table 3.

As far as the hermitian symmetric spaces are concerned, the Dynkin diagram of group H can be obtained from that of group G by the following procedure: first, we delete one of the nodes (say, with Dynkin number one) from the Dynkin diagram of G and then replace it by $U(1)$. The $\mathcal{G}$-representation Ξ to be constructed by putting a unit on this node and zero elsewhere has the same dimension as the one of the cohomology ring of the coset G/H. With k instead of the unit, we get the representation $\Xi_{(k)}$ instead, whose weight is given by the product of k with the weight of Ξ. It is now the outcome of a straightforward inspection of the KS models that the chiral primary states in the corresponding LG model are given by $\Xi_{(k)}$, whose $U(1)$ charges are proportional to ρ_G [155]. In particular, eq. (5.106) now reproduces the central charges of the KS models! The dimensions of the chiral rings $\mathcal{K}$, given by eqs. (6.45) and (6.46), also agree with those following from the LG models. For example, taking the coset $SU(n+1)/SU(n) \times U(1)$ at kevel k yields for the dimension of the associated chiral ring the value

$$\mu = \frac{(n+k)!}{n!k!}, \qquad (6.49)$$

which exactly coincides with the dimension of the k-fold symmetric tensor product of the $SU(n+1)$ fundamental representations. The $U(1)$ charges of the $N = 2$ SCA are proportional to the action of $\rho_G \cdot H$ on the module [155].

We conclude that some of the KS coset models, although not all of them, can be identified with the LG models in the sense that they both lead to the same chiral rings.

---

[3]It is our convention that $d_j = 1$ for any $U(1)$ factor.

The LG superpotential seems to be dictated by the cohomological properties of a given coset manifold. A full understanding of this relation is, however, yet to be developed.

Exercises

VI-5 ▷ (*) Calculate the central charges of the level-one KS models collected in Table 1, when $\hat{k} = 1$.

VI-6 ▷ Check the relation (5.106) for the data in Tables 1 and 2.

VI.4 Gauged WZNW theories

A large class of CFT's based on cosets G/H can be described in terms of the *gauged* WZNW theories [199, 200, 201, 202, 203, 204, 205]. In this section we review this approach. The WZNW action (4.55) defined on a group manifold G is a good starting point for this construction,

$$kI(g) = \frac{k}{8\pi} \int d^2\sigma \operatorname{tr}\left(\partial_m g \partial^m g^{-1}\right) + \frac{k}{12\pi} \int \operatorname{tr}\left(dg g^{-1}\right)^3 . \qquad (6.50)$$

Let H be an anomaly-free vector subgroup of $G_L \times G_R$ which is the global symmetry of the action (6.50). Let $\tilde{H}$ be Lie algebra of the Lie group H. We assume that the Lie algebra $\mathcal{G}$ of G admits an $\operatorname{ad}(\tilde{H})$-invariant orthogonal decomposition having the form [4]

$$\mathcal{G} = \tilde{H} \oplus M , \quad [\tilde{H}, \tilde{H}] \subset \tilde{H} , \quad [\tilde{H}, M] \subset M , \qquad (6.51)$$

where $M \equiv \mathcal{G} \setminus \tilde{H}$; $g \in G$, $h \in H$, and $m \in G/H$. A Lie algebra $\mathcal{G}$-valued field J can always be written as the sum, $J = J_H + J_M$, w.r.t. the decomposition (6.51).

We are now in a position to consider the modified WZNW action, where the subgroup H is gauged. This means that we add the minimal coupling with the H-valued gauge fields and make the action to be invariant under the gauge transformations (with the σ-dependent parameter λ) having the form

$$g \to \lambda g \lambda^{-1} , \quad \lambda \in H ,$$

$$A_m \to \partial_m \lambda \cdot \lambda^{-1} + \lambda A_m \lambda^{-1} . \qquad (6.52)$$

---

[4]For simplicity, the embedding index of H in G is assumed to be one.

In the light-cone variables $\sigma_\pm$, the gauged WZNW action takes the form [5]

$$kI(g, A) = kI(g) + \frac{k}{4\pi}\text{tr}\int d^2\sigma \left\{A_+\partial_-gg^{-1} - A_-g^{-1}\partial_+g + A_+gA_-g^{-1} - A_-A_+\right\} ,$$
(6.53)

where the light-cone components of the gauge field $A_\pm$ in the adjoint representation of H have been introduced. It is useful to parametrize the gauge fields in terms of the H-group elements as [199, 200, 201, 202, 203, 204, 205]

$$A_- = \partial_-\tilde{h}\tilde{h}^{-1} , \quad A_+ = \partial_+hh^{-1} .$$
(6.54)

The convenience of this parametrization becomes obvious after exploiting the Polyakov-Wiegmann identity

$$I(gh) = I(g) + I(h) - \frac{1}{4\pi}\int d^2\sigma\, \text{tr}\left[g^{-1}\partial_+g\partial_-hh^{-1}\right] ,$$
(6.55)

which allows us to rewrite the action (6.53) to the form

$$I(g, A) = I(h^{-1}g\tilde{h}) - I(h^{-1}\tilde{h}) ,$$
(6.56)

where the vector gauge invariance (6.52) is manifest.

The gauge fields in eq. (6.53) are non-propagating. They play the role of Lagrange multipliers forcing the H-currents to vanish in the classical gauged WZNW theory. To see how the action (6.53) describes the G/H coset structure, let us consider its equations of motion. Varying the action w.r.t. the gauge fields yields

$$(g^{-1}\nabla_+g)_H = 0 , \quad (\nabla_-g \cdot g^{-1})_H = 0 ,$$
(6.57)

where the new gauge-covariant derivatives $\nabla_m = \partial_m - [A_m, \;]$ have been introduced. The subscript H above means projection onto the subspace H of G in the sense of eq. (6.51). Eq. (6.57) is only gauge-covariant when the commutation relations (6.51) are fulfilled. The equation of motion for the g-field reads

$$\nabla_-(g^{-1}\nabla_+g) = F ,$$
(6.58)

while the first equation (6.57) implies $F = 0$. Hence, both connections A_+ and A_- are, in fact, on-shell trivial. Exploiting the gauge invariance (6.52) now allows us to choose

---

[5]This form of the action is fixed by the gauge invariance, when requiring the vanishing of the H-current in addition (see below).

the gauge $A_{\pm} = 0$ in the classical theory, in which the equations of motion are simplified to the form [169]

$$(g^{-1}\partial_+ g)_H = 0 , \quad (\partial_- g \cdot g^{-1})_H = 0 ,$$

$$\partial_-(g^{-1}\partial_+ g)_M = 0 , \Longleftrightarrow \partial_+(\partial_- g \cdot g^{-1})_M = 0 . \tag{6.59}$$

The M-currents are therefore non-vanishing on-shell, while the H-currents do vanish on-shell, as they should. In the quantum theory, the situation is a bit more complicated since the change of variables (6.54) induces the Jacobian in the partition function of the gauged WZNW theory,

$$Z = \int [dg][dA_+][dA_-] \exp\left[-kI(g, A)\right] , \tag{6.60}$$

which then takes the form

$$Z = \int [dg][dh][d\tilde{h}] \det D_+ \det D_- \exp\left[-kI(h^{-1}g\tilde{h})\right] \exp\left[kI(h^{-1}\tilde{h})\right] , \tag{6.61}$$

where

$$\det D_{\pm} = \det\{\partial_{\pm} - [A_{\pm}, \ ]\} . \tag{6.62}$$

The conventional QFT techniques can now be applied to calculate the determinants, [6] with the result [170, 171]

$$\det D_+ \det D_- = \exp\left[c_H I(h^{-1}\tilde{h})\right] \det \partial_+ \det \partial_- , \tag{6.63}$$

where the quadratic Casimir operator eigenvalue c_H for the adjoint representation of H, and the chiral determinants $\det \partial_{\pm}$ have been introduced. The latter can be conveniently represented as the (Gaussian) integrals over the Faddeev-Popov (FP) ghost fields $b_{\pm}$ of (conformal) dimension one, and $c, \bar{c}$ of dimension zero, all in the adjoint representation of H. Changing the variables $(h^{-1}g\tilde{h}) \to g$ and fixing the gauge $\tilde{h} = 1$, (or, equivalently, $A_- = 0$) which does not produce any new FP-ghosts (exercise # VI-7), one finally arrives at the following expression for the gauge-fixed partition function (after changing the variable h^{-1} back to h) [203, 204, 205]:

$$Z = \int [dg][dh][db_+][db_-][dc][d\bar{c}] \exp\left[-kI(g)\right] \exp\left[(k + c_H)I(h)\right]$$

$$\times \exp\left(-\text{tr}\int d^2\sigma \, b_+ \partial_- c\right) \exp\left(-\text{tr}\int d^2\sigma \, b_- \partial_+ \bar{c}\right) . \tag{6.64}$$

---

[6]One should use a UV-regulator preserving the gauge invariance, and divide the partition function by the (infinite) gauge volume.

The partition function (6.64) factorizes into three conformally invariant pieces, which are actually coupled to each other by some constraints [203, 204, 205]. A simple way to get the constraints is to introduce an *external* gauge field $B_\pm$ and make use of the parametrization similar to that of eq. (6.54):

$$B_+ = \partial_+ q \cdot q^{-1} , \quad B_- = \partial_- p \cdot p^{-1} , \tag{6.65}$$

where p, q are in the fundamental representation of H. The identity

$$\int [dg] \exp\left[-kI(g, B)\right] = \int [dg] \exp\left[-kI(g)\right] \exp\left[kI(q^{-1}p)\right] \tag{6.66}$$

implies that the naively B-dependent partition function is actually independent on B:

$$Z(B_+, B_-) = Z(0) \exp\{[k - (k + c_H) + c_H]I(q^{-1}p)\} = Z(0) , \tag{6.67a}$$

or, equivalently,

$$\left.\frac{\delta Z(B)}{\delta B_-}\right|_{B=0} = \left\langle J_+^{\text{tot}}(\sigma_+)\right\rangle = 0 , \tag{6.67b}$$

and similarly for J_-^{tot}. [7] Being a function of the holomorphic coordinate $z = \exp(i\sigma_+)$, the current J_+^{tot} takes the form [205]

$$J^{i,\text{tot}}(z) = k\text{tr}(t^i g^{-1}\partial_z g) - (k + c_H)\text{tr}(t^i h^{-1}\partial_z h) - if^{ijk}b_z^j c^k$$

$$\equiv J^i(z) + \tilde{J}^i(z) + J_{\text{gh}}^i(z) , \tag{6.68}$$

where t^i, $i = 1, \ldots, |H|$, are the generators of H in the adjoint representation. The relevant OPE's read:

$$J^i(z) J^j(w) \sim if^{ijk}\frac{J^k(w)}{z - w} + \frac{k\delta^{ij}}{(z - w)^2} ,$$

$$\tilde{J}^i(z)\tilde{J}^j(w) \sim if^{ijk}\frac{\tilde{J}^k(w)}{z - w} - \frac{(k + c_H)\delta^{ij}}{(z - w)^2} ,$$

$$J_{\text{gh}}^i(z) J_{\text{gh}}^j(w) \sim if^{ijk}\frac{J_{\text{gh}}^k(w)}{z - w} + \frac{c_H d^{ij}}{(z - w)^2} . \tag{6.69}$$

The level and the SS-constructed central charge, associated with the current $J^{i,\text{tot}}(z)$, obviously vanish, while the current itself represents the first-class constraint in the gauged WZNW theory. Though we cannot impose it as the operator constraint (i.e. in the strong sense) in the Fock space of states (exercise # VI-8), this constraint can

---

[7]We restrict ourselves to the holomorphic dependence.

be imposed in the weak sense, which is quite usual in QFT, namely, $J_m^{\text{tot}} |\text{phys}\rangle = 0$ for $m \geq 0$, where $J^{\text{tot}}(z) = \sum_{m=-\infty}^{\infty} J_m^{\text{tot}} z^{-m-1}$.

The holomorphic stress tensor for the theory (6.64) reads

$$T(z) = \frac{1}{2k + c_G} : J^a(z) J^a(z) : - \frac{1}{2k + c_H} : \tilde{J}^a(z) \tilde{J}^a(z) : - : b_z^i \partial_z c^i :$$

$$\equiv T^G + \tilde{T}^H + T^{\text{gh}} , \tag{6.70}$$

where $a = 1, \ldots, |G|$. The OPE's (6.69) imply the Virasoro OPE's

$$T^G(z) T^G(w) \sim \frac{c(G, k)}{2(z - w)^4} + \frac{2T^G(w)}{(z - w)^2} + \frac{\partial_w T^G(w)}{z - w} ,$$

$$\tilde{T}^H(z) \tilde{T}^H(w) \sim \frac{c(H, -k - c_H)}{2(z - w)^4} + \frac{2\tilde{T}^H(w)}{(z - w)^2} + \frac{\partial_w \tilde{T}^H(w)}{z - w} ,$$

$$T^{\text{gh}}(z) T^{\text{gh}}(w) \sim \frac{-2|H|}{2(z - w)^4} + \frac{2T^{\text{gh}}(w)}{(z - w)^2} + \frac{\partial_w T^{\text{gh}}(w)}{z - w} , \tag{6.71}$$

where, in accordance with eq. (4.38), the WZNW central charge is given by

$$c(G, k) = \frac{2k|G|}{2k + c_G} . \tag{6.72}$$

Hence, the total central charge of the gauged WZNW theory is

$$c^{\text{tot}} = \frac{2k|G|}{2k + c_G} + \frac{2(-k - c_H)|H|}{2(-k - c_H) + c_H} - 2|H|$$

$$= \frac{2k|G|}{2k + c_G} - \frac{2k|H|}{2k + c_H} , \tag{6.73}$$

in agreement with the central charge of the GKO construction for the cosets! However, the stress tensors $T(z)$ above and $T^{\text{GKO}} \equiv T^{G/H} = T^G - T^H$ differ since

$$T = T^G - T^H + T' \equiv T^{\text{GKO}} + T' , \tag{6.74}$$

where

$$T' = \frac{: J^i J^i :}{2k + c_G} - \frac{: \tilde{J}^i \tilde{J}^i :}{2k + c_H} - : b_z^i \partial_z c^i : . \tag{6.75}$$

The stress tensor T' commutes with T^{GKO}, and its central charge is zero. We know from Ch. I that any unitary highest weight representation of $c = 0$ is trivial. Hence, given a unitary physical spectrum in the gauged WZNW theory, the stress tensor T' should be trivial too. The actual proof of unitarity is far from being obvious for this model, but, nevertheless, it can be done by using the BRST methods (sect. XII.5), because of the first-class constraints only in the theory [205]. The physical states are given by the cohomology $\text{Ker}Q/\text{Im}Q$ of BRST charge Q, where $Q\,|\text{phys}\rangle = 0$ but $|\text{phys}\rangle \neq Q\,|*\rangle$, and $Q^2 = 0$. An explicit expression for the BRST charge of the gauged WZNW model follows according to the general rule in eq. (12.157). The resulting BRST charge appears to be commuting with T^{tot}, T^{GKO} and T' individually. T' can therefore be recognized as a BRST *exact* operator, since there exists an operator $X(z)$ satisfying

$$T' = \{Q, X(z)\}\,, \tag{6.76}$$

which makes the difference between $T(z)$ and T^{GKO} to be irrelevant in the matrix elements between the physical states. In particular, the states created by T' all have zero norm and they are orthogonal to any physical state. Hence, they actually decouple from the physical sector, and one gets

$$T(z)\,|\text{phys}\rangle = T^{\text{GKO}}(z)\,|\text{phys}\rangle\;. \tag{6.77}$$

The decoupling of the negative norm states and ghosts from the physical spectrum of the gauged WZNW theory can actually be proved [205] for the case of an abelian subgroup H. The abelian unitarity provides the strong evidence that the non-abelian case should also be the same in this respect. The supersymmetrization of this field-theoretical construction is straightforward, and it follows along the similar lines [169, 204].

Exercises

VI-7 ▷ Check that the gauge $\bar{h} = 1$ does not produce the FP-ghosts in the partition function of the gauged WZNW theory. *Hint:* use the (left) invariance of the Haar measure on a group.

VI-8 ▷ (*) Verify that the vanishing condition for the operator $J^{\text{tot}}(z)$ in the quantum gauged WZNW theory cannot be consistently imposed in the strong (operatorial) sense.

VI-9 ▷ Directly calculate an algebra satisfied by the stress tensor in eq. (6.70), and find the corresponding central charge.

VI.5 Felder's (BRST) approach to coset models

The Felder (or BRST) approach (see sects. II.9 and III.5) can be further generalized
to the CFT's based on cosets (G/H), by using a free field resolution of the G-based
WZNW theory and a 'projection' of its Fock space onto the subspace corresponding to
the coset G/H [206]. We already know from the previous sections about performing the
projections onto irreps in free field Fock space representations (sect. II-9), and about
computing the chiral blocks on a torus (sect. III-5) or even on arbitrary Riemann surface
[99, 100, 102, 103, 104]. The whole BRST approach is based on the key observation [66]
that an irrep (submodule) L_Λ of Fock space F_Λ is characterized by a BRST *complex*
$(\mathbf{F}_\Lambda, d) \equiv \left\{ (F_\Lambda^{(i)}, d^{(i)}) \right\}$, $F_\Lambda \subset F_\Lambda^{(0)}$, where

$$\ldots \longrightarrow F_\Lambda^{(-1)} \xrightarrow{d^{(-1)}} F_\Lambda^{(0)} \xrightarrow{d^{(0)}} F_\Lambda^{(1)} \xrightarrow{d^{(1)}} F_\Lambda^{(2)} \longrightarrow \ldots , \tag{6.78}$$

and

$$H^i(d) \equiv \frac{\mathrm{Ker} d^{(i)}}{\mathrm{Im} d^{(i-1)}} = \delta^{i,0} \Lambda_A . \tag{6.79}$$

In CFT, such BRST complex is called a (Felder's) **resolution** of the irreducible module
L_Λ in terms of the Fock space modules. The chiral blocks are given by the vacuum
expectation values for products of the chiral vertex operators $V : L_\Lambda \to L_{\Lambda'}$. They can
be formally computed, when one can find a collection of maps (i.e. the screened vertex
operators) $V^{(i)} : F_\Lambda^{(i)} \to L_{\Lambda'}^{(i)}$, satisfying $V^{(0)}\big|_{L_\Lambda} = V$, $d'^{(i)}V^{(i)} = V^{(i+1)}d^{(i)}$, so that

$$
\begin{array}{ccccccccc}
\ldots & \longrightarrow & F_\Lambda^{(-1)} & \xrightarrow{d^{(-1)}} & F_\Lambda^{(0)} & \xrightarrow{d^{(0)}} & F_\Lambda^{(1)} & \xrightarrow{d^{(1)}} & F_\Lambda^{(2)} & \longrightarrow & \ldots \\
& & \downarrow{\scriptstyle V^{(-1)}} & & \downarrow{\scriptstyle V^{(0)}} & & \downarrow{\scriptstyle V^{(1)}} & & \downarrow{\scriptstyle V^{(2)}} & & \\
\ldots & \longrightarrow & F_{\Lambda'}^{(-1)} & \xrightarrow{d'^{(-1)}} & F_{\Lambda'}^{(0)} & \xrightarrow{d'^{(0)}} & F_{\Lambda'}^{(1)} & \xrightarrow{d'^{(1)}} & F_{\Lambda'}^{(2)} & \longrightarrow & \ldots
\end{array}
\tag{6.80}
$$

The chiral blocks on a torus are just traces of products of the vertex operators
$\mathcal{O} = V_1 \cdots V_n$ (sect. III-5),

$$\mathrm{Tr}_{L_\Lambda} \mathcal{O} = \sum_{i \in \mathbf{Z}} (-1)^i \mathrm{Tr}_{F_\Lambda^{(i)}} \mathcal{O}^{(i)} . \tag{6.81}$$

It is worthy to emphasize again here some known facts about the group construction
(Ch. IV). Given an AKM algebra $\hat{\mathcal{G}}_k$, its free field representation can be constructed
in terms of the scalar fields $\phi^i(z)$, $i = 1, \ldots, \mathrm{rank}\, G$, and the spin-1 bosonic fields
$\beta^\alpha(z), \gamma^\alpha(z)$ to be defined for every positive root $\alpha \in \Delta_+$ of G, and having the free
propagators

$$\left\langle \phi^i(z)\phi^j(w) \right\rangle = -(\alpha_i, \alpha_j) \ln(z - w) ,$$

$$\left\langle \gamma^\alpha(z)\beta^{\alpha'}(w)\right\rangle = \frac{\delta^{\alpha\alpha'}}{z-w}\ ,\tag{6.82}$$

where $\{\alpha_i\}$ denote simple roots of $\mathcal{G}$. Given any vector $\alpha = \sum_i m_i\alpha_i$, one can formally associate a field $\phi^\alpha(z) = \sum_i m_i\phi^i(z)$ with it. Cartan subalgebra (CSA) generators in a Chevalley basis take the form

$$h_i(z) = i\sqrt{k+\bar{h}_G}\,\partial\phi^i(z) + \sum_{\alpha\in\Delta_+}(\alpha_i,\alpha) : \gamma^\alpha(z)\beta^\alpha(z): \ ,\tag{6.83}$$

whereas the SS-constructed stress tensor of central charge (4.38) reads in terms of the free fields as follows:

$$T(z) = -\tfrac{1}{2} : \partial\phi(z)\cdot\partial\phi(z) : -\alpha_0\rho\cdot i\partial\phi(z) - \sum_{\alpha\in\Delta_+} : \beta^\alpha(z)\partial\gamma^\alpha(z): \ .\tag{6.84}$$

The free Fock spaces F_Λ are labeled in this case by the scalar field zero modes p^i to be defined for the highest-weight vector $|\Lambda\rangle$,

$$p^i\,|\Lambda\rangle = \alpha_0(\Lambda,\alpha_i)\,|\Lambda\rangle\ ,\quad \alpha_0^2 = (k+\bar{h}_G)^{-1}\ .\tag{6.85}$$

The free field resolutions $F_\Lambda^{(i)}$ of L_Λ, in accordance with eq. (6.78), are given by [206]

$$F_\Lambda^{(i)} = \oplus_{\{w\in\hat{W}|\bar{l}(w)=i\}}F_{w*\Lambda}\ ,\tag{6.86}$$

where the Fock space $F_{w*\Lambda}$ of the highest weight $w*\Lambda = w(\Lambda+\rho)-\rho$ has been introduced, with $\bar{l}(w)$ being the 'modified length' of an element w belonging to the affine Weyl group $\hat{W}$ of $\hat{\mathcal{G}}$, whose explicit form is not needed for what follows (see, however, refs. [207, 208, 209, 210]). The generalization of this construction to the cosets was suggested in ref. [206], and it is briefly reviewed below.

As far as the *finite*-dimensional ordinary Lie groups and algebras are concerned, given an irreducible highest-weight module L_Λ^G of G, we can always decompose it w.r.t. H into the H- and Z-modules, where Z is the *centralizer* of H in G,

$$L_\Lambda^G = \oplus_{\Lambda'}L_{\Lambda,\Lambda'}^Z L_{\Lambda'}^H\ ,\tag{6.87}$$

and the sum goes over a finite set of dominant integral weights of H. The dimensions of spaces $L_{\Lambda,\Lambda'}^Z$ are just the multiplicities with which the H-weight Λ' appears in L_Λ^G. Having the resolution of L_Λ^G given by eq. (6.78), we want to project it onto a subcomplex which would be a resolution of $L_{\Lambda,\Lambda'}^Z$. Since the H-singular vectors (those annihilated by n_+^H) clearly form a subcomplex, the zeroth cohomology of d to be restricted on the space of states with the integral H-weight Λ' appears to be isomorphic to $L_{\Lambda,\Lambda'}^{Z}$. What is not obvious is to show that the non-zeroth cohomology of this subcomplex is trivial elsewhere, as is required by the definition of a resolution (see ref. [206] for the proof based on the theory of categories).

The actual projection can be accomplished by the use of BRST techniques [206]. Assuming, for simplicity, that the positive roots Δ_+^H of H are identified with a subset of Δ_+^G, as well as the corresponding CSA elements, and introducing ghosts $\{c^{-\alpha}, b^{\alpha'}\} = \delta^{\alpha\alpha'}, \alpha, \alpha' \in \Delta_+^H$, we construct the nilponent BRST operator (sect. XII.5) as follows:

$$Q = \sum_{\alpha \in \Delta_+^H} c^{-\alpha} e^{\alpha} - \frac{1}{2} \sum_{\alpha,\beta,\gamma \in \Delta_+^H} f^{\alpha\beta}{}_{\gamma} c^{-\alpha} c^{-\beta} b^{\gamma} , \qquad (6.88)$$

where the raising $(+)$ operators e^{α} and the structure constatns $f^{\alpha\beta}{}_{\gamma}$ of n_+^H have been introduced. Let F_{gh} be the Fock space built on the ghost vacuum $|0\rangle_{\mathrm{gh}}$, so that $b^{\alpha} |0\rangle_{\mathrm{gh}} = 0$ for any $\alpha \in \Delta_+^H$. The replacement $F_{w*\Lambda}$ by $F_{w*\Lambda} \otimes F_{\mathrm{gh}}$ extends the G-complex to the double complex graded by the ghost number $N_{\mathrm{gh}} = \sum c^{-\alpha} b^{\alpha}$: $F_{\mathrm{gh}} = \oplus F_{\mathrm{gh}}^n$. The Q-cohomology on the space $F_{w*\Lambda} \otimes F_{\mathrm{gh}}^{(0)}$ is isomorphic to the full set of the H-singular vectors in $F_{w*\Lambda}$, and it is trivial on $F_{w*\Lambda} \otimes F_{\mathrm{gh}}^{(n)}$ when $n \neq 0$ [206]. Therefore, the resolution of $L_{\Lambda,\Lambda'}^{Z}$ can be identified with this extended complex to be restricted to the states of H-weight Λ'.

The *infinite*-dimensional generalization of this construction to AKM algebras goes along the similar lines. Given an irreducible highest-weight module L_Λ^G of $\hat{G}$, one can decompose it into the irreducible $\hat{H}$- and $\hat{G}/\hat{H}$-modules, just like that in eq. (6.87),

$$L_\Lambda^G = \oplus_{\Lambda'} L_{\Lambda,\Lambda'}^{G/H} L_{\Lambda'}^H , \qquad (6.89)$$

where the sum now goes over a finite set of integrable weights of $\hat{H}$. The associated AKM character is defined by

$$\mathrm{ch}_{\Lambda,\Lambda'}^{G/H} = \mathrm{Tr}_{L_{\Lambda,\Lambda'}^{G/H}} q^{L_0^G - L_0^H - c/24} , \qquad (6.90)$$

where $c = c^G - c^H$. The module $L_{\Lambda,\Lambda'}^{G/H}$ can be represented by the set of vectors in L_Λ^G which are the $\hat{H}$ highest weight vectors of weight Λ'. A projection onto the H-singular states gives the subcomplex, and the cohomology of this subcomplex it trivial everywhere except of $F_\Lambda^{(0)}$, similarly to the finite-dimensional case. Having introduced ghosts for each generator in the set

$$\hat{n}_+^H \equiv \{e_n^{\alpha}, \alpha \in \Delta_+^H, n \geq 0\} \cup \{e_n^{-\alpha}, \alpha \in \Delta_+^H, n > 0\} \cup \{h_n^i, i = 1,\ldots,l', n > 0\} , \quad (6.91)$$

we construct the nilpotent BRST operator,

$$\hat{Q} = \sum_{\Delta_+^H} \left(\sum_{n \geq 0} c_{-n}^{-\alpha} e_n^\alpha + \sum_{n > 0} c_{-n}^\alpha e_n^{-\alpha} \right) + \sum_{i=1}^{l'} \sum_{n > 0} c_{-n}^i h_n^i$$

$$-\frac{1}{2} \sum_{a,b,c} \sum_{m,n} f_{abc} c_{-m}^a c_{-n}^b b_{m+n}^c , \tag{6.92}$$

where f_{abc} are the structure constants of H. Since $\hat{Q}$ does commute with d, we get the double complex again, whereas the $\hat{Q}$-cohomology at the zeroth ghost number is isomorphic to the subspace of the $\hat{H}$ singular vectors. Elsewhere, the $\hat{Q}$-cohomology is supposed to be trivial, although there seems to be no proof yet, but only some explicit examples [206] supporting the resolution conjecture.

The simplest non-trivial example is provided by parafermions and their higher-rank generalizations, when choosing $H = U(1)^l$ where l is rank of G, which corresponds to the coset $G/U(1)^l$ (sect. 1). We find it convenient to bosonize the (β, γ) system first, by writing (Ch. III)

$$\gamma^\alpha = \eta^\alpha e^{-i\omega^\alpha} = e^{i(\chi^\alpha - \omega^\alpha)} , \quad \beta^\alpha = \partial \xi^\alpha e^{i\omega^\alpha} = \partial \left(e^{-i\chi^\alpha} \right) e^{i\omega^\alpha} , \tag{6.93}$$

in terms of the spin-1 fermionic fields η, ξ, and the symplectic boson ω. The Fock space of the (η, ξ) system is *not* isomorphic to that of the (β, γ) system, since the bosonization (6.93) introduces the additional degrees of freedom associated with zero mode ξ_0^α. The equivalence can be maintained after removing the zero modes ξ_0^α and identifying the momentum operators of χ^α and ω^α [35].

After the bosonization, the CSA currents (6.83) for a simple and simply-laced Lie group G of rank l take the form

$$h_i(z) = i\sqrt{k + \tilde{h}_G} \partial \phi^i(z) - \sum_{\alpha \in \Delta_+} (\alpha_i, \alpha) i \partial \omega^\alpha . \tag{6.94}$$

The realization of $\widehat{U(1)}_l$ in terms of l bosonic fields is simply $h_i(z) = i\sqrt{k}\partial\tilde{\phi}^i(z)$, which suggests us to make the following redefinition:

$$\sqrt{k}\tilde{\phi}^i = \sqrt{k + \tilde{h}_G}\phi^i - \sum_{\alpha \in \Delta_+} (\alpha_i, \alpha)\omega^\alpha . \tag{6.95}$$

We also need to define the coset fields which are orthogonal to the fields introduced above. It is not difficult to construct them in this case as follows [8]

$$\sqrt{k}\hat{\omega}^\alpha = \sqrt{k}\omega^\alpha + \left(\frac{\sqrt{k+\bar{h}_G} - \sqrt{k}}{\bar{h}_G}\right) \sum_{\beta \in \Delta_+} (\alpha, \beta)\omega^\beta - \phi^\alpha \ ,$$

$$\hat{\chi}^\alpha = \chi^\alpha \ . \tag{6.96}$$

The stress tensor of the theory under consideration,

$$T(z) =: -\tfrac{1}{2}\partial\phi \cdot \partial\phi - \sqrt{\frac{1}{k+\bar{h}_G}} i\partial^2\phi^\rho$$

$$+ \sum_{\alpha \in \Delta_+} \left[\tfrac{1}{2}(i\partial\chi^\alpha)^2 - \tfrac{1}{2}i\partial^2\chi^\alpha - \tfrac{1}{2}(i\partial\omega^\alpha)^2 + \tfrac{1}{2}i\partial^2\omega^\alpha\right] : \ , \tag{6.97}$$

can now be decomposed as [206]

$$T^G = T^H(\tilde{\phi}) + T^{G/H}(\hat{\chi}, \hat{\omega}) \ , \tag{6.98}$$

where

$$T^H =: -\tfrac{1}{2}\partial\tilde{\phi} \cdot \partial\tilde{\phi} : \ ,$$

$$T^{G/H} =: \sum_{\alpha \in \Delta_+} \left[\tfrac{1}{2}(i\partial\hat{\chi}^\alpha)^2 - \tfrac{1}{2}i\partial^2\hat{\chi}^\alpha - \tfrac{1}{2}(i\partial\hat{\omega}^\alpha)^2\right.$$

$$\left.+ \tfrac{1}{2}i\partial^2\hat{\omega}^\alpha - \left(\frac{\sqrt{k+\bar{h}_G} - \sqrt{k}}{\bar{h}_G\sqrt{k+\bar{h}_G}}\right) (\rho, \alpha)i\partial^2\hat{\omega}^\alpha\right] \ . \tag{6.99}$$

The *screening* vertex operators in this approach take the form [209, 210]

$$V_i = \bar{e}_i(\chi, \omega) \exp\left[-i\sqrt{\frac{1}{k+\bar{h}_G}}\phi^i\right] \ , \quad i = 1, \ldots, l \ , \tag{6.100}$$

where $\bar{e}_i$ is generically a sum of ω-dependent terms of the type $\exp(i\sum n_\alpha\omega^\alpha)$, where $n_\alpha \in \mathbf{Z}$ and $\sum n_\alpha\alpha = \alpha_i$. Being functions of the linear combination

$$\sum n_\alpha\omega^\alpha - \sqrt{\frac{1}{k+\bar{h}_G}}\phi^i \ , \tag{6.101}$$

the screening vertex operators are actually independent on $\tilde{\phi}$, i.e. $V_i = V_i(\hat{\chi}, \hat{\omega})$.

It is usually convenient to decompose fields in the coset models into parallel (i.e. belonging to the subgroup H) fields $\Phi_\parallel$ and perpendicular (i.e. belonging to the coset

---

[8]In accordance with the notation used in ref. [206], our coset quantities carry hats, whereas those of $H = U(1)^l$ carry tildes instead.

G/H) fields $\Phi_\perp$, which is the obvious generalization of the decomposition familiar for finite-dimensional Lie algebras. The $\hat{G}$ screening operators only depend on the orthogonal fields, whereas the $\hat{H}$-currents can be entirely written in terms of $\Phi_\parallel$'s. This implies that the restriction to the Fock spaces $F_\Lambda^\perp \equiv F_\Lambda(\Phi_\perp)$ forms a subcomplex,

$$\cdots \longrightarrow F_{\Lambda,\lambda}^{(-1)\perp} \xrightarrow{d^{(-1)}} F_{\Lambda,\lambda}^{(0)\perp} \xrightarrow{d^{(0)}} F_{\Lambda,\lambda}^{(1)\perp} \xrightarrow{d^{(1)}} F_{\Lambda,\lambda}^{(2)\perp} \longrightarrow \cdots , \qquad (6.102)$$

where

$$F_{\Lambda,\lambda}^{(i)\perp} = \oplus_{\{w \in \hat{W} | \hat{l}(w)=i\}} \left(\oplus_{\{n_\alpha \in \mathbb{Z} | w*\Lambda - \lambda = \sum n_\alpha \alpha\}} F_{w*\Lambda}^{\{n_\alpha\}\perp} \right) , \qquad (6.103)$$

$\{n_\alpha\}$ are zero modes of ω_α, and λ is a fixed H-isospin. The projection is achieved by restricting the sum over n_α as shown above. The desired resolution is finally obtained after removing the degrees of freedom associated with the zero modes ξ_0^α, which were introduced in the bosonization procedure (6.93). This can be done by *extending* the complex (6.102) and introducing the intertwining operators $\eta_0^\alpha = \oint exp(i\hat{\chi}^\alpha)$ as follows [206]:

$$
\begin{array}{ccccccc}
0 & & 0 & & 0 & & \\
\downarrow & & \downarrow & & \downarrow & & \\
\cdots \longrightarrow & F_{\Lambda,\lambda}^{(-1)\perp} & \xrightarrow{d^{(-1)}} & F_{\Lambda,\lambda}^{(0)\perp} & \xrightarrow{d^{(0)}} & F_{\Lambda,\lambda}^{(1)\perp} & \xrightarrow{d^{(1)}} \cdots \\
& \downarrow \eta_0^\alpha & & \downarrow \eta_0^\alpha & & \downarrow \eta_0^\alpha & \\
\cdots \longrightarrow & \cdots & \longrightarrow & \cdots & \longrightarrow & \cdots & \longrightarrow \cdots \\
& \downarrow & & \downarrow & & \downarrow &
\end{array}
\qquad (6.104)
$$

The cohomology of this complex is concentrated in the Fock space $F_{\Lambda,\lambda}^{(0)\perp}$ which is isomorphic to the irreducible module $L_{\Lambda,\lambda}$ by construction.

The power of the resolution technique can be illustrated by deriving the AKM character of the irreducible parafermion module $L_{\Lambda,\lambda}$, representing the chiral partition function on a torus [206]. The highest-weight vector in the Fock space to be derived from $F_{\Lambda,\lambda}^{\{n_\alpha\}\perp}$ by taking m steps with the intertwining operator η_0^α has the (conformal) dimension

$$\tfrac{1}{2}(m+n_\alpha)(m+n_\alpha+1) - \tfrac{1}{2}n_\alpha(n_\alpha+1) = \tfrac{1}{2}m(m+1) + n_\alpha m , \qquad (6.105)$$

where the zero-mode relation $L_0^\chi = \tfrac{1}{2}p_\chi(p_\chi+1)$ has been used. The χ-trace over the η_0^α directions of the complex contains the alternating sum over zero modes, which implies the factor

$$\phi_n = \sum_{m \geq 0} (-1)^m q^{\frac{1}{2}m(m+1)+nm} , \qquad (6.106)$$

according to eq. (6.105). A calculation of the standard alternating sum over the affine Weyl group $\hat{W}$ of $\hat{\mathcal{G}}$ in the general character formula goes as usual (sect. XII.2), and it results, in this case, in the following character for the irreducible parafermionic module:

$$\text{ch}_{\Lambda,\lambda} = \frac{q^{-c/24}}{\prod_{n\geq 1}(1-q^n)^{2|\Delta+|}} \sum_{w\in\hat{W}} (-1)^{\hat{l}(w)} q^{\Delta_{w*\Lambda,\lambda}} \sum_{\{n_\alpha\in\mathbb{Z}|w*\Lambda-\lambda=\sum n_\alpha\alpha\}} \left(\prod_\alpha \phi_{n_\alpha}\right) , \quad (6.107)$$

where (cf sect. XII.2)

$$\Delta_{\Lambda,\lambda} = \frac{(\Lambda,\Lambda+2\rho)}{2(k+\bar{h}_G)} - \frac{|\lambda|^2}{2k} , \quad \rho = \tfrac{1}{2}\sum_{\alpha\in\Delta_+}\alpha . \quad (6.108)$$

Eq. (6.108) is to be compared with other known representations of the parafermionic character in terms of the so-called Kač-Peterson string functions [211] or Hecke modular forms [212]. This leads to the non-trivial identity [206]

$$\frac{1}{\prod_{n\geq 1}(1-q^n e^{2\pi i\theta\cdot\alpha})(1-q^{n-1}e^{-2\pi i\theta\cdot\alpha})} = \frac{1}{\prod_{n\geq 1}(1-q^n)^2} \sum_{n\in\mathbb{Z}} e^{-2\pi in\theta\cdot\alpha}\phi_n . \quad (6.109)$$

Other examples of Felder's resolution for the non-trivial cosets G/H with a non-abelian subgroup H can be found in ref. [206]. It should be noticed here that a free field resolution of a given coset CFT is not unique.

Exercise

VI-10 ▷ Verify that the transformation $(\chi,\omega,\phi) \to (\hat{\chi},\hat{\omega},\tilde{\phi})$ defined by eqs. (6.94) and (6.95) is orthogonal, i.e.

$$\left\langle \tilde{\phi}^i(z)\tilde{\phi}^j(w) \right\rangle = -(\alpha_i,\alpha_j)\ln(z-w) ,$$

$$\left\langle \hat{\chi}^\alpha(z)\hat{\chi}^{\alpha'}(w) \right\rangle = -\delta^{\alpha\alpha'}\ln(z-w) ,$$

$$\left\langle \hat{\omega}^\alpha(z)\hat{\omega}^{\alpha'}(w) \right\rangle = \delta^{\alpha\alpha'}\ln(z-w) . \quad (6.110)$$

Check that the other two-point functions vanish.

VI.6   Generalized affine-Virasoro construction

The SS and GKO constructions discussed in the previous sections, can be further generalized [47, 213, 214, 215]. The construction begins with the currents $J_a(z)$ of an AKM algebra $\hat{\mathcal{G}}$, satisfying the commutation relations

$$[J_a^m, J_b^n] = if_{ab}{}^c J_c^{m+n} + mG_{ab}\delta_{m+n,0} , \quad (6.111)$$

in terms of the structure constants $f_{ab}{}^c$ of Lie algebra $\mathcal{G}$ and the *generalized* Killing metric G_{ab} to be defined below. Given the currents, one considers the general stress tensor

$$L(z) = L^{ab}(z) : J_a(z)J_b(z) : , \qquad (6.112)$$

which is bilinear in the currents. The stress tensor (6.112) is required to satisfy the Virasoro algebra

$$[L^m, L^n] = (m - n)L^{m+n} + \frac{c}{12}m(m^2 - 1)\delta_{m+n,0} . \qquad (6.113)$$

Eq. (6.113) is the equation on the tensor $L^{ab}(z)$. The latter is sometimes called the *inverse inertia tensor*, in analogy with the spinning top [47]. Eq. (6.113) implies, in particular

$$[L^m, J_a^n] = -n\left(2G_{ab}L^{be} + f_{ab}{}^d L^{bc}f_{cd}{}^e\right)J_e^{m+n} - 2if_{ab}{}^d L^{bc}L_{cd}^{m+n} , \qquad (6.114)$$

and

$$L^{ab} = 2L^{ac}G_{cd}L^{db} - L^{cd}L^{ef}f_{ce}{}^a f_{df}{}^b - L^{cd}f_{ce}{}^f f_{df}{}^{(a}L^{b)e} , \qquad (6.115a)$$

$$c = 2G_{ab}L^{ab} . \qquad (6.115b)$$

This construction is not restricted to semi-simple or compact Lie groups. Eq. (6.115a) is known as the **Virasoro master equation** [47, 213, 214, 215].

Let $\mathcal{G} = \bigoplus_I \mathcal{G}_I$ be a semi-simple Lie algebra with dual Coxeter number $\tilde{h}_I = Q_I/\psi_I^2$, and $k_I = 2\tilde{k}_I/\psi_I^2$ the level of the corresponding AKM algebra. One has [213]

$$T_a = \bigoplus_I T_a^I , \quad G_{ab} = \bigoplus_I \tilde{k}_I \eta_{ab}^I ,$$

$$f_{ac}{}^d f_{bd}{}^c = -\bigoplus_I \text{tr}\,(T_a^I T_b^I) = -\bigoplus_I Q_I \eta_{ab}^I , \qquad (6.116)$$

where $(T_a)_c^b = -if_{ab}{}^c$ are the Lie algebra generators in the adjoint representation, and η_{ab}^I is the Killing metric of $\mathcal{G}_I$.

Is is not difficult to see that the SS construction (sect. IV-2) is a particular solution to the Virasoro master equation. The SS construction is reproduced by assigning

$$L^{ab}_{(\mathcal{G})} = \bigoplus_I \frac{\eta_I^{ab}}{2\tilde{k}_I + Q_I} , \quad c_{(\mathcal{G})} = \sum_I \frac{k_I \dim \mathcal{G}_I}{k_I + \tilde{h}_I} . \qquad (6.117)$$

A classification of all solutions to the Virasoro master equation (6.115a) is still unknown. Nevertheless, it is possible to make some general statements about all of its solutions [214]. First, given any solution L^{ab} to the Virasoro master equation, then so is

$$L^{ab} \rightarrow \tilde{L}^{ab} = L^{ab}_{(\mathcal{G})} - L^{ab} , \quad c \rightarrow \tilde{c} = c_{(\mathcal{G})} - c . \qquad (6.118)$$

This generalizes the GKO coset construction, where we had $L = L_{(\tilde{H})}$ for a subalgebra $\tilde{H}$ of $\mathcal{G}$, so that $\tilde{L} = L_{(G/H)}$.

Second, the Virasoro master equation is invariant under the *local* group transformations in the adjoint representation,

$$L^{ab} \rightarrow L^{cd}\Omega_c{}^a\Omega_d{}^b , \tag{6.119}$$

where $\Omega_a{}^c G_{cd}\Omega_b{}^d = G_{ab}$ and $\Omega(z) \in G$. This allows us to geometrically interpret the Virasoro master equation as an equation describing an Einstein-like system on the group manilod G in terms of the left-invarint metric $g_{ij} = e_i^a L_{ab} e_j^b$. [9] Introducing the left-right invariant metric $g_{ij}^{(\mathcal{G})}$ associated with the SS construction, the master equation can be rewritten to the form [214]

$$\hat{R}_{ij} + g_{ij} = g_{ij}^{(\mathcal{G})} , \quad c = \dim \mathcal{G} - 4R , \tag{6.120}$$

where $\hat{R}$ stands for the the Ricci tensor with specific torsion squared terms, and R is the scalar curvature to be constructed in terms of the metric g_{ij} (see ref. [214] for more details.)

One could go even further, and generalize the 'generalized' affine-Virasoro construction by adding the linear terms as follows [214]

$$L(z) = L^{ab}(z) : J_a(z)J_b(z) : +D^a(z)\partial J_a(z) + d^a(z)J_a(z) , \tag{6.121}$$

with the coefficients $L^{ab}(z)$, $D^a(z)$ and $d^a(z)$ to be determined. The generalized Virasoro master equation and the associated central charge then take the form (e.g., when $d^a = 0$ for simplicity) [214]

$$L^{ab} = 2L^{ac}G_{cd}L^{db} - L^{cd}L^{ef}f_{ce}{}^a f_{df}{}^b - L^{cd}f_{ce}{}^f f_{df}{}^{(a}L^{b)e} + if_{cd}{}^{(a}L^{b)c}D^d ,$$

$$D^a\left(2G_{ab}L^{be} + f_{ab}{}^d L^{bc}f_{cd}{}^e\right) = D^e , \quad c = 2G_{ab}\left(L^{ab} - 6D^aD^b\right) . \tag{6.122}$$

These equations can also be geometrically interpreted after rewriting them to the form of Einstein-Maxwell-like equations [214],

$$\hat{R}_{ij} + g_{ij} + i\mathcal{L}_D g_{ij} = g_{ij}^{(\mathcal{G})} , \quad D^i F_{ij} = iD^i\mathcal{L}_D g_{ij} ,$$

$$c = \dim \mathcal{G} - 4R + 6(F^2 - D^2) , \tag{6.123}$$

---

[9]Early Latin letters are used for tangent space indices, while middle Latin letters are world indices. The matrix e_i^a stands for a 'vielbein'.

where $\mathcal{L}_D$ is the Lie derivative in D-direction, and $F_{ij} = \partial_{[i}D_{j]}$. It is yet unclear, whether there are any other interesting solutions to the master equations beyond the known conformal deformations by conformal operators of dimension one. The supersymmetric form of the generalized affine-Virasoro construction is quite similar [215].

Chapter VII

W Algebras

Since the Virasoro symmetry plays the fundamental role in CFT, we should also investigate its possible extensions, including non-linear generalizations, which may serve as the chiral algebras in CFT. One of the natural ways is to explore bosonic and fermionic *higher-spin* extensions of the Virasoro algebra, which are generically called the W **algebras**. Like any other additional symmetry in CFT, the W symmetries are useful for further investigations of RCFT's, as well as for finding exact solutions to the 2d QFT's having these symmetries. In Chs. IV and V we already discussed the AKM and supersymmetric *linear* extensions of the Virasoro algebra. The W algebras are of the more general type, with *multi-linear* OPE's. These algebras go beyond the scope of Lie and AKM algebras but, nevertheless, they share some of their basic properties, e.g. the associativity. In this Chapter, the brief account of the W algebras is given. We concentrate on ideology and examples. We start with the simplest non-trivial W_3 algebra of Zamolodchikov, then we discuss its generalizations and their relevance for 2d CFT. The highest-weight representations and free field constructions of some W algebras are given. The W algebras can be considered as quantum versions of certain classical Hamiltonian structures appearing in the theory of KdV-type equations (Ch. XI). The powerful technique achieved by Drinfeld and Sokolov, and known as the *classical Hamiltonian reduction*, provides the tools for explicit constructions of the classical Hamiltonian structures associated with Lie and AKM algebras. The W algebras can be constructed by quantizing these classical structures. This method is known as the *quantum Drinfeld-Sokolov (DS) reduction*. We discuss general features of the DS reduction. The Lagrangian QFT's associated with the quantum DS reduction are the so-called *Toda* field theories arising from the constrained WZNW models (sect. XI.3).

VII.1 W_3 algebra and its generalizations

The W_3 algebra was first introduced by Zamolodchikov [130]. In addition to the Virasoro field $T(z)$, it contains a primary field $W(z)$ of dimension (spin) 3. Their mode expansions are $(n, m \in \mathbf{Z})$

$$T(z) = \sum_n L_n z^{-n-2} , \quad W(z) = \sum_n W_n z^{-n-3} . \tag{7.1}$$

The operators L_n satisfy the Virasoro algebra (1.78) of central charge c. whereas one has, in addition

$$[L_n, W_m] = (3n - n - m)W_{n+m} , \tag{7.2}$$

in accordance with eq. (1.66). The nontrivial commutator of the Laurent modes of the W field reads as follows [130]:

$$[W_m, W_n] = \frac{c}{360}m(m^2 - 1)(m^2 - 4)\delta_{m+n,0}$$

$$+(m - n)\left[\frac{1}{15}(m + n + 3)(m + n + 2) - \frac{1}{6}(m + 2)(n + 2)\right]L_{m+n}$$

$$+\beta(m - n)\Lambda_{m+n} , \tag{7.3}$$

where

$$\beta = \frac{16}{22 + 5c} , \quad \Lambda_m = \sum_n (L_{m-n}L_n) - \frac{3}{10}(m + 3)(m + 2)L_m . \tag{7.4}$$

In terms of the OPE's, the W_3-algebra commutation relations can be rewritten to the form

$$T(z)T(w) = \frac{c/2}{(z - w)^4} + \frac{2T(w)}{(z - w)^2} + \frac{\partial T(w)}{z - w} + \left[\Lambda(w) + \frac{3}{10}\partial^2 T(w)\right] + \dots ,$$

$$T(z)W(w) \sim \frac{3W(w)}{(z - w)^2} + \frac{\partial W(w)}{z - w} ,$$

$$W(s)W(w) \sim \frac{c/3}{(z - w)^6} + \frac{2T(w)}{(z - w)^4} + \frac{\partial T(w)}{(z - w)^3}$$

$$+\frac{1}{(z - w)^2}\left[2\beta\Lambda(w) + \frac{3}{10}\partial^2 T(w)\right] + \frac{1}{z - w}\left[\beta\partial\Lambda(w) + \frac{1}{15}\partial^3 T(w)\right] , \tag{7.5}$$

where

$$\Lambda(w) =: T(w)T(w) : -\frac{3}{10}\partial^2 T(z) . \tag{7.6}$$

The leading *finite* term, which is the normally ordered product of T's, has been added to the singular terms of the OPE in the first line of eq. (7.5), in order to make it explicit that the stress tensor T itself can be used to define the composite field Λ.

The W_3 algebra is clearly *not* a Lie-type algebra because of the quadratic terms appearing on the r.h.s. of eq. (7.5). However, its commutation relations are actually fixed by the 'Jacobi identities'. The linear commutation relations could be achieved when considering Λ as the new field. This would imply, however, the necessity to introduce additional commutation relations as well as new higher-spin generators without end. The different opportunity is to interpret Λ as the composite field and then ask for a closure in the *enveloping* algebra of T and W. To a large extent, the form of any W algebra is fixed by the conformal Ward identities, which severely restrict operator products of any two Virasoro primaries. The OPE's (7.5) are totally fixed by requiring the crossing symmetry of the correlation functions or the associativity (i.e. the 'Jacobi identities') [130]. In particlular, the four-point function of the W-fields takes the form

$$\langle W(z_1)W(z_2)W(z_3)W(z_4)\rangle = f(x)(z_1 - z_3)^{-6}(z_2 - z_4)^{-6} , \qquad (7.7)$$

where (see eq. (1.49) for the definition of x variable)

$$f(x) = \frac{c^2}{9}\left[\frac{1}{x^6} + \frac{1}{(1-x)^6} + 1\right]$$

$$+2c\left[\frac{1}{x^4} + \frac{1}{(1-x)^4} + \frac{1}{x^3} + \frac{1}{(1-x)^3} - \frac{1}{x} + \frac{1}{1-x}\right]$$

$$+c\left(\frac{9}{5} + \frac{32}{5}\frac{16}{(22+5c)}\right)\left[\frac{1}{x^2} + \frac{1}{(1-x)^2} + \frac{2}{x} + \frac{2}{1-x}\right] , \qquad (7.8)$$

whose crossing symmetry can be explicitly verified (exercise # VII-1). The central charge c of the W_3 algebra given above is arbitrary. Like the stress tensor T, the composite field Λ is quasi-primary.

Given the AKM currents $J^a(z)$ associated with an AKM algebra $\hat{\mathcal{G}}_k$, one can try to generalize the SS construction to the case of the W-algebras [216]. The simplest example is provided by the AKM algebra $\hat{\mathcal{G}} = A_2^{(1)}$ which has two independent Casimir invariants of order 2 and 3. Quite generally, the third-order Casimir invariant of the underlying representation of Lie algebra $\mathcal{G}$ takes the form

$$W^{(3)}(z) = \frac{1}{6(k+3)}\sqrt{\frac{6}{5(2k+3)}}d_{abc} : J^a(z)J^b(z)J^c(z) : , \qquad (7.9)$$

where d_{abc} is the totally symmetric traceless $\mathcal{G}$-invariant tensor. Eq. (7.9) defines the $W^{(3)}$ field of spin 3, having the following OPE with itself [216]

$$W^{(3)}(z)W^{(3)}(w) \sim \frac{c/3}{(z-w)^6} + \frac{2T(w)}{(z-w)^4} + \frac{\partial T(w)}{(z-w)^3}$$

$$+\frac{1}{(z-w)^2}\left[2\beta\Lambda(w)+\frac{3}{10}\partial^2 T(w)+R^{(4)}\right]$$

$$+\frac{1}{z-w}\left[\beta\partial\Lambda(w)+\frac{1}{15}\partial^3 T(w)+\tfrac{1}{2}\partial R^{(4)}\right] \tag{7.10}$$

with the central charge $c = 8k/(k+3)$, where the new primary field $R^{(4)}$ appears. In general, the new field $R^{(4)}$ does not vanish and, hence, the operator algebra on the space spanned by the original fields T and $W^{(3)}$ does not close. [1] Fortunately, in the case of level $k = 1$, it can be shown [216] that the field $R^{(4)}$ actually corresponds to a null state, and it decouples from the rest of the algebra. Without this field, we are just left with the W_3 algebra of central charge $c = 2$. The similar constructions can be applied to all simply-laced classical Lie algebras, in order to get the corresponding W algebras, when $k = 1$ [131].

The $N = 1$ *supersymmetric* W_3 algebra is obtained from the $N = 1$ super-Virasoro algebra of eq. (5.4) by adding a super-primary field $W(z)$ of dimension $5/2$ [217]. This implies the OPE

$$T(z_1)W(z_2) \sim \frac{5\theta_{12}}{2z_{12}^2}W(z_2) + \frac{1}{2z_{12}}\mathbf{D}_2 W(z_2) + \frac{\theta_{12}}{z_{12}}\partial W(z_2) . \tag{7.11}$$

The crucial OPE for $W(z_1)W(z_2)$ is fixed by the condition of associativity under the 'minimal' assumption that the superconformal family members can only appear on the r.h.s. The TWW 'Jacobi identity' then ultimately fixes this OPE, in the form

$$W(z_1)W(z_2) \sim \frac{1}{z_{12}^5}\frac{c}{15} + \frac{\theta_{12}}{z_{12}^4}T(z_2) + \frac{1}{3z_{12}^3}\mathbf{D}_2 T(z_2) + \frac{2\theta_{12}}{3z_{12}^3}\partial T(z_2)$$

$$\frac{\theta_{12}}{z_{12}^2}\frac{1}{(4c+21)}\left[(c+\tfrac{5}{2})\partial^2 T(z_2) + 22T\mathbf{D}_2 T(z_2)\right]$$

$$+\frac{1}{z_{12}}\frac{1}{(4c+21)(10c-7)}[(36c+2)\mathbf{D}_2 T(z_2)\mathbf{D}_2 T(z_2)$$

$$+\frac{4}{3}(2c^2-c-37)\mathbf{D}_2\partial^2 T(z_2) - (4c-166)T\partial T(z_2)]$$

$$+\frac{\theta_{12}}{z_{12}}\frac{1}{(4c+21)(10c-7)}[(112c-160)T\mathbf{D}_2\partial T(z_2)$$

$$+(2c^2-29c+3)\partial^3 T(z_2) + (144c+8)\mathbf{D}_2 T\partial T(z_2)] . \tag{7.12}$$

The final check comes from the WWW 'Jacobi identity', which yields [131]

$$[W,\{W,W\}] = \partial^2\Psi(z) , \tag{7.13}$$

---

[1]The generalized algebras to be constructed this way are called the *Casimir algebras* [131].

where the field $\Psi(z)$ takes the form [131]

$$\Psi(z) = \frac{1}{9(4c+21)(10c-7)}\left[3(2c-83)T\partial DW(z) + 12(18c+1)\mathbf{D}T\partial W(z)\right.$$

$$\left. -6(2c-83)\partial T\mathbf{D}W(z) - 15(18c+1)\partial \mathbf{D}TW(z) + (2c^2 - 29c+3)\partial^3 W(z)\right] . \quad (7.14)$$

Therefore, the 'Jacobi identities' are, in general, not satisfied. The only hope, which still remains, is to take a look at the case where the field $\Psi(z)$ actually becomes the superprimary field corresponding to a null state. This happens only at specific values of central charge, namely $c = 10/7$ and $c = -5/2$, and, hence, only for these central charge values the super-W_3 algebra does exist [131]. The same conclusion comes from requiring the crossing symmetry of the 4-point correlation functions [217]. Zamolodchikov's (bosonic) W_3 algebra is a subalgebra of the super-W_3 algebra for $c = 10/7$. The $N = 2$ super-W_3 algebra can be constructed along the similar lines [218].

More general W-algebras $W(2, s_2, s_3, \ldots, s_n)$ comprising a stress tensor of dimension 2 and primary fields of conformal dimensions (spins) $s_2, s_3, \ldots, s_n$, $s_i \geq 3$, can also be constructed [219, 220, 221]. The W_3 algebra of Zamolodchikov corresponds to the $W(2,3)$, while the $W(2,3,\ldots,n)$ are generically called the W_n algebras. As an explicit example, we present the $W(2,4)$ algebra whose relevant OPE's take the form [220, 221]

$$W(z)W(w) \sim \frac{c/4}{(z-w)^8} + \frac{2T(w)}{(z-w)^6} + \frac{\partial T(w)}{(z-w)^5}$$

$$+\frac{1}{(z-w)^4}\left[\frac{3}{10}\partial^2 T(w) + 2\gamma\Lambda(w)\right] + \frac{1}{(z-w)^3}\left[\frac{1}{15}\partial^3 T(w) + \gamma\partial\Lambda(w)\right]$$

$$+\frac{1}{(z-w)^2}\left[\frac{1}{84}\partial^4 T(w) + \frac{5}{18}\gamma\partial^2\Lambda(w) + \frac{24}{\mu}(72c+13)\Omega(w)\right.$$

$$\left. -\frac{1}{6\mu}(95c^2 + 1254c - 10904)P(w)\right]$$

$$+\frac{1}{z-w}\left[\frac{1}{560}\partial^5 T(w) + \frac{1}{18}\gamma\partial^3\Lambda(w) + \frac{12}{\mu}(72c+13)\partial\Omega(w)\right.$$

$$\left. -\frac{1}{12\mu}(95c^2 + 1254c - 10904)\partial P(w)\right]$$

$$+C_{44}^4\left\{\frac{1}{(z-w)^4}W(w) + \frac{1}{2(z-w)^3}\partial W(w) + \frac{1}{(z-w)^2}\left[\frac{5}{36}\partial^2 W(w) + \frac{28}{3(c+24)}H(w)\right]\right.$$

$$\left. +\frac{1}{z-w}\left[\frac{1}{36}\partial^3 W(w) + \frac{14}{3(c+24)}\partial H(w)\right]\right\} , \quad (7.15)$$

where

$$\Lambda(w) = \; : T(w)T(w) : -\frac{3}{10}\partial^2 T(w) \; ,$$

$$\Omega(w) = \; : \Lambda(w)T(w) : -\frac{3}{5} : \partial^2 T(w)T(w) : -\frac{1}{28}\partial^4 T(w) \; ,$$

$$P(w) = \tfrac{1}{2}\partial^2\Lambda(w) - \frac{9}{5} : \partial^2 T(w)T(w) : +\frac{3}{70}\partial^4 T(w) \; ,$$

$$H(w) = \; : T(w)W(w) : -\frac{1}{6}\partial^2 W(w) \; ,$$

(7.16a)

and

$$\gamma = \frac{21}{22+5c} \; , \qquad \mu = (5c+22)(2c-1)(7c+68) \; ,$$

$$(C_{44}^4)^2 = \frac{54}{\mu}(c+24)(c^2-172c+196) \; .$$

(7.16b)

These explicit formulae are enough to emphasize a complexity of such algebras. Other explicit examples of the non-linear $W(2, s_2, s_3, \ldots, s_n)$ algebras were also discovered [219, 220, 221], as well as their supersymmetric extensions [222, 223]. [2] All of them can be divided into two categories: those ('*generic*') which allow arbitrary values of their central charge c, and those ('*exotic*') which only exist at some (finite number of) values of c [131]. For instance, the algebras $W(2, n)$ are generic for $n = 3, 4, 6$, but they are exotic otherwise. The 'minimal' W_n-superalgebras are all exotic [131].

The different class of the W algebras is formed by the so-called $w_\infty, W_\infty, W_{1+\infty}$ and $W_\infty(\lambda)$ algebras [224, 225, 226, 227, 228, 229]. They include arbitrarily high spins, but they are all, in fact, *linear* algebras, that sharply distinquishes them from the algebras introduced above. Some of these W_∞-type algebras have very clear geometrical meaning, being related to the higher-dimensional gauge field theories such as the (super)-self-dual Yang-Mills theories and the self-dual (super)gravities in four dimensions (sects. XI.4 and XI.5). The simplest algebra among them is the w_∞ algebra [224], which is the algebra of the *area-preserving* diffeomorphisms of a two-dimensional cylinder. It comprises the generators $w_m^{(s)}$ of all spins $s = 2, 3, 4, \ldots$, and it has the commutation relations

$$[w_m^{(s)}, w_n^{(t)}] = [(t-1)m - (s-1)n]\, w_{m+n}^{(s+t-2)} \; .$$

(7.17)

Though the w_∞ algebra contains the (classical) conformal algebra (1.16) as a subalgebra, the Virasoro algebra cannot be extended this way. Hence, the w_∞ should be considered as a classical W algebra. Its quantum version is the so-called W_∞-algebra introduced

---

[2] See ref. [131] and references therein for a review and more details.

in refs. [226, 227]. Denoting V_m^i the generators of this W_∞-algebra, $s = i + 2$, its commutation relations take the form [226, 227]

$$[V_m^i, V_n^j] = \sum_{l \geq 0} g_{2l}^{ij}(m, n) V_{m+n}^{i+j-2l} + c_i(m) \delta^{ij} \delta_{m+n} , \qquad (7.18)$$

where the structure constants $g_{2l}^{ij}(m, n)$ and the central terms $c_i(m)$ are to be fixed by the 'Jacobi identities'. This results in [226, 227]

$$c_i(m) = m(m^2 - 1)(m^2 - 4) \cdots (m^2 - (i + 1)^2) c_i ,$$

$$g_{2l}^{ij}(m, n) = \frac{1}{2(l + 1)!} \phi_l^{ij} N_l^{ij}(m, n) , \qquad \text{(no sum!)} , \qquad (7.19)$$

where the central charges c_i and the constants ϕ_l^{ij}, $N_l^{ij}(m, n)$ are

$$c_i = \frac{2^{2i-3} i! (i + 2)!}{(2i + 1)!! (2i + 3)!!} c , \qquad (7.20)$$

and

$$N_l^{ij} = \sum_{k=0}^{l+1} (-1)^k \binom{l + 1}{k} [i + 1 + m]_{(l+1-k)} [i + 1 - m]_k [j + 1 + n]_{(l+1-k)} [j + 1 - n]_{(l+1-k)} ,$$

$$\phi_l^{ij} = \sum_{k \geq 0} \frac{(-\frac{1}{2})_k (\frac{3}{2})_k (-\frac{l}{2} - \frac{1}{2})_k (-\frac{l}{2})_k}{k! (-i - \frac{1}{2})_k (-j - \frac{1}{2})_k (i + j - l + \frac{5}{2})_k} , \qquad (7.21)$$

with $[x]_n \equiv \Gamma(x + 1)/\Gamma(x + 1 - n)$, and $(x)_n \equiv \Gamma(x + n)/\Gamma(x)$.

The $W_{1+\infty}$-algebra contains, in addition, a generator of spin 1 [228], whereas the $W_\infty(\lambda)$-algebras represent the whole family parametrized by a deformation parameter λ [229]. Their supersymmetric generalizations are also known [229, 230].

There also exist the so-called *twisted* and *projected* W algebras, which can be constructed by exploiting the discrete symmetries (automorphisms) $\mathbf{Z}_2$ or $\mathbf{Z}_3$ of the underlying AKM algebras, and projecting these algebras by integer modding [231, 232, 233]. Given so many W algebras, one might ask about an existence of the *universal* W_∞-like algebra, from which all the others would follow by reductions and/or truncations [234]. In particular, it can be shown that the W_n algebras *at* $c = -2$ can actually be derived from the W_∞-algebra by reductions [235]. It was suggested [236] that the *non-linear* deformation $\hat{W}_\infty(k)$ of the W_∞-algebra, which is the chiral algebra of the non-compact coset model $\widehat{SL(2, \mathbf{R})}_k / \widehat{U(1)}$, should play the role of a 'universal' W algebra for all the W_n algebras based on Lie algebras A_{n-1}. Similar universal W algebras are expected to exist for the other series of W algebras based on classical Lie algebras.

The additional motivation for studying the W algebras in CFT was given in Ch. IV. These algebras can in fact be identified with the *chiral* algebras of the specific conformal models posessing extended symmetries. In particular, the W_3 algebra originated as the underlying symmetry of the critical 3-state Potts model of central charge $c = 4/5$. This model appears in the unitary discrete series of the minimal models [130]. Similarly, the W_n algebras are associated with the statistical models having Z_n symmetry. In fact, the two modular-invariant partition functions are known to exist for the 3-state Potts model. One of them corresponds to the exceptional modular invariant defined in eq. (4.117), in terms of the Virasoro characters χ_h with the highest weights (dimensions) $h = 0, 1/15, 2/5, 2/3, 7/5, 3$. The fusion rules of this model take the form [12]

$$[\phi_{(3,0)}] \times [\phi_{(2/5,*)}] = [\phi_{(7/5,*)}] \,, \quad [\phi_{(3,0)}] \times [\phi_{(7/5,*)}] = [\phi_{(2/5,*)}] \,,$$

$$[\phi_{(3,0)}] \times [\phi_{(2/3,2/3)}] = [\phi_{(2/3,2/3)}] \,, \quad [\phi_{(3,0)}] \times [\phi_{(1/15,1/15)}] = [\phi_{(1/15,1/15)}] \,, \quad (7.22)$$

where the star stands for either 2/5 or 7/5. Eqs. (4.117) and (7.22) indicate the presence of the extended symmetry generated by the spin-3 primary fields $\phi_{(3,0)}(z)$ and $\phi_{(0,3)}(\bar{z})$. In particular, the states $|0\rangle$ and $\phi_{(3,0)}(z)|0\rangle$ should then belong to the same representation, whereas the partition function Z should be expressible in terms of the characters $\hat{\chi}$ associated with the extended chiral algebra generated by $T(z)$ and $\phi_{(3,0)}(z)$, i.e. with the W_3 algebra, and this does happen indeed: see eq. (4.118). Notably, when written in terms of the extended characters, the partition function is of diagonal form, representing the simplest W_3 modular invariant. Loosely speaking, the bigger the chiral algebra, the 'more diagonal' the partition function would be. Therefore, it is the clear advantage to have the maximal chiral algebra for the conformal model under consideration.

The similar phenomenon takes place in the minimal SCFT of central charge $c = 10/7$, which has the same central charge as one of the super-W_3 algebras introduced above. Being written in terms of the $N = 1$ SCA Neveu-Schwarz and Ramond characters χ^{NS} and χ^{R}, resp., the corresponding partition function is known to be of non-diagonal form, namely [225]

$$Z_1 = \frac{1}{4} \sum_{s=1,\text{odd}}^{13} \left(\left| \chi_{1s}^{\text{NS}} + \chi_{5s}^{\text{NS}} + \chi_{7s}^{\text{NS}} + \chi_{11s}^{\text{NS}} \right|^2 + \left(\chi^{\text{NS}} \to \tilde{\chi}^{\text{NS}} \right) + \left| \chi_{4s}^{\text{R}} + \chi_{8s}^{\text{R}} \right|^2 \right) \,, \quad (7.23)$$

where the subscripts indicate the corresponding highest-weight states. Eq. (7.23) can be rewritten in terms of the super-W_3 algebra characters $\text{ch}^{\text{NS,R}}$ and $\bar{\text{ch}}^{\text{NS}}$ as follows [237]:

$$Z_1 = \tfrac{1}{2} \left(\left| \text{ch}_0^{\text{NS}} \right|^2 + \left| \text{ch}_{1/14}^{\text{NS}} \right|^2 + \left| \text{ch}_{5/14}^{\text{NS}} \right|^2 + \left| \text{ch}_{1/7,+}^{\text{NS}} \right|^2 + \left| \text{ch}_{1/7,-}^{\text{NS}} \right|^2 \right) + \tfrac{1}{2} \left(\text{ch}^{\text{NS}} \to \bar{\text{ch}}^{\text{NS}} \right)$$

$$+\frac{1}{8}\left(\left|\mathrm{ch}_{1/14}^{\mathrm{R}}\right|^2+\left|\mathrm{ch}_{5/14}^{\mathrm{R}}\right|^2+\left|\mathrm{ch}_{3/2}^{\mathrm{R}}\right|^2+\left|\mathrm{ch}_{9/14,+}^{\mathrm{R}}\right|^2+\left|\mathrm{ch}_{9/14,-}^{\mathrm{R}}\right|^2\right), \tag{7.24}$$

which is of the diagonal form! [3] These results are extendable to the W_n algebras at $c=\frac{2(n-1)}{(n+2)}$, and to the minimal super-W_n algebras at $c_n=\frac{(3n+1)(n-1)}{2(2n+1)}$ as well [237].

The GKO-type coset construction can also be promoted to the level of the W algebras, and this provides, in fact, the very powerful method for explicit constrictions of the highest-weight representations of the W algebras, and of the corresponding modular-invariant partition functions [131, 186]. For instance, taking the coset

$$\frac{\widehat{SU(n)}_k\otimes\widehat{SU(n)}_1}{\widehat{SU(n)}_{k+1,\mathrm{diag}}}, \tag{7.25}$$

results in the unitary representation of the W_n algebra [131, 186] of central charge

$$c=(n-1)\left[1-\frac{n(n+1)}{(k+n)(k+n+1)}\right]. \tag{7.26}$$

The stress tensor T is provided by the SS construction, $T=T_k+T_1-T_{k+1}$, $T_k\sim:J^aJ^a:$, whereas the $W^{(l)}$-fields, $l=3,4,\ldots,n$, are given by the products of the J-currents, corresponding to the Casimir invariants of order l (similarly to that in eq. (7.9)). [4]

As far as the minimal super-W_n algebras are concerned, there exists the similar (supersymmetric) GKO-type coset construction which is based on the coset

$$\frac{\widehat{SU(n)}_1\otimes\widehat{SU(n)}_n}{\widehat{SU(n)}_{n+1,\mathrm{diag}}}. \tag{7.27}$$

This results in the unitary representation of central charge

$$c_{\mathrm{s}}=(n-1)\left[1-\frac{n(n+1)}{2n(2n+1)}\right]. \tag{7.28}$$

When $n=3$, eq. (7.28) yields $c=10/7$, as it should.

Exercise

\# VII-1 ▷ (*) Verify the crossing symmetry of the 4-point function defined by eqs. (7.7) and (7.8).

---

[3] The additional subscripts $\pm$ in eq. (7.24) denote the sign of the W_0-eigenvalue.
[4] See sect. 4 for more about the W coset construction.

VII.2   Free field approach

The free-field construction in the Feigin-Fuchs-type approach, which was initiated in sect. II.6 and then developed further in sect. IV.4, can also be applied to the W_n algebras [238, 239]. Our presentation in this section is based on the Lectures of Bilal [240]. The basic idea is to construct $n-1$ fields $T, W^{(3)}, W^{(4)}, \ldots, W^{(n)}$ out of $n-1$ *free* fields ϕ^i, $i = 1, \ldots, n-1$.

First we need, however, to generalize the DF construction (sects. II.6 and IV.4) to the case of general Lie algebras. Given $n-1$ quantized scalar free fields ϕ^i, $i = 1, \ldots, n-1$, with the 2-point correlation functions

$$\left\langle \phi^i(z)\phi^j(w) \right\rangle = -\delta^{ij}\ln(z-w) , \qquad (7.29)$$

their mode expansions are given by

$$\phi^j(z) = \phi_0^j - ia_0^j \ln z + i \sum_{n \neq 0} a_n^j \frac{z^{-n}}{n} , \qquad (7.30)$$

where

$$[a_n^j, a_m^l] = \delta^{jl} n \delta_{m+n,0} , \quad [\phi_0^j, a_0^l] = i\delta^{jl} . \qquad (7.31)$$

The 'Hilbert space' of states [5] in this case is clearly the direct sum of the Fock spaces, $a_{-n_1}^{j_1} \cdots a_{-n_r}^{j_r} |\alpha\rangle$, each one being built upon the vacuum $|\alpha\rangle$, $a_0^j |\alpha\rangle = \alpha^j |\alpha\rangle$, where α^j is the eigenvalue of the zero-mode a_0^j. These Fock spaces are related by the operator $\exp(i\beta \cdot \phi_0)$, so that

$$\exp(i\beta \cdot \phi_0) |\alpha\rangle = |\alpha + \beta\rangle . \qquad (7.32)$$

The space of states associated with these free fields is obviously bigger than a representation space of the W_n algebra we are looking for, because of the null states and symmetries associated with them. Nevertheless, this is the good starting point for a free field construction of the W_n fields.

Eq. (2.79) suggests us to represent the (holomorphic part of) stress tensor $T(z)$ as follows

$$T(z) = -\tfrac{1}{2} : \partial\phi(z) \cdot \partial\phi(z) : +2i\alpha_0\rho \cdot \partial^2\phi(z) , \qquad (7.33)$$

---

[5]The hermiticity conditions $a_n^\dagger = -a_{-n}$, which are implied here at $n \neq 0$, lead to the indefinite inner product.

where a (constant) $(n-1)$-component vector ρ, as well as another constant α_0 (for a later convenience) have been introduced. The associated central charge is given by (sect. II.6)

$$c = (n-1) - 48\alpha_0^2\rho^2 . \tag{7.34}$$

The Virasoro primary fields of free field system are known to be represented by the vertex operators $V_\alpha(z) =: \exp(i\alpha\phi)(z) :$ of conformal dimension (sect. II.6)

$$h_2(\alpha) = \tfrac{1}{2}\alpha \cdot (\alpha - 4\alpha_0\rho) = \tfrac{1}{2}(\alpha - 2\alpha_0\rho)^2 - 2\alpha_0^2\rho^2 . \tag{7.35}$$

These vertex operators connect different Fock spaces, while the a_0-eigenvalue becomes shifted, $a_0 \to a_0 + \alpha$, in particular.

To introduce a free field representation of the W algebras having $A_{n-1} \sim su(n)$ as their underlying Lie algebras, [6] we need a basis in the $(n-1)$-dimensional 'target space' of our free fields. The convenient choice is provided by the simple roots or the fundamental weights of the A_{n-1} (sect. XII.2), when the vector ρ introduced above is chosen to be the Weyl vector to be defined by eq. (12.27). The Virasoro central charge then takes the form

$$c = (n-1)[1 - 4\alpha_0^2 n(n+1)] . \tag{7.36}$$

To construct correlation functions, it is useful to introduce the Feigin-Fuchs currents of conformal dimension one,

$$V_\pm^j(z) \equiv V_{\alpha_\pm e_j}(z) =: \exp[i\alpha_\pm e_j \cdot \phi](z) : , \tag{7.37}$$

where the parameters $\alpha_\pm$ are subject to the condition (sect. II.6)

$$\alpha_\pm^2 - 2\alpha_0\alpha_\pm = 1 , \tag{7.38}$$

in accordance with eq. (7.35). Therefore, we get $2(n-1)$ Feigin-Fuchs operators having the form

$$Q_\pm^j = \oint dw\, V_\pm^j(w) , \tag{7.39}$$

with the appropriately chosen closed contours of integration (sect. IV.4). The relevant properties of the Feigin-Fuchs operators are

$$T(z)V_\pm^j(w) \sim \frac{V_\pm^j(w)}{(z-w)^2} + \frac{\partial V_\pm^j(w)}{z-w} = \partial_w\left(\frac{V_\pm^j(w)}{z-w}\right) , \tag{7.40}$$

---

[6]The cases of $B_n \sim so(2n+1)$ and $D_n \sim so(2n)$ can be considered similarly [131]. In our notation, we often use the real forms of classical (complex) Lie algebras.

and

$$\oint dw\, T(z) V_{\pm}^{j}(w) = 0 \ , \quad [L_n, Q_{\pm}^{j}] = 0 \ . \tag{7.41}$$

The stress tensor (7.33) is known to be derivable from the free scalar action S_0, modified by the 'background charge' contribution (sect. II.6) and having the form

$$S = S_0 + \frac{i\alpha_0}{\pi} \int d^2 z\, \sqrt{g} R^{(2)} \rho \cdot \phi \ . \tag{7.42}$$

After the change of variables $\phi^j \to \phi^j + c^j$ with arbitrary constants c^j, one finds the Ward identity for the correlators of the primary fields,

$$\left\langle \prod_j V_{\alpha_j} \right\rangle = \left\langle \prod_j V_{\alpha_j} \right\rangle \exp \left(i \sum_j \alpha_j \cdot c - \frac{i\alpha_0}{\pi} \int d^2 z\, \sqrt{g} R^{(2)} \rho \cdot c \right) \ , \tag{7.43}$$

which implies the charge conservation,

$$\sum_j \alpha_j = \frac{\alpha_0}{\pi} \int d^2 z\, \sqrt{g} R^{(2)} \rho = 2(2 - 2g)\alpha_0 \rho \ , \tag{7.44}$$

for the non-vanishing correlators on a genus-g Riemannian surface. For the sphere we find, in particular

$$\sum_j \alpha_j = 4\alpha_0 \rho \ . \tag{7.45}$$

The conservation rule (7.44) generalizes that in eq. (2.92) to non-trivial Lie algebras and to higher genus as well. As was explained in sect. II.6, the role of the Feigin-Fuchs screening operators to be inserted into the correlation functions is to adjust a balance of charge, in accordance with the condition (7.44) or (7.45). The screening charges have vanishing conformal dimensions but, nevetherless, they can carry non-vanishing charges. Inserting m_i screening vertex operators V_{-}^{i} and m_i' screening vertex operators V_{+}^{i} gives a balance-of-charge equation in the form

$$\sum_j \alpha_j + \sum_{i=1}^{n-1} (m_i \alpha_- + m_i' \alpha_+) e_i = 4\alpha_0 \rho \ . \tag{7.46}$$

We are now prepared to construct the primary fields $W^{(k)}(z)$ having dimensions (spins) $k = 3, \ldots, n$, and forming a closed (in the non-linear sense, but without any explicit dependence on ϕ's) algebra, out of the free fields $\phi^i(z)$ at our disposal. The obvious ansatz for $W^{(k)}(z)$ would be the most general linear combination of various products of $(\partial^m \phi)^n$, each term having k derivatives in total. The coefficients of this ansatz could then be determined from the relevant OPE's. In practice, however, this way only seems to be tractable for the simplest case of the W_3 algebra, because of a complexity of calculations. Fortunately, the trials which were actually done for the W_3 case resulted in a more 'clever' ansatz. Therefore, instead of repeating former trials, we are in a better position to use the 'clever' ansatz from the very beginning. The structure of this ansatz is actually dictated by the weights of $su(n)$, whereas the ansatz inself has the form of a 'Lax operator', namely [238, 239]

$$(2i\alpha_0)^n \mathcal{D}_n =: \prod_{\mu=1}^{n} [2i\alpha_0 \partial_z + h_\mu \cdot \partial\phi(z)] :$$

$$\equiv: [2i\alpha_0 \partial_z + h_n \cdot \partial\phi(z)] \cdots [2i\alpha_0 \partial_z + h_2 \cdot \partial\phi(z)][2i\alpha_0 \partial_z + h_1 \cdot \partial\phi(z)] : . \qquad (7.47)$$

Expanding this differential operator to the form

$$\mathcal{D}_n = \partial_z^n + \sum_{k=1}^{n} (2i\alpha_0)^{-k} u_k(z)\partial_z^{n-k} , \qquad (7.48)$$

which is called a *quantum Miura transformation*, [7] one finds, in particular

$$u_1 = \sum_\mu h_\mu \partial\phi = 0 , \qquad (7.49)$$

where eqs. (12.29) and (12.30) have been used. Similary, the u_2 reads

$$u_2 = \sum_{\mu>\nu} : h_\mu \cdot \partial\phi h_\nu \cdot \partial\phi : +2i\alpha_0 \sum_\mu (n-\mu)h_\mu \cdot \partial^2\phi . \qquad (7.50)$$

The first term on the r.h.s. of this equation can be rewritten as

$$\sum_{\mu>\nu} = \tfrac{1}{2}\sum_{\mu\neq\nu} = \tfrac{1}{2}\sum_{\mu,\nu} - \tfrac{1}{2}\sum_{\mu=\nu} = \tfrac{1}{2}\left(\sum_\mu\right)^2 - \tfrac{1}{2}\sum_{\mu=\nu} = 0 - \tfrac{1}{2}\sum_{\mu=\nu} . \qquad (7.51)$$

Eq. (12.31) now implies

$$u_2 = -\tfrac{1}{2} : \partial\phi(z) \cdot \partial\phi(z) : +2i\alpha_o\rho \cdot \partial^2\phi(z) = T(z) , \qquad (7.52)$$

where T is the stress tensor! Next, one finds

$$u_3 = \sum_{\mu>\nu>\sigma} : h_\mu \cdot \partial\phi h_\nu \cdot \partial\phi h_\sigma \cdot \partial\phi : +2i\alpha_0 \sum_{\mu>\nu} : h_\mu \cdot \partial\phi(n-1-\nu)h_\nu \cdot \partial^2\phi :$$

$$+2i\alpha_0 \sum_{\nu>\mu} : (n-\nu)h_\nu \cdot \partial^2\phi h_\mu \cdot \partial\phi : +(2i\alpha_0)^2 \sum_\mu \binom{n-\mu}{2} h_\mu \cdot \partial^3\phi . \qquad (7.53)$$

---

[7]The *classical Miura transformations* are known in the theory of KdV-type equations [241].

Computing its OPE with the stress tensor of eq. (7.52) yields that the u_3 is a quasi-primary field and is an 'almost' primary field of spin 3, modulo some terms depending on u_3 and ∂u_2 on the r.h.s. The simple deformation

$$W^{(3)} = u_3 - \frac{n-2}{2}(2i\alpha_0)\partial u_2 \; , \tag{7.54}$$

finally yields the true primary field of spin 3, which can be straightforwardly verified. The OPE of this field with itself results in Zamolodchikov's W_3 algebra (in the different normalization for the $W^{(3)}$ field, in fact). The construction of the $W^{(k)}$ field follows the similar lines, and it results in

$$W^{(k)} = u_k + \beta \partial u_{k-1} + \gamma \partial^2 u_{k-2} + \delta : u_2 u_{k-2} : + \dots \; , \tag{7.55}$$

with some calculable numerical coefficients $\beta, \gamma, \delta, \dots$. The validity of this procedure for any k, as well as the closure of the OPE algebra (in the non-linear sense) was argued in ref. [242]. The proof only exists in the classical case, when disregaring all the normal ordering [243, 244]. The nature of the apparent singularities in this construction, at certain isolated values of the central charge, is yet to be explained.

The highest-weight state of highest weight $\Delta_2(\alpha)$ is obtained by applying the vertex operator V_α to the $SL(2, \mathbf{C})$-invariant vacuum $|0\rangle$ (sect. II.6),

$$\lim_{z \to 0} V_\alpha(z) |0\rangle =: \exp(i\alpha \cdot \phi_0) : |0\rangle = |\alpha\rangle \; . \tag{7.56}$$

Since any non-negative mode of $W^{(k)}$ contains the non-negative modes a_m^j annihilating the vacuum $|0\rangle$, all the modes $W_m^{(k)}$ at $m \geq 0$ will also annihilate this vacuum. Hence, the states (7.56) are also the highest weight states of the W_n-algebra:

$$W_n^{(k)} |\alpha\rangle = 0 \qquad n > 0 \; ,$$

$$W_0^{(k)} |\alpha\rangle = \tilde{\Delta}_k(\alpha) |\alpha\rangle \qquad k = 2, 3. \dots, n \; , \tag{7.57}$$

where $W^{(2)} \equiv T$. The eigenvalues $\tilde{\Delta}_k(\alpha)$ simply depend on the eigenvalues $\Delta_k(\alpha)$ of the u_k, in accordance with eqs. (7.54) and (7.55); for example, $\tilde{\Delta}_2(\alpha) = \Delta_2(\alpha)$ and $\tilde{\Delta}_3(\alpha) = \Delta_3(\alpha) + (n-2)(2i\alpha_0)\Delta_2(\alpha)$. To compute $\Delta_k(\alpha)$, one applies the (operator-valued) differential operator $\mathcal{D}_n$, defined by eqs. (7.47) and (7.48), to the highest-weight state $|\alpha\rangle$, and one keeps only the most singular terms, *viz.*

$$(2i\alpha_0)^n \left[\partial^n + \sum_{k=2}^{n} \frac{\Delta_k(\alpha)}{(2i\alpha_0)^k z^k} \partial^{n-k} \right] |\alpha\rangle$$

$$\sim \left(2i\alpha_0\partial - i\frac{h_n \cdot a_0}{z}\right) \cdots \left(2i\alpha_0\partial - i\frac{h_1 \cdot a_0}{z}\right) |\alpha\rangle . \qquad (7.58)$$

Both sides of this equation are still the differential operators. Being applied to the monomials z^j, $j = 0, 1, \ldots, n-2$, they lead to a linear system of $(n-2)$ differential equations, having the form

$$\sum_{k=0}^{j} (2i\alpha_0)^k \frac{j!}{(j-k)!} \Delta_{n-k}(\alpha) = i^n \prod_{m=1}^{n} [2\alpha_0(j - m + 1) - h_m \cdot \alpha] , \qquad (7.59)$$

and whose solution is given by

$$\Delta_k(\alpha) = (-i)^k \sum_{\mu_1 > \mu_2 > \ldots \mu_k} \prod_{m=1}^{k} [h_{\mu_m} \cdot \alpha + 2\alpha_0(k - m)] . \qquad (7.60)$$

The correspondence between the vector α and the set of dimensions $\Delta_k(\alpha)$ is, in fact, $n!$-to-one, which implies that all vertex operators V_α with the same $\Delta_k(\alpha)$ have to be identified. Indeed, the r.h.s. of eq. (7.59) is not going to change if the replacement $\alpha \to \tilde{\alpha}$ would be equivalent to a permutation of the factors in the product, i.e. if

$$h_m \cdot \alpha + 2\alpha_0 m = h_{\pi(m)} \cdot \tilde{\alpha} + 2\alpha_0 \pi(m) , \qquad (7.61a)$$

or, equivalently, since $h_m \cdot \rho = \frac{n+1}{2} - m$,

$$h_m \cdot (\alpha - 2\alpha_0 \rho) = h_{\pi(m)} \cdot (\tilde{\alpha} - 2\alpha_0 \rho) , \qquad (7.61b)$$

where π is a permutation of $(1, 2, \ldots, n)$, and ρ is the Weyl vector. Defining the map $\alpha \to \alpha^*$ by

$$h_m \cdot \alpha^* = -h_{n+1-m} \cdot \alpha , \quad \text{for} \quad m = 1, 2, \ldots, n , \qquad (7.62)$$

one finds, in particular, that $\rho^* = \rho$, and $h_m \cdot [(4\alpha_0\rho - \alpha^*) - 2\alpha_0\rho] = h_{n+1-m} \cdot (\alpha - 2\alpha_0\rho)$. Hence, one gets

$$\Delta_k(4\alpha_0\rho - \alpha^*) = \Delta_k(\alpha) , \qquad (7.63)$$

for any k (*cf* sect. II.6).

Since the hermiticity of T implies $a_0^\dagger = 4\alpha_0\rho - a_0$, there are states with negative norm in our 'Hilbert space', while all the representations of the W algebra we are considering, are in fact, non-unitary. Like the ordinary CFT's (Ch. II), they may actually contain unitary representations for certain values of central charge. The relevant representations of the W algebra are just those which are the most degenerate, and which have $\alpha = -\alpha_- \lambda - \alpha_+ \lambda'$, where λ and λ' are the highest dominant weights (see eq. (7.78) below).

Since $2\alpha_0 = \alpha_- + \alpha_+$, one gets $\alpha - 2\alpha_0\rho = -\alpha_-(\lambda + \rho) - \alpha_+(\lambda' + \rho)$. Eq. (7.61b) now tells us that for the $n!$ equivalent representatives of α, both $\tilde{\lambda}+\rho$ and $\tilde{\lambda}'+\rho$ are obtained from $\lambda+\rho$ and $\lambda'+\rho$, resp., by applying the same element of the Weyl group. If α_+/α_- is *rational*, $\alpha_+/\alpha_- = -\alpha_+^2 = -q/p$, one has $\alpha - 2\alpha_0\rho = [p(\lambda + \rho) - q(\lambda' + \rho)]/\sqrt{pq}$. Therefore, α is actually labeled not by both weights λ and λ', but by their combination $\Lambda = p(\lambda + \rho) - q(\lambda' + \rho)$ only. This implies in fact many more transformations $(\lambda, \lambda') \rightarrow (\tilde{\lambda}, \tilde{\lambda}')$ resulting in the Weyl transformation of Λ, and, therefore, more identifications for the vertex operators V_α have to be done. This qualitatively explains the origin of the unitary minimal models for the W algebras.

To actually construct these degenerate representations, one should find null states in the Verma module $\mathcal{V}_\Delta$ spanned by the vectors $\prod_{k=2}^{n} W_{-n_r^k}^{(k)} |\Delta\rangle$ with $n_r^k > 0$, where $|\Delta\rangle$ is the highest weight state,

$$W_n^{(k)} |\Delta\rangle = 0 \ \ n > 0 \ , \quad W_0^{(k)} |\Delta\rangle = \tilde{\Delta}_k |\Delta\rangle \ , \tag{7.64}$$

and then dicregard all the submodules which they generate (Ch. II). Taking the W_3 algebra for definiteness, the null state $|\chi_1\rangle$ at level 1 is given by

$$|\chi_1\rangle = \left(2\Delta_2 W_{-1}^{(3)} - 3\tilde{\Delta}_3 L_{-1}\right)|\Delta\rangle \ , \tag{7.65}$$

provided

$$9\tilde{\Delta}_3^2 = 2\Delta_2^2 \left(\frac{32}{22 + 5c}(\Delta_2 + \tfrac{1}{5}) - \tfrac{1}{5}\right) \ . \tag{7.66}$$

Since the correlation functions of primary fields with the null vector have to vanish (sect. II.4), one gets this way some differential equations for them. The completely degenerate representations with $2n - 2$ null vectors imply enough differential constrains to completely determine all of their correlation functions.

The formal construction of null states is based on the following property of the screening vertex operators (7.37), which have the charges $\alpha_\pm$ satisfying eq. (7.38) [239]:

$$W^{(k)}(z)V_\pm^j(w) = \partial_w(\ldots) + O(1) \ , \tag{7.67a}$$

or, equivalently,

$$[W^{(k)}(z), Q_\pm^j] = 0 \ , \tag{7.67b}$$

because of eq. (7.39). To prove eq. (7.67), it is enough to check that the differential operator $\mathcal{D}_n$ introduced in eq. (7.48) does commute with $Q_\pm^j$, i.e. one has $\mathcal{D}_n(z)V_\pm^j(w) = \partial_w(\ldots) + O(1)$. Since the definition (7.47) and the identities $h_\mu = \lambda_\mu - \lambda_{\mu-1}$, $e_j \cdot h_\mu =$

$\delta_{j,\mu} - \delta_{j+1,\mu}$, the only non-zero contractions (with $\mathcal{D}_n$) of $e_j \cdot \phi(w)$ in $V_{\pm}^j(w)$ come from only two factors in $\mathcal{D}_n$, namely, from those with $\mu = j+1$ and $\mu = j$. The relevant Wick-theorem 'propagator' takes the form

$$\langle h_\mu \cdot \partial\phi(z) e_j \cdot \phi(w) \rangle = (\delta_{j,\mu} - \delta_{j+1,\mu}) \frac{-1}{z-w} \, , \tag{7.68}$$

and the proof can now be completed as follows:

$$: [2i\alpha_0\partial_z + h_{j+1} \cdot \partial\phi(z)][2i\alpha_0\partial_z + h_j \cdot \partial\phi(z)] : V_{\pm}^j(w)$$

$$\sim: \left[2i\alpha_0\partial_z \frac{-i\alpha_\pm}{z-w} + h_{j+1} \cdot \partial\phi(z) \frac{-i\alpha_\pm}{z-w} + h_j \cdot \partial\phi(z) \frac{i\alpha_\pm}{z-w} + \frac{\alpha_\pm^2}{(z-w)^2} \right] : V_{\pm}^j(w)$$

$$\sim: \left[\frac{-2\alpha_0\alpha_\pm + \alpha_\pm^2}{(z-w)^2} + \frac{\partial_w}{z-w} \right] : V_{\pm}^j(w) \sim \partial_w \left[\frac{V_{\pm}^j(w)}{z-w} \right] \, , \tag{7.69}$$

where eqs. (7.38) and (12.30) have been used, and only the singular terms have been taken into account. It follows

$$W_m^{(k)} Q_{\pm}^j |\Delta(\beta)\rangle = Q_{\pm}^j W_m^{(k)} |\Delta(\beta)\rangle = 0 \tag{7.70}$$

for all $m > 0$. Hence, the vectors

$$\left| \chi_{\pm}^j(\beta) \right\rangle \equiv Q_{\pm}^j |\Delta(\beta)\rangle \tag{7.71}$$

are the highest weight states. Simultaneously, since $|\Delta(\beta)\rangle = V_\beta(0)|0\rangle$, one finds

$$\left| \chi_{\pm}^j(\beta) \right\rangle = \oint dz \, : e^{i\alpha_\pm e_j \cdot \phi(z)} :: e^{i\beta \cdot \phi(0)} : |0\rangle$$

$$= \oint dz \, z^{\alpha_\pm e_j \cdot \beta} : e^{i\alpha_\pm e_j \cdot \phi(z) + i\beta \cdot \phi(0)} : |0\rangle \, . \tag{7.72}$$

The r.h.s. of eq. (7.72) is only well-defined provided

$$\alpha_\pm e_j \cdot \beta = -\bar{l}_j - 1 \, , \quad \text{where } \bar{l}_j = 0, 1, 2, \dots \, , \tag{7.73}$$

when the contour integral in eq. (7.72) makes sense, and it yields

$$\left| \chi_{\pm}^j(\beta) \right\rangle = \frac{1}{\bar{l}_j!} \partial_z^{\bar{l}_j} : e^{i\alpha_\pm e_j \cdot \phi(z) + i\beta \cdot \phi(0)} : |0\rangle \Big|_{z=0} \, . \tag{7.74}$$

Therefore, the state $\left| \chi_{\pm}^j(\beta) \right\rangle$ is both a primary and a secondary state, i.e. it is a null state or a singular vector. Let $\beta = 4\alpha_0\rho - \alpha^* - \alpha_\pm e_j$, so that $\left| \chi_{\pm}^j(\beta) \right\rangle$ becomes the null state for the Verma module associated with the highest weight $\Delta(4\alpha_0\rho - \alpha^*) = \Delta(\alpha)$.

Inserting this β into eq. (7.73) and using eq. (7.38), one finds $\alpha_\pm e_j \cdot \alpha^* = \bar{l}_j - 1$ and $e_j \cdot \alpha^* = e_{n-j} = \alpha_\mp (1 - \bar{l}_j)$, and finally $(l_{n-j} = \bar{l}_j)$

$$e_{n-j} \cdot \alpha = \alpha_\mp (1 - l_{n-j}) , \tag{7.75}$$

where $l_{n-j} = \bar{l}_j$. When $l_{n-j} = 0$, $\left| \chi_\pm^j (\beta) \right\rangle$ is not a null state but a highest weight state $|\Delta(\alpha)\rangle$, so that we should only consider positive integer values for l_{n-j}.

Eq. (7.75) for the null vectors can be generalized as

$$\left| \chi_\pm^{j,\bar{l}'} (\beta) \right\rangle = \oint \cdots \oint V_\pm^j \cdots V_\pm^j |\Delta(\beta)\rangle , \tag{7.76}$$

with $\bar{l}'_j$ screening operators [239]. They will be still well-defined and non-vanishing provided

$$\alpha_\pm^2 (\bar{l}'_j - 1) + \alpha_\pm e_j \cdot (4\alpha_0 \rho - \alpha^* - \alpha_\pm e_j) = -\bar{l}'_j - 1 , \tag{7.77a}$$

or, equivalently,

$$e_{n-j} \cdot \alpha = \alpha_\mp (1 - l_{n-j}) + \alpha_\pm (1 - \bar{l}'_{n-j}) , \tag{7.77b}$$

where $l_{n-j} = \bar{l}_j$ and $l'_{n-j} = \bar{l}'_j$. The completely degenerate representations parametrized by α are the solutions to eq. (7.77) for all values of j. The general solution reads

$$\alpha = \sum_{j=1}^{n-1} [(1 - l_j)\alpha_- + (1 - \bar{l}'_j)\alpha_+]\lambda_j , \tag{7.78a}$$

or, equivalently,

$$\alpha = 2\alpha_0 \rho - \sum_{j=1}^{n-1} (l_j \alpha_- + \bar{l}'_j \alpha_+)\lambda_j , \tag{7.78b}$$

where $l_j, \bar{l}'_j = 1, 2, \ldots$. Since $\lambda = \sum_j (l_j - 1)\lambda_j$ and $\lambda' = \sum_j (\bar{l}'_j - 1)\lambda_j$ are the highest weights of $su(n)$, this can be rewritten to the form

$$\alpha = -\alpha_- \lambda - \alpha_+ \lambda' . \tag{7.78c}$$

Therefore, the highest weights of the A_{n-1}-based W_n algebra are labeled by pairs (λ, λ') of the $su(n)$ highest weights. The construction does not guarantee that *any* completely degenerate representation can be derived this way, but various consistency checks indicate that it should be true [239].

The full structure of W modules could, in principle, be derived from the analysis of the Kač determinant (Ch. II). This determinant is the function of central charge c and highest weight λ. Being a polynomial with respect to λ, the Kač determinant is fixed by its zeroes, which correspond to the null states. A free-field Fock space representation is most suitable for the explicit construction of a W module [46], and this ultimately leads to the same condition (7.78) for the existence of infinitely many null states in the W-RCFT's corresponding to the completely degenerate representations [131]. Some explicit formulae for the Kač determinants associated with the highest-weight modules of the W_n-algebras are available [245, 246]. Unfortunately, it is extremely difficult to extract the information about unitary representations of W-algebras from them.

The fusion rules follow from evaluating the 3-point functions [240]. The charge conservation for the 3-point function

$$\left\langle V_{4\alpha_0\rho-\alpha}V_\beta V_\gamma Q^+\cdots Q^-\cdots\right\rangle . \tag{7.79}$$

implies

$$4\alpha_0\rho - \alpha + \beta + \gamma + \sum_{i=1}^{n-1}(m_i\alpha_- + m_i'\alpha_+)e_i = 4\alpha_0\rho . \tag{7.80}$$

Eq. (7.78) tells us to take $\alpha = -\alpha_-\lambda_\alpha - \alpha_+\lambda_\alpha'$, and similarly for β and γ. The charge conservation condition now takes the simple form

$$\lambda_\alpha = \lambda_\beta + \lambda_\gamma - r , \quad \lambda_\alpha' = \lambda_\beta' + \lambda_\gamma' - r' , \tag{7.81}$$

where $r = \sum m_i e_i$ and $r' = \sum m_i' e_i$ are *positive* or *zero* roots! Eq. (7.81) can be recognized as a part of the branching rules for the decomposition of the tensor product of two $su(n)$ representations with the highest weights λ_β and λ_γ into representations with highest weights λ_α. They can be written down in the form of restrictions 'from below' as follows:

$$\lambda_\alpha \leq \lambda_\beta + \lambda_\gamma , \quad \lambda_\alpha' \leq \lambda_\beta' + \lambda_\gamma' . \tag{7.82}$$

The similar analysis for the same 3-point function, but now in the two different forms, namely

$$\left\langle V_{\alpha^*}V_{4\alpha_0\rho-\beta^*}V_\gamma Q^+\cdots Q^-\cdots\right\rangle \tag{7.83a}$$

and

$$\left\langle V_{\alpha^*}V_\beta V_{4\alpha_0\rho-\gamma^*}Q^+\cdots Q^-\cdots\right\rangle , \tag{7.83b}$$

implies

$$\lambda_\alpha^* \geq \lambda_\beta^* - \lambda_\gamma , \quad \lambda_\alpha'^* \geq \lambda_\beta'^* - \lambda_\gamma' , \tag{7.84a}$$

and

$$\lambda_\alpha^* \geq \lambda_\gamma^* - \lambda_\beta , \quad \lambda_\alpha'^* \geq \lambda_\gamma'^* - \lambda_\beta' , \tag{7.84b}$$

respectively. Putting all together, one finds the fusion rules

$$[\Phi_{\lambda_\beta \lambda'_\beta}] \times [\Phi_{\lambda_\gamma \lambda'_\gamma}] = \sum_{\substack{\lambda^*_\alpha \geq \lambda^*_\beta - \lambda_\gamma \\ \lambda^*_\alpha \geq \lambda^*_\gamma - \lambda_\beta}}^{\lambda_\alpha \leq \lambda_\beta + \lambda_\gamma} \sum_{\substack{\lambda'^*_\alpha \geq \lambda'^*_\beta - \lambda'_\gamma \\ \lambda'^*_\alpha \geq \lambda'^*_\gamma - \lambda'_\beta}}^{\lambda'_\alpha \leq \lambda'_\beta + \lambda'_\gamma} [\Phi_{\lambda_\alpha \lambda'_\alpha}] \; . \tag{7.85}$$

When $\alpha_+^2 = q/p$ is a rational number, one has

$$\alpha_- = -\frac{p}{\sqrt{qp}} \; , \quad \alpha_+ = \frac{q}{\sqrt{qp}} \; , \quad 2\alpha_0 = \frac{q-p}{\sqrt{qp}} \; , \tag{7.86}$$

and

$$\alpha - 2\alpha_0 \rho = \frac{1}{\sqrt{pq}} \sum_{i=1}^{n-1} (pl_i - p'l'_i)\lambda_i \; , \tag{7.87}$$

where eq. (7.78b) has been used. There are many transformations of the parameters which do not change α. Eq. (7.87) already implies some of them, namely, $l_i \to l_i + rq$ and $l'_i \to l'_i + rp$ with integer r. There are, in fact, more of them, *viz.*

$$\lambda \to (q-2)\rho - \lambda^* \; , \quad \lambda' \to (p-2)\rho - \lambda'^* \; , \tag{7.88}$$

yielding $\alpha = -\alpha_- \lambda - \alpha_+ \lambda' \to -(q\alpha_- + p\alpha_+)\rho + 2(\alpha_- + \alpha_+)\rho + \alpha_- \lambda^* + \alpha_+ \lambda'^* = 4\alpha_0 \rho - \alpha^*$, which is the symmetry of the Δ_k. Altogether, they imply some additional restrictions 'from above', namely

$$\lambda_\alpha \geq \lambda_\beta + \lambda_\gamma - (q-2)\rho \; , \quad \lambda_\alpha \leq \lambda_\beta - \lambda^*_\gamma + (q-2)\rho \; , \quad \lambda_\alpha \leq \lambda_\gamma - \lambda^*_\beta + (q-2)\rho \; , \tag{7.89}$$

which ultimately result in the closure of the chiral algebra on a *finite* number of primary fields,

$$\mathcal{A} = \bigoplus_{\sum l_i \leq q-1} \bigoplus_{\sum l'_i \leq p-1} \Phi_{\lambda(l), \lambda'(l')} \; . \tag{7.90}$$

These models are called the *W-minimal* models. The fusion rules (7.85), in combination with those following from eq. (7.89), are incomplete in this case. The central charge of the (p, q) *W*-minimal models is given by

$$c_{pq}^n = (n-1)\left[1 - n(n+1)\frac{(p-q)^2}{pq}\right] \; , \tag{7.91}$$

which generalizes the first line of eq. (2.62) for the Virasoro minimal models with $n = 2$.

The simple illustration to the fusion rules of the *W*-minimal models is provided by the 3-state Potts model ($n = 3$, $p = 4$, $q = 5$) [240]. This model has six *W*-primary fields: the identity **1**, two spin fields σ and σ^+ of dimension $1/15$, two parafermionic fields ψ and ψ^+ of dimension $2/3$, and energy density ε of dimension $2/5$. The corresponding states are created from a vacuum by the vertex operators $V_{-\alpha_-\lambda-\alpha_+\lambda'}$, whose dominant highest weights λ and λ' are given by:

$$
\begin{aligned}
\mathbf{1} &: \quad (0,0) \cong (2\lambda_1, \lambda_1) \cong (2\lambda_2, \lambda_2) \ , \\
\sigma &: \quad (\lambda_1, 0) \cong (\lambda_1 + \lambda_2, \lambda_1) \cong (\lambda_2, \lambda_2) \ , \\
\sigma^+ &: \quad (\lambda_2, 0) \cong (\lambda_1, \lambda_1) \cong (\lambda_1 + \lambda_2, \lambda_2) \ , \\
\psi^+ &: \quad (2\lambda_1, 0) \cong (2\lambda_2, \lambda_1) \cong (0, \lambda_2) \ , \\
\psi &: \quad (2\lambda_2, 0) \cong (0, \lambda_1) \cong (2\lambda_1, \lambda_2) \ , \\
\varepsilon &: \quad (\lambda_1 + \lambda_2, 0) \cong (\lambda_2, \lambda_1) \cong (\lambda_1, \lambda_2) \ .
\end{aligned}
\tag{7.92}
$$

On the one hand, when choosing $(2\lambda_1, \lambda_2)$ to represent ψ and $(0, \lambda_2)$ to represent ψ^+, eqs. (7.85) and (7.89) imply

$$
\psi \times \psi^+ \cong (2\lambda_1, \lambda_2) \times (0, \lambda_2) = (2\lambda_1, 2\lambda_2) + (2\lambda_1, \lambda_1) \cong \omega + 1 \ ,
\tag{7.93a}
$$

where ω is a possible new field. On the other hand, when choosing $\psi \cong (2\lambda_2, 0)$ and $\psi^+ \cong (0, \lambda_2)$, one gets

$$
\psi \times \psi^+ \cong (2\lambda_2, 0) \times (0, \lambda_2) = (2\lambda_2, \lambda_2) \cong \mathbf{1} \ .
\tag{7.93b}
$$

The consistency between these two equations implies

$$
\psi \times \psi^+ \cong \mathbf{1} \ .
\tag{7.93c}
$$

Hence, the fusion of ψ with ψ^+ cannot generate the new field. Similarly, one finds for the rest of the fusion rules

$$
\sigma \times \sigma \cong (\lambda_1, 0) \times (\lambda_1, 0) = (2\lambda_1, 0) + (\lambda_2, 0) \cong \psi^+ + \sigma^+ \ ,
$$

$$
\psi \times \psi \cong (2\lambda_2, 0) \times (0, \lambda_1) = (2\lambda_2, \lambda_1) \cong \psi^+ \ .
\tag{7.94}
$$

The complete fusion rules are closed on the six primary fields introduced above. In particular, one can now see how the general equation (7.90) for the W-minimal models can be fulfilled in this case, namely

$$
\sum_i l_i \le 4 = q - 1 \ , \qquad \sum_i l_i' \le 3 = p - 1 \ .
\tag{7.95}
$$

Exercises

VII-2 ▷ Verify that eq. (7.60) correctly reproduces the previously obtained result (7.35) for $\Delta_2(\alpha)$.

VII-3 ▷ (*) Verify that the star-operation introduced in eq. (7.62) preserves the lengths of vectors, i.e.

$$(\alpha^* - 2\alpha_0\rho^*)^2 = (\alpha^* - 2\alpha_0\rho)^2 = (\alpha - 2\alpha_0\rho)^2 \ , \tag{7.96}$$

and $\Delta_2(\alpha^*) = \Delta_2(\alpha)$. *Hint*: use the $*$-transformations of the simple roots and weights, $e_j \cdot \alpha^* = (h_j - h_{j+1}) \cdot \alpha^* = e_{n-j} \cdot \alpha$, $e_j^* = e_{n-j}$, and $\lambda_j^* = \lambda_{n-j}$.

VII-4 ▷ (*) Prove that

$$|\alpha\rangle^\dagger = \langle 4\alpha_0\rho - a_0| \ . \tag{7.97}$$

What are the hermiticity properties of the modes $W_m^{(k)}$?

VII-5 ▷ Use eq. (7.97) to fix the relative charges α of the vetrex operators V_α in a non-vanishing two-point correlation function. Compare the result with that in eq. (7.45). Check that the non-vanishing two-point correlator takes the form

$$\left\langle V_{\Delta(\alpha^*)} V_{\Delta(\alpha)} \right\rangle \neq 0 \ . \tag{7.98}$$

VII-6 ▷ (*) Consider the fusion rules (7.84) in the case of the W_3 algebra, and take $\lambda_\beta' = \lambda_\gamma' = 0$ and $\lambda_\beta = \lambda_\gamma = \lambda_1$ as the highest weight of the **3** representation of $su(3)$. Do they follow the well-known branching rules for the $su(3)$ tensor products? *Hint*: use the highest weights of the $su(3)$-irreps of dimensions 3 and 6, and Fig. 22.

VII-7 ▷ (*) Derive the fusion rules (7.94) for the 3-state Potts model.

VII.3   Quantum Drinfeld-Sokolov reduction

The lessons from the explicit free-field constructions of W algebras in the previous section can be generalized in the form of the so-called **quantum DS reduction** [247, 248], which gives the most general known procedure for constructing various W algebras by relating them to ordinary Lie and AKM algebras. It turns out that the W algebras can be considered as the quantum versions of the so-called *Gel'fand-Dikii* algebras [249] known in the theory of KdV-type equations [250]. [8] The Gel'fand-Dikii algebras are usually

---

[8]If the reader is not familiar with the KdV theory, he is invited to consult sect. XI.1. before reading this paragraph.

realized in terms of the Poisson brackets associated with the Hamiltonian structure of Lax operators in the theory of integrable equations of the KdV-type (sect. XI.1). For instance, eq. (11.11) shows that the (second) Poisson bracket associated with the KdV equation results in the classical version of the Virasoro algebra which is the simplest W algebra! Moreover, the Lax representation (11.12) of the KdV equation defines the 3-rd order differential operator $Q \equiv W^{(3)}$, whose Fourier components together with those of the KdV field u form the Gel'fand-Dikii algebra (w.r.t. the Poisson bracket) which is exactly the semiclassical limit of the commutation relations of the W_3 algebra in eqs. (7.2) and (7.3) [251, 252]. Replacing the Poisson brackets by the commutators, and the classical phase space by a quantum algebra module, one gets the W_3 algebra and some of its irreducible representation, resp. In fact, the irreps of the Virasoro algebra can be extracted from the irreps of the $SL(2, \mathbf{R})$ current algebra by imposing certain constraints on the latter, when using the BRST formalism [247]. This can also be generalized to the W algebras. In particular, the highest weights of the completely degenerate representations of the W_n algebras can be expressed in terms of the highest weights of the $SL(n, \mathbf{R})$ current algebra, as was already noticed in the previous sect. 2.

The KdV theory admits generalizations by means of the *classical Drinfeld-Sokolov reduction* of natural Hamiltonian structures connected with AKM algebras [250]. The classical Hamiltonian (DS) reduction consists of two steps. The first step is imposing a certain constraint on the Hamiltonian generators of a gauge group, in order to reduce the phase space. The second step is to introduce a new phase space, comprising the orbits of the gauge group in the subspace defined by the constraint, as well as a new Poisson bracket in the reduced phase space. This leads to the generalized KdV-type equations and the higher-order Lax operators in the Hamiltonian form, which are associated with the original AKM algebra. The quantum DS reduction goes further. First, the BRST ghosts and the BRST charge associated with the gauge fixing are introduced. Second, one notices that the resulting BRST cohomology turns out to be isomorphic to an irrep of the W algebra. The proof [247] is based on the fact that both the W_n algebra and the $SL(n, \mathbf{R})$ current algebra ($\sim sl(n)$) can be realized in terms of free fields (sect. 2). Though all free field realizations are highly redundant, their BRST cohomologies are actually isomorphic to the W_n or $SL(n, \mathbf{R})$ current algebra irreps, while their BRST charges are equivalent modulo BRST trivial operators.

To give a simple example [247, 252], let us consider the level-k current algebra $\widehat{sl(2)}_k$

having the form

$$J^+(z)J^-(w) \sim \frac{2}{z-w}J^3(w) + \frac{k}{(z-w)^2} \; ,$$

$$J^3(z)J^\pm(w) \sim \frac{\pm 1}{z-w}J^\pm(w) \; , \quad J^3(z)J^3(w) \sim \frac{k/2}{(z-w)^2} \; , \qquad (7.99)$$

with the SS-constructed stress tensor

$$T_{sl(2)}(z) = \frac{1}{k+2}\sum \; : J^a(z)J^a(z) : \; . \qquad (7.100)$$

In this case, we want to put the constraint

$$J^-(z) = 1 \; . \qquad (7.101)$$

All the currents $J^{3,\pm}(z)$ are the conformal fields of dimension 1 w.r.t. the stress tensor (7.100). Therefore, in order to have the constraint to be consistent with conformal invariance, we should *deform* the stress tensor in such a way that the conformal dimension of the current $J^-(z)$ vanishes. The improved tensor reads [252]

$$T_{\text{impr}}(z) = T_{sl(2)}(z) - \partial J^3(z) \; , \qquad (7.102)$$

and it has the central charge

$$c_{sl(2)}^{(k)} = \frac{3k}{k+2} - 6k = 15 - \frac{6}{k+2} - 6(k+2) \; . \qquad (7.103)$$

In addition, for a degenerate representation of the Virasoro algebra, eq. (2.62) gives $c_{\text{Vir}} = 1 - 6(p-q)^2/(pq) = 13 - 6q/p - 6p/q$. Substituting $p/q = k+2$, we find

$$c_{\text{Vir}}^{(k)} = 13 - \frac{6}{k+2} - 6(k+2) = c_{sl(2)}^{(k)} - 2 \; . \qquad (7.104)$$

The difference between $c_{sl(2)}^{(k)}$ and $c_{\text{Vir}}^{(k)}$ is independent on k, and it is exactly equal to the contribution of the ghost system (b,c) of $\lambda = 1$ to the conformal anomaly (sect. I.6). The conformal ghost system naturally emerges as a result of the constraint (7.101) in the BRST formalism.

Let us now consider an irreducible module $\mathcal{H}_{sl(2)}^{(k)}$ of the $SL(2,\mathbf{R})$ current algebra and the Fock space of the ghost system $\mathcal{H}_{b,c}$. The BRST operator defined by [247]

$$Q_{\text{BRST}} = \oint \frac{dz}{2\pi i}(J^-(z) - 1)c(z) \qquad (7.105)$$

is nilpotent, $Q_{\text{BRST}}^2 = 0$. The BRST cohomology w.r.t. this charge,

$$H_{Q_{\text{BRST}}}\left(\mathcal{H}_{sl(2)}^{(k)} \otimes \mathcal{H}_{b,c}\right) = \text{Ker}(Q_{\text{BRST}})/\text{Im}(Q_{\text{BRST}}) \; , \qquad (7.106)$$

is actually isomorphic to the irreducible module $\mathcal{H}_{\mathrm{Vir}}^{(k)}$ of the Virasoro algebra of central charge $c = c_{sl(2)}^{(k)} - 2$:

$$H_{Q_{\mathrm{BRST}}}\left(\mathcal{H}_{sl(2)}^{(k)} \otimes \mathcal{H}_{b,c}\right) \cong \mathcal{H}_{\mathrm{Vir}}^{(k)} . \qquad (7.107)$$

The proof of the quantum Hamiltonian (DS) reduction represented by eq. (7.107) can be found in ref. [247]. In our example, the classical DS-reduced phase space is the space of orbits (in the classical phase space restricted by the constraint (7.101)) generated by the action of $J^-(z)$ via the Poisson bracket. The quantum physical subspace is defined by the constraint $Q_{\mathrm{BRST}} |\mathrm{phys}\rangle = 0$ in the total 'Hilbert space' of the quantized theory, whereas the *reduced* Hilbert space is the space of orbits of the BRST charge [247]. It is now natural to expect an existence of the quantum Hamiltonian DS reduction for the $SL(n, \mathbf{R})$ current algebras of $n \geq 3$, which should lead to the W_n algebras. See ref. [247] for more non-trivial examples of the DS reduction.

In the most general terms, the idea of the quantum DS reduction is to begin with an AKM algebra $\hat{\mathcal{G}}$, its subalgebra $\hat{\mathcal{G}}'$ and a one-dimensional representation χ_{DS} of $\hat{\mathcal{G}}'$, and then impose the first class-constraints $g \sim \chi(g)$ for any $g \in \hat{\mathcal{G}}'$ by means of the BRST procedure. The cohomology of the BRST operator on a set of the (normally-ordered) currents, ghost fields and their derivatives gives an algebraic structure known in the mathematical literature as the *Hecke algebra*. The only non-trivial cohomology over there is the zeroth cohomology (*cf* sects. II.9 and III.5), which forms a subalgebra and may have a W algebra structure for the suitable choice of the triple $(\hat{\mathcal{G}}, \hat{\mathcal{G}}', \chi_{\mathrm{DS}})$ [131]. The quantum DS reduction also provides the tools to map AKM modules to the corresponding W modules [253]. The quantum Hamiltonian DS construction of the W algebras is formally invariant under the reflections in the root space forming the Weyl group, which nicely justifies the use of the letter W to denote those algebras [131].

A tractable case (in the Chevalley basis) [254] is given by the principal embedding of an $sl(2)$ subalgebra into a Lie algebra $\mathcal{G}$, the former being characterized by the root $\hat{\alpha} = \tilde{\rho}/(\tilde{\rho}, \tilde{\rho})$, where $\tilde{\rho}$ is the dual Weyl vector, [9] so that $(\tilde{\rho}, \alpha) = 1$ for any *simple* root α of $\mathcal{G}$. With respect to this $sl(2)$ subalgebra, the adjoint representation of $\mathcal{G}$ decomposes into the $sl(2)$ representations of spin e_i, $i = 1, \ldots, l$, where $\{e_i\}$ are known as the *exponents* of $\mathcal{G}$, and $l = \mathrm{rank}(\mathcal{G})$. [10] We should then expect an appearance of the $W(s_1, \ldots, s_l)$ algebra ($s_i = e_i + 1$) from the quantum DS reduction, in agreement with the classical

---

[9]For the simply-laced Lie algebras one has $\tilde{\rho} = \rho$.

[10]See sect. XII.2 for the relevant information about classical Lie algebras.

result [250]. Let $e_\alpha(z)$ be the currents corresponding to the positive roots, and

$$\chi_{\text{DS}}(e_\alpha) = \begin{cases} 1 & \text{for simple roots ,} \\ 0 & \text{otherwise ,} \end{cases} \tag{7.108}$$

the constraints. Having introduced the ghost fields $(b_\alpha(z), c_\alpha(z))$ for each positive $\mathcal{G}$-root $\alpha \in \Delta_+$, we can define the BRST operator in the form $Q = \oint j_{\text{BRST}}(z) = Q_0 + Q_1$,

$$Q_0 = \oint \frac{dz}{2\pi i} : \left\{ \sum_{\alpha \in \Delta_+} c_\alpha(z) e_\alpha(z) - \frac{1}{2} \sum_{\alpha,\beta,\gamma \in \Delta_+} f^{\alpha\beta\gamma} b_\gamma(z) c_\alpha(z) c_\beta(z) \right\} : , \tag{7.109}$$

and

$$Q_1 = -\oint \frac{dz}{2\pi i} \sum_{\alpha \in \Delta_+} : c_\alpha(z) \chi_{\text{DS}}(e_\alpha(z)) : , \tag{7.110}$$

where the structure constants $f^{\alpha\beta\gamma}$ have been introduced. By construction, one gets

$$Q^2 = Q_0^2 = Q_1^2 = \{Q_0, Q_1\} = 0 . \tag{7.111}$$

The Virasoro generators now read as follows:

$$T(z) = T_{\text{SS}}(z) + \tilde{\rho} \cdot \partial h(z) + T_{\text{gh}}(z) , \tag{7.112}$$

where the conformal fields $h_i(z)$ correspond to the CSA elements of $\mathcal{G}$, the field $T_{\text{SS}}(z)$ is the SS-constructed stress tensor, and

$$T_{\text{gh}}(z) = \sum_{\alpha \in \Delta_+} [(\tilde{\rho}, \alpha) - 1) : b_\alpha \partial c_\alpha : (z) + (\tilde{\rho}, \alpha) : \partial b_\alpha c_\alpha : (z)] . \tag{7.113}$$

The ghost contribution is fixed by requiring the BRST current $j_{\text{BRST}}(z)$ to be a primary field of dimension 1, so that $[Q, T(z)] = 0$.

The central charge associated with the stress tensor (7.104) is given by

$$c = \frac{k \dim \mathcal{G}}{k + \tilde{h}} - 12k |\tilde{\rho}|^2 - 2 \sum_{\alpha \in \Delta_+} \left[6(\tilde{\rho}, \alpha)^2 - 6(\tilde{\rho}, \alpha) + 1 \right]$$

$$= l - 12 |\alpha_+ \rho + \alpha_- \tilde{\rho}|^2 , \tag{7.114}$$

where $\alpha_- \alpha_+ = -1$ and $\alpha_- = -\sqrt{k + \tilde{h}}$.

The construction of other W generators is not so simple. The key point, however, is the observation that the BRST cohomology can be described as the *centralizer* of the set of screening Feigin-Fuchs charges $\oint \tilde{Q}_i^+(z)$, $i = 1, \ldots, l$, on the space of polynomials (and derivatives thereof) in $i\partial\phi^i(z)$, in terms of free scalar fields $\phi^i(z)$, i.e. in the associated free-field Fock spaces $\mathcal{F}_\Lambda$ labeled by the vacuum eigenvalues (highest weights) Λ of the scalar zero modes [248]. The screening currents take the form

$$\tilde{Q}_i^+(z) =: \exp[-i\alpha_+\alpha_i \cdot \phi(z)] := c_{\alpha_i}(z) , \qquad (7.115)$$

where the scalar fields $\phi(z)$ have come after bosonization of the field

$$\tilde{h}_i(z) = \alpha_+ \left[h_i(z) + \sum_{\alpha \in \Delta_+} (\alpha, \tilde{\alpha}_i) : b_\alpha c_\alpha : (z) \right] , \qquad (7.116)$$

where $\tilde{h}_i(z) = \tilde{\alpha}_i \cdot \partial\phi(z)$.

It can be shown that the zeroth cohomology of the BRST charge Q_0 is generated by the fields $\tilde{h}_i(z)$ and $c_{\alpha_i}(x)$, which allows us to identify $c_{\alpha_i}(z)$ with the Feigin-Fuchs currents. In the previous section, the W generators were commuting with the Feigin-Fuchs operators, and it helped to explicitly construct the W generators. In particular, the Virasoro generator was identified with the conformal operator of dimension two, whereas all the operators (7.115) were primary and of dimension one w.r.t. it. The unique solution given in the previous section takes the form

$$T(z) = -\tfrac{1}{2} : \partial\phi(z)\partial\phi(z) : -(\alpha_+\rho + \alpha_-\tilde{\rho}) \cdot i\partial^2\phi(z) , \qquad (7.117)$$

and it leads to the following conformal dimensions of the Fock vacua:

$$h_\Lambda = \tfrac{1}{2}[\Lambda, \Lambda + 2(\alpha_+\rho + \alpha_-\tilde{\rho})] . \qquad (7.118)$$

A formal description of a W algebra as the centralizer of the screening operators can be obtained in terms of a resolution of a $\hat{\mathcal{G}}$-module L, when using the isomorphism between fields and states in the Hilbert space $\mathcal{H}$ where the W generators act [131]. The proper resolution $(C^{(i)}\mathcal{H}^g, d^{(i)})$ of the Hilbert space $\mathcal{H}^g$ of $\hat{\mathcal{G}}$ has to be constructed in terms of free-field Fock spaces, provided the differentials $d^{(i)}$ commute with generators of $\hat{\mathcal{G}}$. By definition, a resolution $(C^{(i)}L, d^{(i)})$ of a $\hat{\mathcal{G}}$-module L is a complex $d^{(i)} : C^{(i)}L \to C^{(i+1)}L$, where $d^{(i+1)}d^{(i)} = 0$, provided the cohomology of this complex is exactly the module L itself, i.e. $H_d^{(i)}(L) = \delta_{i,0}L$ (sects. II.9, III.5 and VI.5). In the resolution terms, the free-field Fock spaces are labeled by the elements of the Weyl group W of $\mathcal{G}$, namely

$$C^{(i)}\mathcal{H}^g = \bigoplus_{\{w \in W | l(w) = i\}} \mathcal{F}_{w\rho-\rho}^{\alpha\beta\gamma} , \qquad (7.119)$$

where the bosonic free-field Fock spaces $\mathcal{F}_\Lambda^{\phi\beta\gamma} = \mathcal{F}_{\alpha+\Lambda}^\phi \otimes \mathcal{F}^{\beta\gamma}$ have been introduced. The first-order bosonic fields $(\beta^a(z), \gamma^a(z))$ represent the conformal ghost systems (sect. I.6) of dimension one, which are to be associated with each positive root $\alpha \in \Delta_+$ of $\mathcal{G}$. The fields $\phi^i(z)$ are just l scalar fields. It can be shown that eq. (7.119) gives the resolution $(C^{(i)}\mathcal{H}^g, \tilde{d}^{(i)})$ of the W-algebra Hilbert space $\mathcal{H}^g$ indeed, in the form [253]

$$C^{(i)}\mathcal{H}^g = \bigoplus_{\{w \in W | l(w) = i\}} \mathcal{F}_{\alpha+(w\rho-\rho)}^\phi , \qquad (7.120)$$

where the collection of the Feigin-Fuchs charges $\oint \bar{s}_i^+(z)$ plays the role of the zeroth-order differential $\tilde{d}^{(0)}$ acting as $\mathcal{F}_0^\phi \to \mathcal{F}_{-\alpha+\alpha_i}^\phi$, and the intersection of its kernels is isomorphic to $\mathcal{H}^g$. This description leads to an identification of the W algebra with the centralizer of the Feigin-Fuchs screening operators. As far as the W-minimal representations are concerned, this resolution takes the form [253]

$$C^{(i)}L_\lambda = \bigoplus_{\{w \in \tilde{W} | \tilde{l}(w) = i\}} \mathcal{F}_{w*(\lambda'\alpha_+ + \lambda\alpha_-)} , \qquad (7.121)$$

where $w * \lambda = w(\lambda + \rho) - \rho$. The free-field resolution allows us to calculate characters of the highest-weight W-modules. In part, the characters of the W-minimal modules L_λ take the form [255]

$$\text{ch}_{L_\lambda}(q) = \frac{1}{\eta(\tau)^l} \sum_{w \in \tilde{W}} \epsilon(w) q^{\frac{1}{2pq}[p(\lambda+\tilde{\rho}) - qw(\lambda'+\rho)]^2} . \qquad (7.122)$$

Most of the results about the quantum DS reduction carry over to the supersymmetric case, when using the Lie superalgebras and super-AKM algebras instead of the bosonic ones. The quantum Miura transformation can also be supersymmetrized. For example, the $N = 1$ supersymmetric 'Lax operator', generalizing the bosonic one in eq. (7.47), takes the form (in $N = 1$ superspace) [11] [131]

$$R_n(Z) = \sum_{k=0}^{2n+1} U_{k/2}(Z)(\alpha_- D)^{2n+1-k}$$

$$=: [(\alpha_- D + h_1 \cdot iD\Phi(Z)] \cdots [(\alpha_- D + h_{2n+1} \cdot iD\Phi(Z)] : , \qquad (7.123)$$

where the $A_n^{(1)}$-weights h_μ have been introduced in the previous section. The Feigin-Fuchs screening operators in superspace take the simple form, namely, $: \exp(i\alpha_+\alpha_j \cdot \Phi) :$. The

---

[11] The $N = 1$ superspace coordinates are $Z = (z, \theta)$, where θ is a single anticommuting coordinate; a superfield takes the form $\Phi^i(Z) = \phi^i(z) + i\theta\psi^i(z)$, whereas the superspace covariant derivative is given by $D_Z = \partial_\theta + \theta\partial_z$ in this case.

leading singular terms in the OPE of the 'Lax operator' (7.123) with a screening operator come from the following terms

$$[(\alpha_- D_1 + h_j \cdot iD\Phi(Z_1))][(\alpha_- D_1 + h_{j+1} \cdot iD\Phi(Z_1))]\, e^{i\alpha + \alpha_j \cdot \Phi(Z_2)}$$

$$= (-1)^{j+1} D_2 \left[\frac{\theta_{12}}{Z_{12}} e^{i\alpha + \alpha_j \cdot \Phi(Z_2)} \right] , \qquad (7.124)$$

where $Z_{12} = z_1 - z_2 - \theta_1\theta_2$, $\theta_{12} = \theta_1 - \theta_2$ (Ch. V) and $\alpha_+ \alpha_- = -1$. All the operators $U_{k/2}(Z)$ with $k = 0, \ldots, 2n+1$ are in the centralizer of the supersymmetric screening operators. In particular, the generators $U_0(Z)$ and $U_{1/2}(Z)$ appear to be constants, while $U_1(Z)$ and $U_{3/2}(Z)$ form an $N = 2$ SCA of central charge $c = 3n[1 - (n+1)\alpha_-^2]$. This leads to the supersymmetric W algebras.

Exercise

VII-8 (*) Verify that the total BRST stress tensor, having the form $T_{\text{tot}} = T_{\text{impr}} + (\partial b)c$ and acting on the space $\mathcal{H}_{sl(2)}^{(k)} \otimes \mathcal{H}_{b,c}$, commutes with the BRST charge (7.105), but it is not BRST exact: $T_{\text{tot}}(z) \neq \{Q_{\text{BRST}}, *\}$. *Hint:* examine descendants of the vacuum state $|0\rangle_{sl(2)} \otimes |0\rangle_{b,c}$ at grade 2.

VII.4 W coset construction

The W-minimal (p,q)-models are unitary when $q = p + 1$, since they can be derived from the generalized coset construction [216] based on

$$\frac{\widehat{SU(n)}_k \otimes \widehat{SU(n)}_1}{\widehat{SU(n)}_{k+1,\text{diag}}} , \qquad p = k + n . \qquad (7.125)$$

Let J_n^a be the generators of $\widehat{su(n)}_k$, and $\tilde{J}_n^a$ the generators of $\widehat{su(n)}_1$. The diagonal subalgebra $\widehat{su(n)}_{k+1,\text{diag}}$ is then generated by their linear combinations, $J_n^a + \tilde{J}_n^a$. Having introduced the SS-constructed stress tensors for all of them as

$$T_k(z) = \frac{1}{2(k+n)} \sum_a : J^a(z)J^a(z) : ,$$

$$T_1(z) = \frac{1}{2(1+n)} \sum_a : \tilde{J}^a(z)\tilde{J}^a(z) : , \qquad (7.126)$$

$$T_{k+1}(z) = \frac{1}{2(k+1+n)} \sum_a : [J^a + \tilde{J}^a](z)[J^a + \tilde{J}^a](z) : ,$$

with the central charges

$$c_k = \frac{k(n^2 - 1)}{k + n} \, ,$$

$$c_1 = \frac{(n^2 - 1)}{1 + n} = n - 1 \, , \tag{7.127}$$

$$c_{k+1} = \frac{(k + 1)(n^2 - 1)}{k + 1 + n} \, ,$$

resp., one easily finds (sect. VI.1) that the coset stress tensor

$$T(z) = T_k(z) + T_1(z) - T_{k+1}(z) \tag{7.128}$$

commutes with both $J^a + \tilde{J}^a$ and T_{k+1}, and it has the central charge

$$c = c_k + c_1 - c_{k+1} = (n - 1) \left[1 - \frac{n(n + 1)}{(k + n)(k + 1 + n)} \right] \equiv c_{p,p+1}^n \, , \tag{7.129}$$

in exact agreement with eq. (7.91), when $p = k + n$. As far as the spectra of conformal dimensions are concerned, one obtains

$$\Delta_2^k = \frac{\lambda' \cdot (\lambda' + 2\rho)}{2(k + n)} \, , \quad \Delta_2^1 = \frac{\epsilon \cdot (\epsilon + 2\rho)}{2(1 + n)} \, , \quad \Delta_2^{k+1} = \frac{\lambda \cdot (\lambda + 2\rho)}{2(k + 1 + n)} \, , \tag{7.130}$$

where λ', ϵ and λ are the highest weights of $su(n)$. Unitarity imposes the contraints (Ch. II)

$$\lambda' \cdot \psi \le k \, , \quad \lambda \cdot \psi \le k + 1 \, , \quad \epsilon \cdot \psi \le 1 \, , \tag{7.131}$$

where ψ is the highest $su(n)$-root normalized to $\psi^2 = 2$. Eq. (7.131) implies $\epsilon = 0$ or $\epsilon = \lambda_i$. The crucial identity is

$$\Delta_2^k + \Delta_2^1 - \Delta_2^{k+1} = \Delta_2(\alpha = -\alpha_- \lambda - \alpha_+ \lambda') \bmod \mathbf{Z} \, , \tag{7.132}$$

where $\Delta_2(\alpha) = \frac{1}{2} \alpha \cdot (\alpha - 4\alpha_0 \rho)$, $\alpha_-^2 = \frac{p}{p+1} = \frac{k+n}{k+n+1}$ and $\alpha_+^2 = \frac{k+1+n}{k+n}$, in accordance with the results of sect. 2. To prove eq. (7.132), one notices [216, 240] that the defining relations $\alpha_- \alpha_+ = -1$ and $\alpha_+ \alpha_+ = 2\alpha_0$ imply

$$\alpha_- = -\frac{p}{\sqrt{p(p + 1)}} \, , \quad \alpha_+ = \frac{p + 1}{\sqrt{p(p + 1)}} \, , \quad 2\alpha_0 = \frac{1}{\sqrt{p(p + 1)}} \, , \tag{7.133}$$

and, hence,

$$\alpha_- = -2p\alpha_0 \, , \quad \alpha_+ = 2(p + 1)\alpha_0 \, . \tag{7.134}$$

It follows for the r.h.s. of eq. (7.132) that

$$\Delta_2(\alpha) = \frac{1}{2}(\lambda \alpha_- + \lambda' \alpha_+) \cdot (\lambda \alpha_- + \lambda' \alpha_+ + 4\alpha_0 \rho)$$

$$= \frac{1}{2p(p+1)}[(p+1)\lambda' - p\lambda] \cdot [(p+1)\lambda' - p\lambda + 2\rho] , \qquad (7.135)$$

whereas the l.h.s. yields

$$\Delta_2^k + \tfrac{1}{2}(\lambda - l')^2 - \Delta_2^{k+1}$$

$$= \frac{(p+1)\lambda' \cdot (\lambda' + 2\rho) - p\lambda \cdot (\lambda + 2\rho) + p(p+1)(\lambda^2 + \lambda'^2 - 2\lambda \cdot \lambda')}{2p(p+1)}$$

$$= \frac{(p+1)^2\lambda'^2 + p^2\lambda^2 - 2p(p+1)\lambda \cdot \lambda' + 2(p+1)\lambda' \cdot \rho - 2p\lambda \cdot \rho}{2p(p+1)}$$

$$= \frac{[(p+1)\lambda' - p\lambda] \cdot [(p+1)\lambda' - p\lambda + 2\rho]}{2p(p+1)} = \Delta_2(\alpha) . \qquad (7.136)$$

Therefore, we find

$$\Delta_2(\alpha) - \left(\Delta_2^k + \Delta_2^1 - \Delta_2^{k+1}\right) = -\Delta_2^1 + \tfrac{1}{2}(\lambda - l')^2 \equiv \delta . \qquad (7.137)$$

Clearly, $\lambda' + \epsilon - \lambda = r$, where r is a root. It follows $\tfrac{1}{2}(\lambda - l')^2 = \tfrac{1}{2}(\epsilon - r)^2 = \tfrac{1}{2}\epsilon^2 + m$, where $m = r^2/2 - \epsilon \cdot r$ is an integer. In its turn, this implies that $\delta = \tfrac{1}{2}\epsilon^2 - \frac{\epsilon \cdot (\epsilon + 2\rho)}{2(n+1)} + m$. Since we have, in addition, $n\epsilon^2 - 2\epsilon \cdot \rho = 0$ for $\epsilon = 0$ or $\epsilon = \lambda_i$, according to the first equality in eq. (12.28), we finally get $\delta = m$, thus completing the proof of eq. (7.132). The value of δ is not necessarily zero, which means that the W-primaries are not necessarily to be constructed solely from the AKM primaries.

We conclude that the W coset construction gives correct values for central charges and conformal dimensions of the unitary W-minimal models. The actual construction of W_n-generators exists in terms of the totally symmetric current products which are related to the order-n Casimir invariants, like that in eq. (7.9) [216]. More about the quantum DS reduction and the W coset construction, can be found in refs. [131, 256]. The construction of the classical W (or Gel'fand-Dikii) algebras and the Lagrangian realization of the associated symmetries in the Toda field theories are further discussed in sects. XI.2 and XI.3, resp. The Lagrangian approach is based on the conformal reduction of the *non-compact* WZNW models. Therefore, it gives the direct way to see the $\widehat{SL(n, \mathbf{R})}$ AKM symmetry, which is the CFT symmetry hidden by the quantum Hamiltonian reduction.

Chapter VIII

Conformal Field Theory and Strings

The theory of **strings** and **superstrings** (see refs. [70, 128, 257, 258, 259, 260, 261, 262, 263, 264, 265, 266, 267, 268, 269, 270, 271, 272, 273, 274, 275, 276, 277, 278, 279, 280, 281, 282, 283, 284, 285, 286, 287, 288, 289, 290, 291, 292, 293, 294, 295, 296, 297] and references therein [1]) describes (quantized) relativistic 1-dimensional objects (strings) propagating in a D-dimensional space-time. The very idea of string theory originated from numerous efforts to describe quark confinement and to construct a theory of quantum gravity, or a unified quantum theory of all fundamental physical interactions. A better understanding of string theory and, especially, of its non-perturbative aspects, which are believed to be crucial for physical applications, is yet to be developed. Therefore, we are not going to discuss physical content of string theory, if any (see the cited literature), but we want to discuss instead the strong connection which exists between (super)strings and (S)CFT's. We explain the origin of (super)conformal symmetries in string theories, and outline the main consequences. Our way of doing is based on the BRST quantization. The (S)CFT is relevant in describing and classifying (super)string vacua, and the (S)CFT techniques are very efficient in (super)string perturbation theory as well. The power of the (S)CFT tools becomes even more transparent in describing a propagation of strings and superstrings in non-trivial space-time backgrounds, or in addressing the (super)string compactification. The relatively unknown topics of the extended fermionic strings and of the W strings to be connected with the W algebras are also included. In this Chapter, we consider only the *critical* strings and superstrings (see, however, Chs. IX and X, as far as the *non-critical* (super)strings are concerned).

---

[1] We give here the references on books, lecture notes and review papers only.

VIII.1   Bosonic strings in $D \leq 26$

A string propagating in a D-dimensional space-time [2] sweeps out a two-dimensional surface called a string **world-sheet**. Let X^μ, $\mu = 0, 1, \ldots, D - 1$, be the space-time coordinates, while (σ^0, σ^1) the two-dimensional coordinates parametrizing a world-sheet. A string motion is therefore described by the set of D functions $X^\mu(\sigma^0, \sigma^1)$ specifying the space-time position of each point of the string. The σ^1 may be interpreted as the coordinate labeling points along the string, whereas the σ^0 may be identified with the proper time, although this is not actually necessary.

Strings can be either *closed* or *open*. For definiteness, we confine ourselves to the closed strings, which satisfy the boundary condition $X^\mu(\sigma^0, 0) = X^\mu(\sigma^0, \pi)$. Though a physical description of strings implies the use of Minkowski signatures for a world-sheet and a space-time, we apply the Wick rotations to get Euclidean formulation of the theory, in order to make contact with CFT and the theory of Riemann surfaces. The coordinate functions $X^\mu(\sigma)$ describing an embedding of Riemann surface Σ (representing a string world-sheet) into Euclidean spacetime $\mathbf{R}^D$, can then be equally considered as the set of D two-dimensional scalar fields to be defined on Σ.

An action governing string dynamics should be independent on the way how a world-sheet Σ is parametrized. This property is called the *reparametrization invariance*. This invariance is nothing but the usual 2d general coordinate invariance, of course, so that the standard way of implementing it is just to couple the scalar fields $X^\mu(\sigma)$ with the 2d 'gravity' to be described by a metric $g_{\alpha\beta}(\sigma)$. The minimal coupling results in the action known as **Polyakov's string action** having the Euclidean form

$$S_\mathrm{P} = \frac{1}{4\pi\alpha'} \int_\Sigma d^2\sigma \sqrt{g} g^{\alpha\beta} \partial_\alpha X^\mu \partial_\beta X_\mu \, , \tag{8.1}$$

where g is the determinant of the 2d metric $g_{\alpha\beta}$, and α' is the string constant (sometimes called the '*Regge slope parameter*' also). The action (8.1) respects the global Poincaré symmetry in the target space (i.e. in the spacetime). It is an important property of the Polyakov string action (8.1) that it is not only invariant w.r.t. the 2d diffeomorphisms, but also under the local **Weyl** transformations

$$g_{\alpha\beta} \to \Lambda(\sigma) g_{\alpha\beta} \, , \tag{8.2}$$

which do not move the points of Σ. The Weyl symmetry (8.2) is anomalous in quantum theory, whereas the 2d reparametrization invariance can be safely used to bring the 2d

---

[2] The space-time is temporarily assumed to be flat.

metric to the conformally-flat form called the *orthonormal* or *conformal gauge*,

$$g_{\alpha\beta} = \delta_{\alpha\beta} e^{2\phi(\sigma)} \ . \tag{8.3}$$

In this gauge, a *conformal factor* $\phi(\sigma)$ disappears from the classical Polyakov string action. As we already know from sect. I.1, the metric (8.3) preserves its form under the conformal transformations, which are nothing but the analytic transformations in the conformal coordinates $(z, \bar{z})$, where $z = \sigma^0 + i\sigma^1$, on the Euclidean string worldsheet. Therefore, the string theory has the conformal symmetry which arises as the remnant of the reparametrizational symmetry in the conformal gauge. This observation is not related, in fact, with the specific choice of a string action made in eq. (8.1). The Polyakov string action can be generalized in many ways while maintaining the underlying conformal symmetry, so that it is actually the CFT that provides us with the appropriate framework to deduce general action-independent properties of strings, and thus avoid to analyze various string actions separately. It is just the approach that we are going to use throughout this Chapter.

In the conformal gauge, the action (8.1) is just the action of D *free* scalar fields, which can be easily quantized (sect. I.5). The *conformal* (or *Weyl*) anomaly is given by central charge in the quantum conformal (Virasoro) algebra. This anomaly is, of course, equal to D for D free scalar fields (as usual, we concentrate on a holomorphic dependence of the fields). In the BRST quantized string theory, we should also take into account the bosonic FP-ghost fields associated with the covariant gauge-fixing of the 2d reparametrization invariance. They are given by a particular (b, c) conformal ghost system (sect. I.6) of spin $\lambda = 2$, and they are known as the *reparametrization ghosts*. Hence, in acccordance with eq. (1.129), they contribute -26 to the Virasoro central charge. This immediately yields the celebrated result: the quantized bosonic string is anomaly-free only in the **critical dimension** of $D = 26$. The zero modes of string scalar fields are identified with spacetime coordinates. The physical space-time has to have the Minkowski signature, which is dictated by the *no-ghost theorem* [46, 298] requiring the number of time-like directions of the critical target space be the same as that of the (b, c) ghost system, i.e. one in both cases.

The critical space-time dimension of $D = 26$ is actually true provided that all the string scalar fields are free and uncompactified. At the same time, nothing forces us to assume that. Indeed, there are many CFT's with different values of central charge, so that we can compose some of them in order to compensate the only relevant ghost contribution of $c_{\mathrm{gh}} = -26$. In particular, the space-time dimension can be adjusted to be equal to four, by using only four uncompactified free scalar fields as above, while taking the rest of the central charge ($c_{int} = 22$) from some other CFT to be representing the *internal* string degrees of freedom. This is precisely the underlying philosophy which leads to the so-called *four-dimensional strings* [299, 300, 301, 302, 303, 304, 305]. The central charge value of $c_{int} = 22$ is not the only condition to impose on CFT describing the internal sector. Among the other fundamental physically-motivated requirements are (i) unitarity, and (ii) a discrete bounded-from-below spectrum of the 'Hamiltonian' L_0. In particular, this implies that the integrable highest-weight unitary representations of the Virasoro algebra are the only ones to consider. A physical content of a four-dimensional string model crucially depends on a choice of CFT describing the internal sector. Because of an enormous variety of CFT's to be suitable for this role, there is very considerable amount of uncertainty in any physical prediction to be made from the string theory. This observation explains why we are not going to discuss here any possible physical applications of strings and superstrings, since they inevitably would be highly speculative. This also explains a need for developing a non-perturbative approach to string theory. [3]

Let us now generalize the Polyakov string action. First, from the two-dimensional viewpoint, the action (8.1) can be considered as a particular minimal coupling of conformal matter represented by free scalar fields to the two-dimensional gravity. In QFT, the most general reparametrization-invariant action has to be considered, as is normally required by renormalization. Since the 2d Euler characteristic and the 'cosmological' term are the only possible modifications of this type, the 'generalized' (in the 2d sense) string action reads [306]

$$S_B = S_P + \lambda \chi(\Sigma) + \mu_0^2 \int_\Sigma d^2z \sqrt{g} , \qquad (8.4)$$

where λ and μ_0^2 are constants. The second term in eq. (8.4) is a 2d topological invariant, while the third one explicitly breaks the Weyl invariance. The Weyl invariance can however be restored in the critical dimension $D = 26$, if we want to insist on it. Away from the critical dimension, the Weyl anomaly will be represented by the Liouville theory (Ch. X), which is the effective theory of the quantum 2d gravity. The Liouville potential is an exponential of the conformal factor ϕ, which just appears due to the last term in eq. (8.4) to be written in the conformal gauge (8.3).

Second, since the Polyakov string action also has a spacetime interpretation, there is another way of its geometrical generalization towards a *non-linear sigma-model* (NLSM)

---

[3]The matrix models (Ch. X) originated, in part, from the attempts to find a solution to this issue.

[307]. Having required the NLSM renormalizability in addition, we are able to write down the most general reparametrization-invariant local NLSM action in the form [308, 309, 310, 311]

$$S_{\text{NLSM}} = \frac{1}{4\pi\alpha'} \int_\Sigma d^2z \sqrt{g} g^{\alpha\beta} \partial_\alpha X^\mu \partial_\beta X^\nu G_{\mu\nu}(X) + \frac{1}{4\pi\alpha'} \int_\Sigma d^2z \, \varepsilon^{\alpha\beta} \partial_\alpha X^\mu \partial_\beta X^\nu B_{\mu\nu}(X)$$

$$+ \frac{1}{4\pi} \int_\Sigma d^2z \sqrt{g} R^{(2)} \Phi(X) - \frac{1}{\pi\alpha'} \int_\Sigma d^2z \sqrt{g} T(X) \; ; \tag{8.5}$$

where a symmetric *spacetime* metric $G_{\mu\nu}(X)$, an antisymmetric *Kalb-Ramond* field $B_{\mu\nu}(X)$, a *dilaton* field $\Phi(X)$, and a *tachyon* field $T(X)$ have been introduced (and, in fact, defined this way). The action (8.5) describes a string propagation in a space-time background of these fields, which are simultaneously in one-to-one correspondence with the lowest string excitation modes. The dilaton contribution to eq. (8.5) explicitly breaks down the *classical* Weyl invariance of the string action, but the Weyl invariance is supposed to be restored after taking into account quantum corrections. [4] The Weyl invariance then implies the conformal invariance. For simplicity, we ignore the tachyon ($T = 0$) in what follows.

The quantized NLSM of eq. (8.5) is conformally-invariant provided $\langle T_\alpha{}^\alpha \rangle = 0$, where $T_\alpha{}^\beta$ is the two-dimensional stress tensor. On dimensional and symmetry grounds, the trace can be represented in terms of finite NLSM composite operators as

$$2\pi T_\alpha{}^\alpha = \bar\beta^G_{\mu\nu} \sqrt{g} g^{\alpha\beta} \partial_\alpha X^\mu \partial_\beta X^\nu + \bar\beta^B_{\mu\nu} \varepsilon^{\alpha\beta} \partial_\alpha X^\mu \partial_\beta X^\nu + \bar\beta^\Phi \sqrt{g} R^{(2)} \; , \tag{8.6}$$

where the '*Weyl-anomaly coefficients*' $\bar\beta^G_{\mu\nu}$, $\bar\beta^B_{\mu\nu}$ and $\bar\beta^\Phi$ are some local functions of the fields $G_{\mu\nu}$, $B_{\mu\nu}$ and Φ. They almost coincide with the NLSM renormalization group (RG) β-functions $\beta^G_{\mu\nu}$, $\beta^B_{\mu\nu}$ and β^Φ, resp., while the differences are given by the world-sheet total derivative terms coming in a passage from integrated into local expressions [312, 313, 314, 315]. The RG β-functions of NLSM are subject to the ambiguities caused by field reparametrizations, whereas the 'Weyl-anomaly coefficients' are actually invariant w.r.t. them.

The NLSM RG β-functions can be explicitly calculated in the lowest orders of α' on a sphere (genus $h = 0$), by using the conventional quantum perturbation theory for the NLSM (see, e.g., refs. [316, 317, 320, 318, 319] for details, as well as for higher-order generalizations), and they take the form

$$\beta^G_{\mu\nu} = R_{\mu\nu} - H^2_{\mu\nu} + 2\nabla_\mu \nabla_\nu \Phi + O(\alpha') \; ,$$

---

[4]This explains the absence of the dimensional string constant α', playing the role of Planck constant, in the dilaton term of eq. (8.5).

$$\beta_{\mu\nu}^B = 2\nabla_\rho H_{\mu\nu}^\rho - 4(\nabla_\rho\Phi)H_{\mu\nu}^\rho + O(\alpha') \,,$$

$$\frac{\beta^\Phi}{\alpha'} = \frac{1}{\alpha'}\frac{(D-26)}{48\pi^2} + \frac{1}{16\pi^2}\left\{4(\nabla\Phi)^2 - 4\nabla^2\Phi - R + \frac{1}{3}H^2\right\} + O(\alpha') \,, \qquad (8.7)$$

where the Ricci tensor $R_{\mu\nu}$ and the scalar curvature R defined in terms of the D-dimensional metric G, and the totally antisymmetric field strength $H_{\mu\nu\rho} \equiv \frac{3}{2}\partial_{[\mu}B_{\nu\rho]}$ have been introduced. In addition, we use here the following book-keeping definitions:

$$H_{\mu\nu}^2 \equiv H_{\mu\lambda\rho}H_\nu^{\lambda\rho} \,, \quad H^2 \equiv H_{\mu\nu\rho}H^{\mu\nu\rho} \,. \qquad (8.8)$$

Therefore, the vanishing of all the RG β-functions of NLSM leads to the effective equations of motion for the background fields, which generalize the Einstein equations of general relativity. In their exact (to all orders in α') form, they are just the conditions of the conformal invariance for the NLSM in question. Given the exact conditions, the NLSM represents a CFT.

The central charge of this NLSM-realized CFT is of special interest. To relate it with the NLSM 'Weyl-anomaly coefficients', let us first note that a *constancy* of the dilaton β-function follows from the vanishing of other RG β-functions, in the lowest order of α' [317]. This consistency condition should be the case, since the conformal anomaly is essentially (up to a constant normalization factor) given by the dilaton $\bar{\beta}$-function. One finds from eq. (8.7) that

$$0 = \nabla^\mu\left[R_{\mu\nu} - H_{\mu\nu}^2 + 2\nabla_\mu\nabla_\nu\Phi\right]$$

$$= \nabla_\mu\left[-2(\nabla\Phi)^2 + 2\nabla^2\Phi + \frac{1}{2}R - \frac{1}{6}H^2\right] = 0 \,, \qquad (8.9)$$

where the Riemannian Bianchi identity has been used. To identify the (appropriately normalized) 'on-shell' dilaton $\bar{\beta}$-function with the central charge of the corresponding CFT, a similar consistency condition has to be true in general. The general theorem is called the **Curci-Paffuti** relation [321], which generically reads

$$\partial_\mu\bar{\beta}^\Phi = \mathcal{O}_\mu^G\bar{\beta}^G + \mathcal{O}_\mu^H\bar{\beta}^H \,, \qquad (8.10)$$

where $\mathcal{O}_\mu^{G,H}$ are some field-dependent first-order *linear* differential operators, whose explicit form can be calculated in perturbation theory. The original proof of eq. (8.10) in ref. [321] is based on the detailed structure of NLSM renormalization, and it is rather complicated. Fortunately, to understand the origin of this result, there exists the simple argument due to Polchinski who noticed, by analyzing the first two variations of the

NLSM partition function Z w.r.t. the conformal factor ϕ in the conformal gauge (8.3), that

$$\frac{\delta \ln Z}{\delta \phi(z)} \propto \langle T_\alpha{}^\alpha(z) \rangle \propto \langle \bar\beta^\Phi(z) R^{(2)}(z) \rangle \ , \tag{8.11}$$

and, hence,

$$\frac{\delta^2 \ln Z}{\delta \phi(z) \delta \phi(z')} \propto \langle \bar\beta^\Phi(z) \Box_z \delta(z - z') \rangle \ . \tag{8.12}$$

Because of the symmetry of the l.h.s. of eq. (8.12) w.r.t. the exchange $z \leftrightarrow z'$, it follows

$$\langle \left[\bar\beta^\Phi(z) - \bar\beta^\Phi(z') \right] \Box_z \delta(z - z') \rangle = 0 \ , \tag{8.13}$$

which seems to be enough to conclude that $\bar\beta^\Phi(z) = \bar\beta^\Phi(z')$. The stronger argument is given at the end of this section (see, however, ref. [321] for the proof).

One can get additional insights into general features of the NLSM-based CFT from the partial results about the NLSM RG β-functions in eq. (8.7), by questioning the existence of an action from which these β-functions could follow as the equations of motion. This action is known in string theory as the *low-energy string effective action*. First, let us rewrite the on-shell equations (8.7) to the more conventional form used in general relativity, namely [317]

$$0 = \beta^G_{\mu\nu} + 8\pi^2 g_{\mu\nu} \frac{\beta^\Phi}{\alpha'} = \left(R_{\mu\nu} - \tfrac{1}{2} G_{\mu\nu} R \right) - T^{\text{matter}}_{\mu\nu} \ ,$$

$$0 = \beta^B_{\mu\nu} = 2\nabla_\rho H^\rho_{\mu\nu} - 2(\nabla_\rho \Phi) H^\rho_{\mu\nu} \ ,$$

$$0 = 8\pi^2 \frac{\beta^\Phi}{\alpha'} + \tfrac{1}{2} G^{\mu\nu} \beta^G_{\mu\nu} = 2(\nabla \Phi)^2 - \nabla^2 \Phi - \tfrac{1}{3} H^2 \ , \tag{8.14}$$

where the spacetime matter stress-energy tensor $T^{\text{matter}}_{\mu\nu}$ has been introduced,

$$T^{\text{matter}}_{\mu\nu} \equiv H^2_{\mu\nu} - \frac{1}{6} G_{\mu\nu} H^2 - 2\nabla_\mu \nabla_\nu \Phi + 2G_{\mu\nu} \nabla^2 \Phi - 2G_{\mu\nu} (\nabla \Phi)^2 \ . \tag{8.15}$$

It is now easily verified (exercise # VIII-4) that the equations (8.14) do follow as the equations of motion from the following action in the target space (spacetime) [317]:

$$I^{\text{eff}} \propto \int d^D X \sqrt{-G} e^{-2\Phi} \left[-R + 4(\nabla \Phi)^2 + \frac{1}{3} H^2 \right] \ . \tag{8.16}$$

The exponential factor in eq. (8.16) is a direct consequence of eq. (8.5) and the definition of the quantum string effective action, since the zero-mode of the dilaton field is multiplied by the Euler characteristic $\chi(\Sigma)$ in eq. (8.5), while $\chi = 2$ for the Riemann sphere. In general, contributions to the string effective action from the string world-sheets of genus-h should come with the factor $\exp[-2(1-h)\Phi]$. The action (8.16) takes the conventional form after the Weyl transformation of space-time metric as

$$G_{\mu\nu} \to \exp\left[\frac{4\Phi}{D-2}\right] G_{\mu\nu} \ . \tag{8.17}$$

The equivalent action results in

$$I^{\text{eff}} \propto \int d^D X \sqrt{-G} \left\{ -R + \frac{4}{D-2}(\nabla\Phi)^2 + \frac{1}{3}\exp\left[-\frac{8\Phi}{D-2}\right] H^2 \right\} \ , \tag{8.18}$$

and it yields the bosonic terms in the gravitational sector of the *Chapline-Manton action* [322], describing the ten-dimensional $N=1$ supergravity interacting with the $N=1$ supersymmetric Yang-Mills matter, when $D=10$. This result is relevant for the ten-dimensional critical superstrings, since two-dimensional fermions do not contribute to the one-loop RG β-functions of the supersymmetric NLSM.

One may ask whether the Weyl invariance conditions in a general two-dimensional NLSM, or even in an arbitrary 2d QFT, can always be derived as the equations of motion from some action? The existence proof is usually referred to as **Zamolod-chikov's c-theorem** [323]. The c-theorem is valid under the three basic assumptions: (i) reparametrizational invariance, (ii) renormalizability and (iii) locality of two-dimensional QFT under consideration. The first condition implies the conservation law for the QFT stress-energy tensor in the form $\partial_\alpha(\sqrt{g}T^{\alpha\beta}) = 0$, and this allows us to choose the conformal gauge; the second condition ensures that the stress-energy tensor is finite; while the third condition leads to the positivity of QFT, which is expressed by the positivity of the function $E(G)$ to be defined below.

The proof of the c-theorem goes as follows [323, 324]. Consider a two-dimensional [5] reparametrization-invariant and renormalizable QFT characterized by an action $S = \int d^2z \mathcal{L}$, and let $\{G^i\}$ be a set of all its (renormalized) couplings, while β^i the corresponding RG β-functions, i.e.

$$\sqrt{g}T_{\alpha\beta} \equiv \frac{\delta S}{\delta g^{\alpha\beta}} \ , \quad \beta^i \equiv \frac{dG^i}{dt} \ , \tag{8.19}$$

where t is a RG parameter. It should be stressed that we are now dealing with a 2d renormalizable QFT *away* from its critical RG fixed point. The QFT under consideration

---

[5]It is not known yet whether the statement of the c-theorem can be extended to dimensions higher than two [325, 326].

becomes a CFT at a RG fixed point G_0^i, where $\beta^i(G_0) = 0$. Using the conformal coordinates $(z, \bar{z})$, and the notation $T(z, \bar{z}) = \sqrt{g}\, T_{zz}$, $\Theta(z, \bar{z}) = \sqrt{g}\, T_{z\bar{z}}$, let us define the following real functions of the renormalized couplings $\{G\}$:

$$C(G) = 2z^4 \, \langle T(z, \bar{z}) T(0, 0)\rangle|_{z\bar{z}=1} \ ,$$

$$D(G) = z^3 \bar{z} \, \langle T(z, \bar{z}) \Theta(0, 0)\rangle|_{z\bar{z}=1} \ ,$$

$$E(G) = z^2 \bar{z}^2 \, \langle \Theta(z, \bar{z}) \Theta(0, 0)\rangle|_{z\bar{z}=1} \ , \tag{8.20}$$

where $z\bar{z} = 1$ has been chosen to be a normalization 'point'. Because of the conservation and the non-renormalizability of the stress-energy tensor $(\partial_i \equiv \partial/\partial G^i)$,

$$\frac{dT_{\alpha\beta}}{dt} = \frac{\partial T_{\alpha\beta}}{\partial t} + \beta^i \partial_i T_{\alpha\beta} = 0 \ , \tag{8.21}$$

we have the relations

$$\tfrac{1}{2}\beta^i \partial_i C = 3D - \beta^i \partial_i D \ , \quad \beta^i \partial_i D - D = 2E - \beta^i \partial_i E \ . \tag{8.22}$$

Hence, the function $I(G)$ defined by

$$I(G) \equiv C(G) - 4D(G) - 6E(G) \ , \tag{8.23}$$

satisfies the equation

$$\frac{dI(G)}{dt} = \beta^i \partial_i I(G) = -12E(G) \ . \tag{8.24}$$

Eq. (8.24) implies that the function $I(G)$ monotonically decreases under the RG flow, since the function $E(G)$ is positively definite, according to its definition in eq. (8.20). The analogue of eq. (8.6) near the critical point takes the form

$$\Theta = \beta^i \Lambda_i \ , \tag{8.25}$$

and it is valid on-shell. The normally-ordered composite operators Λ_i are defined from the QFT action as $\Lambda_i = \partial_i \mathcal{L}$. It follows

$$E(G) = \beta^i \beta^j \mathcal{G}_{ij} \ , \tag{8.26}$$

in accordance with eq. (8.20), where the symmetric and positively definite 'metric' $\mathcal{G}_{ij}$ in the space of couplings has been introduced,

$$\mathcal{G}_{ij} = z^2 \, \bar{z}^2 \, \langle \Lambda_i(z, \bar{z}) \Lambda_j(0, 0)\rangle\Big|_{z\bar{z}=1} \ . \tag{8.27}$$

Eq. (8.24) now yields

$$\beta^i \left(\partial_i I + 12 \mathcal{G}_{ij} \beta^j \right) = 0 \ . \tag{8.28}$$

Since the positivity of the metric $\mathcal{G}_{ij}$, the stationarity of the Zamolodchikov action $I(G)$, $\partial_i I(G) = 0$, means the criticality, $\beta^i = 0$, according to eq. (8.28). Simultaneously, one gets the existence proof for the action itself. The critical value of the action $I(G)$ yields the central charge of the corresponding CFT. The criticality condition $\beta^i = 0$ does not, however, imply $\partial_i I = 0$. This would be true only under an anticipated existence of the *off-shell* relation

$$\partial_i I = K_{ij} \beta^j \tag{8.29}$$

with an invertible matrix K_{ij}. It was suggested in ref. [323] that the '*K-matrix*' exists, and it might even be the same as the 'metric' $\mathcal{G}_{ij}$. Explicit perturbative calculations up to three-loop order show, in particular, the presence of derivatives in the K-operator w.r.t. fields, thus indicating its very non-trivial nature [327, 328, 329].

Having taken into account the total derivative terms making a difference between the β's and $\bar{\beta}$'s in NLSM, the relation similar to that in eq. (8.28) can be established in terms of the 'Weyl-anomaly coefficients', after appropriate changes in the meaning of the Λ_i-operators in eq. (8.25): they now have to be the normally-ordered finite operator products resulting from the renormalization of the NLSM composites, $\Lambda_i \to [\Lambda_i]$. The Curci-Paffuti relation can now be justified on the basis of Zamolodchikov's c-theorem [324]. Let us introduce the new function

$$H_i(G) = z^3 \bar{z} \left\langle T(z, \bar{z})[\Lambda_i] \right\rangle \big|_{z\bar{z}=1} \ , \tag{8.30}$$

so that $\bar{\beta}^i H_i = D$. The RG equation generically reads

$$\frac{d[\Lambda_i]}{dt} = (P_i{}^j - \partial_i \beta^j)[\Lambda_j] \equiv \Gamma_i{}^j[\Lambda_j] \ , \tag{8.31}$$

where the 'matrix' $P_i{}^j$ arises due to total derivative terms on the world-sheet. [6] It follows

$$\beta^k \partial_k H_i - \Gamma_i{}^k H_k - H_i = 2\mathcal{G}_{ik}\bar{\beta}^k + \Gamma_i{}^k \mathcal{G}_{kn}\bar{\beta}^n - \beta^j \partial_j \mathcal{G}_{in}\bar{\beta}^n - \mathcal{G}_{ij}\beta^n \partial_n \bar{\beta}^j \ , \tag{8.32}$$

and, hence,

$$\beta^i \partial_i D + \beta^i \partial_i (H_j \beta^j) = D + 2E - \beta^i \partial_i E + \left\{ \beta^i \partial_i \bar{\beta}^j - \bar{\beta}^i \partial_i \beta^j - \bar{\beta}^i P_i{}^j \right\} \left(H_j + \mathcal{G}_{jn}\bar{\beta}^n \right) \ . \tag{8.33}$$

---

[6]See the notice made after eq. (8.6).

Eq. (8.33) is to be compared with eq. (8.22). This immediately implies the vanishing of the last term in eq. (8.33),i.e. the vanishing coefficient in curly brackets, which is just the Curci-Paffuti relation. Therefore, this relation is a corrollary of the three basic assumptions about the relevant two-dimensional QFT's listed above.

Most of the particular constructions introduced in the previous Chapters (minimal models, characters, partition functions, free field representations, group and coset constructions, modular invariance) can be applied to study the four-dimensional strings. The string 'Hamiltonian' L_0 is always a sum of a space-time term and an internal term. The (one-loop) string partition function is therefore factorized into a product of space-time and internal contributions, the similar structure being shared by a string space of physical states too. The modular invariance is a strong condition in string theory, which follows from an invariance of the string theory w.r.t. *all* world-sheet reparametrizations, including the topologically non-trivial (i.e. 'global' or 'big') 2d diffeomorphisms.

Exercises

\# VIII-1 ▷ (*) Eliminate the metric $g_{\alpha\beta}$ from Polyakov's string action (8.1) by using its algebraic equations of motion. Prove that this gives rise to the *Nambu-Goto* string action which is proportional to world-sheet area [330, 331],

$$S_{\text{NG}} = \frac{1}{2\pi\alpha'} \int_\Sigma d^2\sigma \sqrt{h(X)} \,, \qquad (8.34)$$

where $h(X)$ is the determinant of the induced metric,

$$h_{\alpha\beta}(X) = \partial_\alpha X^\mu \partial_\beta X_\mu \,. \qquad (8.35)$$

\# VIII-2 ▷ (*) Calculate the normal-ordering constant arising in the definition of the quantum string Hamiltonian L_0. This constant is called a string '*intercept*'. Construct the BRST operator Q of the bosonic string in the Polyakov formulation, when following the rules of sect. XII.5. Derive the value of the critical dimension from the BRST nilpotency condition $Q^2 = 0$.

\# VIII-3 ▷ Verify that the matter stress-energy tensor $T_{\mu\nu}^{\text{matter}}$ defined by eq. (8.15) satisfies the conservation law $\nabla^\mu T_{\mu\nu}^{\text{matter}} = 0$ on the matter equations of motions, as it should.

\# VIII-4 ▷ Derive eq. (8.16) from eqs. (8.14) and (8.15). Calculate the *improved* matter stress-energy tensor from the action (8.18).

VIII.2   Supersymmetry and picture-changing

The Polyakov string action is the good starting point for world-sheet supersymmetrization of the bosonic string, while maintaining its Poincaré invariance in the target space. The relevant action is constructed by minimal coupling of the two-dimensional (N-extended) supergravity to an (N-extended) 2d scalar 'matter' multiplet. The case of $N = 1$ supersymmetry leads to the *fermionic* string, also known as the **Neveu-Schwarz-Ramond** (NSR) model [332, 333, 334, 335, 336]. The *extended* fermionic strings with $N > 1$ world-sheet supersymmetry are considered in sect. 3.

To fix NSR string action, let us begin with the bosonic string theory to be defined on a *Minkowski* world-sheet, since 2d MW spinors can be introduced there. [7] Since the (conformally) gauge-fixed bosonic string action is free, it can be easily supersymmetrized in two dimensions to the form

$$S_{\text{g.-f.}}^{\text{NSR}} = -\frac{1}{2\pi} \int d^2\sigma \left(\partial_\alpha X^\mu \partial^\alpha X_\mu + i\overline{\Psi}^\mu \rho^\beta \partial_\beta \Psi_\mu \right) , \qquad (8.36)$$

where D two-dimensional Majorana spinor fields Ψ^μ have been introduced, $\overline{\Psi} = \Psi^{\mathrm{T}} \rho^0$. The two types of boundary conditions for the world-sheet fermions, resulting in the two (NS and R) different sectors of the fermionic string theory have been already discussed in sect. I.5. In eq. (8.36), the matrices ρ^α stand for 2d Dirac matrices, and they have an explicit representation in the form

$$\rho^0 = \begin{pmatrix} 0 & -1 \\ 1 & 0 \end{pmatrix} , \quad \rho^1 = \begin{pmatrix} 0 & 1 \\ 1 & 0 \end{pmatrix} . \qquad (8.37)$$

The action (8.36) is invariant w.r.t. rigid two-dimensional $N = 1$ supersymmetry transformations having the form

$$\delta X^\mu = i\bar{\eta}\Psi^\mu , \quad \delta\Psi^\mu = (\rho^\alpha \partial_\alpha X^\mu)\eta , \qquad (8.38)$$

with spinor Grassmannian parameter η. In the Minkowski space-time, the (covariant) quantization of the theory (8.36) leads to the additional (compared to the bosonic string) non-physical propagating degrees of freedom because of the Ψ^0-field component. In the

---

[7]The Minkowski formulation of Polyakov's string is obtained by the 'reverse' Wick rotation from the Euclidean formulation. The formal differences between the Minkowski string action and the action in eq. (8.1) amount to the signs in front of the action and under the square root of the 2d metric determinant. We also choose the string constant to be $\alpha' = 1/2$, just for simplicity.

bosonic string theory, the similar problem with the X^0-field component was resolved by the local (conformal) invariances of the theory, resulting in the (Virasoro) constraints on the bosonic string physical states. In the fermionic case, we therefore need an additional local fermionic invariance, whose remnant in appropriate (superconformal) gauge would lead to the superconformal invariance. The symmetry in question is just the local 2d *supersymmetry*. All we need is therefore to couple the action (8.36) to two-dimensional supergravity, in order to get full fermionic string theory action. The two-dimensional $N = 1$ supergravity is described in components by a 2d metric $g_{\alpha\beta}$ or a *zweibein* e^b_β, and its superpartner – a 2d Majorana '*gravitino*' field χ_β. The zweibein provides a local basis on a string world-sheet. This basis consists of two vectors $\vec{e}^{(b)}$ having the components e^b_β, and it has the following properties:

$$g_{\beta\gamma} = e^b_\beta \eta_{bc} e^c_\gamma , \quad e^2 = [\det(e^b_\beta)]^2 = -\det(g_{\beta\gamma}) , \tag{8.39}$$

where e^β_b is the inverse zweibein. The full 2d locally supersymmetric action can be constructed either by 'trial and error' (this is known as the *Noether procedure*) or by superspace methods. The result takes the form [337, 338, 339]

$$S^{\text{NSR}} = -\frac{1}{2\pi} \int d^2\sigma \, e \left\{ g^{\alpha\beta} \partial_\alpha X^\mu \partial_\beta X_\mu + i\overline{\Psi}^\mu e^\beta_b \rho^b \partial_\beta \Psi_\mu \right.$$
$$\left. + 2ie^\gamma_c e^\beta_b \overline{\chi}_\gamma \rho^b \rho^c \Psi^\mu (\partial_\beta X_\mu + \frac{i}{2}\overline{\Psi}_\mu \chi_\beta) \right\} . \tag{8.40}$$

The NSR string action (8.40) is invariant under the *local $N = 1$ supersymmetry* transformations

$$\delta X^\mu = i\overline{\eta}\Psi^\mu , \quad \delta\Psi^\mu = e^\beta_b \rho^b (\partial_\beta X^\mu + i\overline{\lambda}^\mu \chi_\beta)\eta ,$$
$$\delta e^b_\beta = -2i\overline{\eta}\rho^b \chi_\beta , \quad \delta\chi_\beta = -D_\beta \eta \equiv -\partial_\beta \eta + \frac{1}{2}\omega_\beta \rho^0 \rho^1 \eta , \tag{8.41}$$

with anticommuting spinor parameter $h(\sigma)$. The 2d Lorentz connection ω_β reads [337, 338]

$$\omega_\beta = \frac{1}{2}\varepsilon^{ab} e^\alpha_a \left[\partial_\beta e_{b\alpha} - \partial_\alpha e_{b\beta} + e^\gamma_b e^c_\beta \partial_\gamma e_{c\alpha} \right] + 2ie\varepsilon_{\beta\alpha}\overline{\chi}^\alpha e^\gamma_b \rho^b \chi_\gamma . \tag{8.42}$$

The action (8.40) is invariant, in addition, under the 2d reparametrizations and the 2d local Lorentz rotations, as well as under the local Weyl transformations (with parameter $\Lambda(\sigma)$) having the form

$$\delta X^\mu = 0 , \quad \delta\Psi^\mu = -\frac{1}{2}\Lambda\Psi^\mu , \quad \delta e^b_\beta = \Lambda e^b_\beta , \quad \delta\chi_\beta = \frac{1}{2}\Lambda\chi_\beta , \tag{8.43}$$

and, finally, under the local superconformal (fermionic) transformations (with anticommuting spinor parameter $\pi(\sigma)$) of the form

$$\delta X^\mu = \delta\Psi^\mu = \delta e^b_\beta = 0 , \quad \delta\chi_\beta = e^b_\beta \rho_b \pi . \tag{8.44}$$

All the local symmetries of the NSR theory are just enough to (locally) gauge away all the supergravity fields. However, the symmetries (8.43) and (8.44), representing together the $N = 1$ super-Weyl invariance, are, in fact, anomalous in quantum theory. The rest of the local symmetries are non-anomalous, and they can therefore be used to define the *superconformal gauge*, in which the supergravity fields are given by a super-Weyl transformation of their flat values. In the superconformal gauge, the reparametrization invariance and local supersymmetry are fixed, which causes us to introduce the fermionic reparametrization ghosts (b, c) of $\lambda = 2$, as well as the bosonic superconformal ghosts (β, γ) of $\lambda = 3/2$. The latter are just a particular (β, γ) conformal ghost system introduced in sect. I.6. The NSR matter contributes $(1 + \frac{1}{2})D$ into the central charge c (or just D to the central charge $\hat{c}$), whereas all the ghosts together contribute $-26 + 11 = -15$ to the c (or just -10 to the $\hat{c}$). The total central charge vanishes when $D = 10$, which is the critical dimension of the NSR model. Again, we are not forced to represent the $N = 1$ matter by the uncompactified free fields, e.g. we could use only four scalar $N = 1$ supermultiplets whose bosonic zero modes would represent the space-time coordinates (they contribute $4 + 2 = 6$ to the central charge c), whereas take the rest (the 'internal' sector) to be represented by a SCFT of the central charge $c = 9$. This is known as the *superstring compactification*.

From the SCFT point of view, the holomorphic sector of the NSR model in a flat ten-dimensional space-time is described by ten free scalar fields $x^\mu(z)$ and their fermionic superpartners $\psi^\mu(z)$, with the OPE's

$$x^\mu(z)x^\nu(w) \sim -\eta^{\mu\nu}\ln(z - w) , \quad \psi^\mu\psi^\nu \sim -\frac{\eta^{\mu\nu}}{z - w} , \tag{8.45}$$

where the Minkowski tensor $\eta^{\mu\nu} = \text{diag}(-, +, +, \dots, +)$ in $1 + 9$ dimensions has been introduced. It is convenient to unify the NSR fields into superfields as

$$\mathbf{D}X^\mu(\mathbf{z}) \equiv \psi^\mu(z) + \theta\partial x^\mu(z) , \quad C(\mathbf{z}) \equiv c(z) + \theta\gamma(z) , \quad B(\mathbf{z}) \equiv \beta(z) + \theta b(z) , \tag{8.46}$$

where the superspace notation of sect. V.1 has been used. The total superfield stress tensor takes the form

$$T(\mathbf{z}) = T_{\text{matter}}(\mathbf{z}) + T_{\text{gh}}(\mathbf{z}) , \tag{8.47}$$

where

$$T_{\text{matter}} =: -\tfrac{1}{2}\mathbf{D}X^\mu\mathbf{D}X_\mu : , \tag{8.48a}$$

and

$$T_{\text{gh}} =: -C(\partial B) + \tfrac{1}{2}(\mathbf{D}C)(\mathbf{D}B) - \frac{3}{2}(\partial C)B : . \tag{8.48b}$$

The ghost superfields C and B have dimensions -1 and $3/2$, resp., w.r.t. the stress tensor T or T_{gh}.

The NSR-string BRST operator reads in superspace as

$$Q_{\text{BRST}} = -\oint \frac{dz}{2\pi i} \int d\theta \; : C \left[T_{\text{matter}} + \tfrac{1}{2} T_{\text{gh}} \right] : \, , \tag{8.49}$$

and it is nilpotent in $D = 10$ (exercise # VIII-8).

Quite generally, there is the equivalence between a BRST cohomology of *states* in an (extended by ghosts) Fock space and a BRST cohomology of *operators* creating these states. The operatorial BRST cohomology is defined by the operators (anti)commuting with the BRST charge *modulo* BRST (anti)commutators of the form $[Q_{\text{BRST}}, *\}$. Physical (on-shell) string vertex operators (anti)commute with the BRST charge by definition. They also have definite conformal dimensions, which allow us to integrate them over a Riemann surface Σ or a proper superspace, consistently with the (super)conformal invariance (see also exercises # VIII-6 and # VIII-7). Being a representative of the BRST cohomology, any (chiral) physical vertex operator creating a physical state of on-shell momentum k^μ contains (in components) a universal factor : $e^{ik \cdot x(z)}$: which may be multiplied, in general, by fermionic fields $\psi(z)$ and/or derivatives of bosonic $x(z)$-fields, contracted with some polarization tensors in a way to be consistent with all the symmetries of a given string theory. The physical vertex operators also carry definite total ghost number U_{tot} which is associated with the SCFT ghosts. The total ghost number U_{tot} is a sum of ghost numbers for each conformal ghost field, the latter being defined as $\text{gh}(c) = -\text{gh}(\bar{c}) = -\text{gh}(b) = \text{gh}(\bar{b}) = 1$ and quite similarly for other ghost systems.

The bosonization of the superconformal ghosts (β, γ) in eq. (3.73), in terms of the fields ρ and χ, actually introduces more conformal degrees of freedom and an additional ghost charge. Let $q_{\rho,\chi,\sigma}$ be charges assosiated with the bosonized fields: $q_\rho \left[e^{l\rho} \right] = l$, $q_\chi \left[e^{m\chi} \right] = m$ and $q_\sigma \left[e^{n\sigma} \right] = n$. For instance, the Fock space of the (β, γ) ghost system corresponds to $q_\rho + q_\chi = 0$ (called the [0]-'*picture*'), while $U_{\beta,\gamma} = U_\rho$. Therefore, the bosonization of the superconformal ghosts in eq. (3.73) extends the (already BRST-extended) Fock space by infinitely many copies, which can be labeled by a charge combination $\tilde{\pi} \equiv q_\rho + q_\chi$. The allowed values of $\tilde{\pi}$ are half-integral in NSR model, while the parameter $\tilde{\pi}$ itself is called **picture**. Hence, a picture value does not necessarily need to be zero, in general. The total ghost number now reads $U_{\text{tot}} = q_\sigma + q_\rho$.

A total ghost number and a picture are to be specified in order to find a solution to the superstring BRST cohomology problem. A 'clever' choice of picture is also practically important in describing superstring interactions in terms of superconformally invariant correlation functions of superstring vertex operators. For instance, given an $N = 1$ primary superfield $\Phi(\mathbf{z})$ of dimension $h = \frac{1}{2}$, the superfield operator [8]

$$V_{(1)}(\mathbf{z}) \equiv C\mathbf{D}\Phi - \tfrac{1}{2}(\mathbf{D}C)\Phi \equiv \tilde{V}_{(1)}(z) + \theta V_{(1)}(z) \tag{8.50a}$$

of dimension $h = 0$ represents a BRST cohomology class (exercise # VIII-9) in the [0]-picture and with the total ghost number $U_{\text{tot}} \equiv U_{b,c} + U_{\beta,\gamma} = 1$ indicated by the subscript (1). Its BRST cohomology copy with the total ghost number $U_{\text{tot}} = 2$ takes a slightly more complicated form, namely

$$V_{(2)}(\mathbf{z}) \equiv C\mathbf{D}(\partial C\Phi) - \tfrac{1}{2}(\mathbf{D}C)(\partial C\Phi) . \tag{8.50b}$$

The two different BRST cohomology copies are related by the descent equations [9]

$$Q_{\text{BRST}}(\Phi) = \mathbf{D}V_{(1)} , \quad Q_{\text{BRST}}(\partial C\Phi) = \mathbf{D}V_{(2)} . \tag{8.51}$$

Unfortunately, the NSR-string physical vertex operators cannot be made 'local' w.r.t. the θ-dependence (i.e. in our superspace) for an *arbitrary* picture. In particular, an integration over θ cannot be replaced by a multiplication with $\gamma(z)$ at $\theta = 0$, because of the discrepancy in conformal dimensions: $h[d\theta] = \frac{1}{2} \neq h[\gamma] = -\frac{1}{2}$.

We only know how to deal with the *picture-changing* in components where $\Phi(\mathbf{z}) \equiv \tilde{\phi}(z) + \theta\phi(z)$ and $C(\mathbf{z}) \equiv c(z) + \theta\gamma(z)$, so that, for example, $\tilde{V}_{(1)}(z) = (c\phi - \frac{1}{2}\gamma\tilde{\phi})(z)$ and $V_{(1)}(z) = (\frac{1}{2}\gamma\phi - c\partial\tilde{\phi} - \frac{1}{2}(\partial c)\tilde{\phi})(z)$. In particular, the picture-changing [35] from the [0]-picture considered above to the [−1]-picture is realized by the replacement $\theta \to \delta(\gamma)$, so that

$$V_{[-1]}(z) \equiv \delta(\gamma)\tilde{V}_{[0]}(z) =: e^{-\rho} : \tilde{V}_{[0]}(z) , \tag{8.52}$$

where a subscript now indicates a picture. We find, for instance, that $\delta(\gamma)\tilde{V}_{(1),[0]} = c\delta(\gamma)\phi - \frac{1}{2}\gamma\delta(\gamma)\tilde{\phi} = c\delta(\gamma)\phi \equiv V_{(0),[-1]}$.

The substitution (8.52) is consistent with the conformal properties of all fields under consideration, but it has to be replaced by some other rules for a different picture-changing. The superstring BRST cohomology is known to take the simple and natural form in the (*canonical*) [−1]-picture for the NS sector, and in the (*canonical*) [−1/2]-picture (or in the [−3/2]-picture which is mirror-symmetric) for the R sector [128, 283,

---

[8]We use the notation: $V(\mathbf{z}) = \tilde{V}(z) + \theta V(z)$ and $V(\mathbf{z}) = \int d\theta V(\mathbf{z})$.

[9]The BRST charge Q_{BRST} commutes with the picture-changing.

288, 293]. The NS physical states in the $[-1]$-picture receive the simple description (in components) as

$$|\text{phys}; [-1]\rangle_{\text{NS}} = c(0)\delta(\gamma)(0)\phi(0)\,|0\rangle_{\text{NS}}\ , \tag{8.53}$$

where the entire ghost content is hidden in the factor $c\delta(\gamma) = \delta(c)\delta(\gamma)$, and $|0\rangle_{\text{NS}}$ is the formal NS Fock space vacuum. It is now natural to introduce the NS physical ground state $|\Omega\rangle$ as

$$|\Omega\rangle_{\text{NS}} \equiv \delta(C)(0)\,|0\rangle_{\text{NS}} \equiv \delta(c)(0)\delta(\gamma)(0)\,|0\rangle_{\text{NS}}\ . \tag{8.54}$$

The NS- and R-sectors of the NSR-model are distinguished by the boundary conditions (1.104), which are valid for all μ values because of the Lorentz symmetry. The $N = 1$ SCA generators can be easily constructed in terms of field bilinears (exercise # VIII-10). It follows from eq. (8.45) that the zero modes of the field $\psi^\mu(z)$ in the R-sector satisfy the Clifford algebra, $\{\psi_0^\mu, \psi_0^\nu\} = -\eta^{\mu\nu}$. Hence, they have to be identified with the Dirac matrices in *ten* dimensions as

$$\psi_0^\mu = \frac{1}{\sqrt{2}}\Gamma^\mu\ , \quad \text{where}\ \ \{\Gamma^\mu, \Gamma^\nu\} = -2\eta^{\mu\nu}\ . \tag{8.55}$$

We conclude that the NS-states are all space-time *bosons*, whereas the R-states have to be space-time *fermions*. [10]

Let $S^\alpha(z)$ be a *spin field* connecting the NS- and R-sectors, so that $|\alpha\rangle_{\text{R}} = S^\alpha(0)\,|0\rangle_{\text{NS}}$. When being applied to a 2d spinor field $\psi^\mu(z)$, the spin field $S^\alpha(z)$ changes its boundary condition from the NS-type to the R-type and vice versa. Hence, it has to be a twist field of (conformal) dimension $1/16$ (sect. I.5). The spin field obviously has to be non-local in terms of ψ's, which implies a non-local nature of the underlying SCFT, where the NS- and R-sectors are related. In addition, a space-time fermion vertex operator has to be ghost-dependent, since eq. (5.2) implies (without ghosts) $G_0^2 = L_0 - \frac{1}{16}\hat{c} = L_0 - \frac{5}{8}$, and, hence, the naive ghost-independent fermion vertex would be of dimension $5/8$ instead of 1.

Before going any further, we need to introduce some more facts about the representation theory of the Lorentz algebra $so(1,9)$ and about the assosiated AKM algebra $\widehat{so(1,9)}$. The space-time Lorentz symmetry of the NSR-model is generated by the (anti-hermitean) currents

$$j^{\mu\nu}(z) \equiv -: \psi^\mu(z)\psi^\nu(z): , \tag{8.56}$$

---

[10]In ten dimensions, spinors can be of either Majorana-, or Weyl- or Majorana–Weyl (MW)-type. A MW spinor S^α has $2^{10/2-1} = 16$ real components. The opposite chirality is distinguished by dots over spinor indices.

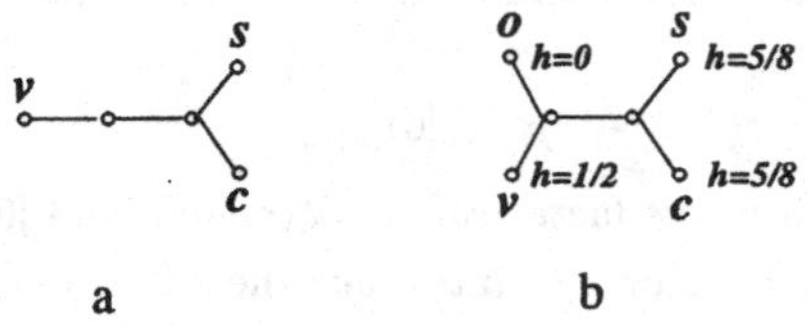

Fig. 17. The Dynkin diagrams for $D_5 \sim so(1,9)$ (a),
and its affine Kac-Moody extension $\hat{D}_5$ (b).

which are antisymmetric in their space-time indices. They form the adjoint representation of the $so\overline{(1,9)}$ of level $k = 1$,

$$j^{\mu\nu}(z)j^{\rho\lambda}(w) \sim \frac{\eta^{\mu[\lambda}\eta^{\rho]\nu}}{(z-w)^2} + \frac{1}{z-w}\left[\eta^{\nu\rho}j^{\mu\lambda} - \eta^{\nu\lambda}j^{\mu\rho} - \eta^{\mu\rho}j^{\nu\lambda} + \eta^{\mu\lambda}j^{\nu\rho}\right] . \qquad (8.57)$$

The SS-construction (sect. IV.2) in terms of the currents (8.56) yields the stress-tensor

$$T_\psi = -\frac{1}{4(D-1)} : j^{\mu\nu}j_{\mu\nu} := \tfrac{1}{2} : \psi^\mu\partial\psi_\mu : , \qquad (8.58)$$

where the dual Coxeter number $\tilde{h}_{SO(1,9)} = C_2/2 = D - 2 = 8$ and the quantum equivalence have been used.

The Dynkin diagrams for the Lie algebra $D_5 \sim so(1,9)$ and its AKM-extension $\hat{D}_5$ are given in Fig. 17, where the relevant irreps are indicated too. All finite-dimensional representations of $so(1,9)$ or, equally, of $so(10)$, can be assigned in a 5-dimensional weight space. [11] In particular, the vector (v), the positive-chirality spinor (s) and the negative-chirality spinor (c) irreps of $so(1,9)$ correspond to the weights

$$\lambda_v : \quad (\pm1,0,0,0,0) , \quad (0,\pm1,0,0,0) , \quad \ldots \quad , \quad (0,0,0,0,\pm1) ,$$
$$\lambda_{s,c} : \quad (\pm\tfrac{1}{2},\pm\tfrac{1}{2},\pm\tfrac{1}{2},\pm\tfrac{1}{2},\pm\tfrac{1}{2}) , \qquad (8.59)$$

with an *even* number of pluses in the case of (s), and an *odd* number of pluses in the case of (c).

The OPE's for the products of currents (8.56) with the fermionic fields and the spin fields read as follows:

$$j^{\mu\nu}(z)\psi^\rho(w) \sim \frac{1}{z-w}\eta^{\rho[\nu}\psi^{\mu]}(w) ,$$

---

[11] See sect. XII.2 for an introduction to the theory of Lie and AKM algebras.

$$j^{\mu\nu}(z)S^{\alpha}(w) \sim \frac{1}{z-w}\frac{1}{4}[\Gamma^{\mu},\Gamma^{\nu}]^{\alpha}{}_{\beta}S^{\beta}(w) \; ,$$

$$j^{\mu\nu}(z)S^{\dot{\alpha}}(w) \sim \frac{1}{z-w}\frac{1}{4}[\Gamma^{\mu},\Gamma^{\nu}]^{\dot{\alpha}}{}_{\dot{\beta}}S^{\dot{\beta}}(w). \tag{8.60}$$

On general symmetry and consistency grounds, the OPE's for the fusion of the free fermions with the spin fields, as well as those for the spin fields alone, take the form (exercise # VIII-11)

$$\psi^{\mu}(z)S^{\alpha}(w) \sim \frac{1}{\sqrt{2}}(z-w)^{-1/2}(\Gamma^{\mu})^{\alpha}{}_{\dot{\beta}}S^{\dot{\beta}}(w) \; ,$$

$$S^{\alpha}(z)S^{\dot{\beta}}(w) \sim (z-w)^{-5/4}C^{\alpha\dot{\beta}} + \frac{1}{4}(z-w)^{-1/4}(\Gamma_{\mu}\Gamma_{\nu}C)^{\alpha\dot{\beta}}\,\psi^{\mu}\psi^{\nu}(w) \; ,$$

$$S^{\alpha}(z)S^{\beta}(w) \sim \frac{1}{\sqrt{2}}(z-w)^{-3/4}(\Gamma_{\mu}C)^{\alpha\beta}\,\psi^{\mu}(w) \; , \tag{8.61}$$

where the ten-dimensional charge conjugation matrix $C^{\alpha\dot{\beta}}$ has been introduced.

A nice description of spin fields and fermion-vertex operators can be obtained via bosonization [35, 91]. Given an AKM algebra $\hat{\mathcal{G}}$ in the Cartan-Weyl basis (12.32), the fields

$$H^{i}(z) = \sum_{i\in\mathbf{Z}}H^{i}_{n}z^{-n-1} \; , \quad E^{\alpha}(z) = \sum_{i\in\mathbf{Z}}E^{\alpha}_{n}z^{-n-1} \; , \tag{8.62}$$

can be bosonized in terms of free scalars $\phi^{i}(z)$ as

$$H^{i}(z) = \partial\phi^{i}(z) \; , \quad i = 1,2,\ldots,\text{rank}\,\mathcal{G} \; ,$$

$$E^{\alpha}(z) =: e^{\alpha\cdot\phi(z)}: c_{\alpha} \; , \tag{8.63}$$

where the *Klein cocycle* operators c_{α} have been explicitly introduced to take care about statistics, and the dot product means the summation over CSA indices i. In the particular case of $\mathcal{G} = so(1,9)$, after making the Wick rotation to $so(10)$ and introducing the linear combinations

$$f^{\pm i} \equiv \frac{1}{\sqrt{2}}\left(\psi^{2i-1}\mp i\psi^{2i}\right) \; , \quad i = 1,2,\ldots,5 \; , \tag{8.64}$$

we can bosonize the latter as

$$f^{\pm i}(z) =: e^{\pm\phi^{i}(z)}: c_{\pm i} =: e^{\lambda_{v}\cdot\phi(z)}: c_{v} \; . \tag{8.65}$$

We now define spin fields in the form

$$S^{\alpha}(z) =: e^{\lambda_{s}\cdot\phi(z)}: c_{\alpha} \; , \quad S^{\dot{\alpha}}(z) =: e^{\lambda_{c}\cdot\phi(z)}: c_{\dot{\alpha}} \; . \tag{8.66}$$

The Klein operator c_i also gets the explicit form as [91]

$$c_i = (-1)^{\sum_{j=1}^{i-1} \oint \frac{dz}{2\pi i} \partial \phi^j} = (-1)^{\sum_{j<i} \phi_0^j} = e^{i\pi \sum_{j<i} \phi_0^j} \ . \tag{8.67}$$

The boundary conditions for the superconformal ghosts (β, γ) are periodic in the NS sector, and are antiperiodic in the R sector. The exchange NS$\leftrightarrow$R is performed by using other spin-fields having the form

$$\Sigma^{\pm}(z) \equiv: e^{\pm \frac{1}{2} \rho(z)} : \ , \tag{8.68}$$

and dimensions $h(\Sigma^+) = -5/8$ and $h(\Sigma^-) = 3/8$, resp. (sect. III.3). The ghost field Σ^- 'completes' the operator S^α to the full dimension-one physical vertex operator $S^\alpha \Sigma^-$, which intertwines the BRST-constructed NS- and R-sectors of superstring [340, 341]. Another interesting combination $S^\alpha \Sigma^+$ has vanishing dimension.

The natural dimension-one candidate for the simplest fermion vertex operator describing an emission of the Ramond ground-state fermion, with space-time momentum k^μ and a (Dirac) wave function $u_\alpha(k)$ is given by

$$V_{[-1/2]}(z) = \bar{u}_{\overset{\bullet}{\alpha}} S^{\overset{\bullet}{\alpha}} : e^{-\rho/2} e^\sigma e^{ik \cdot x} : (z) \equiv \bar{u}_{\overset{\bullet}{\alpha}} c V_{[-1/2]}^{\overset{\bullet}{\alpha}}(z) \tag{8.69}$$

in the $[-1/2]$-picture. It anticommutes with the NSR Q_{BRST} charge on-shell, where $k^2 = \bar{u}_{\overset{\bullet}{\alpha}} (\Gamma^\mu)^{\overset{\bullet}{\alpha}}{}_\beta k_\mu = 0$.

The fusion of two on-shell fermion vertex operators takes the form [12]

$$V_{[-1/2]}(u, k, z) V_{[-1/2]}(v, p, w) \sim (z-w)^{-1+k \cdot p} \frac{1}{\sqrt{2}} (\bar{u} \Gamma^\mu v) : \psi_\mu e^{-\rho} e^{i(k+p) \cdot x} : (w)$$

$$\equiv (z-w)^{-1+k \cdot p} V_{[-1]}(\bar{u} \Gamma^\mu v, k+p, w) \ , \tag{8.70}$$

and this obviously *mixes* the pictures because the OPE is a picture-additive operation.

The vertex operator algebra of the NSR model is therefore 'picture-graded'. The only exception is a subalgebra of *bosonic* physical vertex operators in the [0]-picture, while a half-integer picture is needed to introduce physical fermions. This explains the need for picture-changing in describing the space-time fermionic structure and the space-time supersymmetry (see below) of the NSR-model.

---

[12] The first argument of a vertex operator $V(u, k, z)$ means polarization, the second one is its space-time momentum, while the third one is the holomorphic argument on a Euclidean string world-sheet.

A *picture-changing operator* X increases a total ghost-number and a picture by one. In its explicit form, it reads [35, 91]

$$X(z) = \{Q_{\mathrm{BRST}}, \xi(z)\} \; . \tag{8.71}$$

A simple calculation yields $X = X_0 + X_1 + X_2$, where (see sect. III.3 for the notation and the bosonization rules for the superconformal ghosts)

$$X_0 = c\partial\xi \; , \quad X_1 = \tfrac{1}{2} : e^\rho \psi \cdot \partial x : , \quad X_2 = -\frac{1}{4} : \partial\eta e^{2\rho}b : -\frac{1}{4} : \partial(\eta e^{2\rho}b) : \; . \tag{8.72}$$

A normal-ordering ambiguity in this definition of a picture-changing operator has been fixed by demanding $[Q_{\mathrm{BRST}}, X(z)] = 0$.

The picture-changing operator $X(z)$ is actually *not* a BRST trivial operator, as one might naively conclude from its definition as a BRST (anti)commutator. It is, in fact, non-trivial in the space of physical states, since the BRST cohomology in the extended (by the non-physical zero mode of $\xi(z)$) Fock space is *smaller* than the physical BRST cohomology (without that zero mode). Notably, the *inverse* picture-changing operator also has the simple form [13]

$$Y(z) \equiv X^{-1}(z) =: 2c\partial\xi e^{-2\rho} : (z) = 2 : e^{\sigma-2\rho+\chi}\partial\chi(z) : \; . \tag{8.73}$$

We are now in a position to introduce the on-shell generators of *space-time* supersymmetry in the NSR-string model as the BRST-invariant integrated fermion vertex operators of zero-momentum, having the form [35, 91]

$$Q_-^{\dot\alpha} = \oint \frac{dz}{2\pi i} \, V_{[-1/2]}^{\dot\alpha}(z) = \oint \frac{dz}{2\pi i} \, S^{\dot\alpha} : e^{-\rho/2} : (z) \equiv \oint \frac{dz}{2\pi i} \, \theta_-^{\dot\alpha}(z) \; ,$$

$$Q_+^{\dot\alpha} = \oint \frac{dz}{2\pi i} \, V_{[+1/2]}^{\dot\alpha}(z) = \oint \frac{dz}{2\pi i} \, \partial x^\mu (\Gamma_\mu)^{\dot\alpha}{}_\beta S^\beta : e^{+\rho/2} : (z) \equiv \oint \frac{dz}{2\pi i} \, \theta_+^{\dot\alpha}(z) \; . \tag{8.74}$$

It follows [14]

$$\{Q_-^{\dot\alpha}, V_{[-1/2] \text{ or } [+1/2]}(u, k, z)\} = V_{[-1] \text{ or } [0]}\left(\epsilon_\mu = [\Gamma_\mu u]^{\dot\alpha}, k, z\right) \; ,$$

$$\{Q_-^{\dot\alpha}, V_{[-1] \text{ or } [0]}(\epsilon_\mu, k, z)\} = V_{[-1/2] \text{ or } [+1/2]}\left([i(\Gamma \cdot k)(\Gamma \cdot \epsilon)]^{\dot\alpha}{}_{\dot\beta}, k, z\right) \; . \tag{8.75}$$

---

[13] We ignore complications related with possible zero modes of X.

[14] We use the notation ϵ_μ for space-time polarization vectors, while $(u^\alpha, u^{\dot\alpha})$ stand for space-time (Dirac) spinor wave functions.

The space-time supersymmetry algebra reads

$$\{Q_-^{\dot\alpha}, Q_-^{\dot\beta}\} = \frac{1}{\sqrt{2}}(\Gamma_\mu C)^{\dot\alpha\dot\beta} \oint \frac{dz}{2\pi i} \psi^\mu : e^{-\rho} :\equiv \frac{1}{\sqrt{2}}(\Gamma_\mu C)^{\dot\alpha\dot\beta} P_{[-1]}^\mu , \tag{8.76}$$

and similarly for $Q_+^{\dot\alpha}$. The current (space-time translation) operator $P_{[-1]}^\mu(z)$ in the charge $P_{[-1]}^\mu \equiv \oint \frac{dz}{2\pi i} P_{[-1]}^\mu(z)$ just introduced fulfils the OPE's

$$P_{[-1]}^\mu(z)P_{[-1]}^\mu(w) \sim \frac{\eta^{\mu\nu}}{(z-w)^2} : e^{-2\rho} : ,$$

$$P_{[-1]}^\mu(z)Q_-^{\dot\alpha}(w) \sim \frac{1}{\sqrt{2}(z-w)}(\Gamma^\mu)^{\dot\alpha}{}_\beta : S^\beta e^{-3\rho/2} : . \tag{8.77}$$

This implies that the space-time supersymmetry algebra *cannot* be closed on a finite number of operators, or, equivalently, it needs picture-changing. The current $P_{[-1]}^\mu$ in the [0]-picture takes the form

$$P_{[0]}^\mu(z) = X \cdot P_{[-1]}^\mu(z) = \partial x^\mu(z) , \tag{8.78}$$

which is the space-time translation operator indeed. In addition, we find

$$\{Q_-^{\dot\alpha}, Q_+^{\dot\beta}\} = (\Gamma_\mu C)^{\dot\alpha\dot\beta} P_{[0]}^\mu . \tag{8.79}$$

The non-closure of the full operator algebra is already clear from eq. (8.77), or the OPE

$$P_{[0]}^\mu(z)Q_+^{\dot\alpha}(w) \sim \frac{1}{(z-w)^2}(\Gamma^\mu)^{\dot\alpha}{}_\beta : S^\beta e^{+\rho/2} : . \tag{8.80}$$

Stated differently, the space-time supersymmetry algebra only closes *modulo* picture-changing.

The space-time supersymmetry in the spectrum of the NSR-string states only exists after a GSO-projection (sect. III.1), while an action of the supersymmetry charge is only well-defined on the GSO-projected states. The corresponding 2d 'fermion-number' (or *G-parity*) operator $(-1)^F$ can also be bosonized as

$$(-1)^F = \exp\left[i\pi(\sum_{i=1}^5 \oint \frac{dz}{2\pi i}\partial\phi^i + \oint \frac{dz}{2\pi i}\partial\rho)\right] \equiv \exp\left[i\pi(\sum_{i=1}^6 \oint \frac{dz}{2\pi i}\partial\hat\phi^i)\right] , \tag{8.81}$$

where adding the additional field $\rho \equiv \phi^6$ corresponds to extending the root system of D_5 by one *time-like* root α^0 (or, equivalently, to extending the root system of D_4 by two light-like roots): this results in the Lorentzian AKM algebra $\hat D_{5,1}$ containing $\hat D_5$ as a subalgebra. The G-parity is equal to $(+1)$ for the representations (0) and (s), and (-1) for (v) and (c). The Lorentzian weight lattice $\Gamma_{5,1}$ is integral and self-dual [299, 342].

The GSO-projected NSR-model having the space-time supersymmetry is called the **superstring**. In terms of theta functions or CFT characters (sect. III.1), the equality of the bosonic and fermionic physical states at each mass level of the superstring formally follows from the Riemann identity

$$\theta_3^4(0|\tau) = \theta_2^4(0|\tau) + \theta_4^4(0|\tau) \ . \tag{8.82}$$

In the *light-cone* gauge, the space-time Lorentz symmetry of supersting is reduced to its 'little' subgroup $SO(8)$. There are no ghosts there, while bosonization can still be used to introduce spin fields and supersymmetry charges, in the way very similar to the one discussed so far [343]. The supersymmetry algebra in the light-cone gauge closes without picture-changing, while the GSO-projection can be easier understood on a basis of the *triality symmetry* of the $D_4 \sim so(8)$ Dynkin diagram (Fig. 21). Each of the three $so(8)$-irreps (v), (s) and (c) has dimension 8, and their weights are equal. They are all related in fact by the Weyl group automorphisms.

The philosophy of superstring compactification follows, in essence, the lines of the bosonic case (sect. 1). One can even combine, in principle, a (properly compactified) bosonic string construction in the holomorphic (left-moving) sector with a fermionic string construction in the anti-holomorphic (right-moving) sector, which leads to the very useful physical concept of space-time supersymmetric **heterotic** strings [344, 345, 346, 267]. Of course, one still needs to sum over all spin structures in the superstring partition function in a modular-invariant way [73, 347, 348]. For the phenomenologically and field-theoretically appealing $N = 1$ space-time supersymmetry in a superstring theory compactified down to four dimensions, the (rigid) $N = 2$ superconformal symmetry on a superstring world-sheet is sufficient [349, 350]. This explains the relevance of the $N = 2$ minimal unitary SCFT's and of the $N = 2$ Landau-Ginzburg models (sect. V.4) for the superstring compactification. In particular, when using a tensor product of n exactly solvable $N = 2$ minimal unitary models to represent the left-moving internal sector of superstring [351], the central charge of the corresponding $N = 2$ SCFT has to satisfy the relation

$$c = \sum_{i=1}^n \frac{3k_i}{k_i + 2} = 9 \ . \tag{8.83}$$

where eq. (5.13) has been used. There exists a variety of the superstring-inspired 'quasi-phenomenological' models developed towards making contact with supersymmetrical extensions of the Standard Model. The reader can easily find numerous discussions of these issues in the physical literature. The role of the underlying extended superconformal symmetry in all these studies is to get exactly solvable SCFT's for the four-dimensional superstring vacua, which could then be used to *derive* a number (three?) of the space-time chiral fermionic generations, hopefully *determine* the low-energy gauge group, and ultimately *predict* the spectrum of particles (of the minimal supersymmetric Standard Model?) with realistic lifetimes, etc. A full (still lacking) classification of all CFT's and SCFT's of central charges 22 and 9, resp., is clearly needed for any attempt to find satisfactory solutions to these problems.

Exercises

VIII-5 ▷ Show that the commutator of two rigid supersymmetry transformations defined in eq. (8.38) yields a two-dimensional translation.

VIII-6 ▷ Prove that the *bosonic* string BRST charge annihilates the state $c\phi(0)\,|0\rangle$ created by the operator $\phi(z)$ which is a primary field of dimension $h = 1$. *Hint*: use the ghost part of the BRST charge in the bosonized form (3.92), and integrate by parts.

VIII-7 ▷ Consider an action of the *bosonic* string BRST charge Q on a primary field ϕ_h of dimension h, and prove the descent equations

$$Q(\phi_1) = \partial(c\phi_1)\,, \quad Q(c\phi_h) = (1-h)c(\partial c)\phi_h\,. \tag{8.84}$$

VIII-8 ▷ Prove a nilpotency of the NSR string BRST operator (8.49) in $D = 10$. The naive BRST supercurrent $:C\left[T_{\text{matter}} + \frac{1}{2}T_{\text{gh}}\right]:$ is not, in fact, a super-primary field (check it!). Show that this naive form can be improved to get a dimension-1/2 superprimary field, by adding the total derivative $-\frac{3}{4}:\mathbf{D}\left(C(\mathbf{D}C)B\right):$.

VIII-9 ▷ Given an $N = 1$ primary superfield $\Phi(\mathbf{z})$ of dimension $h = 1/2$, verify that the superfield operator $V_\Phi \equiv C\mathbf{D}\Phi - \frac{1}{2}(\mathbf{D}C)\Phi$ is BRST-invariant, and it is not a BRST (anti)commutator.

VIII-10 ▷ Construct the $N = 1$ SCA generators (see eq. (5.1)) in terms of bilinears of the free 'matter' fields introduced in eq. (8.45). Show that these fields transform under the rigid $N = 1$ conformal supersymmetry as follows:

$$[G_{-1/2}, x^\mu(z)] = \psi^\mu(z)\,, \quad [G_{-1/2}, \psi^\mu(z)] = \partial x^\mu(z)\,, \tag{8.85}$$

so that $G^2_{-1/2} = L_{-1}$. *Hint*: consider the stress tensor and supercurrent of the NSR-model in the superconformal gauge.

VIII-11 ▷ Derive the OPE's in eq. (8.61) from the bosonization rules.

VIII-12 ▷ Complete the space-time supersymmetry algebra of the operators introduced in eqs. (8.74) and (8.78) by showing that

$$[P^\mu_{(0)}, P^\nu_{(0)}] = [P^\mu_{(0)}, Q^{\dot\beta}_-] = 0 \ . \tag{8.86}$$

VIII.3 Extended fermionic strings

The theories of $N = 2$ and $N = 4$ fermionic (or '*spinning*') strings are the quite natural extensions of the conventional ($N = 1$) NSR model: the latter has the *gauged* $N = 1$ superconformal symmetry on a string world-sheet, while the former have the $N = 2$ or $N = 4$ extended *gauged* superconformal symmetry. A string theory always provides a bridge between the two-dimensional (world-sheet) and the target (space-time) concepts, as well as between the corresponding symmetries. One such relation we already noticed in the previous section, namely, the correspondence between the $N = 2$ *rigid* superconformal symmetry on a superstring world-sheet and the $N = 1$ space-time sypersymmetry of effective four-dimensional field theory resulting from the superstring (Calabi-Yau) compactification.

The $N = 2$ SCA is given by eq. (5.11). The so-called '*small*' $N = 4$ SCA (sect. V.1) contains a *complex doublet* of supersymmetry charges $G_i(z)$, belonging to the fundamental representation of the internal symmetry group, which can be either the *compact* group $SU(2)$ or the *non-compact* group $SU(1,1)$. The most general '*large*' linear $N = 4$ SCA has the $\widehat{su(2)}_{k+} \oplus \widehat{su(2)}_{k-} \oplus \widehat{u(1)}$ AKM subalgebra. The 'large' $N = 4$ SCA is parametrized by two KM levels $k^\pm$, $c = 12k^+k^-/(k^+ + k^-)$, and it includes four free fermions in addition to the canonical generators [352, 353]. The internal symmetry generators $J^I(z)$ of the 'small' $N = 4$ SCA form instead the AKM subalgebra $\widehat{su(2)}_k$ or $\widehat{su(1,1)}_k$, whose KM 'level' k is fixed by central charge, $c = 6k$. All linear SCA's of $N > 4$ are known to contain *subcanonical* charges [15] needed to close an algebra [147]. The case of $N = 4$ is special, due to the relevant group decompositions

$$SO(4) \cong SU(2) \otimes SU(2) \ , \quad SO(2,2) \cong SU(1,1) \otimes SU(1,1) \ . \tag{8.87}$$

---

[15] By 'subcanonical' charges [147] we mean charges of dimensions other than 2, 3/2 or 1. The canonical charges are just charges for conformal symmetry, supersymmetry and internal symmetry, resp.

The 'large' $N = 4$ internal symmetry generators can therefore be restricted to be *self-dual* (SD) or *anti-self-dual* (ASD), which just leads to the 'small' $N = 4$ SCA with the canonical charges only.

The bridge between fermionic strings and SCA's is provided by free field realisations of the latter. The convenient free field representation of the N-extended SCA is given by a *supercurrent* multiplet of the N-extended Polyakov's string action in the superconformal gauge. A locally N-supersymmetric string action can be constructed by coupling the N-extended supergravity to an N-extended scalar multiplet in two dimensions. In the $N = 2$ case, the simplest action (in components) is known as the *Brink-Schwarz* (BS) string action [354] [16]

$$S_{\text{BS}} = \frac{1}{\pi} \int d^2\xi\, e \left\{ \frac{1}{2} h^{\alpha\beta} \partial_\alpha Z^+ \cdot \partial_\beta Z^- + \frac{i}{2} \overline{\Psi}^+ \cdot \rho^\alpha \overleftrightarrow{D}_\alpha \Psi^- + A_\alpha \overline{\Psi}^+ \cdot \rho^\alpha \Psi^- \right.$$
$$\left. + \left(\partial_\alpha Z^+ + \overline{\Psi}^+ \chi_\alpha^- \right) \cdot \overline{\chi}_\beta^+ \rho^\alpha \rho^\beta \Psi^- + \left(\partial_\alpha Z^- + \overline{\chi}_\alpha^+ \Psi^- \right) \cdot \overline{\Psi}^+ \rho^\beta \rho^\alpha \chi_\beta^- \right\} , \tag{8.88}$$

where a set of D *complex* scalar $N = 2$ matter multiplets is coupled to the $N{=}2$ world-sheet supergravity multiplet comprising a real zweibein $e_\alpha^a(\xi)$ with the metric $h_{\alpha\beta} = \eta_{ab} e_\alpha^a e_\beta^b$, a complex gravitino field $\chi_\alpha^\pm(\xi)$, and a real $U(1)$ gauge field $A_\alpha(\xi)$. The gravitational spinor covariant derivative [355] is denoted by D_α. Dot products stand for contractions with a flat target space metric $\eta_{\mu\nu}$.

Via BRST quantization in the $N{=}2$ superconformal gauge, the fields of the $N{=}2$ string on the Euclidean world-sheet become free, so that they can be decomposed into their holomorphic and anti-holomorphic parts as usual. Lower-case ψ will be used to denote the chiral parts of Ψ. The ghost systems appropriate for the $N{=}2$ string are [357, 358]:

- the reparametrization ghosts (b, c), an anticommuting pair of free world-sheet fermions with conformal dimensions $(2, -1)$.

- the two-dimensional $N{=}2$ supersymmetry ghosts (β^i, γ^i) or $(\beta^\mp, \gamma^\pm)$, two commuting pairs of free world-sheet fermions with conformal dimensions $(\frac{3}{2}, -\frac{1}{2})$.

- the $U(1)$ ghosts $(\tilde{b}, \tilde{c})$, an anticommuting pair of free world-sheet fermions with conformal dimensions $(1, 0)$.

---

[16] We find it convenient to use here purely imaginary (Majorana) representation for 2d Dirac matrices ρ^α, $\alpha = 0, 1$: $\rho^0 = \sigma_2$, $\rho^1 = i\sigma_1$, $\rho^3 = \rho^0\rho^1 = \sigma_3$, where σ_i, $i = 1, 2, 3$, are Pauli matrices, and $\{\rho^\alpha, \rho^\beta\} = -2\eta^{\alpha\beta}$, with Minkowski metric $\eta = diag(-, +)$. Dirac conjugation (denoted by bar) $\overline{\Psi} \equiv \Psi^T \rho^0$ does *not* include complex conjugation (denoted by superscript '-').

To calculate the critical dimension of the $N = 2$ fermionic string, the easiest way is to use the conformal anomaly counting [356, 359],

$$
\begin{array}{ccccccc}
(b,c) & & (\beta^i,\gamma^i) & & (\bar{b},\bar{c}) & & (Z^i,\psi^i)^\mu \\
-26 & + & 2\cdot 11 & + & -2 & + & 2\cdot D\cdot(1+\tfrac{1}{2}) \,,
\end{array}
\tag{8.89}
$$

whose vanishing yields $D = 2$, i.e. two *complex* dimensions of signature $(2, 2)$ or $(4, 0)$ [360]. The choice of *two* time-like directions is consistent with the no-ghost theorem [46, 298] requiring this number to match with the number of time-like directions due to the fermionic ghosts (the (b, c) and $(\bar{b}, \bar{c})$ conformal ghost systems contribute by one). It is remarkable that a calculation of the $N = 2$ string partition function on a torus results in the *absence* of 'massive' physical string modes, because of the cancellations between the non-zero modes of the $N = 2$ conformal matter and that of the $N = 2$ ghosts [361, 362, 363]. Therefore, the critical $N = 2$ string theory should be equivalent to a local quantum *field* theory (QFT) in $2 + 2$ dimensions, describing $N = 2$ string ground states and their interactions. [17]

The simplest way to determine this effective QFT is to use the gauge-fixed $N = 2$ string action in the complex $U(1, 1)$ notation ($a = 1, 2$),

$$
S_{\text{g.-f.}}^{N=2} = \frac{1}{\pi}\int d^2z\, d^2\theta\, d^2\bar{\theta}\, \eta_{a\bar{b}} X^a \bar{X}^{\bar{b}} \equiv \frac{1}{\pi}\int d^4Z\, K_0(X,\bar{X}) \,,
\tag{8.90}
$$

and the vertex operator describing an emission of the $N = 2$ string bosonic ground state of momentum $k^{\underline{a}} = (k^a, \bar{k}^{\bar{a}})$,

$$
V_{\text{c}} = \frac{\kappa}{\pi} : \exp[i(k\cdot\bar{X} + \bar{k}\cdot X)] : \,,
\tag{8.91}
$$

written in terms of the $N = 2$ chiral scalar superfields $X(Z, \bar{Z}) = X(z, \bar{z}, \theta, \bar{\theta})$, with coupling constant κ. The $N = 2$ string tree scattering amplitudes are the first ones to be calculated. In particular, the 3-point tree takes the form [362, 363]

$$
A_3 \sim \left\langle V_{\text{c}}|_{\theta=0}(0)\cdot\int d^2\theta\, d^2\bar{\theta}\, V_{\text{c}}(1)\cdot V_{\text{c}}|_{\theta=0}(\infty) \right\rangle
$$

$$
= \kappa\left(k_1\cdot\bar{k}_2 - \bar{k}_1\cdot k_2\right)^2 \equiv \kappa c_{12}^2 \,,
\tag{8.92}
$$

where the $N = 2$ super-Möbius invariance has been used to fix three points on the $N = 2$ super-Riemannian sphere. Note that A_3 is non-vanishing and apparently non-covariant

---

[17]The case of $4+0$ dimensions does not seem to be interesting since the Euclidean on-shell momentum condition leads to zero momentum.

w.r.t. the natural 'Lorentz' group $SO(2,2)$ in $2+2$ dimensions. A calculation of the 4-point function yields [362, 363]

$$A_4 \sim \int d^2z \left\langle V_c|_{\theta=0}(0) \cdot \int d^2\theta d^2\bar\theta\, V_c(z) \cdot \int d^2\theta d^2\bar\theta\, V_c(1) \cdot V_c|_{\theta=0}(\infty) \right\rangle$$

$$= \frac{\kappa^2}{16\pi} \int d^2z \left| \frac{1}{(1-z)^2}t(t+2) + \frac{c_{12}c_{34}}{z} + \frac{c_{23}c_{41}}{1-z} \right|^2 |z|^{-s}\,|1-z|^{-t}$$

$$= \frac{\kappa^2}{\pi} F^2 \frac{\Gamma(1-s/2)\Gamma(1-t/2)\Gamma(1-u/2)}{\Gamma(s/2)\Gamma(t/2)\Gamma(u/2)} , \tag{8.93}$$

where s, t, u are the Mandelstam variables, $s = -(k_1 \cdot \bar k_2 + \bar k_1 \cdot k_2)$, *etc.*, and

$$F \equiv 1 - \frac{c_{12}c_{34}}{su} - \frac{c_{23}c_{41}}{tu} = 0 . \tag{8.94}$$

The vanishing kinematical factor F in the amplitude (8.93) is actually needed for consistency: otherwise, the massive poles of this amplitude would indicate the existence of the massive modes known to be absent in the $N = 2$ string spectrum [362, 363]. For the same reason, all the other trees A_n, at $n > 4$, should vanish too. The bosonic QFT reproducing this scattering behavior was constructed in refs. [362, 363]. It is the theory of **self-dual gravity** (SDG) in the *Plebanski* form [364]:

$$I_{\text{SDG}} = \int d^4x \left(\frac{1}{2}\eta^{a\bar b}\partial_a\Phi\partial_{\bar b}\Phi + \frac{\kappa}{3}\Phi\partial_a\partial_{\bar a}\Phi\varepsilon^{ab}\varepsilon^{\bar a\bar b}\partial_b\partial_{\bar b}\Phi \right) . \tag{8.95}$$

The $N = 2$ string NS-like 'scalar' Φ should therefore be identified with a deformation of the flat Kähler potential, $K = K_0 + 2\kappa\Phi$. The equations of motion in the Plebanski theory (8.95) follow from the SDG equations, $^*R = R$, in terms of the 4d curvature tensor R to be constructed from the metric $g_{a\bar b} = \partial_a\partial_{\bar b}K$, and its dual $^*R$, *after* a gauge-fixing *and* choosing a complex structure. Note that the SDG equations are explicitly $SO(2,2)$ covariant. [18] Hence, the $N = 2$ string 'scalar' represents a self-dual 'graviton' of helicity $+2$ and of 'spin' 2 to be defined w.r.t. the *little group* of the 'Lorentz' group $SO(2,2)$ [368]. The $N = 2$ *open* strings are quite similar [369], and they give rise to the $(2+2)$-dimensional **self-dual Yang-Mills** (SDYM) theory in the so-called *Siegel-Parkes* form (with only cubic self-interaction) [370, 371, 372]) as the effective QFT representing the tree S-matrix of open $N = 2$ fermionic strings. The SDYM field equations of motion, $^*F = F$, in terms of the Yang-Mills field strength F and its dual $^*F$, are also covariant w.r.t. $SO(2,2)$. The same conclusions follow from an analysis of RG fixed points of the relevant $N = 2$ supersymmetric 2d NLSM's [373, 374].

---

[18]See refs. [365, 366, 367], as regards the covariant actions for self-dual field theories.

Our notation, in what follows in this section, is now going to be adjusted to $2 + 2$ dimensions. Target space indices (internal and Lorentz) always appear as superscripts. The complex bosonic and fermionic matter fields are split either into real and imaginary components, $Z^\mu = Z^{2\mu} + iZ^{3\mu}$ and $\Psi^\mu = \Psi^{2\mu} + i\Psi^{3\mu}$, or into holomorphic and antiholomorphic parts, $(Z^\mu, Z^{\mu*}) = (Z^{+\mu}, Z^{-\mu})$ and $(\Psi^\mu, \Psi^{\mu*}) = (\Psi^{+\mu}, \Psi^{-\mu})$, where $\mu = 0, 1$. These fields may also be grouped into light-cone combinations,

$$
\begin{aligned}
Z^{i\pm} &= Z^{i0} \pm Z^{i1} \quad , \qquad \Psi^{i\pm} = \Psi^{i0} \pm \Psi^{i1} \qquad i = 2, 3 \\
Z^{a\pm} &= Z^{a0} \pm Z^{a1} \quad , \qquad \Psi^{a\pm} = \Psi^{a0} \pm \Psi^{a1} \qquad a = \pm ,
\end{aligned}
\tag{8.96}
$$

with respect to the 2d Lorentz index $\mu = 0, 1$. Lower-case Greek indices refer to a 2d Minkowski space, while the lower-case Latin indices specify the real and imaginary components of the complex fields, and $a = \pm$ denote their holomorphic and antiholomorphic parts. To avoid confusing the same numerical values, an unusual range for the lower-case Latin indices is taken, so that $\{i\mu\} = \{20, 21, 30, 31\}$. We frequently suppress inessential Lorentz indices.

The critical string action (8.88) should be supplemented by boundary conditions. The simplest *Ooguri-Vafa* boundary conditions, which were implicitly assumed above, are [362, 363]

$$
Z(\pi) = Z(0) \quad , \qquad \Psi(\pi) = (+\,\text{or}\,-)\Psi(0) , \tag{8.97}
$$

jointly for all components. They are the only ones which allow us to keep the single-valuedness of the bosonic matter fields and, hence, yield $\mathbf{C}^{1,1}$ as the consistent $(2+2)$-dimensional background 'spacetime' for $N=2$ string propagation. This choice of un-twisted boundary conditions only deals with untwisted line bundles and their square roots (spin bundles) to define fermions, just as for the $N=1$ string. The two possible signs in eq. (8.97) are common for all the world-sheet spinors, and correspond to the usual NS-R distinction familiar from the $N=1$ case. In other words, untwisted states reside in the (NS,NS) or (R,R) sectors, where the two factors refer to the boundary conditions of (Ψ^2, Ψ^3). The relevant example of twisted boundary conditions was first considered by Mathur and Mukhi (M-M) in ref. [375]. In the holomorphic basis, their (M-M) global twist reads

$$
Z^\pm(\pi) = Z^\mp(0) \quad , \qquad \Psi^\pm(\pi) = (+\,\text{or}\,-)\Psi^\mp(0) . \tag{8.98}
$$

Unlike the other possible twists, the M-M twist still allows massless neutral ground states among physical states [376]. Fields of integral spin are now allowed to pick up signs, i.e. be *double*-valued, just as for fields of half-integral spin. The complete monodromy behavior is fixed by the signs picked up by the components of Z, in addition to an overall sign between Z and Ψ related to the NS-R distinction. Twisted states comprise (NS,R) and (R,NS).

Interpreting the 'internal' index i as part of a (2+2)-dimensional 'extended space-time' label $i\mu$, the extended spacetime Lorentz symmetry for the $N=2$ string should be $SO(2,2)$ [359]. Although the *gauge-fixed* $N=2$ string action has this symmetry (in the matter part), the interaction terms in the *gauge-invariant* action (8.88) break one of the two $SU(1,1)$ factors of $SO(2,2)$ (see eq. (8.87)) down to $U(1) \otimes \mathbf{Z}_2$, with $\mathbf{Z}_2$ representing the M-M twist. The global symmetry of the $N=2$ string action is only part of the 'hidden' $SO(2,2)$ symmetry. The latter actually exhibits itself in the *twistor* space, since the space of all complex structures is exactly isomorphic to the coset $SU(1,1)/U(1)$! Given the 'extended Lorentz' symmetry, we are able to distinguish between 'bosons' and 'fermions' in the target space, since the $SO(2,2)$ group has representations of continuous spin and its little group is non-trivial. [19]

In order to construct fermionic vertex operators we make use of chiral bosonization. The reparametrization ghosts (b,c) are expressed as

$$c \cong e^{+\sigma}, \quad b \cong e^{-\sigma}, \quad \text{with} \quad \sigma(z)\,\sigma(w) \sim +\ln(z-w)\,. \tag{8.99}$$

Similarly, for the $U(1)$ ghosts $(\tilde{b}, \tilde{c})$ one has

$$\tilde{c} \cong e^{+\tilde{\sigma}}, \quad \tilde{b} \cong e^{-\tilde{\sigma}}, \quad \text{with} \quad \tilde{\sigma}(z)\,\tilde{\sigma}(w) \sim +\ln(z-w)\,. \tag{8.100}$$

For the complex fields, bosonization depends on the basis. In the real basis,

$$\psi^{i+}(z)\,\psi^{j-}(w) \sim \frac{2\,\delta^{ij}}{z-w} \quad \text{and} \quad \gamma^i(z)\,\beta^j(w) \sim \frac{\delta^{ij}}{z-w} \tag{8.101}$$

can be represented by [20]

$$\psi^{i\pm} \cong \sqrt{2}\,e^{\pm\phi^i}\,, \qquad \gamma^i \cong \eta^i e^{\varphi^i}\,, \qquad \beta^i \cong e^{-\varphi^i}\partial\xi^i\,, \tag{8.102}$$

$$\phi^i(z)\phi^j(w) \sim +\delta^{ij}\ln(z-w)\,, \quad \varphi^i(z)\varphi^j(w) \sim -\delta^{ij}\ln(z-w)\,, \quad \xi^i(z)\eta^j(w) \sim \frac{\delta^{ij}}{z-w}\,,$$

---

[19]The little group of $U(1,1)$ is trivial.

[20]Normal ordering is always suppressed, as well as cocycle operators.

where four scalar bosons ϕ^i and φ^i with values in $i(\mathbf{R}/2\pi)$ have been introduced. The auxiliary anticommuting (η^i, ξ^i) conformal system of spin $(1,0)$ may also be bosonized as

$$\xi^i \cong e^{+\theta^i} , \quad \eta^i \cong e^{-\theta^i} , \quad \text{with} \quad \theta^i(z)\,\theta^j(w) \sim +\delta^{ij}\ln(z-w) . \tag{8.103}$$

Though the 'solitons' $e^{\pm\varphi^i}$ are outside the monomial field algebra of (β^i, γ^i), one finds that [95]

$$e^{+\varphi^i} \cong \delta(\beta^i) , \quad e^{-\varphi^i} \cong \delta(\gamma^i) , \quad \xi^i \cong \Theta(\beta^i) , \quad \eta^i \cong \delta(\gamma^i)\partial\gamma^i . \tag{8.104}$$

Finally, we introduce real spin fields with helicity index $\pm$ as

$$S^{i\pm} \cong e^{\pm\frac{1}{2}\phi^i} . \tag{8.105}$$

Their action on the vacuum state implements the M-M twist for the fermions and flips the boundary conditions of a given ψ^i.

If one does not perform the M-M twist, the holomorphic basis,

$$\psi^{\pm+}(z)\,\psi^{\mp-}(w) \sim \frac{4}{z-w} , \qquad \gamma^+(z)\,\beta^-(w) \sim \frac{2}{z-w} , \tag{8.106}$$

invites the alternative bosonization,

$$\psi^{\pm+} \cong 2e^{+\phi^\pm} , \quad \psi^{\mp-} \cong 2e^{-\phi^\pm} , \quad \gamma^\pm \cong \eta^\pm e^{+\varphi^\pm} , \quad \beta^\mp \cong e^{-\varphi^\pm}\partial\xi^\mp , \tag{8.107}$$

$$\phi^\pm(z)\phi^\pm(w) \sim +\ln(z-w) , \quad \varphi^\pm(z)\varphi^\pm(w) \sim -\ln(z-w) , \quad \xi^\mp(z)\eta^\pm(w) \sim \frac{2}{z-w} .$$

In particular, the holomorphic spin fields are

$$S^{\pm+} \cong e^{+\frac{1}{2}\phi^\pm} \quad \text{and} \quad S^{\mp-} \cong e^{-\frac{1}{2}\phi^\pm} . \tag{8.108}$$

The two bosonization schemes are related by *non-local* field redefinitions, as becomes clear by comparing, for example, the expressions for the local matter $U(1)$ current,

$$\begin{aligned} J &= \tfrac{i}{2}\,\psi^2 \cdot \psi^3 = -\tfrac{i}{2}\left(e^{+\phi^2-\phi^3} + e^{-\phi^2+\phi^3}\right) \\ &= -\tfrac{1}{4}\,\psi^+ \cdot \psi^- = \tfrac{1}{2}(\partial\phi^+ - \partial\phi^-) . \end{aligned} \tag{8.109}$$

One reads off that $e^{q\phi^\pm}$ has $U(1)$ charge $e = \pm q$. Moreover, $\xi^\pm$ and $\eta^\pm$ are not linear combinations of ξ^i and η^i, as is the case for ψ, γ and β.

Here and in what follows we use the general results of Chs. II and III for free chiral bosons $\rho \in \{\sigma, \tilde{\sigma}, \phi, \varphi, \theta\}$ with a background charge $\tilde{Q}_\rho$,

$$h\left[e^{q\rho}\right] = \frac{\epsilon}{2}q(q - \tilde{Q}), \qquad \rho(z)\,\rho(w) \sim \epsilon \ln(z - w), \qquad \epsilon = \pm 1,$$

$$\tilde{Q}_\sigma = 3, \qquad \tilde{Q}_{\tilde{\sigma}} = 1, \qquad \tilde{Q}_\phi = 0, \qquad \tilde{Q}_\varphi = -2, \qquad \tilde{Q}_\theta = 1, \tag{8.110}$$

where the factor ϵ takes into account statistics. [21] It follows that the spin fields S twisting ψ and χ have conformal dimension $h{=}1/8$. Similarly, the fields $e^{-\frac{1}{2}\varphi}$ twisting β and γ have conformal dimension $h{=}3/8$.

As we already know from the previous section, the bosonization of the superconformal ghosts enormously enlarges the set of local fields. Indeed, for the real as well as for the holomorphic scheme, the extended Fock space based on φ and θ is a direct sum of $(\mathbf{Z} \times \mathbf{Z})$ copies ('pictures') of the original (β, γ) Fock space, each labelled by two picture numbers (either π_i or $\pi_\pm$) which are the eigenvalues of

$$\Pi_i = -\oint[\beta^i\gamma^i + \eta^i\xi^i] = \oint[-\partial\varphi^i + \partial\theta^i] \qquad i = 2, 3 \tag{8.111}$$

or

$$\Pi_\pm = -\tfrac{1}{2}\oint[\beta^\pm\gamma^\mp + \eta^\pm\xi^\mp] = \oint[-\partial\varphi^\mp + \partial\theta^\mp], \tag{8.112}$$

respectively. Obviously, any polynomial only in β and γ has $\pi{=}0$, whereas $e^{q_i\varphi^i}$ carries $\pi_i{=}q_i$ and $e^{q\varphi^\mp}$ has $\pi_\pm{=}q$. A further subtlety, also mentioned in the previous section, is that the extended Fock space does not contain the constant zero modes of ξ, since only derivatives of these fields appear in eqs. (8.102) and (8.107).

To describe the $N{=}2$ string Fock space of states, we define the formal super-$SL(2, \mathbf{C})$-invariant vacuum state $|0\rangle$ in the untwisted NS sector, with vanishing momentum, conformal dimension and local $U(1)$ charge. States in the (R,R) sector may then be created by the joint action of two spin fields, $S^2 S^3$ or $S^+ S^-$. In order to generate the twisted sectors (R,NS) or (NS,R), one acts with a single spin field, S^2 or S^3, respectively. However, this is not sufficient. One also had to utilize the $\mathbf{Z}_2$-*twist* fields $t^{i\mu}(z)$, whose role is to flip the boundary conditions for $Z^{i\mu}$ [376]. The twist fields generically act as

$$t^{i\mu}(z)\,\partial Z^{j\nu}(w) \sim \frac{\eta^{\mu\nu}\,\delta^{ij}}{\sqrt{z - w}}\,\tilde{t}(w), \tag{8.113}$$

---

[21] $\epsilon = -1$ for $\rho = \varphi$, and $\epsilon = +1$ otherwise.

where $\tilde{t}$ is another (excited) twist field. Twist fields act trivially on ψ. We assume that all twist fields are Virasoro primaries. The conformal dimension of $t^{i\mu}$ is $h=1/16$, and their OPE reads

$$t^{i\mu}(z) \ t^{j\nu}(w) \ \sim \ -\frac{\eta^{\mu\nu} \ \delta^{ij}}{(z-w)^{1/8}} \ . \tag{8.114}$$

We find it convenient to define the composites

$$t^i \ \equiv \ t^{i0}t^{i1} \ \text{(no sum!)} \quad , \qquad t^i_+ \ \equiv \ e^{\tilde{\sigma}/2} \, t^i \ , \tag{8.115}$$

where t^i_+ have vanishing conformal dimension.

The $N=2$ string BRST charge is

$$Q_{\text{BRST}} \ = \ \oint_0 \frac{dz}{2\pi i} \, j_{\text{BRST}}(z) \ \equiv \ \oint j_{\text{BRST}} \ , \tag{8.116}$$

where the dimension-one BRST current reads

$$\begin{aligned} j_{\text{BRST}} \ &= \ c\hat{T} + bc\partial c + \gamma^2 G + \gamma^3 \overline{G} + \tilde{c}\hat{J} \\ &- (\gamma^2\gamma^2 + \gamma^3\gamma^3)b + 2i(\gamma^2\partial\gamma^3 - \gamma^3\partial\gamma^2)\tilde{b} + \tfrac{3}{4}\partial[c(\beta^2\gamma^2 + \beta^3\gamma^3)] \end{aligned} \tag{8.117b}$$

in the real basis, or

$$\begin{aligned} j_{\text{BRST}} \ &= \ c\hat{T} + bc\partial c + \tfrac{1}{2}\gamma^- G^+ + \tfrac{1}{2}\gamma^+ G^- + \tilde{c}\hat{J} \\ &- \gamma^+\gamma^- b + (\gamma^-\partial\gamma^+ - \gamma^+\partial\gamma^-)\tilde{b} + \tfrac{3}{8}\partial[c(\beta^+\gamma^- + \beta^-\gamma^+)] \end{aligned} \tag{8.117b}$$

in its holomorphic form. Here, we use the notation

$$\hat{T} \ = \ T_{\text{tot}} - T_{b,c} \quad \text{and} \quad \hat{J} \ = \ J_{\text{tot}} - \partial(\tilde{b}c) \ , \tag{8.118}$$

where $T_{b,c} = -2b\partial c - (\partial b)c$, and we introduced the full (BRST-invariant) stress tensor T_{tot} and the $U(1)$ current J_{tot} as

$$\begin{aligned} T_{\text{tot}} \ &= \ \{Q_{\text{BRST}}, b\} \ = \ T + T_{b,c} - \tilde{b}\partial\tilde{c} - \tfrac{3}{2}(\beta^2\partial\gamma^2 + \beta^3\partial\gamma^3) - \tfrac{1}{2}(\gamma^2\partial\beta^2 + \gamma^3\partial\beta^3) \\ &= \ T + T_{b,c} - \tilde{b}\partial\tilde{c} - \tfrac{3}{4}(\beta^+\partial\gamma^- + \beta^-\partial\gamma^+) - \tfrac{1}{4}(\gamma^+\partial\beta^- + \gamma^-\partial\beta^+) \ , \\ J_{\text{tot}} \ &= \ \{Q_{\text{BRST}}, \tilde{b}\} \ = \ J + \partial(\tilde{b}c) - \tfrac{i}{2}(\beta^2\gamma^3 - \beta^3\gamma^2) \\ &= \ J + \partial(\tilde{b}c) + \tfrac{1}{4}(\beta^+\gamma^- - \beta^-\gamma^+) \ . \end{aligned} \tag{8.119}$$

Here T, G and J are the $N=2$ string (matter) currents without ghosts, *viz.*

$$\begin{aligned} T \ &= \ -\tfrac{1}{2}\left(\partial Z^i \cdot \partial Z^i - \psi^i \cdot \partial\psi^i\right) \ = \ -\tfrac{1}{2}\partial Z^+ \cdot \partial Z^- + \tfrac{1}{4}\psi^+ \cdot \partial\psi^- + \tfrac{1}{4}\psi^- \cdot \partial\psi^+ \ , \\ G \ &= \ \delta^{ij}\partial Z^i \cdot \psi^j \quad , \qquad\qquad\qquad G^+ \ = \ \partial Z^- \cdot \psi^+ \ , \\ \overline{G} \ &= \ \varepsilon^{ij}\partial Z^i \cdot \psi^j \quad , \qquad\qquad\qquad G^- \ = \ \partial Z^+ \cdot \psi^- \ , \\ J \ &= \ \tfrac{i}{4}\varepsilon^{ij}\psi^i \cdot \psi^j \ = \ \tfrac{i}{2}\psi^2 \cdot \psi^3 \ = \ -\tfrac{1}{4}\psi^+ \cdot \psi^- \ . \end{aligned} \tag{8.120}$$

Given any vertex operator V of the type cV_1, where V_1 is conformal primary of dimension one and built without (b, c) ghosts, one easily verifies that

$$\left[c\hat{T} + bc\partial c \right](z)\, V(w) \sim \text{regular} . \tag{8.121}$$

Furthermore, if V_1 is of the type $V_{1,0}$, where $V_{1,0}$ has vanishing local $U(1)$ charge and does not contain $(\tilde{b}, \tilde{c})$ ghosts, one also has

$$\tilde{c}\hat{J}(z)\, V(w) \sim \text{regular} . \tag{8.122}$$

$h(V_{1,0})=1$ and $e(V_{1,0})=0$ are just the physical state conditions of on-shell momentum and charge neutrality. Twisted states are, however, not covered by this argument since they require a twisting of $(\tilde{b}, \tilde{c})$ by $e^{\tilde{\sigma}/2}$. Replacing cV_1 by $c\partial cV_1$, or $V_{1,0}$ by $\tilde{c}V_{1,0}$, creates additional BRST-closed states which are, however, BRST-exact when off-shell or non-neutral. Finally, if $V_{1,0} \in (NS, NS)$ has the form $e^{-\varphi^2-\varphi^3}W$, with a neutral, zero-dimensional, and ghost-free W satisfying $G(z)W(w) \sim O(\frac{1}{z-w}) \sim \overline{G}(z)W(w)$, BRST invariance of V is guaranteed. For (R, R) states, one needs $V_{1,0} = e^{-\frac{1}{2}\varphi^2-\frac{1}{2}\varphi^3}W'$, with a ghost-free W', and $G(z)W'(w) \sim O(\frac{1}{\sqrt{z-w}}) \sim \overline{G}(z)W'(w)$ yields Dirac equations. More generally, the bosonic string constraints T_{tot} and J_{tot} annihilate a given physical state $|\text{phys}\rangle$ only in the generalised Siegel gauge [22]

$$b_0 |\text{phys}\rangle = \tilde{b}_0 |\text{phys}\rangle = 0 . \tag{8.123}$$

The constraints associated with the local $U(1)$ symmetry are absent for twisted states, where $\tilde{b}_0$ does not exist.

As representatives of the massless physical states in the *twisted* $N=2$ string, let us choose the following BRST-invariant (and BRST non-trivial) chiral vertex operators (multiple sign choices are correlated)

$$\begin{aligned}
(NS, NS) : &\quad \Phi^\pm = c\, e^{-\varphi^2-\varphi^3}\, e^{-\frac{i}{2}k^{2\mp}Z^{2\pm}} , \\
(R, R) : &\quad \Upsilon^\pm = c\, e^{-\frac{1}{2}\varphi^2-\frac{1}{2}\varphi^3}\, S^{2\pm}S^{3\pm}\, e^{-\frac{i}{2}k^{2\mp}Z^{2\pm}} , \\
(NS, R) : &\quad \Xi^\pm = c\, e^{-\varphi^2-\frac{1}{2}\varphi^3}\, S^{3\pm}t_+^3\, e^{-\frac{i}{2}k^{2\mp}Z^{2\pm}} , \\
(R, NS) : &\quad \Lambda^\pm = c\, e^{-\frac{1}{2}\varphi^2-\varphi^3}\, S^{2\pm}t_+^3\, e^{-\frac{i}{2}k^{2\mp}Z^{2\pm}} .
\end{aligned} \tag{8.124}$$

The first four (untwisted) operators just represent the NS-type Ooguri-Vafa boson (Φ) and its R-type bosonic partner (Υ), while the other four (twisted) operators correspond to the M-M chiral fermions (Ξ, Λ). Since the abelian gauge field has been

---

[22] The fermionic constraints G_{tot} are visible in the canonical pictures $\pi = -1, -\frac{1}{2}$.

twisted, no states are related by spectral flow. The twisted string supports a merely $(1+1)$-dimensional kinematics, i.e. $k^3=0$, which turns massless states into either 'space-time' left-movers ($k^{2+}=0$, the upper sign choice) or right-movers ($k^{2-}=0$, the lower sign choice). It is straightforward to check that all the chiral vertex operators introduced in eq. (8.124) do belong to the $N=2$ string BRST cohomology [376].

The untwisted string, in contrast, lives in $2+2$ dimensions so that massless states only satisfy

$$-2\,k^+ \cdot k^- \;=\; k^{++}k^{--} + k^{+-}k^{-+} \;=\; 0 \quad . \tag{8.125}$$

In this case, $N=2$ supersymmetry allows only (NS,NS) and (R,R) sectors, and holomorphic representations of massless physical states are

$$
\begin{aligned}
(NS,NS): \quad & \Phi \;=\; c\,e^{-\varphi^- -\varphi^+}\,e^{ik\cdot Z}\,, \\[4pt]
(R,R): \quad & \Upsilon \;=\; c\,e^{-\frac{1}{2}\varphi^- -\frac{1}{2}\varphi^+}\left(k^{\pm -}S^{-+}S^{++} \pm ik^{\pm +}S^{+-}S^{--}\right)e^{ik\cdot Z}\,,
\end{aligned}
\tag{8.126}
$$

where the choice of sign for Υ is irrelevant since the two expressions agree via eq. (8.125), as long as no $k^{a\pm}$ vanishes. Spectral flow [363] identifies Φ with Υ, leading us back to the single massless scalar boson.

To compute correlation functions with the BRST-invariant chiral vertex operators introduced in eqs. (8.124) and (8.126), we need to discuss the picture-changing operations and GSO projections in $N=2$ string theory. The BRST cohomology problem is simplified by identifying grading operators. For the $N=2$ string, these are

- the total ghost charge
  $$
  \begin{aligned}
  U &= -\oint[bc + \tilde{b}\tilde{c} + \beta^2\gamma^2 + \beta^3\gamma^3] = \oint[\partial\sigma + \partial\tilde{\sigma} - \partial\varphi^2 - \partial\varphi^3] \\
  &= -\oint[bc + \tilde{b}\tilde{c} + \tfrac{1}{2}\beta^+\gamma^- + \tfrac{1}{2}\beta^-\gamma^+] = \oint[\partial\sigma + \partial\tilde{\sigma} - \partial\varphi^+ - \partial\varphi^-]
  \end{aligned}
  $$

- the picture charges $\Pi_i = -\oint[\beta^i\gamma^i + \eta^i\xi^i] = \oint[-\partial\varphi^i + \partial\theta^i]$ $i=2,3$
  or $\Pi_\pm = -\tfrac{1}{2}\oint[\beta^\pm\gamma^\mp + \eta^\pm\xi^\mp] = \oint[-\partial\varphi^\mp + \partial\theta^\mp]$

- the full bosonic constraints T_{tot} and J_{tot}

of which only U does not commute with Q_{BRST}. Accordingly, it suffices to separately investigate simultaneous eigenspaces of the commuting set $\{U, \Pi, L_0^{\text{tot}}, J_0^{\text{tot}}\}$, labelled by $\{u, \pi, h, e\}$. From $L_0^{\text{tot}} = \{Q_{\text{BRST}}, b_0\}$ and $J_0^{\text{tot}} = \{Q_{\text{BRST}}, \tilde{b}_0\}$ it readily follows that non-trivial cohomology only exists for $h = e = 0$. [23] At this point, no restriction on

---

[23] Note, however, that J_0^{tot} and e are not defined for twisted states.

the values of u or π arises, so that an infinity of (massless) physical vertex operators is anticipated. Like in the $N=1$ string, however, BRST cohomology classes differing by integral values of u or π correspond to the same physical state, if their spacetime properties agree. Hence, we should identify physical states with equivalence classes of BRST cohomology classes under integral total ghost and picture number changes.

For the evaluation of string amplitudes, the collection (8.124) of representatives is not sufficient. Consider a correlation function of exponential operators $e^{q(k)\rho}$ for one of the ghost systems listed in eq. (8.110). The tree-level ghost number selection rules [24]

$$\sum_k q(k) = -\tilde{Q} \tag{8.127}$$

yield, in particular,

$$\sum_k \pi(k) = -2 \quad \text{and} \quad \sum_k u(k) = 0 \tag{8.128}$$

for non-zero tree-level correlations of vertex operators. These constraints generally require the use of vertex operators in several pictures and total ghost sectors.

Let us begin with the representatives of eq. (8.124) which have $u(\Phi, \Upsilon, \Xi, \Lambda) = (-1, 0, 0, 0)$; their pictures can be read off the φ^i charges. We are looking for a map to equivalent vertex operators which changes the u and/or π_i values. For the simplest BRST class in our list, represented by Φ with $(u, \pi_2, \pi_3) = (-1, -1, -1)$, the total ghost number u may be increased from -1 to 0, 0, or $+1$, without changing the pictures, upon replacing the factor c by $c\partial c$, $\bar{c}c$, or $\bar{c}c\partial c$, respectively. This reflects the vacuum degeneracy due to the anticommuting ghost zero modes in the untwisted sector, and leaves the generalized Siegel gauge. A way to lower u to -2 is exchanging the factor c by a chiral integration $\int dz$, delocalizing the resulting vertex $V^\int$. The other vertex operators mentioned above, Υ, Ξ and Λ, have 'partners' in non-zero ghost numbers as well. In general, however, there does not seem to exist a simple recipe to derive them directly from one another and establish an explicit equivalence. Nevertheless, from the argument just given one should expect in a given picture that localized untwisted vertex operators come in four types and localized twisted ones in two.

None of the above changes the picture numbers π_i. The existence of the doubly infinite set of pictures is related to the superconformal ghost algebra. Fortunately, we do

---

[24] Since the constant zero modes of ξ are not part of the Fock space, the θ charges must sum to zero instead of one.

not need to repeat the cohomology analysis for each picture, since in this case an explicit equivalence relation is known. More precisely, the picture-changing operations X^i shift $\pi_j \to \pi_j + \delta_j^i$ while commuting with Q_{BRST}, b_0 and L_0^{tot}, and raising u by one. The latter fact makes the total ghost number impractical to label picture-equivalence classes of vertex operators, so we introduce instead the real picture-invariant combination

$$V_r = U - \Pi_2 - \Pi_3 = -\oint [bc + \tilde{b}\tilde{c} - \eta^2\xi^2 - \eta^3\xi^3] = \oint [\partial\sigma + \partial\tilde{\sigma} - \partial\theta^2 - \partial\theta^3] , \quad (8.129)$$

with eigenvalues

$$v_r = u - \pi_2 - \pi_3 . \qquad (8.130)$$

The v_r-charges in a tree-level correlation function must sum to 4. The non-trivial BRST cohomology actually exists only for $v_r = 1, 2, 2', 3$ in the untwisted sectors, [25] and for $v_r = \frac{3}{2}, \frac{5}{2}$ in the twisted ones, with the indicated multiplicities, and omitting $V^{\int}$ [376, 377].

Let us display the picture of a vertex operator by parenthesized subscripts and its v_r-charge by a parenthesized superscript, $V_{(\pi_2,\pi_3)}^{(v_r)}$. Picture-changing in the real basis then proceeds as

$$V_{(\pi_2+1,\pi_3)} = [Q_{\text{BRST}}, \xi^2 V_{(\pi_2,\pi_3)}\} =: X^2 \cdot V_{(\pi_2,\pi_3)}$$
$$V_{(\pi_2,\pi_3+1)} = [Q_{\text{BRST}}, \xi^3 V_{(\pi_2,\pi_3)}\} =: X^3 \cdot V_{(\pi_2,\pi_3)} , \qquad (8.131)$$

where, in the absence of normal ordering between ξ^i and V, one defines the real picture-changing operators

$$X^i(z) \equiv \{Q_{\text{BRST}}, \xi^i(z)\} \qquad i = 2, 3 . \qquad (8.132a)$$

Analogously, in the holomorphic basis,

$$X^{\pm}(z) \equiv \{Q_{\text{BRST}}, \xi^{\pm}(z)\} \qquad (8.132b)$$

shifts π_+ or π_- by one unit, while leaving the eigenvalues

$$v_h = u - \pi_+ - \pi_- \qquad (8.133)$$

of

$$V_h = U - \Pi_+ - \Pi_- = -\oint [bc + \tilde{b}\tilde{c} - \tfrac{1}{2}\eta^-\xi^+ - \tfrac{1}{2}\eta^+\xi^-] = \oint [\partial\sigma + \partial\tilde{\sigma} - \partial\theta^+ - \partial\theta^-] \quad (8.134)$$

unchanged. Again (apart from discrete states), physical states reside in $v_h = 1, 2, 2', 3$ sectors only.

---

[25] The two BRST classes with $v = 2$ are being distinguished by using a prime.

By construction, the X are BRST invariant but *not* BRST trivial due to the lack of the zero modes of ξ in the bosonization formulae. The picture-changing operators just introduced take the following explicit form:

$$X^2 = c\partial\xi^2 + e^{\varphi^2}\left[G + \tfrac{i}{2}\bar{c}\beta^3 + 2i(\partial\bar{b})\gamma^3 + 4ib\partial\gamma^3\right] + 2e^{2\varphi^2}b\partial\eta^2 + \partial(e^{2\varphi^2}b)\eta^2 \,,$$

$$X^3 = c\partial\xi^3 + e^{\varphi^3}\left[\bar{G} - \tfrac{i}{2}\bar{c}\beta^2 - 2i(\partial\bar{b})\gamma^2 - 4ib\partial\gamma^2\right] + 2e^{2\varphi^3}b\partial\eta^3 + \partial(e^{2\varphi^3}b)\eta^3 \,,$$

$$\tag{8.135a}$$

$$X^+ = c\partial\xi^+ + e^{\varphi^-}\left[G^+ + 2\partial\bar{b}\gamma^+ + 4\bar{b}\partial\gamma^+ - 2b\gamma^+\right] \,,$$

$$X^- = c\partial\xi^- + e^{\varphi^+}\left[G^- - 2\partial\bar{b}\gamma^- - 4\bar{b}\partial\gamma^- - 2b\gamma^-\right] \,.$$

$$\tag{8.135b}$$

It is clear that $X^\pm$ are not just linear combinations of X^i. The two types of picture-changing operators differ in two respects. The holomorphic version does not contain $e^{2\varphi}$ terms, and it also lacks any $\bar{c}$ dependence. The latter means that $X^\pm$ commute with J_0^{tot}, whereas X^i do so only modulo BRST-exact terms. One can easily check that $h[X] = 0$ and $u[X] = 1$, in particular.

Since the picture-changing X establishes an equivalence of cohomology classes, its inverse Y can only be well-defined modulo BRST-trivial terms and may, like X itself, possess a BRST-trivial non-zero kernel. It seems to be natural to require, in the real form, that

$$[Q_{\text{BRST}}, Y^i] = 0 \qquad \text{and} \qquad Y^2(z)\, X^2(w) \sim 1 \sim Y^3(z)\, X^3(w) \tag{8.136}$$

but do not constrain the mixed products. The quantum numbers of Y^i are determined as $(h, u, \pi_j) = (0, -1, -\delta_{ij})$. A simple ghost number analysis shows that this leaves only a single candidate for each φ^i ghost charge value below -1. One then finds the unique solution [378],

$$Y^3 = \sum_{\substack{k=2 \\ k \text{ even}}}^{\infty} \left(\prod_{\ell=1}^{k-1} \ell!\right)^{-1} Y_k^3 = Y_2^3 + \frac{1}{12}Y_4^3 + \frac{1}{34560}Y_6^3 + \dots \,, \tag{8.137}$$

where

$$Y_k^3 = c\left(\gamma^2\right)^{k-2} \partial^{k-1}\xi^3 \dots \partial^2\xi^3\partial\xi^3\, e^{-k\varphi^3} \qquad k \geq 2 \tag{8.138}$$

satisfy

$$Y_k^3(z)\, X^3(w) \sim \delta_{k2} + O(z-w) \,. \tag{8.139}$$

The first term,

$$Y_2^3 = c\,\partial\xi^3\, e^{-2\varphi^3} \,, \tag{8.140}$$

is identical with the inverse picture-changing operator of the $N=1$ string, but fails to be BRST invariant for the $N=2$ string. This failure is corrected by adding an infinite series of Y_k^3, with $k = 4, 6, 8, \ldots$, in eq. (8.137). The formal series (8.137) can be summed to the non-local expression [378]

$$Y^3(w) = -\sin \oint_w [\gamma^2 \beta^3 - \gamma^3 \beta^2] \cdot Y_1^3(w) = i \sinh(2 \, \mathrm{ad} J_0^{\mathrm{tot}}) \cdot Y_1^3(w) \qquad (8.141)$$

where we introduced

$$Y_1^3 = -c\xi^2 e^{-\varphi^2 - \varphi^3} = c(\gamma^2)^{-1} \delta(\gamma^3) \qquad (8.142)$$

and understand the action on Y_1^3 as a power series of iterated commutators. Of course, a mirror image expression emerges for Y^2. Note that the Y^i are pure ghost operators and do not contain any matter fields. The formal sum in eq. (8.137) is no longer contained in the local bosonized field algebra. Thus, $Y^i \cdot V$ may not correspond to a state even in the extended Fock space, unless only a finite number of terms from eq. (8.137) contribute.

The massless vertex operators in eq. (8.124) or (8.126) may be written as

$$\Phi \equiv V_{(-1,-1)}^{(1)} \quad , \quad \Upsilon \equiv V_{(-\frac{1}{2},-\frac{1}{2})}^{(1)} \quad , \quad \Xi \equiv V_{(-1,-\frac{1}{2})}^{(\frac{3}{2})} \quad , \quad \Lambda \equiv V_{(-\frac{1}{2},-1)}^{(\frac{3}{2})} \, , \qquad (8.143)$$

with additional helicity indices in the twisted string.

The principal tools, which are essential for string amplitudes, are picture-changing (which changes π but not v) and fusion, $V_1(z)V_2(w) \sim V_3(w)$ (which is additive in π and v). To simplify the notation, let us suppress the ubiquitous factor $e^{ik \cdot Z}$. The masslessness condition reads

$$k^2 \cdot k^2 + k^3 \cdot k^3 = -(k^{2+}k^{2-} + k^{3+}k^{3-}) = -\tfrac{1}{2}(k^{++}k^{--} + k^{+-}k^{-+}) = 0 \, . \qquad (8.144)$$

Since the untwisted vertex operators of the twisted model are obtained from restricting the momenta of the real vertex operators of the untwisted string, it is convenient to discuss the full (2+2)-dimensional vertex operators for Φ and Υ in the real representation.

For the Ooguri-Vafa boson, Φ, it is straightforward to work upwards in pictures as we already know $V_{(-1,-1)}^{(v)}$ for $v = 0, 1, 2, 2', 3$. For example, some $v_r=1$ vertex operators are [26]

$$
\begin{aligned}
V_{(-1,-1)}^{(1)} &= c\, e^{-\varphi^2 - \varphi^3} \\
V_{(\ 0,-1)}^{(1)} &= c\, e^{-\varphi^3}(k\cdot\psi) + \gamma^2 e^{-\varphi^3} + \tfrac{i}{2}\bar{c}c\partial\xi^3 e^{-2\varphi^3} \\
V_{(-1,\ 0)}^{(1)} &= c\, e^{-\varphi^2}[k\cdot\psi] + \gamma^3 e^{-\varphi^2} - \tfrac{i}{2}\bar{c}c\partial\xi^2 e^{-2\varphi^2} \\
V_{(\ 0,\ 0)}^{(1)} &= c\,\{[k\cdot\partial Z] - [k\cdot\psi](k\cdot\psi)\} - \gamma^2[k\cdot\psi] + \gamma^3(k\cdot\psi) \, ,
\end{aligned}
\qquad (8.145)
$$

---

[26] See ref. [378] for more details.

where we have abbreviated

$$(k\cdot\psi) \equiv i\delta^{ij} k^i \cdot \psi^j \qquad \text{and} \qquad [k\cdot\psi] \equiv i\epsilon^{ij} k^i \cdot \psi^j \,, \tag{8.146}$$

and similarly for $(k\cdot\partial Z)$ and $[k\cdot\partial Z]$.

Next, let us consider the spectral-flow partner Υ in the pictures $\pi_i = -\frac{1}{2}$ and $-\frac{3}{2}$. Rather than solving the fermionic constraints in the canonical $(-\frac{1}{2}, -\frac{1}{2})$ picture, it proves more straightforward to begin with the subcanonical $(-\frac{3}{2}, -\frac{3}{2})$ picture. Applying X^2 and X^3 one obtains for $v_r = 1$

$$V^{(1)\pm}_{(-\frac{3}{2},-\frac{3}{2})} \;=\; c\, e^{-\frac{3}{2}\varphi^2 - \frac{3}{2}\varphi^3} S^{2\mp} S^{3\mp}$$

$$V^{(1)\pm}_{(-\frac{1}{2},-\frac{3}{2})} \;=\; c\, e^{-\frac{1}{2}\varphi^2 - \frac{3}{2}\varphi^3} [k^{2\mp} S^{2\pm} S^{3\mp} - ik^{3\mp} S^{2\mp} S^{3\pm}] + \frac{i}{2}\tilde{c}c\partial\xi^3 e^{-\frac{1}{2}\varphi^2 - \frac{5}{2}\varphi^3} S^{2\mp} S^{3\mp}$$

$$V^{(1)\pm}_{(-\frac{3}{2},-\frac{1}{2})} \;=\; c\, e^{-\frac{3}{2}\varphi^2 - \frac{1}{2}\varphi^3} [k^{3\mp} S^{2\pm} S^{3\mp} + ik^{2\mp} S^{2\mp} S^{3\pm}] - \frac{i}{2}\tilde{c}c\partial\xi^2 e^{-\frac{5}{2}\varphi^2 - \frac{1}{2}\varphi^3} S^{2\mp} S^{3\mp}$$

$$V^{(1)\pm}_{(-\frac{1}{2},-\frac{1}{2})} \;=\; c\, e^{-\frac{1}{2}\varphi^2 - \frac{1}{2}\varphi^3} [k^{\pm\mp} S^{2\pm} S^{3\pm} \pm k^{\pm\pm} S^{2\mp} S^{3\mp}] \,,$$

$$\tag{8.147}$$

where the last expression obtains after division by $k^{\mp\mp}$. The two sign choices are BRST-equivalent, which may be checked for $V_{(-\frac{3}{2},-\frac{3}{2})}$ but is obvious from

$$k^{-\pm} V^{(1)+}_{(-\frac{1}{2},-\frac{1}{2})} \;\propto\; +k^{-\pm}[k^{+-} S^{2+} S^{3+} + k^{++} S^{2-} S^{3-}]$$

$$= \;\pm k^{+\pm}[k^{-+} S^{2-} S^{3-} - k^{--} S^{2+} S^{3+}] \;\propto\; \pm k^{+\pm} V^{(1)-}_{(-\frac{1}{2},-\frac{1}{2})} \,, \tag{8.148}$$

as long as all $k^{a\pm}$ are generically non-zero. Each picture-changing operator by X^i essentially raises the φ^i charge and applies one of two Dirac operators, $\slashed{k} \equiv \delta^{ij} k^i \cdot \Gamma^j$ or $\overline{\slashed{k}} \equiv \epsilon^{ij} k^i \cdot \Gamma^j$, via

$$\partial Z^i \cdot \psi^j(z) \;\; S^{2\alpha} S^{3\beta} e^{ik\cdot Z}(w) \;\sim\; (z-w)^{-3/2} (-ik^i) \cdot (\Gamma^j S^2 S^3)^{\alpha\beta} e^{ik\cdot Z} \,. \tag{8.149}$$

One may write, for example,

$$k^{\mp\mp} V^{(1)\pm}_{(-\frac{1}{2},-\frac{1}{2})} \;=\; c\, e^{-\frac{1}{2}\varphi^2 - \frac{1}{2}\varphi^3} \left(\slashed{k}\overline{\slashed{k}} S^2 S^3 \right)^{\mp\mp} \,, \tag{8.150}$$

or use $\slashed{k}^+ \slashed{k}^- = -2i\slashed{k}\overline{\slashed{k}}$ for a holomorphic description, with $\slashed{k}^\pm \equiv k^\mp \cdot \Gamma^\pm$. This ensures automatically that $V^{(1)}_{(-\frac{1}{2},-\frac{1}{2})}$ satisfies two Dirac equations,

$$\slashed{k} V^{(1)}_{(-\frac{1}{2},-\frac{1}{2})} = 0 = \overline{\slashed{k}} V^{(1)}_{(-\frac{1}{2},-\frac{1}{2})} \qquad \text{or} \qquad \slashed{k}^\pm V^{(1)}_{(-\frac{1}{2},-\frac{1}{2})} = 0 \,, \tag{8.151}$$

since each $\not{k}$ squares to zero on-shell. Such is not required for the BRST invariance of $V_{(-\frac{3}{2},-\frac{3}{2})}$. As a rule, lower pictures enlarge the image of Q_{BRST} but increase its kernel as well.

In the twisted theory, we must restrict the kinematics to $k^{3+} = 0 = k^{3-}$ which, together with masslessness, implies either

$$k^{++} = 0 = k^{-+} \qquad \text{or} \qquad k^{-+} = 0 = k^{--} . \tag{8.152}$$

Now the upper and lower sign choices in eq. (8.147) are no longer equivalent, since division by $k^{\mp\mp}$ is no longer admissible, so that

$$k^{a+} = 0 \quad \implies \quad k^{++} V^{(1)-}_{(-\frac{1}{2},-\frac{1}{2})} = 0 \quad \text{and} \quad k^{--} V^{(1)+}_{(-\frac{1}{2},-\frac{1}{2})} \propto S^{2+} S^{3+} \tag{8.153}$$

and the opposite in case $k^{a-}=0$. Hence, the two above momentum constraints lead to different BRST classes, i.e. Υ^+ and Υ^-. The helicity index distinguishes $(1+1)$-dimensional 'spacetime' left-movers (Υ^+ with $k^{2+}=0$) and right-movers (Υ^- with $k^{2-}=0$), and it agrees with the spinor helicities in the $(-\frac{1}{2}, -\frac{1}{2})$ picture. The two fermions in the twisted sector, Ξ and Λ, are related by interchanging indices $2 \leftrightarrow 3$ (except for twist fields and momenta). The list of vertex operators for them in the pictures $\pi_2 = -1, 0$ and $\pi_3 = -\frac{3}{2}, -\frac{1}{2}$ can be found in ref. [378].

Let us now turn to the fusion algebra of the localized chiral vertex operators. The relevant information is contained in their various OPEs. More precisely, the fusion of two BRST cohomology classes obtains as the constant piece in the OPE of any two representatives. The resulting vertex operator may be BRST-trivial, as must be the case for all singular terms in the OPE. The BRST cohomology classes are labeled by (v, π_2, π_3) which behave additively under fusion. Since the spectrum of non-trivial v-values is presumably bounded, the fusion of two vertex operators should automatically give a trivial answer if the sum of their v-charges exceeds that range. Ultimately, we like to divide the vertex operator algebra by the picture and ghost number equivalence and arrive at a fusion algebra for the physical excitations. In order to retain a non-trivial result we must keep $v_r \in \{1, \frac{3}{2}, 2, \frac{5}{2}, 3\}$ for localized operators. Since we investigate massless states only, the momenta k and k' of the fusing vertex operators are taken to satisfy $k \cdot k' = 0$.

On the one hand, interactions between the Ooguri-Vafa boson Φ and the Mathur-Mukhi fermions (Λ, Ξ) have to be forbidden, since the relevant OPE's are *non-local*. At vanishing momenta we have, for example,

$$\Phi(z)\,\Lambda^{\pm}(w)\ \sim\ (z-w)^{-1/2}\,c\partial c\,e^{-\frac{3}{2}\varphi^2-2\varphi^3}e^{\pm\frac{1}{2}\phi^2}\,t_+^3\ ,$$

$$\Lambda^{+}(z)\,\Lambda^{-}(w)\ \sim\ (z-w)^{-1/2}\,c\partial c\,e^{-\varphi^2-2\varphi^3}\,\tilde{c}\ ,$$

$$\text{(8.154)}$$

and similarly for $\Xi^{\pm}$.

On the other hand, each set, $(\Phi^+,\Phi^-,\Upsilon^+,\Upsilon^-)$, $(\Xi^+,\Lambda^+,\Upsilon^-)$ or $(\Xi^-,\Lambda^-,\Upsilon^+)$, separately has *local* OPE's among its components. This means, in particular, that an interacting theory can make sense for each of those sets seperately. Eq. (8.154) also shows that if we just want to disregard the Ooguri-Vafa boson Φ in favor of the new fermionic fields, we should restrict ourselves to a definite helicity. This gives us two fermionic generators to play with, for example Λ^+ and Ξ^+. Their fusions must either be trivial or lead to some representatives of Υ^-.

Let us concentrate on the *bosonic* theory whose massless physical spectrum is represented by $\Phi^{\pm}$ and $\Upsilon^{\pm}$. Allowing for the momenta to be (2+2)-dimensional amounts to combining the four states to two, Φ and Υ. A fusion of two Φ fields, as represented in eq. (8.124), with momenta k and k' and $k{\cdot}k'=0$ gives

$$\Phi(z)\,\Phi(w)\ \sim\ -(z-w)^{-1}\,c\partial c\,e^{-2\varphi^2-2\varphi^3}e^{i(k+k')\cdot Z}(w)+\tfrac{1}{2}V^{(2)}_{(-2,-2)}(w)+O(z-w)\ ,\quad\text{(8.155)}$$

where the residue

$$c\partial c\,e^{-2\varphi^2-2\varphi^3}e^{i(k+k')\cdot Z}\ =\ \left\{Q_{\text{BRST}},\,c\,e^{-2\varphi^2-2\varphi^3}e^{i(k+k')\cdot Z}\right\}\qquad\text{(8.156)}$$

is BRST-trivial as expected, but the finite term

$$V^{(2)}_{(-2,-2)}=\left\{Q_{\text{BRST}},\partial c\,e^{-2\varphi^2-2\varphi^3}e^{i(k+k')\cdot Z}\right\}-c\partial c\,i(k-k')\cdot\partial Z\,e^{-2\varphi^2-2\varphi^3}e^{i(k+k')\cdot Z}\quad\text{(8.157)}$$

is not. In fact, consecutive application of X^2 and X^3 reproduces $V^{(2)}_{(-1,-1)}$. The same result may be observed in other pictures, e.g. for

$$V^{(1)}_{(0,-1)}(z)\,V^{(1)}_{(-1,0)}(w)\ \sim\ V^{(2)}_{(-1,-1)}(w)\ ,\qquad\text{(8.158)}$$

showing that fusion of Φ with itself reproduces Φ. For the fusion of Υ with Φ one finds

$$\Upsilon(z)\,\Phi(w)\ \sim\ -c\partial c\,e^{-\frac{3}{2}\varphi^2-\frac{3}{2}\varphi^3}\left(\slashed{k}\overline{\slashed{k}}S^2S^3\right)e^{i(k+k')\cdot Z}(w)\ \propto\ V^{(2)}_{(-\frac{3}{2},-\frac{3}{2})}\ ,\qquad\text{(8.159)}$$

which does not picture-change to zero since the momentum on the right-hand side is $k+k'$. Finally, fusion of Υ with itself yields

$$\Upsilon(z)\,\Upsilon(w)\ \sim\ -c\partial c\,e^{-\varphi^2-\varphi^3}(\text{linear in }kk')\,e^{i(k+k')\cdot Z}(w)\ \propto\ V^{(2)}_{(-1,-1)}\ .\qquad\text{(8.160)}$$

Denoting the picture and ghost number equivalence classes of cohomology classes by square brackets, the physical fusion algebra in the bosonic (untwisted) theory takes the simple form of

$$[\Phi] \cdot [\Phi] = [\Phi] \;, \qquad [\Phi] \cdot [\Upsilon] = [\Upsilon] \;, \qquad [\Upsilon] \cdot [\Upsilon] = [\Phi] \;, \tag{8.161}$$

which is consistent with the only non-zero three-point functions $\langle \Phi\Phi\Phi \rangle$ and $\langle \Phi\Upsilon\Upsilon \rangle$. The spectral flow identifies Φ with Υ. Indeed, a short calculation shows that, at tree-level,

$$\langle \Phi\Phi\Phi \rangle = \left\langle V^{(2)}_{(-1,-1)} V^{(1)}_{(\,0,\,0)} V^{(1)}_{(-1,-1)} \right\rangle = k_2^2 \cdot k_3^3 - k_2^3 \cdot k_3^2 = \tfrac{i}{2} c_{23} \tag{8.162}$$

while

$$\langle \Phi\Upsilon\Upsilon \rangle = \left\langle V^{(2)}_{(-1,-1)} V^{(1)}_{(-\frac{1}{2},-\frac{1}{2})} V^{(1)}_{(-\frac{1}{2},-\frac{1}{2})} \right\rangle = k_2^{--} k_3^{++} - k_2^{-+} k_3^{+-} = c_{23} \tag{8.163}$$

as well, with $c_{ij} \equiv k_i^+ \cdot k_j^- - k_i^- \cdot k_j^+ = -c_{ji}$ being the additional non-trivial $U(1,1)$ (but not $SO(2,2)$!) invariant. The bosonic 'scalar' Φ can be identified with deformation of the Kähler potential [363].

The alternative local algebra of physical vertex operators can be based on Ξ^+ and Λ^+ in the *twisted* theory. Note that for this helicity choice only the k^{2-} component of the momenta $k^{i\mu}$ stays non-zero. The OPE's then read

$$\Xi^+(z)\,\Xi^+(w) \sim \bar{c}c\partial c\, e^{-2\varphi^2-\varphi^3}\, e^{\phi^3}\, e^{-\frac{i}{2}(k^{2-}+k'^{2-})Z^{2+}}(w) \xrightarrow{X^2} 0 \;,$$

$$\Lambda^+(z)\,\Lambda^+(w) \sim \bar{c}c\partial c\, e^{-\varphi^2-2\varphi^3}\, e^{\phi^2}\, e^{-\frac{i}{2}(k^{2-}+k'^{2-})Z^{2+}}(w) \xrightarrow{X^3} 0 \;,$$

$$\Xi^+(z)\,\Lambda^+(w) \sim \bar{c}c\partial c\, e^{-\frac{3}{2}\varphi^2-\frac{3}{2}\varphi^3}\, S^{2+}S^{3+} e^{-\frac{i}{2}(k^{2-}+k'^{2-})Z^{2+}}(w) = V^{(3)-}_{(-\frac{3}{2},-\frac{3}{2})} \xrightarrow{X^2 X^3} 0 \;,$$

$$\tag{8.164a}$$

the last vertex being trivial due to its 'wrong' helicity. Finally,

$$\Upsilon^-(z)\,\Upsilon^-(w) \sim c\partial c\, e^{-\varphi^2-\varphi^3}(k^{-+}k'^{--} + k^{--}k'^{-+})e^{-\frac{i}{2}(k^{2-}+k'^{2-})Z^{2+}}(w) = 0 \tag{8.164b}$$

since $k^{a+} = 0 = k'^{a+}$. Thus, the local fusion algebra of chiral massless vertex operators in the twisted $N=2$ string theory becomes trivial (away from $k\equiv 0$). Since there is no spectral flow in the twisted theory, Υ^- here has nothing to do with the field Φ which is absent. The fusion rules are consistent with the vanishing of the only potentially non-trivial tree-level three-point function,

$$\left\langle \Xi^+\Lambda^+\Upsilon^- \right\rangle = \left\langle V^{(\frac{3}{2})}_{(-1,-\frac{1}{2})} V^{(\frac{3}{2})}_{(-\frac{1}{2},-1)} V^{(1)}_{(-\frac{1}{2},-\frac{1}{2})} \right\rangle = 0 \;, \tag{8.165}$$

simply due to the massless (1+1)-dimensional kinematics, for non-zero momenta. An equivalent new theory can of course be obtained by flipping the helicities of $(\Xi^+, \Lambda^+, \Upsilon^-)$ to $(\Xi^-, \Lambda^-, \Upsilon^+)$. Obviously, the twisted string does not support interactions of its massless excitations.

One may now try to build $N=2$ extended 'spacetime supersymmetry' generators from the two 'spacetime' fermionic vertex operators at vanishing momenta. Taking the integrated versions $V^{(\frac{1}{2})}_{(-\frac{1}{2},-1)}$ and $V^{(\frac{1}{2})}_{(-1,-\frac{1}{2})}$, and replacing $\int$ by a contour integration, we arrive at

$$Q_2^\pm \equiv \oint j_2^\pm(z) = \oint \Lambda^\pm\Big|_{c-\text{omitted},\, k=0} \sim \oint e^{-\frac{1}{2}\varphi^2} e^{-\varphi^3} S^{2\pm}\, t_+^3\,,$$
$$Q_3^\pm \equiv \oint j_3^\pm(z) = \oint \Xi^\pm\Big|_{c-\text{omitted},\, k=0} \sim \oint e^{-\varphi^2} e^{-\frac{1}{2}\varphi^3} S^{3\pm}\, t_+^3\,,$$
(8.166)

where $\oint \equiv \oint_0 \frac{dz}{2\pi i}$, and either $+$ or $-$ should be chosen. The relevant OPE's are

$$j_2^+(z)\, j_2^+(w) \sim (z-w)^{-1}\, \tilde c\, e^{-2\varphi^3}\gamma_\mu^{++}\psi^{2\mu}e^{-\varphi^2}(w)\,,$$
$$j_3^+(z)\, j_3^+(w) \sim (z-w)^{-1}\, \tilde c\, e^{-2\varphi^2}\gamma_\mu^{++}\psi^{3\mu}e^{-\varphi^3}(w)\,,$$
$$j_2^+(z)\, j_3^+(w) \sim (z-w)^{-1}\, \tilde c\, e^{-\frac{3}{2}\varphi^2-\frac{3}{2}\varphi^3}S^{2+}S^{3+}(w)\,,$$
(8.167)

and similarly for $j_2^-(z)\, j_3^-(w)$, whereas

$$j_2^+(z)\, j_2^-(w) \sim O(z-w)^{-3/2}\,, \qquad j_3^+(z)\, j_3^-(w) \sim O(z-w)^{-3/2}\,.$$
(8.168)

The operators $\psi^{2\mu}e^{-\varphi^2}$ and $\psi^{3\mu}e^{-\varphi^3}$ appearing on the r.h.s. of the first two OPE's in eq. (8.167) are the 'spacetime' translation operators in the (-1) picture, just like in the $N=1$ superstring theory. However, this does not ultimately come through since the $N=2$ picture-changing of the *full* operators on the r.h.s. of eq. (8.167) brings the first two to zero. The third operator,

$$P_{23} \equiv \oint \tilde c\, e^{-\frac{3}{2}\varphi^2-\frac{3}{2}\varphi^3}S^{2+}S^{3+}\,,$$
(8.169)

should represent Υ^-. However, picture-changing turns it into zero due to vanishing momenta. Therefore, instead of translation operators we get in fact zeros on the r.h.s. of our 'spacetime supersymmetry' algebra,

$$\{Q_i^+, Q_j^+\} = 0 \qquad i, j = 2, 3\,.$$
(8.170)

These $N=2$ extended 'supersymmetry' generators should rather be interpreted as a kind of 'exterior' derivatives.

Having required world-sheet locality for physical vertex operators, the covariant lattice approach can also be used to establish that only *two* distinct maximal GSO projections are possible for the $\mathbf{Z}_2$-twisted $N=2$ fermionic string [378, 379]. The first one simply retains the untwisted states which, on the massless level, were proved to comprise two 'spacetime' scalars, to be identified via spectral flow. The second projection should lead to an $N=2$ superstring, yet it falls short in three respects. First, the necessary 'spacetime' twist restricts the kinematics from $D=2+2$ to $D=1+1$, trivializing massless interactions. Second, there is one bosonic state 'missing' from the massless spectrum of two 'spacetime' fermions and one 'spacetime' boson. Third, the would-be 'spacetime' supersymmetry' generators are null operators instead of producing 'spacetime' translations. Their geometrical significance is yet to be understood. Witten's index counts the mismatch of bosonic and fermionic ground states. In the twisted $N=2$ string theory, this index is non-vanishing, which implies that the $N=2$ extended 'spacetime supersymmetry' is unbroken.

As regards the four-point function, we already know since the beginning of this section that

$$\langle \Phi\Phi\Phi\Phi \rangle \;=\; \left\langle V^{(2')}_{(-1,-1)}\, V^{(0)}_{(-1,-1)}\, V^{(1)}_{(0,0)}\, V^{(1)}_{(0,0)} \right\rangle \tag{8.171}$$

vanishes due to the special kinematics in 2+2 dimensions, which yields

$$c_{12}c_{34}t + c_{23}c_{41}s - 16stu \;=\; 0 \tag{8.172}$$

for $c_{ij} \equiv k_i^+ \cdot k_j^- - k_i^- \cdot k_j^+$ and $-4s_{ij} \equiv k_i^+ \cdot k_j^- + k_i^- \cdot k_j^+$. Because of spectral flow, the same should hold true for $\langle \Phi\Phi\Upsilon\Upsilon \rangle$ and $\langle \Upsilon\Upsilon\Upsilon\Upsilon \rangle$. Indeed, abbreviating $z_{ij} \equiv z_i - z_j$ and

$$\Sigma_i^\pm \;\equiv\; \left(k_i^{\pm -} S^{-+} S^{++} \pm i k_i^{\pm +} S^{+-} S^{--} \right)(z_i) \,, \tag{8.173}$$

and making use of $s + t + u = 0$, one computes, e.g.

$$
\begin{aligned}
\langle \Upsilon\Upsilon\Upsilon\Upsilon \rangle &= \left\langle V^{(2')}_{(-\frac{1}{2},-\frac{1}{2})}\, V^{(0)}_{(-\frac{1}{2},-\frac{1}{2})}\, V^{(1)}_{(-\frac{1}{2},-\frac{1}{2})}\, V^{(1)}_{(-\frac{1}{2},-\frac{1}{2})} \right\rangle \\[4pt]
&= \int dz_2\; z_{13}z_{14}z_{34} \prod_{i<j} z_{ij}^{-2s_{ij}-\frac{1}{2}}\, \left\langle \Sigma_1^+ \Sigma_2^- \Sigma_3^+ \Sigma_4^- \right\rangle \\[4pt]
&= -\frac{\Gamma(-2s)\Gamma(-2t)}{\Gamma(1-2u)} \left[4(k_1^+ \cdot k_2^-)(k_3^+ \cdot k_4^-)\, t + 4(k_2^+ \cdot k_3^-)(k_4^+ \cdot k_1^-)\, s \right] \\[4pt]
&= 0\,,
\end{aligned}
\tag{8.174}
$$

because the expression in square brackets is equal to eq. (8.172).

The $N = 2$ string amplitudes can be reformulated as amplitudes of $N = 4$ *topological* strings [380]. The topological representation does not include ghosts, and has no ambiguities associated with positions of the picture-changing operators on string world-sheet. In addition, the topological approach turned out to be useful to actually *prove* the vanishing to all orders of all $N = 2$ string n-point correlation functions (with the exception of $n = 3$ at genus zero and one) [380]. The critical $N = 2$ string theory is, in fact, a *topological* conformal field theory, because it has *twisted* $N = 4$ superconformal symmetry [377, 381]. The latter is generated by the $N = 2$ stress-tensor superfield $T(z, \theta^+, \theta^-)$, the BRST supercurrent $j_{\mathrm{BRST}}(z, \theta^+, \theta^-)$, the antighost superfield $B(z, \theta^+, \theta^-)$ and the ghost number supercurrent $BC(z, \theta^+, \theta^-)$, with the vanishing central charge (see exercise # VIII-13 for the notation). [27]

The $N = 4$ generalization of the $N=2$ string action in eq. (8.88) is known as the *Pernici-Nieuwenhuizen* (PN) action [382]

$$S_{\mathrm{PN}} = \frac{1}{\pi} \int d^2\xi \, e \left\{ \frac{1}{2} h^{\alpha\beta} \partial_\alpha X^* \cdot \partial_\beta X + \frac{i}{2} \bar\Psi \cdot \gamma^\alpha \overleftrightarrow{D}_\alpha \Psi \right.$$

$$\left. + A_\alpha^I \left(\bar\Psi_i^c \cdot \gamma^\alpha \sigma_{ij}^I \Psi_j^c \right) + \left[(\partial_\alpha X + \bar\chi_\alpha \Psi) \cdot \left(\bar\Psi \gamma^\beta \gamma^\alpha \chi_\beta \right) + \mathrm{h.c.} \right] \right\} , \quad (8.175)$$

where the fields X, Ψ and χ_α are now considered to be *quaternionic*. This means $X^\mu = X_1^\mu + \mathbf{i} X_2^\mu + \mathbf{j} X_3^\mu + \mathbf{k} X_4^\mu = (X_1^\mu + \mathbf{i} X_2^\mu) + \mathbf{j}(X_3^\mu - \mathbf{i} X_4^\mu) \equiv (X^c)_1^\mu + \mathbf{j}(X^c)_2^\mu$, $(X^\mu)^* = X_1^\mu - \mathbf{i} X_2^\mu - \mathbf{j} X_3^\mu - \mathbf{k} X_4^\mu$, and similarly for Ψ and χ_α. Three (quaternionic) complex structures $\mathbf{i}, \mathbf{j}$ and $\mathbf{k}$ satisfy an algebra: $\mathbf{i}^2 = \mathbf{j}^2 = \mathbf{k}^2 = -1$, $\mathbf{ij} = -\mathbf{ji} = \mathbf{k}$, etc. A quaternionic index can equally be represented as a multi-index (ii'), where $i = 1, 2$ and $i' = 1', 2'$ are connected to the simple factors in the decomposition (8.87). The matrices σ_I stand for $SU(2)$ or $SU(1, 1)$ generators in the fundamental representation.

In the $N = 4$ superconformal gauge, the PN action leads to the standard (holomorphic) constraints defining the $N = 4$ NSR-type string [149],

$$T(z) = \frac{1}{2} : P_{ii'}(z) \cdot P_{ii'}(z) : -\frac{i}{2} : \psi_{ii'}(z) \cdot \partial\psi_{ii'}(z) : ,$$

$$G_i(z) = \psi_{i'}(z) \cdot P_{ii'}(z) ,$$

$$J^I(z) = \frac{i}{2} : \psi_i(z) \cdot (\sigma_{ij}^I \psi_j)(z) : . \quad (8.176)$$

Actually, there are many ways to build $N = 4$ extended fermionic string models, by using the linear $N = 4$ SCA's or their non-linear cousins [353, 383]. For example,

---

[27]See sect. IX.3 for the meaning of a *topological* twist in extended SCFT's. The topological twist effectively means shifting the spins of all fields by half of their $U(1)$ charge.

an analysis of the $N = 4$ constraints (8.176) shows that they are irreducible in the compact case [359], but become *reducible* in the non-compact case [370]. Being a subset of the $N = 4$ constraints, the $N = 2$ constraints already forbid all excited $N = 4$ string physical states but ground states, so that the rest of the $N = 4$ constraints may become redundant. The $N = 2$ string physical states are just the highest weight states of the twisted $N = 4$ SCA [358]. The $N = 4$ fermionic string originally introduced in refs. [147, 149] has the *compact* internal symmetry group $SU(2)$ and the irreducible constraints [359]. It can therefore be quantized along the standard BRST lines, which lead to (formally) negative critical dimension, $D = -2$ [375]. The same conclusion arises from the conformal anomaly counting:

$$
\begin{array}{cccc}
(b,c) & (\beta_i^a, \gamma_a^i) & (\tilde{b}^I, \tilde{c}_I) & (x_{ii'}, \psi_{ii'})^\mu \\
-26 & + \; 2 \cdot 2 \cdot 11 & + \; -3 \cdot 2 & + \; 2 \cdot 2 \cdot D \cdot (1 + \tfrac{1}{2}) \, ,
\end{array}
\tag{8.177}
$$

whose cancellation implies $D = -2$.

The reducibility of the 'non-compact' $N = 4$ string constraints means linear dependence between generators of the gauged $SU(1,1)$ superconformal algebra. A more general (than BRST) covariant scheme known as the *Batalin-Fradkin-Vilkovisky* (BFV) quantization [384, 385, 386] (see sect. XII.5) is therefore needed for a consistent quantization in this case. In particular, eq. (8.177) has to be revised [359].

It may happen that some $N = 4$ non-compact string theory could support the *self-dual* supersymmetry and supergravity in $2 + 2$ dimensions [368, 370, 371] (see sect. XI.5 also). The existence of MW spinors in $2 + 2$ dimensions is crucial for their construction [387, 388, 389]. Notably, in a real (Majorana) representation for the $(2+2)$-dimensional Dirac 4×4 matrices $\Gamma^{\underline{a}}$,

$$
\{\Gamma^{\underline{a}}, \Gamma^{\underline{b}}\} = 2\eta^{\underline{ab}} \, , \quad \eta^{\underline{ab}} = \text{diag}\,(-,-,+,+) \, ,
\tag{8.178}
$$

the Γ_5-matrix is *real*, $\Gamma_5 \equiv \Gamma^1 \Gamma^2 \Gamma^3 \Gamma^4$, $\Gamma_5^2 = 1$. MW spinors transform in the fundamental (real) representation of one of the $SU(1,1) \cong SL(2, \mathbf{R})$ factors in $SO(2,2)$, and, hence, they have only $2/2 = 1$ degree of freedom on-shell. A self-dual vector particle and a MW spinor are naturally united into an $N = 1$ supersymmetric self-dual *vector* multiplet with $1_B \oplus 1_F$ components on-shell. In extended supersymmetry, there exists an $N = 2$ self-dual vector multiplet [387, 388] comprising a self-dual vector, two MW spinors of the same chirality and a real scalar ($2_B \oplus 2_F$ components on-shell). One could naively expect that $N = 4$ SDYM theory would also follow this pattern and contain $4_B \oplus 4_F$ components in its on-shell spectrum, but it turns out *not* to be the case [388]. The

$N = 4$ SDYM theory actually needs twice as many degrees of freedom for its definition, even on-shell, and it has the following on-shell field contents [371, 388]:

$$\left(A_{\underline{a}}{}^I, G_{\underline{ab}}{}^I, \rho^I, \tilde{\lambda}^I, S_{\hat{i}}{}^I, T_{\hat{i}}{}^I\right) \ , \tag{8.179}$$

where $G_{\underline{ab}}$ is anti-symmetric and anti-self-dual, ρ and $\tilde{\lambda}$ are MW spinors of opposite chiralities, $S_{\hat{i}}$ and $T_{\hat{i}}$ are scalars, $\hat{i} = \hat{1}, \hat{2}, \hat{3}$; all fields are real and belong to the adjoint representation of a gauge group G, $I = 1, 2, \ldots, \dim G$. The situation with *self-dual supergravities* (SDSG's) in $2 + 2$ dimensions is quite similar [368, 387, 388]. There exist the N-extended SDSG's up to $N \leq 4$, which contain $N_{\rm B} \oplus N_{\rm F}$ on-shell degrees of freedom, but this is no longer true for the $N > 4$ SDSG's, where the naive number of on-shell degrees of freedom has to be doubled.

The Lagrangian of the $N = 4$ SDYM theory takes the form

$$\mathcal{L}_{N=4 \text{ SDYM}} = -\tfrac{1}{2}G^{\underline{ab}\,I}(F_{\underline{ab}}{}^I - \tfrac{1}{2}\epsilon_{\underline{ab}}{}^{\underline{cd}}F_{\underline{cd}}{}^I) + \tfrac{1}{2}(\nabla_{\underline{a}}S_{\hat{i}}{}^I)^2 - \tfrac{1}{2}(\nabla_{\underline{a}}T_{\hat{i}}{}^I)^2$$

$$+i(\rho^I\sigma^{\underline{a}}D_{\underline{a}}\tilde{\lambda}^I) - if^{IJK}\left[(\tilde{\lambda}^I\alpha_{\hat{i}}\tilde{\lambda}^J)S_{\hat{i}}{}^K + (\tilde{\lambda}^I\beta_{\hat{i}}\tilde{\lambda}^J)T_{\hat{i}}{}^K\right] \ , \tag{8.180}$$

where $\nabla_{\underline{a}}$ and $D_{\underline{a}}$ are the gauge-covariant derivatives, $\sigma^{\underline{a}}$ represent the Dirac matrices in the two-component notation for the $SO(2, 2)$ 'Lorentz' group,

$$\Gamma^{\underline{a}} = \begin{pmatrix} 0 & \sigma^{\underline{a}} \\ \bar{\sigma}^{\underline{a}} & 0 \end{pmatrix} \ , \tag{8.181}$$

and Γ_5 is supposed to be diagonal. The α and β matrices are the second independent set of Dirac matrices for the $SO(4)$ or $SO(2, 2)$ internal symmetry [388]. A similar action exists for the $N = 8$ SDSG too [368]. The action (8.180) is covariant and leads to non-vanishing 3-point trees. Hence, there is non-trivial scattering of covariant objects in this theory. Its quantum loops should all vanish because of the non-renormalization theorem in extended supersymmetry [180]. This actually implies the N-extended *superconformal* invariance in 'space-time' [390] for all *quantum* self-dual 4d field theories:

$$
\begin{array}{ccccc}
SO(2,2) & \cong & SL(2) \otimes SL(2) & \xrightarrow{\text{susy}} & SL(2|N) \otimes SL(2|N) \\
& & \Big\downarrow \text{conf} & & \Big\downarrow \text{s--conf} \\
SO(3,3) & \cong & SL(4) & \xrightarrow{\text{susy}} & SL(4|N)
\end{array}
$$

The 4d (supersymmetric) self-dual field theories appear to be the 'master theories' for all (supersymmetric) 2d integrable models (Ch. XI). Simultaneously, they are the most natural *four-dimensional* generalizations of 2d (S)CFT's.

Exercises

VIII-13 ▷ (*) Let (z, θ^+, θ^-) be the coordinates of $N = 2$ (holomorphic) superspace. The $N = 2$ supercovariant derivatives take the form $\mathbf{D}_\pm = \frac{\partial}{\partial \theta^\pm} + \frac{1}{2}\theta^\mp \partial$ and satisfy an algebra $\{\mathbf{D}_+, \mathbf{D}_-\} = \partial$, $\{\mathbf{D}_+, \mathbf{D}_+\} = \{\mathbf{D}_-, \mathbf{D}_-\} = 0$, where $\partial \equiv \frac{\partial}{\partial z}$. The $N = 2$ SCA generators can be united into a single $N = 2$ superfield $T(z, \theta^+, \theta^-) = J(z) + \theta^+ G_+(z) + \theta^- G_-(z) + \theta^+\theta^- T(z)$, whereas the $N = 2$ ghosts receive a simple description in terms of two $N = 2$ superfields as $C(z, \theta^+, \theta^-) = c(z) + \theta^+\gamma_+(z) + \theta^-\gamma_-(z) + \theta^+\theta^-\tilde{c}(z)$ and $B(z, \theta^+, \theta^-) = \bar{b}(z) + \theta^+\beta_+(z) - \theta^-\beta_-(z) + \theta^+\theta^- b(z)$. Show that the $N = 2$ ghost stress-tensor superfield takes the form (normal ordering is always assumed)

$$T_{B,C} = \partial(CB) - (\mathbf{D}_+ C \mathbf{D}_- B + \mathbf{D}_- C \mathbf{D}_+ B) , \qquad (8.182)$$

whereas the *improved* (in order to be an $N = 2$ primary superfield of $h = q = 0$) $N = 2$ BRST supercurrent is given by

$$j_{\text{BRST}}(z, \theta^+, \theta^-) = C\left(T_{\text{N=2 matter}} + \tfrac{1}{2}T_{B,C}\right) + B\left(\mathbf{D}_+ C \mathbf{D}_- C - C\partial C\right) =$$

$$C\left(T_{\text{N=2 matter}} + \tfrac{1}{2}T_{B,C}\right) - \tfrac{1}{2}\left[\mathbf{D}_-(C\mathbf{D}_+(BC)) + \mathbf{D}_+(C\mathbf{D}_-(BC))\right] . \qquad (8.183)$$

VIII-14 (*) Given an $N = 2$ primary superfield Φ of vanishing conformal dimension and $U(1)$-charge, prove that the operator [377]

$$V(\Phi) \equiv \mathbf{D}_-(C\mathbf{D}_+\Phi) - \mathbf{D}_+(C\mathbf{D}_-\Phi) \qquad (8.184)$$

(in the so-called $[0, 0]$-picture [392]) represents a BRST cohomology class, i.e. it is BRST invariant but it is not a BRST (anti)commutator. *Hint*: use the relations

$$Q_{\text{BRST}}(C) = C\partial C - \mathbf{D}_+ C \mathbf{D}_- C ,$$

$$Q_{\text{BRST}}(\Phi) = -\mathbf{D}_-(C\mathbf{D}_+\Phi) - \mathbf{D}_+(C\mathbf{D}_-\Phi) . \qquad (8.185)$$

VIII-15 Derive the critical dimension and the critical intercept of the $N = 2$ string for the NS-type (untwisted) boundary conditions from the nilpotency of the BRST charge to be defined by eqs. (8.116)–(8.120).

VIII-16 (*) Consider the SDG field equations for a Kähler metric $g_{a\bar{b}} = \partial_a\partial_{\bar{b}}K$, and show that the Ricci tensor vanishes. Given a Kähler potential having the form $K = K_0 + 2\kappa\Phi$, prove that the corresponding field equations of motion follow from the Plebanski action in eq. (8.95).

VIII-17 (*) Show that the SDYM field equations can be put into the form

$$F_{ab} = F_{\bar{a}\bar{b}} = \eta^{a\bar{b}}F_{a\bar{b}} = 0 , \quad \text{where} \quad \eta^{a\bar{b}} == \eta^{a\bar{b}} = \begin{pmatrix} 1 & 0 \\ 0 & -1 \end{pmatrix} . \tag{8.186}$$

VIII.4 W gravity and W strings

The existence of W algebras in CFT (Ch. VII) motivates a construction of W-gravity and of related W-string models. A W-**gravity** is a higher-spin (> 2) extension of the ordinary 2d gravity, with the W algebra playing the role of the Virasoro algebra. This obviously leads to the appearance of higher-spin 2d gauge fields in any theory of W-gravity. Like the ordinary gravity and supergravity theories [390, 391], classical W-gravities can be systematically constructed by 'gauging' the underlying classical [28] W algebras [393, 394]. Again, like in the ordinary 2d gravity and supergravity theories, an analysis can be considerably simplified by choosing a convenient gauge where most of the local symmetries are fixed.

One of the simplest constructions of this (latter) type is a free boson realization of the classical W_3 algebra [395, 396]. One starts with a free action for n scalars Φ^i having the transformation laws

$$\delta_\varepsilon \Phi^i = \varepsilon_+ \partial_- \Phi^i , \quad \delta_\lambda \Phi^i = \lambda_{++} d^{ijk} \partial_- \Phi^j \partial_- \Phi^k , \tag{8.187}$$

where $\partial_- = \partial_z$, $\partial_+ = \partial_{\bar{z}}$, and d^{ijk} is a totally symmetric constant tensor. It is not difficult to check (exercise # VIII-18) that the free action

$$S_0 = \frac{1}{\pi} \int d^2z \left[-\tfrac{1}{2} \partial_+ \Phi^i \partial_- \Phi^i \right] \tag{8.188}$$

is invariant under the symmetries (8.187) with holomorphic parameters $\varepsilon_+ = \varepsilon_+(z)$ and $\lambda_{++} = \lambda_{++}(z)$. These symmetries can be promoted to the full local symmetries (with parameters $\varepsilon_+(z, \bar{z})$ and $\lambda_{++}(z, \bar{z})$) by introducing minimal coupling to the W_3 gauge fields h_{++} and b_{+++} and adjusting their transformation laws. This results in a locally symmetric action having the form [395, 396]

$$S_{\mathrm{H}} = \frac{1}{\pi} \int d^2z \left[-\tfrac{1}{2} \partial_+ \Phi^i \partial_- \Phi^i + \tfrac{1}{2} h_{++} \partial_+ \Phi^i \partial_+ \Phi^i + \tfrac{1}{3} b_{+++} d^{ijk} \partial_+ \Phi^i \partial_+ \Phi^j \partial_+ \Phi^k \right] , \tag{8.189}$$

---

[28] By a *classical W* (or Gel'fand-Dikii) algebra we always mean a W algebra where all central terms (like the central charge) are set to be zero. Given a classical W algebra, the corresponding quantum W algebra may not even exist!

provided the identity

$$d^{k(ij}d^{l)mk} = \delta^{(ij}\delta^{l)m} \qquad (8.190)$$

holds. Remarkably, no more terms beyond those explicitly given in eq. (8.189) are needed. Eq. (8.190) implies that d^{ijk} are structure constants of a *Jordan* algebra of degree 3. Similar free field (chiral) realizations can be constructed for the W_n algebras, although their non-chiral gaugings generically lead to the actions which are non-polynomial in the gauge fields [131, 396].

Similarly to the ordinary 2d gravity, a classical 2d W-gravity does not have physical degrees of freedom (up to moduli), but the situation becomes different after quantization, because of the anomalies associated with some of the local symmetries. The ordinary bosonic string originates from coupling 2d gravity to a conformal matter, the latter being usually represented by scalar fields (sect. 1). Coupling W-gravity to a W-invariant matter CFT naturally leads to the similar notion of a W-**string**. The critical dimension of a string characterizes the central charge value at which 2d gravity decouples from conformal matter. Similarly, as far as the most general W-string is concerned, its critical central charge is given by

$$c = \sum_s 2(-1)^{2s}\left(6s^2 - 6s + 1\right) , \qquad (8.191)$$

where eqs. (1.129) and (1.131) have been used, and the summation goes over spins s of independent W-algebra generators. In the case of the W_3 algebra, one gets, in particular, $c = -26 - 74 = -100$.

To actually construct a W string, its explicit (nilpotent) BRST operator is needed. For the case of W_3, the BRST charge is known [397, 398]. See ref. [399] for the results of investigation of the (complicated) structure of the W_3 string.

Quantum dynamics of the W_3-gravity is governed by the effective action, which is rather difficult to construct. A partial integration over matter fields in the corresponding quantum generating functional yields the induced action (the 'integrated anomaly'), which the W-gravity fields are coupled to. In the conformal gauge, the induced action was shown [400] to have the form of a Toda action (sect. XI.3). Since the W_3-gravity can be obtained by a conformal reduction of the WZNW theory based on the AKM algebra $\widehat{sl}(3)_k$ (sect. XI.3), eq. (7.114) gives us the central charge contribution of the matter sector [131]:

$$c_k = 2 - 24\left(\sqrt{\frac{1}{k+3}} - \sqrt{k+3}\right)^2 = 50 - 24\left(\frac{1}{k+3} + k + 3\right) . \qquad (8.192)$$

The total (vanishing in the critical theory) central charge of the coupled (with the W_3-gravity) system is a sum of the central charge c of the W_3-gravity itself, the contribution (-100) of its ghosts and c_k. Since $c_k = 100 - c_{-(k+6)}$, the required central charge results from the quantum DS reduction (sect. VII.3) of the $\widehat{sl(3)}_{-(k+6)}$ WZNW model. Eq. (8.192) now implies

$$-(k+6) = -\frac{1}{48}\left(50 - c + \sqrt{(c-98)(c-2)}\right) - 3 \ . \tag{8.193}$$

Therefore, we can approximately represent the effective W_3-gravity degrees of freedom by the (quantum DS)-reduced $\widehat{sl(3)}$ WZNW model of level (8.193) [131]. The full answer can be obtained, e.g., by the $1/c$ perturbation theory [401].

The *supersymmetric* W-gravities and super-W strings can also be introduced and studied along the similar lines [399].

Exercises

VIII-18 ▷ (*) Check that the free action (8.188) is invariant under the symmetries (8.187), with holomorphic parameters $\varepsilon_+ = \varepsilon_+(z)$ and $\lambda_{++} = \lambda_{++}(z)$. Calculate the corresponding currents T_{--} and W_{---}, and prove that they are holomorphic on-shell, $\partial_+ T_{--} = \partial_+ W_{---} = 0$, and generate the classical W_3 symmetry.

VIII-19 ▷ Calculate the contribution of a spin-4 field to the critical charge in eq. (8.191).

Chapter IX

Quantum 2d Gravity, and Topological Field Theories

To understand the *non-critical* string theories defined away from their critical dimensions, as well as various 2d conformal field systems coupled to Euclidean quantum 2d gravity, we need to understand better the quantum 2d gravity itself. Since there are no classical propagating degrees of freedom associated with the 2d gravity in two dimensions, its total quantum induced action is given by the conformal anomaly alone, so that any quantum description of this theory should be essentially non-perturbative. The solution to the quantum 2d gravity to be discussed below suggests us to take a look at other solvable field theories in two dimensions. The so-called *topological field theories* (TFT's) are a particularly interesting class of them, which are closely related to CFT's.

2d quantum gravity can be solved either in the *Knizhnik-Polyakov-Zamolodchikov* (KPZ) approach [402, 403] or in the *David-Distler-Kawai* (DDK) approach [404, 405], which are based on different gauges. The induced action of 2d gravity is given by the **Liouville theory** whose consistent quantum definition turns out to be closely related with the 'quantum group' structure. The topological field theories provide us with many examples of solvable field theories, and the information extracted from their studies turns out to be particularly useful in CFT.

Coupling the quantum 2d gravity to 2d conformal field systems results in the nontrivial renormalization of the corresponding CFT operators, which is physically interpreted as their 'gravitational dressing'. This gives rise to a very different critical behaviour to be described by critical exponents (some of them are introduced below).

IX.1   Quantum 2d gravity

The local conformal anomaly of the Polyakov string action (sect. VIII.1) in D spacetime dimensions is given by [1]

$$\frac{1}{\sqrt{g}} g^{\alpha\beta} \frac{\delta W}{\delta g^{\alpha\beta}} = \tfrac{1}{2} g^{\alpha\beta} \langle T_{\alpha\beta} \rangle = \frac{D}{24\pi} \left[R^{(2)} + const. \right] . \tag{9.1}$$

The induced action W is [2]

$$W = \frac{D}{96\pi} \int d^2x d^2x' \sqrt{g(x)}\sqrt{g(x')} R^{(2)}(x) R^{(2)}(x') G(x, x') + const. \int d^2x \sqrt{g(x)} , \tag{9.2}$$

where $G(x, x')$ is Green's function of the 2d scalar Laplace operator. The action W is generically non-local, but it becomes local in the conformal gauge $g_{\alpha\beta} = e^{\phi(x)} \delta_{\alpha\beta}$, where it takes the form of the *Liouville* action

$$W = -\frac{D}{48\pi} \int d^2x \left[\tfrac{1}{2} (\partial_\alpha \phi)^2 + const.\, e^\phi \right] . \tag{9.3}$$

The KPZ approach to 2d gravity is based on the following light-cone (Polyakov) gauge for 2d reparametrizations:

$$ds^2 = dx^+ dx^- + h_{++}(x^+, x^-)(dx^+)^2 , \tag{9.4}$$

where the light-cone coordianes $x^\pm = \frac{1}{\sqrt{2}}(x^1 \pm ix^2)$ have been introduced. In this gauge, one has

$$\delta W = \int d^2x\, T_{--} \delta h_{++} , \quad R^{(2)} = \partial_-^2 h_{++} ,$$

$$\nabla_+ T_{--} \equiv \partial_+ T_{--} - h_{++} \partial_- T_{--} - 2(\partial_- h_{++}) T_{--} = \frac{D}{24\pi} \partial_- R^{(2)} . \tag{9.5}$$

It is now convenient to introduce the new field f defined by [402]

$$\partial_+ f = h_{++} \partial_- f , \tag{9.6}$$

instead of h. In terms of the field f, the induced gravity action in the Polyakov gauge takes the form

$$W[f] = \frac{D}{24\pi} \int d^2x \left[\frac{(\partial_-^2 f)(\partial_+ \partial_- f)}{(\partial_- f)^2} - \frac{(\partial_-^2 f)(\partial_+ f)}{(\partial_- f)^3} \right] . \tag{9.7}$$

---

[1] Another derivation of eq. (9.1) from the proper-time (Schwinger-De Witt) heat-kernel expansion can be found in ref. [39]. No ghost contributions are included.

[2] The dimension (central charge) D is the only matter parameter which is relevant in what follows. In fact, *any* conformal matter could be chosen here.

The Jacobian associated with the change of variables in eq. (9.6) is just the determinant of the differential operator appearing in the reparametrizational variation of h_{++} with infinitesimal parameter $\varepsilon_+(x)$,

$$\delta h_{++} = \nabla_+ \varepsilon_+ = (\partial_+ - h_{++}\partial_-)\varepsilon_+ + (\partial_- h_{++})\varepsilon_+ \ , \quad \delta f = \varepsilon_+ \partial_- f \ , \qquad (9.8)$$

and it coincides with the conventional FP (ghost) determinant. This effectively implies replacing D by $D - 26$ in the expression for the conformal anomaly. The equations of motion can be written in the following form (with $c = D - 26$)

$$\nabla_+ T_{--} = \frac{c}{24\pi}\partial_-^3 h_{++} = 0 \ , \qquad (9.9)$$

and they are equivalent to the ones following from the action (9.7).

The gauge symmetry (9.8) in terms of the *rescaled* field $h_{++} \to (-c/6)h_{++}$ leads to the Ward identities

$$\langle h_{++}(z)h_{++}(x_1)\cdots h_{++}(x_M)\rangle$$

$$= \frac{c}{6}\sum_j \frac{(z^- - x_j^-)^2}{(z^+ - x_j^+)^2}\langle h_{++}(x_1)\cdots h_{++}(x_{j-1})h_{++}(x_{j+1})\cdots h_{++}(x_M)\rangle$$

$$+ \sum_j \left[\frac{(z^- - x_j^-)^2}{z^+ - x_j^+}\frac{\partial}{\partial x_j^-} + 2\frac{z^- - x_j^-}{z^+ - x_j^+}\right]\langle h_{++}(x_1)\cdots h_{++}(x_M)\rangle \ , \qquad (9.10)$$

where the identity

$$\frac{1}{4\pi i}\partial_-^3 \frac{(z^-)^2}{z^+} = \delta^{(2)}(z) \ , \qquad (9.11)$$

has been used. Eq. (9.10) actually gives the reccurence relations defining correlation functions with any number of h_{++}. Similarly, given any conformal field of weight λ (w.r.t. a change of the x^- coordinate),

$$\delta\phi = \varepsilon_+ \partial_- \phi + \lambda(\partial_- \varepsilon_+)\phi \ , \qquad (9.12)$$

one finds [402]

$$\langle h_{++}(z)\phi(x_1)\cdots\phi(x_M)\rangle = \sum_j \left[\frac{(z^- - x_j^-)^2}{z^+ - x_j^+}\frac{\partial}{\partial x_j^-} + 2\lambda\frac{z^- - x_j^-}{z^+ - x_j^+}\right]\langle\phi(x_1)\cdots\phi(x_M)\rangle \ . \qquad (9.13)$$

Because of eq. (9.9), the field $h_{++}(z)$ can be decomposed as

$$h_{++}(z) = j^{(1)} - 2j^{(0)}z^- + j^{(-1)}(z^-)^2 \ . \qquad (9.14)$$

Hence, one finds the relation [402]

$$\langle j^a(z)\phi(x_1)\cdots\phi(x_M)\rangle = \sum_j \frac{l_j^a}{z^+ - x_j^+} \langle \phi(x_1)\cdots\phi(x_M)\rangle , \qquad (9.15)$$

where the operators

$$l_j^{(-1)} = \frac{\partial}{\partial x_j^-} , \quad l_j^{(0)} = x_j^- \frac{\partial}{\partial x_j^-} + \lambda , \quad l_j^{(1)} = (x_j^-)^2 \frac{\partial}{\partial x_j^-} + 2\lambda x_j^- , \qquad (9.16)$$

generate an $SL(2,\mathbf{R})$ current (AKM) algebra, equally represented in terms of the currents $j^a(z)$. This means that the quantum 2d gravity has the hidden affine $SL(2,\mathbf{R})$ symmetry! [3] From the viewpoint of quantum DS reduction (sect. VII.3), the theory under consideration is equivalent to the (conformally) constrained $SL(2,\mathbf{R})$ WZNW model (sect. XI.3).

It follows from the Ward identity (9.10) that the corresponding Ward identity for the currents takes the form

$$\left\langle j^a(z) j^{b_1}(x_1) \cdots j^{b_M}(x_M) \right\rangle$$

$$= \frac{c}{12} \sum_j \frac{\eta^{ab_j}}{(z^+ - x_j^+)^2} \left\langle j^{b_1}(x_1) \cdots j^{b_{j-1}}(x_{j-1}) j^{b_{j+1}}(x_{j+1}) \cdots j^{b_M}(x_M) \right\rangle$$

$$+ \sum_j f^{ab_j c_j'} \frac{\eta_{c_j c_j'}}{z^+ - x_j^+} \left\langle j^{b_1}(x_1) \cdots j^{c_j}(x_j) \cdots j^{b_M}(x_M) \right\rangle , \qquad (9.17)$$

where the $SL(2,\mathbf{R})$ Killing-Kartan metric η_{ab} and structure constants f^{abc} have been introduced, $\eta_{+-} = \frac{1}{2}$, $\eta_{00} = -1$.

The derivation above and the resulting current algebra are, in fact, quite similar to the ones in the case of $SU(2)$ current algebra (Ch. IV). This analogy was extended by Polyakov [402] who postulated the following 'renormalized' equation for the quantum 2d gravity:

$$k\partial_+ f =: h_{++}\partial_- f : +\lambda : (\partial_- h_{++})f : , \qquad (9.18)$$

where the constant k is yet to be determined. Without the coupling to 2d gravity, we would have $k = k_0 \equiv (26 - D)/6$. The consistency of the modified equation (9.18) with the $SL(2,\mathbf{R})$ current algebra implies the relation

$$: h_{++}\partial_- f : +\lambda : (\partial_- h_{++})f := \eta_{ab} : j^a(z)l^b f(z) : , \qquad (9.19)$$

---

[3]A canonical derivation of the $SL(2,\mathbf{R})$ current algebra in the quantum 2d gravity was also given in ref. [247]. The $SL(2,\mathbf{R})$ symmetry allows the geometrical interpretation in terms of the coadjoint orbits of the Virasoro group [406].

which, because of eq. (9.15), leads to differential equations for correlation functions in the form [402]

$$k\frac{\partial}{\partial x_j^+}\langle f(x_1)\cdots f(x_M)\rangle = \sum_{k\neq j}\frac{\eta_{ab}l_j^a l_k^b}{x_j^+ - x_k^+}\langle f(x_1)\cdots f(x_M)\rangle \;. \tag{9.20}$$

The exact (i.e. in the presence of 2d gravity) renormalized scaling dimension λ and the 'level' κ, $\kappa + 2 \equiv -(k + 2)$, of the $SL(2, \mathbf{R})$ AKM algebra are, in general, different from their naive (without the 2d gravity) quantum values λ_0 and $\kappa_0 = -k_0 - 4$, because of the anticipating gravitational dressing. Similarly to the Knizhnik-Zamolodchikov analysis (sect. IV.2), which has led to the differential equation (4.47) for correlators and to eq. (4.52) for scaling dimensions, we can now use the $SL(2, \mathbf{R})$ Ward identities in order to determine the 'dressed' values of the parameters. When using the exact values instead of the naive ones in the $SL(2, \mathbf{R})$-analogue of eq. (4.50), one gets the famous **KPZ equation** for the dressing of scaling dimensions [402, 403]: [4]

$$\lambda - \lambda_0 = \frac{\lambda(\lambda + 1)}{\kappa + 2} = -\frac{\lambda(\lambda + 1)}{k + 2} \;. \tag{9.21}$$

The $\Delta_0 = -\lambda_0$ is the usual conformal dimension of the primary spinless field $\phi_{\lambda_0}(x)$ under consideration, when the 2d gravity is switched off. The $\Delta = -\lambda$ is the conformal (scaling) dimension of the same field in the presence of 2d gravity.

One of the ways to actually calculate the value of k, which should be a universal function of the matter central charge D only, is to consider the simplest conformal matter represented by, say, D Dirac fermions $\chi(x)$ coupled to 2d gravity, and to take advantage of general covariance of the theory. Following ref. [403], consider the Lagrangian

$$\mathcal{L} = \bar{\chi}e_a^\beta\gamma^a\partial_\beta\chi \;, \tag{9.22}$$

where the 2d gravity zweibein $e_a^\alpha(x)$ has been introduced. The theory (9.22) can be rewritten to the form of the conventional gauge theory with the Lagrangian

$$\mathcal{L}_{\text{matter}} = \psi_-(\partial_+ - h_{++}\partial_-)\psi_- + \psi_+(\partial_- - h_{--}\partial_+)\psi_+ \;, \tag{9.23}$$

after the appropriate field redefinitions, *viz.*

$$\psi_- = \sqrt{e_{-+}}\chi_- \;, \quad \psi_+ = \sqrt{e_{+-}}\chi_+ \;,$$

---

[4]This formula will be rederived in the DDK approach in what follows, in this section.

$$h_{++} = \frac{v_{++}}{v_{-+}} , \quad h_{--} = \frac{v_{--}}{v_{+-}} . \tag{9.24}$$

Since the fields in eq. (9.23) decouple, the classical theory under consideration possesses the extended 'chiral' general coordinate invariance w.r.t separate diffeomorphisms for the 'right' and 'left' movers. The extended symmetry is, however, going to be broken down to the initial reparametrizational invariance after quantization (see below). In the Polyakov gauge $h_{--} = 0$, integrating over the fermions yields for induced action

$$W(h_{++}) \propto \log \text{Det}(\partial_+ - h_{++}\partial_-) . \tag{9.25}$$

The general idea of KPZ [403] is to employ *another* (KPZ) gauge having the form

$$h_{--} = h_{--}(x) , \quad h_{+-} = h_{+-}(x) , \tag{9.26}$$

in terms of the *given* functions $h_{--}(x)$ and $h_{+-} = h_{--}(x)$, and to make use of the general covariance which ensures the total quantum stress tensor to vanish. The general covariance of the quantized theory is not spoiled by choosing this gauge, since one could always add a regulator of the form

$$S_{\text{reg}} = \Lambda^{-2} \int d^2 x \, \sqrt{g} (R^{(2)})^2 = \Lambda^{-2} \int d^2 x \, \sqrt{g} (\partial_-^2 h_{++})^2 \tag{9.27}$$

to the classical action, which makes the theory to be convergent and covariant. The invariance conditions in terms of the partition function Z take the form

$$\left. \frac{\delta Z}{\delta h_{--}(x)} \right|_{h_{--}=0} = \left. \frac{\delta Z}{\delta h_{+-}(x)} \right|_{h_{+-}=0} = 0 . \tag{9.28}$$

The total effective action must be supplemeted by the contributions of ghosts, and it should include the vacuum polarization also, due to the coupling to quantum 2d gravity. Fixing the KPZ gauge (9.26) leads to the modified Lagrangian (9.23) in the form

$$\mathcal{L}_{\text{matter+ghosts}} = \mathcal{L}_{\text{matter}} + \eta_{++}\nabla_-\epsilon_- + \zeta(\nabla_+\epsilon_- + \nabla_-\epsilon_+) , \tag{9.29}$$

where the ghosts (ϵ_+, ϵ_-) and the corresponding anti-ghosts (η_{++}, ζ) have been introduced. The vacuum polarization can be accounted by adding a local functional $\Lambda(h_{++}, h_{--}, h_{+-})$ to the effective action,

$$\tilde{W}(h_{++}, h_{--}, h_{+-}) = W(h_{++}) + W(h_{--}) + \Lambda(h_{++}, h_{--}, h_{+-}) , \tag{9.30}$$

and by requiring the invariance of the total action w.r.t. the diffeomorphisms

$$\delta h_{++} = 2\nabla_+\varepsilon_+ , \quad \delta h_{+-} = \nabla_+\varepsilon_- + \nabla_-\varepsilon_+ , \quad \delta h_{--} = 2\nabla_-\varepsilon_- . \tag{9.31}$$

Since eq. (9.28), we actually need to calculate only those terms in Λ,

$$\Lambda \approx \int d^2x \left[h_{--} T_{++}(h_{++}) + h_{+-} \theta(h_{++}) \right] , \tag{9.32}$$

that are linear in h_{--} and h_{+-}. One easily finds

$$T_{++} \propto \tfrac{1}{2} \left[(\partial_- h_{++})^2 - 2h_{++} \partial_-^2 h_{++} \right] + \partial_+ \partial_- h_{++} , \quad \theta \propto \partial_-^2 h_{++} . \tag{9.33}$$

Hence, the general covariance conditions take the form

$$T^{\text{tot}}_{++} \equiv \psi_+ \partial_+ \psi_+ + \eta_{++} \partial_+ \varepsilon_- + \zeta \partial_+ \epsilon_+ + T_{++}(h) = 0 , \quad \theta(h) = 0 . \tag{9.34}$$

Eq. (9.34) is nothing but the requirement of invariance of the theory under the residual gauge transformations in Polyakov's gauge (exercise # IX-1). This implies for the corresponding central charges,

$$c(\psi) \equiv c_{\text{matter}} = D , \quad c(\eta_{++}, \epsilon_-) = -26 , \quad c(\zeta, \epsilon_+) = -2 , \tag{9.35}$$

the relation

$$c_{\text{tot}} = D - 28 + c(h) = 0 . \tag{9.36}$$

To determine $c(h)$, the $SL(2, \mathbf{R})$ current algebra generated by h_{++} with the currents j^a of eq. (9.14) is crucial. The SS-constructed stress tensor associated with these currents takes the form (sect. IV.2)

$$T^{\text{SS}}_{++} = \frac{1}{\kappa + 2} \eta_{ab} : j^a_+ j^b_+ := \frac{1}{2(\kappa + 2)} \left[(\partial_- h_{++})^2 - 2h_{++} \partial_-^2 h_{++} \right] . \tag{9.37}$$

Eqs. (9.33) and (9.34) now imply

$$T_{++}(h) = T^{\text{SS}}_{++} + \partial_+ j^{(0)}_+ . \tag{9.38}$$

It follows

$$c(h) = \frac{3\kappa}{\kappa + 2} - 6\kappa = \frac{3k + 12}{k + 2} + 6(k + 4) . \tag{9.39}$$

The vanishing condition for the total central charge in eq. (9.36) then takes the form

$$D - 28 + \frac{3\kappa}{\kappa + 2} - 6\kappa = 0 , \tag{9.40a}$$

or, equivalently,

$$D = 1 - 6k - \frac{6}{k + 2} . \tag{9.40b}$$

According to this equation, the value of the $SL(2, \mathbf{R})$ AKM 'level' k is obviously different from its naive value $k_0 = (26 - D)/6$. The coincidence is recovered in the weak coupling limit $D \to \infty$ only, where the renormalization effects due to interaction with 2d gravity can be neglected. Eq. (9.40) has the universal form indeed, which makes it applicable to any massless conformal matter of Virasoro central charge D.

Eq. (9.40b) is a quadratic equation w.r.t. k, and it can be solved as

$$k = -\frac{11 + D + \sqrt{(1 - D)(25 - D)}}{12} \, , \tag{9.40c}$$

where the positive value of the square root has been chosen in order to make it compatible with the semiclassical value of k, $k \sim -D/6$. The real solutions for k clearly exist only in the 'weak gravity' regime $D \leq 1$, or when $D \geq 25$. In particular, this applies to the minimal models. A much more physically interesting region is $D < 1$, which presumably corresponds to a 'strong gravity'. At $D > 1$, eq. (9.40) leads to complex values for the 'level' k or κ, which implies, in general, complex values for the critical exponents also (see below). A kind of a phase transition should therefore be expected near $D = 1$ [403].

One of the critical exponents $\gamma(h)$ determines behaviour of the partition function in the large-area limit, and it is called the '*string susceptibility*'. In the case of genus zero, the $\gamma(0) \equiv \gamma$ is defined by

$$Z(A) \propto A^{\gamma - 3} \, , \tag{9.41}$$

where A is the total area of 2d surface. The intuitive physical meaning of the susceptibility as a specifically gravitational parameter makes it natural to identify γ with the solution $-\lambda$ of the KPZ equation (9.21) for $\lambda_0 = 0$, i.e. [403]

$$\gamma = -k - 1 = \frac{D - 1 - \sqrt{(1 - D)(25 - D)}}{12} \, , \tag{9.42}$$

where eq. (9.40) has been used. In the semiclassical limit $D \to -\infty$, eq. (9.42) yields (exercise # IX-2)

$$\gamma \to \frac{D - 19}{6} + 2 \, , \tag{9.43}$$

in exact agreement with the independently derived estimate [407]. For some specific values of D, namely, $d = -2, 0, 1/2$, eq. (9.43) yields $\gamma = -1, -\frac{1}{2}, -\frac{1}{3}$, resp., again in agreement with the known exact solutions of the discretized models of 2d gravity, where the 2d metric is described by random triangulations (Ch. X).

The KPZ solution to quantum 2d gravity described above is only applicable for the sphere (of genus $h = 0$) since it has been derived in the 'light-cone'-like gauges. It is, nevertheless, possible to rederive all the results in the conventional conformal gauge, which is also appropriate for generalizing them to all genera [404, 405]. The good starting point is a Euclidean bosonic string partition function in the form

$$Z = \int \frac{[dg]_g [dX]_g}{\mathrm{Vol(Diff)}} \exp\left[-S_{\mathrm{P}}(X;g) - (\mu_0^2/2\pi) \int_\Sigma d^2\sigma \sqrt{g} \right] , \qquad (9.44)$$

where the (Euclidean) Polyakov string action S_{P} and a bare cosmological term have been introduced ($\alpha' = 1/2$). The integration measure in eq. (9.44) is naturally defined w.r.t. the functional norms

$$\|\delta X\|_g^2 = \int d^2\sigma \sqrt{g}\, \delta X \cdot \delta X , \qquad (9.45a)$$

and

$$\|\delta g\|_g^2 = \int d^2\sigma \sqrt{g}\, (g^{\alpha\gamma} g^{\beta\delta} + const.\, g^{\alpha\beta} g^{\gamma\delta}) \delta g_{\alpha\beta} \delta g_{\gamma\delta} , \qquad (9.45b)$$

which are both invariant under reparametrizations, but are *dependent* on the metric g on a Riemann surface Σ. The action S_{P} is known to be invariant (sect. VIII.1) under the local Weyl transformations of the metric, $g_{\alpha\beta} \to e^\sigma g_{\alpha\beta}$, while the measure in eq. (9.44) is *not*. This is just another manifestation of the conformal anomaly (9.1) and (9.2) in the conformal gauge [408, 409], which now takes the form of the local *Liouville* action [37],

$$[dX]_{e^\sigma g} = \exp\left[\frac{D}{48\pi} S_{\mathrm{L}}(\sigma;g) \right] [dX]_g , \qquad (9.46)$$

where

$$S_{\mathrm{L}}(\sigma;g) = \int d^2\sigma \sqrt{g} \left(\tfrac{1}{2} g^{\alpha\beta} \partial_\alpha \sigma \partial_\beta \sigma + R^{(2)} \sigma + \mu^2 e^\sigma \right) . \qquad (9.47)$$

An arbitrary metric g on a Riemann surface can be parametrized by a conformal factor ϕ_0, a diffeomorphism vector field v_α and moduli $\{\tau\}$ (sect. XII.4) as [5]

$$g = e^{\phi_0} \exp\{v\} \hat{g}(\tau) , \qquad (9.48)$$

where a reference (only moduli-dependent) metric $\hat{g}(\tau)$ has been introduced. Fixing the reparametrizational invariance by the conformal gauge determines the measure in terms of the reparametrizational ghosts to be represented by the conventional (b, c) ghost system of central charge -26. This adds the factor $[db]_g [dc]_g \exp[-S_{\mathrm{gh}}(b,c;g)]$. The ghost action S_{gh} is invariant w.r.t. the Weyl transformations, while the measure is not,

$$[db]_{e^\sigma g} [dc]_{e^\sigma g} = \exp\left[(-26/48\pi) S_{\mathrm{L}}(\sigma;g) \right] [db]_g [dc]_g . \qquad (9.49)$$

---

[5] A finite diffeomorphism can be formally represented by the 'exponentiation' of an infinitesimal general coordinate transformation.

In the critical dimension $D = 26$, the conformal anomaly cancels, while the remaining dependence of the measure upon the Liouville mode (the conformal factor) is usually removed from the partition function by its proper normalization. For a non-critical string theory defined in dimensions $D \neq 26$, the conformal mode has non-trivial dymanics to be represented by the Liouville action (see sect. 2 for more). The Liouville action can therefore be considered as the induced action of the quantum 2d gravity in the conformal gauge. It is the important question to define a path integral measure for this theory.

The naive measure (9.45a) defined w.r.t. the metric g is clearly inappropriate for integrating over ϕ_0 since the g itself is dependent on ϕ_0. The tractable and practically useful definition is provided by the norm

$$||\delta\phi||_{\hat{g}}^2 = \int d^2\sigma \sqrt{\hat{g}} \, (\delta\phi)^2 \ , \tag{9.50}$$

w.r.t. the reference metric $\hat{g}$. This implies, in fact, that all factors in the string partition function measure are to be defined w.r.t. this reference metric, *viz.*

$$[d\phi_0]_g[db]_g[dc]_g[dX]_g = [d\phi]_{\hat{g}}[db]_{\hat{g}}[dc]_{\hat{g}}[dX]_{\hat{g}} J(\phi, \hat{g}) \ , \tag{9.51}$$

where the associated Jacobian $J(\phi, g)$ has been introduced. It is natural to assume that the Jacobian $J(\phi, g)$ can be represented as an exponential of a renormalizable local action to be invariant under the diffeomorphisms w.r.t. the metric $\hat{g}$ (the so-called '*ultra-locality principle*' [81]),

$$J(\phi, \hat{g}) = \exp\left[-S(\phi, \hat{g})\right] \ , \tag{9.52}$$

namely

$$
\begin{aligned}
S(\phi, \hat{g}) &= \frac{1}{8\pi} \int d^2\sigma \sqrt{\hat{g}} \left[\hat{g}^{\alpha\beta}\partial_\alpha\phi\partial_\beta\phi + Q\hat{R}^{(2)}\phi + 4\mu_1^2 e^{\alpha\phi}\right] \ , \\
&= \frac{1}{2\pi} \int d^2 z \left[\partial\phi\bar{\partial}\phi + \frac{1}{4}Q\sqrt{\hat{g}}\hat{R}^{(2)}\phi + \mu_1^2\sqrt{\hat{g}}e^{\alpha\phi}\right] \ ,
\end{aligned}
\tag{9.53}
$$

where the Liouville mode has been rescaled in order to get the standard normalization for the kinetic term. The renormalization coefficients Q, α and μ_1 are to be determined from consistency of the theory (w.r.t. general covariance!) or, equivalently, from the invariance of the theory under the shifts (renormalization group!) [404, 405]

$$\hat{g} \to e^\sigma \hat{g} \ , \quad \phi \to \phi - \sigma/\alpha \ . \tag{9.54}$$

which is equivalent to demanding the metric $g = e^{\alpha\phi}\hat{g}$ invariant. Since a bare parameter μ_0 is at our disposal, it can be chosen to cancel μ_1. The Liouville theory then effectively becomes a free field theory, which makes exact calculations possible. The consistency implies, in particular, the vanishing total central charge, $c_{\rm tot} = c_\phi + D - 26 = 0$, where $c_\phi = 1 + 3Q^2$ because of eqs. (3.62) and (9.53). Hence, we get

$$Q = \sqrt{(25 - D)/3} \, . \tag{9.55}$$

where the choice of sign in front of the square root is not relevant. The consistent value of α follows, when demanding the operator $: e^{\alpha\phi} :$ be a conformal (marginal) tensor of weight $(1, 1)$,

$$\Delta_0(: e^{\alpha\phi} :) = 1 \, , \tag{9.56}$$

which is equivalent to the symmetry (9.54). Eqs. (3.62) and (9.56) now imply the relation $-\frac{1}{2}\alpha(\alpha - Q) = 1$ and, because of eq. (9.55),

$$\alpha = \tfrac{1}{2}Q - \tfrac{1}{2}\sqrt{Q^2 - 8} = -\frac{1}{2\sqrt{3}}\left(\sqrt{25 - D} - \sqrt{1 - D}\right) \, , \tag{9.57}$$

where the negative root has been chosen to get agreement with the semiclassical result (see below). In particular, the *Seiberg bound* $\alpha \leq \frac{1}{2}Q$ is automatically fulfiled, which is, in fact, the necessary condition for an existence of the relevant operator, as was argued in refs. [412, 413] from various viewpoints. When $D > 25$, both Q and α become imaginary. Since the metric has to be real, the field ϕ must be pure imaginary, which, however, implies the non-physical sign in front of the kinetic term in the action (9.53). When $D = 25$, the Liouville mode can be interpreted as a time-like coordinate of the embedding space. In particular, the 'non-critical' bosonic string in 25 Euclidean dimensions is equivalent to the 'critical' bosonic string embedded in $(25 + 1)$-dimensional Minkowski space-time.

To calculate the string susceptibility γ, consider the string partition function at a given area A of Riemann surface, [404, 405]

$$Z(A) = \int [d\phi][dX] e^{-S} \delta\left(\int d^2\sigma \sqrt{\hat{g}}\, e^{\alpha\phi} - A\right) \, . \tag{9.58}$$

The measure was defined to be invariant under constant shifts of the integration variable ϕ, $\phi \to \phi + \rho/\alpha$, so that the scaling law of the partition function $Z(A)$ is entirely determined by the scaling behaviour of the integrand. We have

$$S \to S + Q(1 - h)\rho/\alpha \, , \tag{9.59a}$$

and

$$\delta\left(\int d^2\sigma \sqrt{\hat{g}}\, e^{\alpha\phi} - A\right) \to e^{-\rho} \delta\left(\int d^2\sigma \sqrt{\hat{g}}\, e^{\alpha\phi} - e^{-\rho}A\right) \, . \tag{9.59b}$$

It follows

$$Z(A) = e^{[Q(h-1)/\alpha-1]\rho} Z(e^{-\rho} A) ,\qquad (9.60a)$$

or

$$Z(A) \propto A^{(h-1)Q/\alpha-1} ,\qquad (9.60b)$$

which imply [404, 405]

$$\gamma(h) = \frac{(1-h)}{12}\left[D - 25 - \sqrt{(25 - D)(1 - D)}\right] + 2 ,\qquad (9.61)$$

in agreement with the KPZ equation (9.42) for $h = 0$. In the semiclassical approximation, one finds (exercise # IX-2)

$$\gamma(h) \stackrel{D \to -\infty}{\Longrightarrow} (1 - h)\frac{D - 19}{6} + 2 ,\qquad (9.62)$$

again in agreement with eq. (9.43) for $h = 0$. Comparing eqs. (9.40) and (9.42) with eqs. (9.57) and (9.61) yields the relation between α and κ as follows:

$$\kappa + 2 = -k - 2 = -\frac{2}{\alpha^2} .\qquad (9.63)$$

Given a scalar field Φ of spin zero and conformal dimension (without 2d gravity) Δ_0, its gravitational 'dressing' (i.e. the wave function renormalization) $\Phi \to \Phi e^{\beta\phi}$ in the presence of quantum 2d gravity is fixed by demanding the dressed operator to be 'marginal', i.e. to have conformal dimension $(1, 1)$. [6] This gives the quadratic equation $\Delta_0 - \frac{1}{2}\beta(\beta - Q) = 1$, whose solution is

$$\beta_\pm = -\frac{1}{2\sqrt{3}}\left(\sqrt{25 - D} \mp \sqrt{1 - D + 24\Delta_0}\right) .\qquad (9.64)$$

The gravitational scaling dimension $\Delta = \Delta(\Phi)$ is determined from the expectation value of the 1-point function

$$F_\Phi(A) = Z^{-1}(A) \int [d\phi][dX] e^{-S} \delta\left(\int d^2\sigma\sqrt{\hat{g}}\, e^{\alpha\phi} - A\right) \int d^2\sigma'\sqrt{\hat{g}}\, \Phi e^{\beta\phi} \propto A^{1-\Delta} .\quad (9.65)$$

The scaling behaviour of eq. (9.65) w.r.t. the constant shifts $\phi \to \phi + \rho/\alpha$ yields [404, 405]

$$\Delta_\pm = -\frac{\beta_\pm}{\alpha} = \frac{\pm\sqrt{1 - D + 24\Delta_0} - \sqrt{1 - D}}{\sqrt{25 - D} - \sqrt{1 - D}} ,\qquad (9.66a)$$

---

[6] Only 'marginal' operators can be integrated over a Riemann surface.

or

$$\Delta - \Delta_0 = -\tfrac{1}{2}\alpha^2 \Delta(\Delta - 1) \, , \qquad (9.66b)$$

which is just the KPZ formula (9.21) ! As the simplest application of this formula, consider the Ising model with a fluctuating 2d gravity [403]. The Ising model (sect. II.7) has two Virasoro primary fields, an energy density $\varepsilon \propto \bar\psi_+\psi_-$ of dimension $\Delta_\varepsilon^{(0)} = \tfrac{1}{2}$ and a magnetization σ of dimension $\Delta_\sigma^{(0)} = \tfrac{1}{16}$. Eqs. (9.57) and (9.66) give the different values, namely

$$\Delta_\varepsilon = \frac{2}{3} \, , \quad \Delta_\sigma = \frac{1}{6} \, . \qquad (9.67)$$

In particular, the specific heat and spontaneous magnetization critical exponents, $\tilde\alpha$ and $\tilde\beta$, resp., are given by

$$\tilde\alpha \equiv \frac{1 - 2\Delta_\varepsilon}{1 - \Delta_\varepsilon} = -1 \, , \quad \tilde\beta \equiv \frac{\Delta_\sigma}{1 - \Delta_\varepsilon} = \tfrac{1}{2} \, , \qquad (9.68)$$

in exact agreement with those found in refs. [410, 411] from the matrix models (Ch .X).

Other CFT minimal models in the presence of quantum 2d gravity can be examined in the similar way. The gravitational dressing is described by eq. (9.66), which converts the Kač table (2.62) of the degenerate Virasoro dimensions $\Delta_{(n,m)}^{(0)}$ into the different table [403]

$$\Delta_{(n,m)} == -\frac{1+k}{2} \pm \tfrac{1}{2}[(k+2)n - m] \, , \qquad (9.69)$$

corresponding to the degenerate representations of the $SL(2, \mathbf{R})$ AKM algebra. Since quantum 2d gravity is still described by CFT methods, the DF constructon of correlation functions (sect. II.6) and the bosonization rules (sect. III.3) apply, now in terms of the dressed parameters specified above (exercise # IX-3). After substituting the central charge $D = c(p, q)$ from eq. (2.62) into eq. (9.42), one gets the critical exponent of string susceptibility for the case of the (p, q) minimal model in the following form:

$$\gamma_{(p,q)} = \frac{-2}{p + q - 1} \, . \qquad (9.70a)$$

In particular, for the unitary minimal models with $q = p + 1$, one finds

$$\gamma_{(p,p+1)} = -\frac{1}{p} \, . \qquad (9.70b)$$

Similarly, subsituting eq. (2.62) into eq. (9.55) yields

$$Q(p, q) = \sqrt{\frac{2p}{q}} + \sqrt{\frac{2q}{p}} \, . \qquad (9.70c)$$

A generalization of all these results to the case of SCFT's and quantum 2d super-gravity is straightforward [405]. The SCFT matter fields to be represented by $\hat{D}$ scalar superfields contribute $c_m = \frac{3}{2}\hat{D}$ to central charge, while the reparametrizational and superconformal ghosts contribute -26 and $+11$, resp., $c_{gh} = -15$. The Liouville field ϕ has to be replaced by a superfield $\hat{\phi}$ of central charge $c_{\hat{\phi}} = 1 + 3Q^2 + 1/2$, where $1/2$ appears due to a fermionic superpartner of ϕ. Requiring the total central charge to vanish, $c_{tot} = c_{\hat{\phi}} + c_m + c_{gh} = 0$, yields the relation

$$Q = \sqrt{\frac{9 - \hat{D}}{2}} \ . \tag{9.71}$$

The super-Liouville analogue of the renormalized α-coefficient is determined by demanding the superfield operator $: e^{\alpha\hat{\phi}} :$ be a super-density, i.e. of dimension $(\frac{1}{2}, \frac{1}{2})$. This leads to a quadratic equation of the form $-\frac{1}{2}\alpha(\alpha - Q) = \frac{1}{2}$, whose appropriate solution is

$$\alpha = -\frac{1}{2\sqrt{2}} \left(\sqrt{9 - \hat{D}} - \sqrt{1 - \hat{D}} \right) \ . \tag{9.72}$$

The superstring susceptibility γ_s is determined from the scaling behaviour of the super-symmetric partition function as a function of super-area $\hat{A}$,

$$Z(\hat{A}) = \int [d\hat{\phi}][d\hat{X}] e^{-S} \delta \left(\int d^2\sigma d^2\theta \hat{E} e^{\alpha\hat{\phi}} - \hat{A} \right) \propto \hat{A}^{\gamma_s - 3} \ . \tag{9.73}$$

where the superdeterminant $\hat{E}$ of the reference super-zweibein has been introduced. The scaling argument, similar to the one used above in the bosonic case, gives the equation [405]

$$\gamma_s(h) = (h - 1)Q/\alpha + 2 = \frac{1}{4}(1 - h)\left[\hat{D} - 9 - \sqrt{(9 - \hat{D})(1 - \hat{D})} \right] + 2 \ . \tag{9.74}$$

For a spinless superconformal primary field $\hat{\Phi}$ of dimension Δ_0, the corresponding supergravitationally-dressed operator $: \hat{\Phi} e^{\beta\hat{\phi}} :$ should also be a super-density. This gives the equation $\Delta_0 - \frac{1}{2}\beta(\beta - Q) = \frac{1}{2}$, whose solution is [405]

$$\beta_\pm = -\frac{1}{2\sqrt{2}}\left[\sqrt{9 - \hat{D}} \mp \sqrt{1 - \hat{D} + 16\Delta_0} \right] \ . \tag{9.75}$$

The scaling dimension $\hat{\Delta}$ in the presence of 2d supergravity is defined by the super-analogoue of eq. (9.65): $F_{\hat{\Phi}}(\hat{A}) \propto \hat{A}^{1-\Delta}$. Examining its scaling behaviour under a constant shift of the Liouville superfield, one finds [405]

$$\Delta_\pm = 1 - \frac{\beta_\pm}{\alpha} = \frac{\pm\sqrt{1 - \hat{D} + 16\Delta_0} - \sqrt{1 - \hat{D}}}{\sqrt{9 - \hat{D}} - \sqrt{1 - \hat{D}}} \ . \tag{9.76}$$

Exercises

IX-1 ▷ (*) Prove that the operators T_{++} and θ introduced in eq. (9.33) are the generators of the residual gauge symmetry transformations

$$x^+ \rightarrow a(x^+) , \quad x^- \rightarrow x^-[da/dx^+]^{-1} + b(x^+) , \tag{9.77}$$

which preserve the Polyakov gauge (9.4). *Hint*: consider the infinitesimal transformations with $a(x^+) = x^+ + \varepsilon(x^+)$, $b(x^+) = \eta(x^+)$.

IX-2 ▷ Check the semiclassical corollaries (9.43) and (9.62) from eqs. (9.42) and (9.61), resp.

IX-3 ▷ (*) Consider correlation functions of the dressed primary fields (densities) having the form $\Phi_i(z) =: e^{i\gamma_i \varphi} e^{\beta_i \phi} :$ and defined in terms of a single DF free boson φ (sect. II.6) on a sphere. Let the background charge take the values corresponding to the minimal CFT's. Derive the selection rules for the coefficients β_i and γ_i, which make the correlators non-vanishing. Determine the structure of the Feigin-Fuchs screening operators in terms of φ and ϕ.

IX-4 ▷ Derive the supersymmetric formulae in eqs. (9.71)–(9.76) along the lines of the bosonic case.

IX.2 Liouville theory

Any 2d local field theory coupled to 2d gravity with a metric $g_{\alpha\beta} = e^\phi \hat{g}_{\alpha\beta}$ has the classical conformal symmetry. Indeed, varying its action w.r.t. the conformal factor ϕ gives equation of motion, which immediately implies $T_\alpha{}^\alpha = 0$ or, equivalently, $T_{z\bar{z}} = 0$ in the conformal coordinates. In other words, when a generally covariant 2d local field theory is Weyl-invariant, it is CFT. When, in addition, the 2d conformal 'matter' is represented by an integrable model, it is possible to exactly calculate all its renormalized parameters provided the corresponding quantum field theory is conformally invariant. In the prevous section, the renormalized value of the cosmological constant has been set to zero. When it is not zero, one needs to solve the *quantum* Liouville theory. In this section, we briefly review the classical and quantum Liouville theory [412, 413, 414, 415, 416, 417, 418, 419, 420] [7] and discuss the associated 'quantum group' structure [428, 429, 430, 431].

---

[7]The solution to the quantum Liouville theory in the operator approach was also given in refs. [421, 422, 423, 424]. See refs. [425, 426, 427] for a review.

After choosing a metric $g_{\alpha\beta}$ on a Riemann surface Σ in the form

$$g_{\alpha\beta} = e^{\gamma\phi}\hat{g}_{\alpha\beta} \; , \tag{9.78}$$

with a constant γ and a reference metric $\hat{g}_{\alpha\beta}$, the action of the Liouville theory we consider is given by

$$S_{\rm L}[\phi] = \frac{1}{8\pi}\int d^2z\sqrt{\hat{g}}\left[\hat{g}^{\alpha\beta}\partial_\alpha\phi\partial_\beta\phi + Q\hat{R}^{(2)}\phi + \frac{\mu^2}{\gamma^2}e^{\gamma\phi}\right] \; . \tag{9.79}$$

Requiring conformal invariance of the *classical* Liouville action determines the classical background charge coefficient Q,

$$Q = 2/\gamma \; , \tag{9.80}$$

since the action (9.79) then becomes invariant under the Weyl transformations having the form (exercise # IX-5)

$$\hat{g} \to e^{2\sigma}\hat{g} \; , \quad \gamma\phi \to \gamma\phi - 2\sigma \; . \tag{9.81}$$

This means that the (improved) stress tensor of the classical Liouville theory in conformal coordinates takes the usual DF-type form (sects. II.6 and III.3),

$$T(z) \equiv T_{zz}(z) = -\tfrac{1}{2}(\partial\phi)^2 + \tfrac{1}{2}Q\partial^2\phi \; , \quad T_{z\bar{z}} = 0 \; , \tag{9.82}$$

where the equations of motion have been used. Under the conformal transfromations, $z \to w = f(z)$, the Liouville field transforms as

$$\phi \to \phi - \frac{1}{\gamma}\log\left|\frac{dw}{dz}\right|^2 \; , \tag{9.83}$$

since ϕ is not just an ordinary scalar but a component of the metric. [8] It follows

$$T(z) \to \left(\frac{dw}{dz}\right)^2 T(w) + \frac{1}{\gamma^2}S[w;z] \; , \tag{9.84}$$

where $S[w;z]$ is the Schwartzian derivative defined in eq. (1.83), with $c = 12/\gamma^2$.

The normalization of the Liouville field above is fixed by the kinetic term in eq. (9.79). After the field redefinition $\phi = \gamma^{-1}\varphi$, the constant γ^{-2} appears only in front of the Liouville action provided eq. (9.80). Therefore, it is convenient to identify

$$\gamma^2/4 = \hbar \equiv \frac{3}{25 - D} \; , \tag{9.85}$$

---

[8] The 'improvement' of the Noether stress tensor is due to the second term in the action (9.79), which is responsible for the non-trivial transfromation law in eq. (9.83).

where the vanishing total central charge condition, $c_L + D = 26$ with $c_L = 1 + 3Q^2$, and eq. (9.80) have been used. Since we want to canonically quantize the Liouville theory, let us go back from the complex coordinates $(z, \bar{z})$ to the cylindric coordinates (τ, σ), which are related by eq. (1.34) in Euclidean space, and take $x_\pm = \sigma \pm i\tau$. We are now left with the action [9]

$$S_L = \frac{1}{8\pi\gamma^2} \int d\tau d\sigma \left(\partial_\alpha \varphi \partial^\alpha \varphi + \mu^2 e^\varphi \right) . \tag{9.86}$$

The classical equation of motion takes the form of the *Liouville equation*

$$8\partial_+ \partial_- \varphi = \mu^2 e^\varphi, \tag{9.87}$$

whose general solution can be written as follows:

$$\varphi = \ln \left\{ \frac{16}{\mu^2} \frac{A'(x_+)B'(x_-)}{[A(x_+) - B(x_-)]^2} \right\} , \tag{9.88}$$

where A and B are two arbitrary functions, and prime means the differentiation w.r.t. a given argument. The Liouville field is supposed to be periodic, $\varphi(\tau, \sigma + 2\pi) = \varphi(\tau, \sigma)$. This does not, however, imply the functions A and B must be periodic. Moreover, they are not unique since the solution (9.88) has the obvious $SL(2, \mathbf{R})$ symmetry:

$$A \to TA \equiv \frac{aA + b}{cA + d} , \quad B \to TB \equiv \frac{aB + b}{cB + d} ; \quad T = \begin{pmatrix} a & b \\ c & d \end{pmatrix} \in SL(2, \mathbf{R}) . \tag{9.89}$$

This implies, in general, twisted boundary conditions of the form

$$A(x_+ + 2\pi) = \hat{T}A(x_+) , \quad B(x_- - 2\pi) = \hat{T}B(x_-) , \tag{9.90}$$

since they are consistent with those of φ. [10] There are three sectors to distinguish for the monodromy matrix $\hat{T}$: (i) $\left| \text{tr} \, \hat{T} \right| < 2$ (*elliptic*), (ii) $\left| \text{tr} \, \hat{T} \right| = 2$ (*parabolic*), and (iii) $\left| \text{tr} \, \hat{T} \right| > 2$ (*hyperbolic*). Their canonical forms are

$$\hat{T}_{\text{ell}} = \begin{pmatrix} e^{i\alpha} & 0 \\ 0 & e^{-i\alpha} \end{pmatrix} , \quad \hat{T}_{\text{par}} = \begin{pmatrix} 1 & \alpha \\ 0 & 1 \end{pmatrix} , \quad \hat{T}_{\text{hyp}} = \begin{pmatrix} \alpha & 0 \\ 0 & \alpha^{-1} \end{pmatrix} . \tag{9.91}$$

The fields A and B are not free fields w.r.t. the Poisson brackets to be defined from the Liouville theory (9.86) in a Hamiltonian form. The true action-angle variables

---

[9]Since we do not want to integrate over moduli, we set $\hat{g}_{\alpha\beta} = \delta_{\alpha\beta}$ locally.

[10]The case of trivial monodromy is excluded in the quantum Liouville theory. This is related to the fact that this theory does not have a translationally invariant ground state.

having free-field Poisson brackets can be found by the 'inverse scattering method' [432]. Let us define the new fields as [11]

$$\psi_{-1/2}(x_+) \equiv (A')^{-1/2}(x_+) \, ,$$

$$\psi_{+1/2}(x_+) \equiv A(A')^{-1/2}(x_+) = \psi_{-1/2}(x_+) \int^{x_+} \psi_{-1/2}^{-2}(x'_+)dx'_+ \, . \tag{9.92}$$

The fields $\psi_i(x_+)$ satisfy a 'Schrödinger equation' (or a 'Lax pair') of the form

$$\psi_i'' + \frac{\gamma^2}{2}T(x_+)\psi_i = 0 \, , \quad i = -1/2, +1/2 \, , \tag{9.93}$$

where $T(x_+)$ entirely comes from the Schwartzian derivative, $T(x_+) = \frac{1}{\gamma^2}S[A; x_+]$ (see also exercise # IX-7).

The fields ψ_i transform in the linear spin-$J = 1/2$ representation of $SL(2, \mathbf{R})$, and they are clearly 'chiral', i.e. have a single argument. The general solution to the Liouville equation can now be represented in another form, namely

$$e^{-\frac{1}{2}\varphi} = const. \left[\psi_{-1/2}(x_+)\bar{\psi}_{+1/2}(x_-) - \psi_{+1/2}(x_+)\bar{\psi}_{-1/2}(x_-) \right] \, , \tag{9.94}$$

under the assumption that *arbitrary* chiral functions on the r.h.s. of this equation are so normalized that their Wronskians $(\psi_i'\psi_{i\pm1} - \psi_i\psi_{i\pm1}')$ are all equal to unity [428, 429, 430]. In the classical Liouville theory, arbitrary powers $e^{N\varphi}$ can be simply written down in terms of the chiral fields by the use of eq. (9.94). When the power N is positive, the decomposition becomes infinite. When $2N = -2J$ is a *negative integer*, the number of terms in the decomposition is clearly finite, *viz.*

$$e^{-J\varphi} = const. \sum_{M=-J}^{M=J} (-)^{J-M} \psi_M^{(J)}(x_+)\bar{\psi}_{-M}^{(J)}(x_-) \, , \tag{9.95}$$

where $\psi_M^{(J)}$ transform in the spin-J representation of $SL(2, \mathbf{R})$,

$$\psi_M^{(J)}(x_+) \equiv \sqrt{\binom{2J}{J+M}} \, \psi_{-1/2}^{J-M}(x_+)\psi_{+1/2}^{J+M}(x_+) \, ,$$

$$J_\pm \psi_M^{(J)} = \sqrt{(J \mp M)(J \pm M + 1)}\psi_{M\pm1}^{(J)} \, , \quad J_0\psi_M^{(J)} = M\psi_M^{(J)} \, . \tag{9.96}$$

---

[11] The undergoing construction is supposed to be repeated for the x_--dependent fields $\bar{\psi}_{\pm1/2}$, B and $\bar{T}$, resp.

The fields ψ_i are still not free fields, but they can be 'bosonized' as

$$\psi_i(x_+) \propto \exp\left[-\frac{2}{\gamma}X_i(x_+)\right] \ , \quad P_i(x_+) \equiv X_i'(x_+) \ , \tag{9.97}$$

in terms of some other fields X_i having (with their conjugated 'momenta' P_i) free-field Poisson brackets. This can be simply understood by noticing that, in terms of the free field X, the stress tensor takes the free form (9.82), namely, $T(x_+) = -\frac{1}{2}(X')^2 + \frac{1}{\gamma}X''$. The Euclidean momentum $P_i(x_+)$ can be thought of as a solution to the associated (non-linear) Ricatti equation

$$-\frac{1}{2}P_i^2 + \frac{1}{\gamma}P_i' = T \ . \tag{9.98}$$

From the Hamiltonian viewpoint, the (non-local and non-linear) field transformation from the interacting Liouville field and its conjugated momentum to the set of free fields defined above is a *canonical* (Miura) transformation. The Hamiltonian of the Liouville theory in terms of the fields X and P is a free Hamiltonian. The two sets of free 'phase-space' variables are not independent, but are related by the canonical transformation, so that one of them can be solved in terms of another [412]. The underlying $SL(2,\mathbf{R})$ symmetry is only explicit when using the both sets. The dynamical information about classical interaction is actually encoded in the quite non-trivial Poisson brackets between the fields belonging to different canonical sets. A transformation from the Liouville field to a *single* free field is just a particular example of the so-called *Bäcklund* transformations [433] known in the theory of integrable models, see refs. [421, 422, 423, 424] as far as the Bäcklund transformation in the Liouville theory is concerned.

Quantization of the Liouville theory is performed by imposing the canonical commutation relations on the free fields, which are now considered to be defined on a unit cirle parametrized by σ. Assuming, for simplicity, that a monodromy matrix is diagonalizable, the chiral free fields can be made periodic up to a multiplicative constant,

$$X_j(\sigma) = q_0^{(j)} + p_0^{(j)}\sigma + i\sum_{n\neq 0}\frac{e^{-in\sigma}}{n}p_n^{(j)} \ , \tag{9.99}$$

where p_0 is real in the hyperbolic sector, and is imaginary in the elliptic one. The canonical rule $\{\ ,\ \}_{\text{P.B.}} \to -i[\ ,\ ]$ now implies the commutation relations

$$[p_n^{(j)}, p_m^{(j)}] = n(\gamma^2/4)\delta_{n+m} \ , \quad [q_0^{(j)}, p_0^{(j)}] = i \ . \tag{9.100}$$

Renormalization of the Liouville theory has to be consistent with conformal invariance. It should result in a 'deformation' of the classical CFT introduced above into a quantum CFT. Eq. (9.98) can be used to define the normally ordered [12] stress tensor T and the *new* coupling constant γ of the quantized theory. To avoid a confusion, we now use $\hbar$ instead of the classical γ, and introduce, in addition, the 'quantum deformation' parameter η to be defined by $\gamma \equiv \gamma_{qu} = \eta\gamma_{cl}$. [13] According to the previous section, the quantization of the theory (9.79) will be consistent with the conformal invariance provided the KPZ condition $-\frac{1}{2}\gamma(\gamma - Q) = 1$ or, equivalently, $Q = \gamma + 2/\gamma$ holds, where Q is the classical background charge from eq. (9.80), i.e. $Q = \hbar^{-1/2}$. This determines the 'quantum deformation' parameter as a solution to the equation

$$2\hbar\eta^2 - \eta + 1 = 0 \ . \tag{9.101}$$

The chiral field in eq. (9.97) is replaced after quantization by *two* (because of the two solutions to the quadratic equation) normally ordered chiral operators,

$$\hat{\psi}_i^{(1,2)}(x_+) \propto\ :\exp\left[-Q^{(1,2)}X_i(x_+)\right]\ :\ , \tag{9.102}$$

where Q is the quantum background charge, $Q = 2/\gamma$. These fields are determined as solutions to the quantum analogue of the Schrödinger equation (9.93), which is nothing but the BPZ equation ensuring the decoupling of Virasoro null states (sect. II.4). By construction, products of operators (9.102) generate families of the (normally ordered) chiral operators $\hat{\psi}_M^{(J)}$ which are the true Virasoro primaries. The 'Hilbert' space $\mathcal{H}$, where these operators act, is a direct sum of the Fock spaces $\mathcal{F}(\tilde{\omega})$ built upon the Virasoro highest-weight states $|\tilde{\omega}, 0\rangle$ satisfying the usual conditions in terms of the Fourier modes L_n of the stress tensor T, $L_n |\tilde{\omega}, 0\rangle = 0$, $n > 0$; $(L_0 - \Delta(\tilde{\omega})) |\tilde{\omega}, 0\rangle = 0$. Here, $\tilde{\omega}$ denotes the so-called *quasi-momentum*, which can be restricted to discrete values [428, 429],

$$\tilde{\omega} \equiv \frac{2ip_0^{(1)}}{Q} \ , \quad \Delta(\tilde{\omega}) = \frac{1}{8\hbar} + \tfrac{1}{2}(p_0^{(1)})^2 \ , \tag{9.103}$$

where $\frac{1}{8\hbar}$ results from the normal ordering of T. Eqs. (9.85) and (9.101) imply

$$\eta = \frac{1}{12}\left(25 - D \pm \sqrt{(25 - D)(1 - D)}\right) \ , \tag{9.104}$$

---

[12] The normal ordering here is supposed to be defined w.r.t. a single set of free fields, which makes it necessary to distinguish between the two normal orderings. We are not going to explicitly indicate this distinction in our formulae, unless this leads to a confusion.

[13] The appearance of η can also be understood due to miltiplicative renormalization of a free field wave function [426].

and

$$Q \equiv \sqrt{\frac{2\hat{h}}{\pi}} \ , \quad \hat{h} = \frac{\pi}{12}\left(c_{\mathrm{L}} - 13 \pm \sqrt{(c_{\mathrm{L}} - 25)(c_{\mathrm{L}} - 1)}\right) , \tag{9.105}$$

where the new parameter $\hat{h}$ has been introduced, for a later use. We are therefore restricted to the values in $D \leq 1$ and $c_{\mathrm{L}} \geq 25$, which correspond to the weak coupling with 2d gravity.

The highest weight $\Delta(\tilde{\omega})$ can be rewritten as

$$\Delta(\tilde{\omega}) = \frac{\hat{h}}{4\pi}\left(1 + \frac{\pi}{\hat{h}}\right)^2 - \frac{\hat{h}}{4\pi}\tilde{\omega}^2 . \tag{9.106}$$

Similarly to the classical (elliptic) case, the operators $\hat{\psi}_m^{(J)}$ diagonalize the monodromy matrix, namely

$$\hat{\psi}_m^{(J)}(\sigma + 2\pi) = e^{2i\hat{h}m\tilde{\omega}} e^{2i\hat{h}m^2} \hat{\psi}_m^{(J)}(\sigma) . \tag{9.107}$$

The operators of 'powers of metric', i.e. of the type $e^{-J\gamma\phi}$, are known to play the very special role in quantum 2d gravity, providing gravitational 'dressing' for physical operators. This explains the need to construct such operators in the quantum Liouville theory. According to the previous section, we have

$$\Delta(e^{\alpha\phi}) = -\tfrac{1}{2}\alpha^2 + \tfrac{1}{2}\alpha Q = -\tfrac{1}{2}\left(\alpha - \frac{Q}{2}\right)^2 + \frac{c_{\mathrm{L}} - 1}{24} . \tag{9.108}$$

Hence, the classical dimension $-J$ of the field $e^{-J\gamma\phi}$ is renormalized as

$$\Delta(e^{-J\gamma\phi}) = \frac{J(J+1)}{k+2} - J = -\frac{\hat{h}J(J+1)}{\pi} - J , \tag{9.109}$$

where eqs. (9.40), (9.63) and (9.105) have been used.

After being quantized, the fields ψ become non-commutative operators $\hat{\psi}$, which, under the condition (9.101), satisfy a closed algebra (braiding rules) of the form

$$\hat{\psi}_i(\sigma)\hat{\psi}_j(\sigma') = R_{ij}^{21}(\tilde{\omega}, \sigma - \sigma')\hat{\psi}_j(\sigma')\hat{\psi}_i(\sigma) + R_{ij}^{12}(\tilde{\omega}, \sigma - \sigma')\hat{\psi}_i(\sigma')\hat{\psi}_j(\sigma) . \tag{9.110}$$

When inserted into the correlation functions, eq. (9.110) can be considered as the consequence of the BPZ equation. [14] The quantum operators representing 'powers of metric' would then have the general form

$$e^{-J\gamma\phi(\sigma)} \propto \sum_{m=-J}^{m=J} (-)^{J-m} C_m^{(J)}(\tilde{\omega}) \hat{\psi}_m^{(J)}(\sigma) \hat{\tilde{\psi}}_{-m}^{(J)}(\sigma) , \tag{9.111}$$

---

[14]The Virasoro-Ward identities are, in fact, equivalent to the operator identity (9.110) due to the Wigner-Eckart theorem, as was shown, e.g., in ref. [417], where the explicit form of the R-matrix was found too.

whose coefficients (C) are yet to be determined. The additional requirement, which is quite natural to impose in order to fix the coefficients, is the *locality* of operators on the l.h.s. of eq. (9.111), which just means their commutativity at equal time. [15] Though the operators $\hat{\psi}_M^{(J)}(\sigma)$ are closed w.r.t. the fusion (OPE) and braiding, the coefficients of their R-matrix and their fusion rules are still non-trivially dependent on $\bar{\omega}$, which questions the locality. The basic idea of refs. [428, 429, 430] is to introduce the underlying '*quantum group*' structure, in order to put the braiding and fusion properties of the chiral operators into the universal form dictated by 'quantum groups' (sect. XII.6). This can be made explicit after replacing the $\hat{\psi}_M^{(J)}(\sigma)$ fields (defining a 'DF-type' basis) by their linear combinations $\xi_M^{(J)}(\sigma)$ (defining a 'q-type' basis), in such a way that the braiding and fusion coefficients in the second case would *not* be dependent on $\bar{\omega}$. This can actually be done when the fields $\xi_M^{(J)}(\sigma)$ transform in the irreducible representations of the 'quantum group', which is $U_q(sl(2))$ of $q = \exp[2i\hat{h}] = \exp[\frac{2\pi i}{\kappa+2}]$ in this case, [16] so that the operators $e^{-J\gamma\phi(\sigma)}$ are the 'quantum group' invariants. It seems to be quite natural from the viewpoint of the correspondence principle that the classical $SL(2,\mathbf{R})$ structure is to be replaced by its 'quantum group' counterpart after quantization of the Liouville theory.

Following refs. [428, 429, 430, 431], let us define the new fields as

$$\xi_M^{(J)}(\sigma) = \sum_{m=-J}^{m=J} |J, \bar{\omega})_M^m \, \hat{\psi}_M^{(J)}(\sigma) \,, \quad J \le M \le J \,,$$

$$|J, \bar{\omega})_M^m = \sqrt{\binom{2J}{J+M}} \, e^{i\hat{h}m/2} \sum_t e^{i\hat{h}t(\bar{\omega}+m)}$$

$$\times \binom{J-M}{(J-M+m-t)/2} \binom{J+M}{(J+M+m+t)/2} \,, \tag{9.112}$$

where the t-sum goes over integer values of $\frac{1}{2}(J-M+m-t)$, and the 'quantum'-deformed binomial coefficients and factorials are used, *viz.*

$$\binom{P}{Q} \equiv \frac{\lfloor P \rfloor!}{\lfloor Q \rfloor! \lfloor P-Q \rfloor!} \,, \quad \lfloor n \rfloor! \equiv \prod_{r=1}^{n} \lfloor r \rfloor \,, \quad \lfloor r \rfloor \equiv \frac{\sin(\hat{h}r)}{\sin \hat{h}} \,. \tag{9.113}$$

---

[15] The locality becomes the usual micro-causality, when the theory is formulated in Minkowski space-time [426].

[16] See sect. XII.6 for more about 'quantum groups' and CFT.

In terms of the ξ-fields, the proper definition of the quantum 'powers of metric' is given by [428, 429, 430, 431]

$$e^{-J\gamma\phi(\sigma)} = const. \sum_{M=-J}^{M=J} (-)^{J-M} e^{i\hbar(J-M)} \xi_M^{(J)}(\sigma) \bar{\xi}_{-M}^{(J)}(\sigma) . \tag{9.114}$$

They will be invariant under a 'quantum group' action, when the ξ-fields belong to the irreps of the $U_q(sl(2))$ algebra

$$[J_+, J_-] = \lfloor 2J_3 \rfloor , \quad [J_3, J_\pm] = \pm J_\pm , \tag{9.115}$$

with the co-product generators to be defined as [17]

$$\mathbf{J}_\pm = J_\pm \otimes e^{i\hbar J_3} + e^{-i\hbar J_3} \otimes J_\pm , \quad \mathbf{J}_3 = J_3 \otimes 1 + 1 \otimes J_3 . \tag{9.116}$$

It follows

$$J_3 \xi_M^{(J)} = M \xi_M^{(J)} , \quad J_\pm \xi_M^{(J)} = \sqrt{\lfloor J \mp M \rfloor \lfloor J \pm M + 1 \rfloor} \psi_{M\pm 1}^{(J)} . \tag{9.117}$$

The $\bar{\xi}$-fields are supposed to be introduced and transformed in the similar way. The linear transformation (9.112) can be explicitly inverted as follows:

$$\hat{\psi}_M^{(J)}(\sigma) = \sum_{M=-J}^{M=J} (J, \tilde{\omega}|_m^M \, \xi_M^{(J)}(\sigma) ,$$

$$(J, \tilde{\omega}|_m^M = (2i \sin \hbar \, e^{i\hbar/2})^{2J} (-)^{J+M} e^{i\hbar(J+M)} |J, \tilde{\omega})_{-M}^{-m} \frac{\lfloor \tilde{\omega} - 2m \rfloor (-1)^{J+m}}{E_m^{(J)}(\tilde{\omega})} ,$$

$$E_m^{(J)}(\tilde{\omega}) \equiv \begin{pmatrix} 2J \\ J-m \end{pmatrix} \lfloor \tilde{\omega} - J + m \rfloor_{2J+1} ; \quad \lfloor n \rfloor_N \equiv \lfloor n \rfloor \lfloor n+1 \rfloor \cdots \lfloor n+N-1 \rfloor . \tag{9.118}$$

which, in particular, implies $C_m^{(J)}(\tilde{\omega}) = E_m^{(J)}(\tilde{\omega})/\lfloor \tilde{\omega} + 2m \rfloor$. Moreover, it now becomes possible to exactly calculate the exchange algebra satisfied by the ξ-fields,

$$\xi_{M_1}^{(J_1)} \xi_{M_2}^{(J_2)} = \sum_{N_1 N_2} (J_1, J_2)_{M_1 M_2}^{N_2 N_1} \xi_{M_2}^{(J_2)} \xi_{M_1}^{(J_1)} , \tag{9.119}$$

where the braiding matrix $(J_1, J_2)_{M_1 M_2}^{N_2 N_1}$ is given by a matrix element of the universal R matrix of $U_q(sl(2))$,

$$(J_1, J_2)_{M_1 M_2}^{N_2 N_1} = (<< J_1, M_1| \otimes << J_2, M_2|) \, R \, (|J_1, N_1 >> \otimes |J_2, N_2 >>) ,$$

---

[17]The definition of the tensor product is $(A \otimes B) \left(\xi_{M_1}^{(J_1)}(\sigma) \xi_{M_2}^{(J_2)}(\sigma') \right) \equiv \left(A \xi_{M_1}^{(J_1)}(\sigma) \right) \left(B \xi_{M_2}^{(J_2)}(\sigma') \right)$.

$$R = e^{-2i\hbar J_3 \otimes J_3} \sum_{n=0}^{\infty} \frac{(1 - e^{2i\hbar})^n e^{i\hbar n(n-1)/2}}{\lfloor n \rfloor !} e^{-i\hbar n J_3} (J_+)^n \otimes e^{i\hbar n J_3} (J_-)^n , \qquad (9.120)$$

where the group-theoretical states $|J, M >>$ belonging to the 'spin'-J representation of $U_q(sl(2))$ have been introduced.

The fusion properties of the ξ-fields are determined by the Clebsch-Gordan coefficients $(J_1, M_1; J_2, M_2|J)$ of $U_q(sl(2))$ or, equivalently, the 'quantum group' $3j$-symbols. They lead to the OPE

$$\xi_{M_1}^{(J_1)}(\sigma) \xi_{M_2}^{(J_2)}(\sigma') = \sum_{J=|J_1-J_2|}^{J_1+J_2} \left\{ [d(\sigma - \sigma')]^{\Delta(J)-\Delta(J_1)-\Delta(J_2)} \right.$$

$$\left. (J_1, M_1; J_2, M_2|J) \left(\xi_{M_1+M_2}^{(J)}(\sigma) + \text{descendants} \right) \right\} , \qquad (9.121)$$

where $d(\sigma - \sigma') \equiv 1 - e^{-i(\sigma - \sigma')}$, and the 'anomalous' Virasoro weights $\Delta(J)$ of $\xi_M^{(J)}(\sigma)$ are given by eq. (9.109). The fusion and braiding of the ξ-fields is, therefore, covariant w.r.t. the 'quantum group' action in eq. (9.117). The standard co-product in eq. (9.116), as well as the q-Clebsch-Gordan coefficients, are non-symmetric, but the non-commutativity of the ξ operators precisely cancels this lack of symmetry, which explains the appearance of the universal R matrix in their braiding rules. It is now straightforward (albeit tedious) to check commutativity (i.e. locality) of the operators (9.114), using eqs. (9.119) and (9.120). This means that the 'quantum group' deformation of the classical $SL(2, \mathbf{R})$ symmetry results from consistency of the quantum Liouville theory.

The positive 'powers of metric' (with a negative J above) can be constructed by exploiting the non-trivial formal symmetry of the above construction under the exchange $J \to -J - 1$. It seems to be even legitimate to extend the chiral decomposition with the 'quantum group' structure to a regime of the strongly coupled gravity, i.e. to go beyond the KPZ 'barrier' of $D = 1$. [18] Though complex dimensions will then appear, it can be shown, nevertheless, that for some specific values of D, namely $D = 7, 13, 19$, there exist consistent truncations to *unitary* theories (like the situation with the minimal models). all having operators with real conformal dimensions only [420]. With D free fields to represent the conformal matter, this line of development leads to the so-called *Liouville strings* of Gervais and Bilal [435, 436]. Their supersymmetric generalizations also exist, and they can be constructed along similar lines [437, 438].

---

[18]See ref. [434] for a discussion of the KPZ barrier in the case of 2d quantum gravity.

Exercises

 # IX-5 ▷ (*) Prove the Weyl invariance of the classical Liouville action (9.79) under the transformations (9.81). *Hint:* use the transfromation property of the 2d scalar curvature:

$$R^{(2)}[e^{2\sigma}\hat{g}] = e^{-2\sigma}\left(R^{(2)}[\hat{g}] - 2\hat{\nabla}^2\sigma\right) . \tag{9.122}$$

 # IX-6 ▷ (*) Derive the classical equations of motion in the Liouville theory (9.79), and prove that they describe Riemann surfaces of constant negative curvature $R^{(2)}$.

 # IX-7 [19] ▷ (*) According to the uniformization theorem in the theory of Riemann surfaces (sect. XII.4), a quotient of the upper half plane $\mathbf{C}_+$ by a discrete subgroup $\Gamma \subset SL(2,\mathbf{R})$ gives a general Riemann surface Σ. The standard *Poincaré solution* of the Liouville equation, which is supported by $\mathbf{C}_+$, takes the form

$$ds^2 = \frac{4}{\mu^2}\frac{1}{(\text{Im }z)^2}\,|dz|^2 . \tag{9.123}$$

This solution is invariant under the transformations (9.89) and, hence, it descends to a Euclidean analogue of the metric (9.88) having the form

$$ds^2 = \frac{4}{\mu^2}\frac{\partial A \bar{\partial} B}{[A(z) - B(z)]^2}\,|dz|^2 , \tag{9.124}$$

and defined on $\Sigma = \mathbf{C}_+/\Gamma$. The 'inverse' map $f : \Sigma \to \mathbf{C}_+$ is known as the *uniformizing map*. Show that the solution (9.124) has the stress tensor which is entirely given by the Schwartzian derivative,

$$T(z) = \frac{1}{\gamma^2}S[w; z] , \tag{9.125}$$

where $w = f(z)$.

IX.3 Topological field theories and strings

The general quantum 2d gravity may have *phases* different from the 2d gravity we considered so far. A phase with unbroken general covariance corresponds to the so-called *topological gravity*, which was originally studied in the context of moduli space of Riemann surfaces [440, 441, 442]. Since the quantum 2d gravity can be considered, in some sense, as a perturbation of the topological 2d gravity [443], it is worthy to discuss

---

[19]This exercise I 'borrowed' from ref. [439], p.p. 28–29.

the connection which exists between CFT's and *topological field theories* (TFT's) in general. To set up the framework, we first introduce basic notions of TFT, using the notation of sect. I.1 (see also refs. [444, 445, 446] for a review of TFT's). Our Euclidean spacetime manifold Σ (equipped with a metric $g_{\alpha\beta}$) is temporarily supposed to be d-dimensional and compact. In a moment, we will actually switch to d = 2. This section is based on ref. [445].

The first examples of TFT were introduced in QFT by Schwartz [447] and Witten [448, 449]. Let $\Phi(x)$ be all the fields of a gauge theory, including gauge fields, ghosts and (Nakanishi-Lautrup) multipliers (see sect. XII.5 for the BRST techniques, which are now used as the guidelines for a formal definition of TFT). Let S_{cl} be a classical action of the gauge theory, and S_{tot} the corresponding total action with all the gauge-fixing and ghost terms included. The latter is supposed to be invariant w.r.t. 'BRST'-like fermionic symmetry generated by a nilpotent 'BRST' charge Q. A 'BRST' variation of any functional $\mathcal{O}$ of fields Φ takes the form of a graded commutator $\delta\mathcal{O} = [Q, \mathcal{O}\}$, which is called a 'BRST' commutator. The S_{tot} is supposed to be 'BRST' invariant by construction. The vacuum state is supposed to be annihilated by Q, which implies, in particular, the vanishing VEV of any 'BRST' commutator, $\langle 0|[Q,\mathcal{O}\}|0\rangle \equiv \langle [Q,\mathcal{O}\}\rangle = 0$. The physical subspace $\mathcal{H}_{phys}$ in the total Fock space $\mathcal{H}$ of TFT is supposed to be in a one-to-one correspondence with the cohomology classes of Q, $\mathcal{H}_{phys} = \text{Ker}\,Q/\text{Im}\,Q$. A TFT is now defined by requiring its stress-energy tensor to be Q-exact, i.e.

$$T_{\alpha\beta} = [Q, G_{\alpha\beta}\} , \tag{9.126}$$

where $G_{\alpha\beta}$ is an anticommuting rank-two tensor.

There are two general cases in which eq. (9.126) is automatically satisfied. In the first one, the total action $S_{qu}(\Phi, g)$ can be written as a 'BRST' commutator, $S_{qu} = [Q, G\}$ with some $G(\Phi, g)$. Such theory is called a *cohomological or Witten-type* TFT [446, 448, 449] (see also exercise # IX-8). In the second case, the total action $S_{qu}(\Phi, g)$ takes the form $S_{qu}(\Phi, g) = S_{cl}(\Phi) + [Q, G(\Phi, g)\}$, where $S_{cl}(\Phi)$ is a non-trivial *metric-independent* classical action. Such theory is called a *quantum or Schwartz-type* TFT [446, 447] (see also exercise # IX-9).

The charge Q may actually be a genuine BRST charge, although this is not mandatory. [20] The TFT partition function

$$Z(\Sigma) \equiv \langle 0|0\rangle = \int [d\Phi]\, e^{-S_{tot}} , \tag{9.127}$$

---

[20] In a Schwartz-type quantum TFT, Q is always a BRST operator [446].

is easily seen to be independent on the metric provided the path-integral measure is both 'BRST'- and metric-independent, and eq. (9.126) holds. [21] For the physical correlation functions, eq. (9.126) also implies their metric-independence, [22]

$$\langle T_{\alpha\beta}(x)\Phi_{i_1}(x_1)\cdots\Phi_{i_s}(x_s)\rangle = 0 \ . \tag{9.128}$$

Hence, the TFT partition function is a topological invariant, while all TFT physical correlation functions are constants. All TFT physical states are just ground states (exercise # IX-10). This explains the term 'topological' since TFT only concerns about global properties, and it has no excited physical states.

One could say, a QFT defined over a manifold Σ is called to be topological when it is invariant under arbitrary smooth deformations of any metric $g_{\alpha\beta}$ which one puts on Σ, in the weak sense of eq. (9.126). It should be noticed that eq. (9.126) is actually valid in the strong sense for any theory of matter coupled to gravity because of Einstein equations, or after integrating over the metric. The real meaning of this definition of TFT refers to the field theories which are generally covariant *without* integrating over metric, or even without introducing dynamical metric at all.

The conserved charges

$$P_\alpha = \int d^{d-1}x\, T_{\alpha 0}(x) \ , \quad G_\alpha = \int d^{d-1}x\, G_{\alpha 0}(x) \ , \tag{9.129}$$

associated with $T_{\alpha\beta}$ and $G_{\alpha\beta}$ in eq. (9.126), resp., are related due to the same equation as

$$P_\alpha = [Q, G_\alpha\} \ , \quad [G_\alpha, G_\beta\} = 0 \ . \tag{9.130}$$

Eq. (9.130) resembles the super-Poincaré algebra, where the 'BRST' charge Q is a super-symmetry charge with the 'wrong' spin 0 instead of 1/2. Since Q can be simultaneously interpreted as a 'BRST' and a 'supersymmetry' charge, the associated ghosts can be considered as 'superpartners' to ordinary fields in TFT. This explains, in particular, the absence of dynamical physical excitations in TFT, namely, due to the 'fermion-boson cancellation' phenomenon known in supersymmetric models [446].

Given a physical field $\Phi(x) \equiv \Phi^{(0)}(x)$, one can build up a whole superfield as [445]

$$\Phi(x,\theta) \equiv e^{\theta^\alpha G_\alpha}\Phi(x) = \Phi^{(0)}(x) + \Phi^{(1)}_\alpha(x)\theta^\alpha + \ldots + \Phi^{(d)}_{\alpha_1\cdots\alpha_d}(x)\theta^{\alpha_1}\cdots\theta^{\alpha_d} \ , \tag{9.131}$$

---

[21] This implies, in particular, the absence of conformal anomaly and the vanishing central charge.

[22] We assume that the one-to-one correspondence between physical states and operators still holds in TFT, partly because of we are actually interested in the topological CFT's, see below.

where d Grassmannian anticommuting 'superspace' coordinates θ^α have been introduced. The fields

$$\Phi^{(s)}_{\alpha_1\cdots\alpha_s}(x) = [G_{\alpha_1}, [G_{\alpha_2}, \ldots, [G_{\alpha_s}, \Phi^{(0)}(x)\}\ldots\}\} \qquad (9.132)$$

are totally antisymmetric in their indices (i.e. they are space-time s-forms), and satisfy the *descent equations*

$$d\Phi^{(s)} = [Q, \Phi^{(s+1)}\}, \quad \text{or equivalently,} \quad \mathbf{D}\Phi(x,\theta) = 0, \quad \mathbf{D} \equiv Q - \theta^\alpha \frac{\partial}{\partial x^\alpha}, \qquad (9.133)$$

where Q plays the role of exterior derivative (exercise # IX-11). Eq. (9.133) implies that $\int_C \Phi^{(s)}$ is a physical operator when C is an s-dimensional cycle, $\partial C = 0$, because of Stokes theorem. All such 'BRST'-invariant operators with $s > 0$ are non-local, and represent higher-dimensional analogues of the Wilson loop operators known in QCD. In particular, for a local physical operator $\Phi^{(0)}(x)$, it follows

$$\Phi^{(0)}(x) - \Phi^{(0)}(y) = \int_x^y dx^\alpha [Q, \Phi^{(1)}_\alpha\}, \qquad (9.134)$$

which just means that the TFT correlation functions are to be constants. Non-local physical operators of the form $\int_\Sigma \Phi^{(d)}$ are of special importance, since they can be used to deform the original action S to just another action, $S \to S - t\int_\Sigma \Phi^{(d)}$, whithout spoiling general covariance.

We are now in a position to consider the **topological conformal field theories** (TCFT's) to be defined by combining 2d CFT and 2d TFT. It should be stressed that the TFT axioms already imply the vanishing trace of TFT stress-energy tensor in the weak sense. The CFT axioms require that this trace has to vanish identically, i.e. in the strong sense. To the end of this section, we follow ref. [445].

The TCFT holomorphic stress tensor $T(z)$ is Q-exact, $T(z) = [Q, G(z)\}$, while the TCFT holomorphic 'BRST' current $Q(z)$ corresponds to the (left-moving) part of 'BRST' charge $Q_L \equiv Q$. Since the current has (conformal) dimension one, i.e. of positive energy, it should be Q-exact in TCFT,

$$Q(z) = [Q, J(z)\}, \qquad (9.135)$$

where the new $U(1)$ current $J(z)$ of dimension one has been introduced. As is usual in (S)CFT, all its fields ϕ_i are supposed to have definite conformal dimensions $(h_i, \bar{h}_i)$ and $U(1)$-charges $(q_i, \bar{q}_i)$. In particular, the assignments for the already introduced fields, which are present in any TCFT, are given by

field	$T(z)$	$G(z)$	$Q(z)$	$J(z)$
spin h	2	2	1	1
charge q	0	-1	1	0
$h + \frac{1}{2}q$	2	$\frac{3}{2}$	$\frac{3}{2}$	1

$$(9.136)$$

This field contents is very reminiscent to the one of $N = 2$ SCA (sect. V.1). The actual correspondence càn be established after the twisting

$$T \to T_{N=2\ SCA} = T - \tfrac{1}{2}\partial J\ , \quad h \to h_{N=2\ SCA} = h + \tfrac{1}{2}q\ . \qquad (9.137)$$

Hence, requiring the fields in eq. (9.136) to form a *closed* algebra completely fixes their OPE's due to a uniqueness of the $N = 2$ SCA! This actually yields the so-called *twisted* $N = 2$ SCA with the OPE's

$$T(z)T(w) \sim \frac{2T(w)}{(z-w)^2} + \frac{\partial T(w)}{z-w}\ , \quad J(z)J(w) \sim \frac{d}{(z-w)^2}\ ,$$

$$T(z)Q(w) \sim \frac{Q(w)}{(z-w)^2} + \frac{\partial Q(w)}{z-w}\ , \quad J(z)Q(w) \sim \frac{Q(w)}{z-w}\ ,$$

$$T(z)G(w) \sim \frac{2G(w)}{(z-w)^2} + \frac{\partial G(w)}{z-w}\ , \quad J(z)G(w) \sim \frac{-G(w)}{z-w}\ ,$$

$$T(z)J(w) \sim \frac{-d}{(z-w)^3} + \frac{J(w)}{(z-w)^2} + \frac{\partial J(w)}{z-w}\ ,$$

$$Q(z)G(w) \sim \frac{d}{(z-w)^3} + \frac{J(w)}{(z-w)^2} + \frac{T(w)}{z-w}\ ,$$

$$G(z)G(w) \sim Q(z)Q(w) \sim 0\ , \qquad (9.138)$$

where (constant) d represents the $U(1)$ anomaly. The conformal central charge c has to vanish due to unbroken general covariance required by TFT. Therefore, *any* $N = 2$ SCFT leads to a TCFT after the twisting in accordance to eq. (9.137). The specific features of $N = 2$ SCFT's (such as, e.g., $N = 2$ chiral primary fields Φ_i, chiral rings, $U(1)$ selection rules for correlation functions, etc.) are also shared by $N = 2$ TCFT's. In particular, the free energy $F(t)$ to be defined by the expectation value

$$F(t) = \left\langle \exp\left(\sum_n t_n \int \Phi_i\right)\right\rangle\ , \qquad (9.139)$$

where the sum goes over the chiral ring, satisfies the scaling law (on a sphere) [450]

$$\sum_j (q_j - 1)t_j \frac{\partial}{\partial t_j} F(t) = (d - 3)F(t)\ , \qquad (9.140)$$

which is just the consequence of the $U(1)$ selection rules (sect. V.4) for the $N = 2$ SCFT with chiral primaries Φ_i of charge q_i. [23]

We conclude this section with a few examples of TCFT's. The simplest example is provided by the twisted version of the free field realization of $N = 2$ SCA, comprising a complex boson and a complex fermion (*cf* exercise #V-13), with the Lagrangian representing the $d = 1$ *topological sigma-model*,

$$\mathcal{L} = \partial \bar{X} \bar{\partial} X + \chi \bar{\partial} \psi + \bar{\chi} \partial \bar{\psi} \; , \tag{9.141}$$

where $h(\psi) = 0$, $h(\chi) = 1$, $q(\psi) = -q(\chi) = 1$, and the total central charge vanishes: $c = c_B + c_F = 2 + (-2) = 0$. The non-vanishing BRST-type transformations $\delta = [Q_L, \;\}$ take the form

$$\delta X = \psi, \quad \delta \chi = \partial \bar{X} \; , \tag{9.142}$$

and similarly for $\bar{\delta} = [Q_R, \;\}$. The Lagrangian (9.141) is therefore $(Q_L + Q_R)$-exact, namely

$$\mathcal{L} = [Q_L + Q_R, \chi \bar{\partial} X + \bar{\chi} \partial \bar{X}\} \; . \tag{9.143}$$

As far as the stress tensor is concerned, one finds the similar relation

$$T = \partial \bar{X} \partial X + \chi \partial \psi = [Q_L + Q_R, \chi \partial X\} \; . \tag{9.144}$$

The 'BRST' charge Q takes the form

$$Q \equiv Q_L = \oint \psi \partial \bar{X} = \oint \psi \delta \chi \; . \tag{9.145}$$

The list of fields in eq. (9.136) for this particular case is given by

$$T(z) = \partial \bar{X} \partial X + \chi \partial \psi \; , \quad Q(z) = \psi \partial \bar{X} \; ,$$

$$J(z) = \psi \chi \; , \quad G(z) = \chi \partial X \; , \tag{9.146}$$

whereas the physical states correspond to the operators $\{1, \psi, \bar{\psi}, \psi\bar{\psi}\}$. It could be generalized by considering the twisted version of the $N = 2$ LG-model in eq. (5.101), which is specified by the quasi-homogeneous superpotential $W(X)$ in eq. (5.100).

---

[23] The shift $q \to q - 1$ is due to the fact that the two-forms $\phi_i^{(2)} = G_{-1}\bar{G}_{-1}\phi_i^{(0)}$ actually appear in the expression (9.139) for the free energy, while the shift $d \to d - 3$ is due to the $SL(2, \mathbf{C})$ invariance in the case of sphere.

The critical string theory (Ch. VIII) can also be converted into a topological theory. The untwisted stress tensor $T = T_{\text{matter}} + T_{\text{ghost}}$ and the nilpotent BRST charge $Q = \oint :$ $c(T_{\text{matter}} + \frac{1}{2}T_{\text{ghost}}) :$ of the 26-dimensional bosonic string are both Q-exact: [24]

$$T(z) = [Q, b(z)\} , \quad \Rightarrow \quad G(z) = b(z) ,$$

$$Q(z) = [Q, : bc : (z)\} , \quad \Rightarrow \quad J(z) =: bc : (z) , \tag{9.147}$$

where the anticommuting reparametrizational ghosts (b, c) can be considered as a (b, c) conformal ghost system of $\lambda = 2$ (sect. I.6). The physical operators of the type $\phi^{(0)}$ and $\phi^{(2)}$ can be identified with the *ghost-independent* physical vertex operators $W(z, \bar{z})$ of dimension $(1, 1)$ as (*cf* sect. VIII.2)

$$\phi^{(0)} = c\bar{c}W \equiv V, , \quad \phi^{(2)} = W , \tag{9.148}$$

with the $U(1)$ charges $q = 1$ and $q = 0$, resp. The $U(1)$-selection rule

$$\sum q_n = 3(1 - h) \equiv \frac{3}{2}\chi(\Sigma) \tag{9.149}$$

allows non-vanishing correlators on the sphere $(h = 0)$ where the total background charge can be absorbed, but it does not allow any non-trivial correlator beyond the tree level $(h > 0)$ for the topological strings since all of their local operators have $U(1)$ charge $q = 1$. In the conventional (untwisted) string theory [128], the correlators are not topological but depend on moduli $\{m\}$ of the Riemann surface representing a string world-sheet, and are to be integrated over the moduli space. In the Polyakov prescription [37] for string scattering amplitudes, the Beltrami differentials μ_a (sect. XII.4), which are not BRST-closed, appear inside the modular invariant correlators,

$$\langle \phi_{i_1} \cdots \phi_{i_s} \rangle_h \equiv \int_{\mathcal{M}_h} \prod d^2 m_a \left\langle \int \phi_{i_1} \cdots \int \phi_{i_s} \prod_{a=1}^{3h-3} \int_\Sigma \mu_a b \int_\Sigma \bar{\mu}_a \bar{b} \right\rangle_h . \tag{9.150}$$

The number of insertions in eq. (9.150) is dictated by eq. (1.149), since the anti-ghost field b has charge $q = -1$. The selection rule for the correlators (9.150) is different from that in eq. (9.149), and it takes the form

$$\sum_n q_n = 0 . \tag{9.151}$$

Eq. (9.151) is obviously too restrictive in order to make sense of the topological strings and of the topological gravity. Hence, this naive construction has to be generalized. One

---

[24]This is generally true for any theory coupled to gravity, where metric is a dynamical variable.

of the ways of doing that is to introduce a dynamical metric into TFT, while preserving its 'BRST' symmetry after gauge-fixing [451]. The simplest example is based on the $d = 1$ topological sigma-model (9.141), whose Lagrangian on arbitrary Riemann surface is given by

$$\mathcal{L}_{\text{cov}} = \sqrt{g} g^{\alpha\beta} (\partial_\alpha X \partial_\beta \bar{X} + \chi_\alpha \partial_\beta \psi) , \qquad (9.152)$$

where a 2d metric $g_{\alpha\beta}$ has been introduced. Since we want this metric to be dynamical and the Lagrangian to be Q-invariant, we also need to introduce a 'superpartner' for the metric w.r.t. the Q-supersymmetry. Let it be $\psi_{\alpha\beta}$, together with the transformation laws

$$\delta g_{\alpha\beta} = \psi_{\alpha\beta} , \quad \delta\psi_{\alpha\beta} = 0 . \qquad (9.153)$$

The stress-energy tensor $T_{\alpha\beta}$ should also have a 'superpartner' $G_{\alpha\beta}$ proportional to $\partial\mathcal{L}/\partial\psi_{\alpha\beta}$, where $\mathcal{L}$ is a complete Lagrangian we are looking for. It follows

$$\mathcal{L} = \sqrt{g} g^{\alpha\beta} (\partial_\alpha X \partial_\beta \bar{X} + \chi_\alpha \partial_\beta \psi + \psi_{\alpha\gamma} \chi^\gamma \partial_\beta X) . \qquad (9.154)$$

In the conformal gauge, $g_{\alpha\beta} = e^\phi \delta_{\alpha\beta} , \psi_{\alpha\beta} = \rho e^\phi \delta_{\alpha\beta}$, the commuting 'superconformal' ghosts (β, γ) of $\lambda = -1$ are to be added. They can be interpreted as 'superpartners' to the (b, c) reparametrization ghosts. This would ensure Q-supersymmetry of the theory and the vanishing total central charge. The Q-supersymmetry in the conformal gauge takes the form

$$\delta\phi = \rho , \quad \delta\rho = 0 . \qquad (9.155)$$

The Weyl symmetry can also be fixed by the conditions $\partial\bar{\partial}\phi = \hat{R}^{(2)}$ and $\partial\bar{\partial}\rho = 0$, where $\hat{R}^{(2)}$ is the 2d (world-sheet) curvature w.r.t. a reference metric $\hat{g}_{\alpha\beta}$ on Σ. The total topological string action takes the form

$$S = S_{\text{matter}} + S_{\text{ghost}} + S_{\text{L}} ,$$

$$S_{\text{ghost}} = \int b\bar{\partial}c + \beta\bar{\partial}\gamma + \text{h.c.} ,$$

$$S_{\text{L}} = \int \Lambda(\partial\bar{\partial}\phi - \hat{R}^{(2)}) - \lambda\partial\bar{\partial}\rho , \qquad (9.156)$$

where the Lagrange multipliers Λ and λ have been introduced. The total 'BRST' charge reads [451]

$$Q_{\text{tot}} = Q + Q_{\text{BRST}} , \qquad (9.157)$$

where Q is the nilponent 'BRST' charge corresponding to the 'BRST' symmetry above, while Q_{BRST} is the conventional BRST charge corresponding to the gauged symmetries

generated by the stress tensor T and its fermionic 'superpartner' G: $Q_{\text{BRST}} = \oint c$:
$(T_{\text{m}} + T_{\text{L}} + \frac{1}{2}T_{\text{gh}}) + \gamma(G_{\text{m}} + G_{\text{L}} + \frac{1}{2}G_{\text{gh}})$: . Cohomology of the total 'BRST' charge
determines physical spectrum. The physical operators w.r.t. Q_{tot} can be shown to be of
the form $\mathcal{O}_k \equiv \phi_{i_k} \cdot \sigma_{n_k}$, where ϕ_i are matter fields of charge q_i, and σ_n is the combination
of ghost and Liouville fields, describing gravitational 'dressing' [451],

$$\sigma_n = e^{\frac{2}{3}(n-1)\pi} \left[\bar{\partial}(\tfrac{1}{2}\partial c + c\partial\phi - \text{c.c.}) \right]^n , \quad n = 0, 1, \ldots . \tag{9.158}$$

The topological string theory correlation functions are still defined by eq. (9.150),
after the substitution $b \to G$, but in terms of the 'dressed' operators $\phi_k \to \mathcal{O}_k$, which
leads to the general selection rule

$$\sum_k (n_k + q_{i_k} - 1) = (d - 3)(1 - h) . \tag{9.159}$$

This rule corresponds to the scaling relations in matrix models (Ch. X). For the case of
purely topological gravity $(d = 0)$, eq. (9.159) becomes

$$\langle \sigma_{n_1} \cdots \sigma_{n_s} \rangle_h \neq 0 \Longrightarrow \sum_i (n_i - 1) = 3h - 3 . \tag{9.160}$$

It can be shown that the correlation functions of purely topological gravity are entirely
determined from a boundary of moduli space [451].

Exercises

IX-8 ▷ (*) Show that the semiclasscial partition function Z of a cohomological
Witten-type TFT is exact. *Hint*: introduce the parameter $t \propto 1/\hbar$ in front of the total
TFT action S_{qu} and consider a variation of Z w.r.t. this parameter.

IX-9 ▷ (*) Consider the *Chern-Simons action*,

$$S(A) = \frac{k}{4\pi} \int_\Sigma \text{tr}(A \wedge dA + \frac{2}{3}A \wedge A \wedge A) , \tag{9.161}$$

defined on an oriented *three*-dimensional compact manifold Σ, in terms of a Lie algebra-
valued Yang-Mills gauge field $A(x)$. Show that this theory is a quantum Schwartz-type
TFT.

IX-10 ▷ Prove that the only physical states that can appear in TFT are ground
states with vanishing energy. *Hint*: use eq. (9.126) and the Hamiltonian $H \equiv P_0$ from
eq. (9.129).

\# IX-11 ▷ Prove the TFT descent equations (9.133). *Hint:* use the fact that Q annihilates any physical operator, and eq. (9.130).

\# IX-12 ▷ (*) The first example of TFT [448] was actually obtained by *twisting* the four-dimensional $N = 2$ supersymmetric Yang-Mills theory. [25] Witten's twisting means replacing the chiral set of simple currents, J_L and J_R of $SU(2)_L$ and $SU(2)_R$, resp., in the Euclidean rotation group $SO(4) \cong SU(2)_L \otimes SU(2)_R$ by the 'twisted' currents, say, $J_R \to J_R + J_I$, where J_I are currents of the internal symmetry group $SU(2)$ (i.e., in other words, replacing $SU(2)_R$ by a diagonal subgroup of $SU(2)_R \otimes SU(2)_I$). Accordingly, one gets the decomposition $\mathbf{2} \times \mathbf{2} = \mathbf{1} + \mathbf{3}$ for the $N = 2$ supercharges, $\bar{Q}^I_{\dot\alpha} = Q + Q_i$, where Q is the nilponent 'BRST' charge (i.e. 'zero-spin supersymmetry') in the corresponding four-dimensional $N = 2$ TFT. Show that the axioms of cohomological (Witten-type) TFT are all satisfied in this case.

\# IX-13 ▷ (*) The combined (twisted) truncation and dimensional reduction of Witten's 4d, $N = 2$ TFT (exercise \# IX-12), or the 4d, $N = 4$ SDYM theory (sect. VIII.3), down to two dimensions yields a 2d TFT, with the action [454, 455]

$$S_{2d-top} = \int \mathrm{tr}(\phi \wedge F - 2i\rho\nabla \wedge \psi) , \qquad (9.162)$$

where the Yang-Mills field strength 2-form F, the Yang-Mills exterior covariant derivative ∇, a 'gaugino' 1-form ψ, a scalar field ϕ and a spinor field ρ, all in the adjoint representation of a gauge group, have been introduced. Show that this action is invariant under 'BRST' transformations (with anticommuting parameter ϵ) having the form

$$\delta A = i\epsilon\psi , \quad \delta\rho = \epsilon\phi , \quad \delta\phi = \delta\psi = 0 . \qquad (9.163)$$

---

[25] This work of Witten gives the Lagrangian quantum field theory description of Donaldson's work on a topology of four-manifolds [452, 453].

Chapter X

CFT and Matrix Models

All our considerations in the previous Chapters were based on the 2d conformal *field* theory. Compared to the situation in the general QFT having no symmetries at all, it is the infinite-dimensional conformal symmetry that makes it possible to go beyond the quantum perturbation theory in CFT. Knowing the other ways to extract non-perturbative information is very useful to further justify the CFT results, as well as to expand physical applications of CFT. It is therefore quite worthwhile to look for the alternative approach to CFT. It should also be noticed here that the lack of non-perturbative information is the main stumbling block towards phenomenological applications of the modern theory of strings and superstrings (Ch. VIII). The understanding of the non-perturbative quantum 2d gravity was greatly enhanced due to the remarkable progress achieved by Brézin and Kazakov [456], Douglas and Shenker [457], and Gross and Migdal [458] in 1990. Their approach is based on a *discretization* of a string world-sheet, instead of representing it by a genus-h Riemann surface Σ. This concept naturally leads to the so-called **matrix models** (MM's), in which the all-genus dependence of the string partition function can be analyzed in certain limit. In this Chapter, the basic facts about matrix models are provided, which the emphasis being made on the connection between the MM results and the corresponding facts from CFT and TFT. Our presentation here is based on ref. [459]. Being consistent and self-contained, it does not, however, go beyond the introduction. [1] Nevertheless, the connection to CFT and 2d integrable models can be clearly seen already in this very limited context. The next Chapter is totally devoted to 2d integrable models, in connection with conformal symmetry and 4d self-dual field theories.

---

[1] See refs. [459, 460, 461, 462, 463, 464] for more about the MM's and 2d quantum gravity.

X.1   Why matrix models ?

The central problem of string theory is how to compute the partition function which is given by the sum over all 2d topologies (all genus),

$$Z = \sum_h \int [Dg][dX] e^{-S_{\mathrm{B}}(g,X)} \ , \tag{10.1}$$

where g is a metric on Σ, and $S_{\mathrm{B}} = S_{\text{2d grav}} + S_{\text{matter}}$ is the string action (8.4). The functional integral over metrics in eq. (10.1) has to be properly defined. The natural norm provided by eq. (9.45b) formally supplies a measure in the space of metrics, but the real problem, however, is to integrate over such space. As we already know from sect. IX.1, using the conformal gauge leads to the quantum Liouville theory, which is quite hard to deal with, especially when $h > 0$. [2]

To simplify the situation, one starts with no matter, but the 2d quantum gravity itself, i.e. with the pure theory of surfaces. The corresponding partition function takes the form

$$Z(\beta_0, \gamma_0) = \sum_h Z_h(\beta_0, \gamma_0) \equiv \sum_h \int [Dg] e^{-\beta_0 A + \gamma_0 \chi} \ , \tag{10.2}$$

in terms of the total area $A(\Sigma) = \int_\Sigma \sqrt{g}$ and the Euler characteristic $\chi(\Sigma) = \frac{1}{2\pi} \int_\Sigma \sqrt{g} R = 2 - 2h$ for a genus-h Riemann surface Σ, with arbitrary real parameters β_0 and γ_0. Instead of computing $Z_h(\beta_0, \gamma_0)$ and $Z(\beta_0, \gamma_0)$, one may equally consider the partition functions at fixed area, $Z_h(A)$ and $Z(A)$, since the area-independent results can always be recovered by integrating over A. [3] One has, in particular,

$$Z_h(\beta_0, \gamma_0, A) = \mathrm{Vol}(h, A) \exp[-\beta_0 A(\Sigma) + \gamma_0 \chi(\Sigma)] \ , \tag{10.3}$$

where the γ_0-dependence is now trivial, and the volume $\mathrm{Vol}(h, A)$ is either defined by this formula or it has to be computed independently. The actual evaluation of the path integral uses a finite cutoff (regularization), and the subsequent renormalization of the parameters. Of course, *a priori* there is no guarantee that a sensible renormalized solution exists at all. In addition, there are well-known renormalization ambiguities associated with any solution. In our case, arbitrary cutoff-independent constants can be added to the parameters β_0 and γ_0, which would lead to additional exponential factors

---

[2]According to Polyakov [465], the CFT itself originated from unsuccessful attempts to solve the Liouville theory! [459].

[3]We are going to explicitly indicate only the arguments which are relevant at the moment.

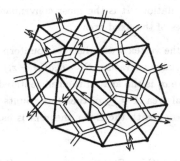

Fig. 18. The random triangulation of a surface. The triangular faces
and the MM quantum 3-point vertices are dual to each other.

in $Z_h(A)$. [4] The so-called '*random triangulation*' method [410, 466, 467] uses a discrete
cutoff, and it deals with triangulations of Σ instead of metrics on Σ, so that the surface
Σ has to be constructed from the triangles (Fig. 18).

Each triangulation in fact determines the metric on Σ. The triangles we consider are
supposed to be equilateral of area ε, where the curvature R is concentrated at the vertices
i. If the surface Σ is covered with a very large number of triangles, the averaging over
all the local singularities is supposed to produce the effective smooth metric on Σ. In
general, it is important to distinguish between a triangulated surface and a triangulated
embedding. Without (conformal) matter fields, we only need to investigate the *intrinsic*
surface properties, and a triangulation is just a special choice of metric on Σ. Therefore,
the first general idea is to consider a discrete version of eq. (10.2) in the form

$$\sum_{\text{genus}} \int [Dg] \;\rightarrow\; \sum_{\text{triangulations}} . \tag{10.4}$$

One may add the attribute '*random*' for the triangulations in question, which is, however,
a bit misleading since it comes from the picture implying that, to average over triangulations, one needs an ensemble of them. In fact, one needs to sum over *all* triangulations
of a given type.

Similarly to an integral over all continuum metrics, which should be understood as
the integral over the conformal classes, i.e. modulo the 2d diffeomorphisms and the
Weyl transformations [37, 128], the discrete sum in eq. (10.4) is meant as a sum over
the topological equivalence classes of triangulations. [5] A class is given by writing down

---

[4] These ambiguities can actually be removed by requiring $Z_h(A)$ to vary as a power of A.

[5] This analogy is motivated by the fact that any two triangles (on a sphere) are conformally equivalent.

the incidence matrix of any representative triangulation. It is the most convenient to represent a class by using only equilateral triangles of the fixed area ε.

Given the number N_i of incident triangles at the vertex i, the discrete counterparts of the infinitesimal volume element $\sqrt{g}$ and the scalar curvature $R = R_i \delta^2(z - z_i)$ are $\sigma_i = N_i \varepsilon/3$ and $R_i = \pi(6 - N_i)/(N_i \varepsilon)$, resp., where σ_i is the area of the cell centered at i in the lattice *dual* to the triangulation. The total area $A = \sum_i \sigma_i$ therefore counts the total number of triangles, $n = A/\varepsilon$. The consistency with the Euler theorem is easily verified:

$$\int \sqrt{g} R \to \sum_i 2\pi(1 - N_i/6) = 2\pi(V - \tfrac{1}{2}F) = 2\pi(V - E + F) = 2\pi\chi , \qquad (10.5)$$

where the total number of vertices V, edges E and faces F of a triangulation has been introduced. They satisfy the obvious relation $3F = 2E$.

Let $V(h, n)$ be the number of equivalence classes of triangulations of Σ (of genus h), with n triangles. The large-n behaviour of $V(h, n)$ is actually under control [468]. Specifically, it is the classical mathematical result [468] that the behaviour of $V(h, n)$ for large n is given by

$$V(h, n) \sim e^{cn} n^{\tilde{\gamma}\chi-1} b_h [1 + O(1/n)] , \qquad (10.6)$$

where the constants c, $\tilde{\gamma}$ and b_h have been introduced. Though the Euler characteristic $\chi = 2 - 2h$ becomes negative for $h > 1$, the $\tilde{\gamma}$ turns out to be negative as well.

Regarding ε as a cutoff when $n = A/\varepsilon$, it makes sense to approximate $\mathrm{Vol}(h, A)$ by $V(h, n)$. Because of $\sum_n \sim \int dA/\varepsilon$ and eq. (10.6), taking the limit $\varepsilon \to 0$ at fixed area A yields the relation

$$\mathrm{Vol}_\varepsilon(h, A) \sim \frac{1}{\varepsilon} e^{cA/\varepsilon} \left(\frac{A}{\varepsilon}\right)^{\tilde{\gamma}\chi-1} b_h [1 + O(\varepsilon)] . \qquad (10.7)$$

It follows

$$Z_{\varepsilon,h}(A) = \mathrm{Vol}_\varepsilon(h, A) \exp[-\beta_0 A + \gamma_0\chi] = \frac{1}{\varepsilon} e^{cA/\varepsilon} \left(\frac{A}{\varepsilon}\right)^{\tilde{\gamma}\chi-1} b_h \exp[-\beta_0 A + \gamma_0\chi]$$

$$= b_h A^{\tilde{\gamma}\chi-1} \exp\left[A(\tfrac{c}{\varepsilon} - \beta_0) - \chi(\tilde{\gamma}\ln\varepsilon - \gamma_0)\right] . \qquad (10.8)$$

The renormalization amounts to adjusting β_0 and γ_0 as the functions of ε in such a way that the function $Z_{\varepsilon,h}(A)$ converges in the limit $\varepsilon \to 0$ to a well-defined function

$Z_h(A)$. This yields $\beta_0 \to \beta_c = c/\varepsilon$ and $\gamma_0 \to \gamma_c = -\tilde{\gamma}\ln(A_0/\varepsilon)$, with the constant A_0 resulting from the renormalization ambiguity, so that

$$Z_h(A) = \frac{1}{A}\left(\frac{A}{A_0}\right)^{\tilde{\gamma}\chi} b_h \ . \tag{10.9}$$

Therefore, the naively 'dominant' term e^{cn} in eq. (10.6) is actually irrelevant because it disappears at criticality after the renormalization !

Absorbing $A_0^{-\tilde{\gamma}\chi}$ into b_h and making the Laplace transformation of $Z_h(A)$ yield

$$Z(\beta_0, \gamma_0) \equiv \sum_h \int_0^\infty dA\, Z_h(A) = \sum_h \int_0^\infty dA\, A^{\tilde{\gamma}\chi - 1} b_h \exp[-(\beta_0 - \beta_c)A + (\gamma_0 - \gamma_c)\chi]$$

$$= \sum_h \frac{\Gamma(\tilde{\gamma}\chi)}{(\beta_0 - \beta_c)^{\tilde{\gamma}\chi}} b_h \exp[(\gamma_0 - \gamma_c)\chi] \ , \tag{10.10}$$

where $Z(\beta_0, \gamma_0)$ now becomes the function of only one (!) 'renormalized' variable alone, viz.

$$Z(z) = \sum_h z^{-\tilde{\gamma}\chi} f_h \ , \quad z \equiv e^{-(\gamma_0 - \gamma_c)/\tilde{\gamma}}(\beta_0 - \beta_c) \ . \tag{10.11}$$

The limit $\beta_0 \to \beta_c$, $\gamma_0 \to \gamma_c$ at fixed z is known as the **double scaling limit**. Since $\tilde{\gamma} < 0$ for each value $h > 1$ of the genus, the partition function $Z_h(\beta)$ in eq. (10.10) diverges at the critical value of the coupling constant β. However, as far as the double scaling limit is concerned, the suppression coming from the explicit γ-dependence is chosen to exactly offset the singularity. This is possible for all genera h *simultaneously*, due to the scaling property in eq. (10.10). Stated differently, the partition function exhibits the singularity, at which one should expect the conformal symmetry, and, hence, the anticipated correspondence with the CFT results obtained in the continuum limit (see below).

The coefficents f_h in the series (10.11) are well-defined and positive. They have the following asymptotical behaviour in the large-h limit:

$$f_h = \Gamma(\tilde{\gamma}\chi) b_h \propto (2h)! \ . \tag{10.12}$$

The sum over genus in eq. (10.11) is divergent, and it is clearly not even Borel-summable [469].

The real advantage of 'random triangulations' comes out after the crucial observation that the integrals in eq. (10.2) can be explicitly calculated in their discretized form (10.4). The basic trick consists of introducing certain matrix integral as the generating functional for the random triangulations. [6] The use of the triangles is not essential, so that they can be replaced by polygons of any type, e.g. by the squares, which is sometimes more convenient. The large-n behaviour of the number of the equivalence classes of polygonizations of a surface with n polygons turns out to be very similar to that in eq. (10.6), with the same $\tilde{\gamma}$ but the different parameters c and b_h [468]. Therefore, there is a good chance to expect that the same theory (characterized by $\tilde{\gamma}$) is obtained in the continuum limit (this is known as the MM *universality*).

First, one recalls the standard Feynman diagrammatic expansion of

$$\ln \int_{-\infty}^{+\infty} \frac{d\varphi}{\sqrt{2\pi}} \exp\left[-\varphi^2/2 - \lambda\varphi^4/4!\right] , \qquad (10.13)$$

in 'zero dimensions', where the 'field' φ takes its values in real numbers, and the fourth power in the interaction potential is chosen just for simplicity. The well-known formal expansion in power series of coupling constant λ can be systematically described by the connected Feynman diagrams of the $\lambda\varphi^4$-theory via the Wick theorem. Being drawn on a compact orientable surface, each such vacuum Feynman diagram with n vertices defines a *dual* graph with n faces (*cf* Fig. 18), where the roles of vertices and faces are interchanged. The connected four-valent Feynman graphs yield therefore dual graphs consisting of tetragons (or squares) and forming something reminiscent to a 'tetragonization' of the surface under consideration. This becomes even more similar after introducing the additional structure into the Feynman graphs by 'thickening' the propagators, i.e. by making them 'fat' so that the ribbon edge orientations are to be preserved when linking together the vertices (Fig. 19). Technically, this corresponds to replacing the scalar φ by an $N \times N$ hermitian matrix $M^i{}_j$. The perturbative contributions to $F \equiv \ln \int dM \exp\left[-\frac{1}{2}\mathrm{tr}\, M^2 - \frac{g}{N}\mathrm{tr}\, M^4\right]$ then take the form

$$\left\langle M^{i_1}{}_{j_1} \cdots M^{i_n}{}_{j_n} \right\rangle \equiv \int dM \, e^{-\mathrm{tr}\, M^2/2} M^{i_1}{}_{j_1} \cdots M^{i_n}{}_{j_n}$$

$$= \frac{\partial}{\partial J^{j_1}{}_{i_1}} \cdots \frac{\partial}{\partial J^{j_n}{}_{i_n}} \int dM \, e^{-\mathrm{tr}\, M^2/2 + \mathrm{tr}\, JM} \Big|_{J=0} = \frac{\partial}{\partial J^{j_1}{}_{i_1}} \cdots \frac{\partial}{\partial J^{j_n}{}_{i_n}} e^{\mathrm{tr}\, J^2/2} \Big|_{J=0} , \qquad (10.14)$$

where the matrix source $J^i{}_j$ and the measure $dM = \prod_i dM^i{}_i \prod_{i<j} d\mathrm{Re}M^i{}_j d\mathrm{Im}M^i{}_j$ have been introduced. The normalization above is fixed by the condition $\int dM e^{-\mathrm{tr}\, M^2/2} = 1$. The hermitian matrix propagators

$$\left\langle M^i{}_j M^k{}_l \right\rangle = \delta^i_l \delta^k_j \qquad (10.15)$$

---

[6]This technique was originally introduced by t'Hooft [470] in his analysis of the large-N limit of the quantum $SU(N)$ gauge theory.

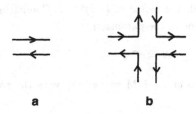

Fig. 19. The graphical representation for the hermitian
matrix propagator (a), and for the hermitian
matrix 4-point vertex (b).

and the matrix four-point vertices represented by the double lines (Fig. 19) yield the
Feynman graphs, with each line being thickened to a ribbon. The two edges of a ribbon
are labeled by i and j corresponding to the indices of the matrices M or J, while N
possible labels of each edge correspond to the N positive values of the matrix index.
Their dual ribbon graphs actually define tetragonizations of Riemann surface, since the
closed internal loops of these graphs uniquely specify locations and orientations of faces.
Indeed, the edges of the double lines join together into circles, and upon filling in these
circles with discs, one obtains the 'tetragonized' Riemann surface Σ. It is clear that
each topological equivalence class of the tetragonizations represented by a 'squarulation'
is obtained from exactly *one* M^4-type Feynman graph (including some singular ones).
Hence, the matrix integral above automatically generates all tetragonizations by the
dualization of the Feynman graphs! Replacing

$$\frac{1}{N}\text{tr}\, M^4 \to \frac{1}{\sqrt{N}}\text{tr}\, M^3 \tag{10.16}$$

in the MM potential generates the triangulations (Fig. 18) instead. After rescaling the
variables $M \to M\sqrt{N}$ in this case, the MM action takes the form $N\text{tr}(-\frac{1}{2}M^2 - g\text{tr}M^3)$.
Since each vertex now contains a factor N, while each propagator (edge) contributes a
factor of N^{-1} and each closed loop (face) gives another factor of N due to the associated
index summation, each Feynman graph has an overall N-dependence given by the factor

$$N^{V-E+F} = N^\chi = N^{2-2h}\,, \tag{10.17}$$

since each triangle was chosen to have the fixed area ε, while the total surface area
is given by $A = n\varepsilon$. The above Feynman rules weigh each triangulation of a surface

(of Euler number χ, with n triangles) with the factor of $(-g)^n N^\chi$. Comparing with eq. (10.3) leads to the following identification of the parameters

$$(-g) = e^{-\beta_0 \varepsilon} , \quad N = e^{\gamma_0} , \tag{10.18}$$

and it implies, in addition, an identification of the MM free energy with the partition function of the quantum 2d gravity.

Eq. (10.18) is enough to identify the continuum limit of Z in eq. (10.2) with that of F after the replacement (10.16). More general MM potentials lead to polygonizations with adjustable mixtures of triangles, tetragons, pentagons, etc. The different 'random polygonizations' still keep eq. (10.18) intact, whereas additional coupling constants simply reflect possible additional degrees of freedom to be represented by conformal matter in the continuum limit [471]. The moral of this story reads: calculating MM integrals of the type

$$e^Z = \int dM \, e^{-\mathrm{tr}\, V(M)} , \tag{10.19}$$

with some function $V(M)$, yields the perturbative definition of quantum 2d gravity, and may yet be able to provide non-perturbative information. Therefore, the first thing to do is to consider the continuum limit of MM's, in order to establish the connection, if any, with the CFT results. One should also find independent ways of calculating the MM integrals (sect. 2).

The MM free energy Z can be expanded in terms of the coupling constants g and N as follows:

$$Z(g, N) = \sum_h N^{2-2h} Z_h(g) = N^2 Z_0(g) + Z_1(g) + N^{-2} Z_2(g) + \dots , \tag{10.20}$$

where eq. (10.18) has been used. Generally speaking, in the large-N limit, only the *planar* contribution $Z_0(g)$ survives. In its turn, $Z_0(g)$ can be perturbatively expanded in the coupling constant g. In accordance with eqs. (10.6) and (10.9), in a vicinity of g_c, it should behave like [7]

$$Z_0(g) \sim \sum_n n^{\gamma-3}(g/g_c)^n \sim (g - g_c)^{2-\gamma} , \quad \bar{\gamma} = \frac{\gamma - 2}{2} , \tag{10.21}$$

where the new constant γ has been introduced. Eq. (10.21) obviously exhibits the critical behaviour when g approaches its critical value g_c, as expected. The continuum limit has

---

[7] This asymptotical behaviour will be derived from an MM integral in the next section.

to correspond to this critical point. This can be actually seen from the expectation value of the surface area as follows:

$$\langle A \rangle = \varepsilon \langle n \rangle = \varepsilon \frac{\partial}{\partial g} \ln Z(g) \sim \frac{\varepsilon}{g - g_c} \ . \tag{10.22}$$

We must rescale the area ε of each triangle to zero in order to obtain a continuum surface. Obviously, this surface can only have a finite area A when simultaneously $g \to g_c$. Intuitively, the perturbation series diverges at the critical point since the integral becomes dominated by the MM Feynman graphs with an infinite number of vertices, and it is just what should be expected to happen in the continuum limit. Of course, this is by no means a proof, and the best we can do is to compare the scaling properties of the partiton function to be calculated by the MM methods with those derived by CFT. Quite natural things to compare are the predictions for the string susceptibility critical exponent γ. On the one hand, it appears in eq. (10.9), $viz.$

$$Z(A) \propto A^{(\gamma-2)\chi/2-1} \ . \tag{10.23}$$

On the other hand, the string susceptibility is quite independently predicted by the quantum Liouville theory (sect. IX.1). In the case of the CFT minimal models, the CFT prediction is given by eq. (9.70b). The case of $p = 2$ corresponds to the pure 2d gravity having $\gamma = -\frac{1}{2}$, while the case of $p = 3$ corresponds to a fermion (or the critical Ising model) coupled to 2d gravity, and having $\gamma = -1/3$. There exists the KPZ barrier at $c_{matter} = 1$ for all these results (sect. IX.1), which seems to indicate a phase transition between the weakly coupled phase and the strongly coupled phase of 2d quantum gravity. The MM computations in the next section confirm the CFT results. This provides the good evidence to believe that the MM's are able to produce sensible continuum results. Much more supporting evidence in favour of such conjecture can be found in the book [439] and in the references therein.

The MM's are capable to provide some all-genus information about the 2d quantum gravity partition function in the double-scaling limit. All the partition function coefficients $Z_h(g)$, not just $Z_0(g)$, should have the critical point at the same critical value $g = g_c$, since a divergence of the perturbation series is a local phenomenon and, hence, it should not depend on the genus which is a global parameter. In accordance with eq. (10.10), eq. (10.21) can be generalized to [8]

$$Z_h(g) \sim \sum_n n^{(\gamma-2)\chi/2-1} (g/g_c)^n \sim (g - g_c)^{(2-\gamma)\chi/2} \ . \tag{10.24}$$

---

[8] This equation will also be derived in the next section.

Hence, the higher-genus contributions with $\chi < 0$ are blowing up when g approaches g_c because of $\gamma < 2$. This shows the way how the double scaling limit works: the large-N suppression at higher genus is compensated by the $g \to g_c$ enhancement, so that they give a coherent contribution together [456, 457, 458]. Eq. (10.11) can be rewritten in terms of the leading singular piece of $Z_h(g)$ in eq. (10.24),

$$Z_h(g) \dot{=} f_h N^\chi (g - g_c)^{(2-\gamma)\chi/2} , \tag{10.25}$$

and of the variable

$$\kappa^{-1} \equiv N(g - g_c)^{(2-\gamma)/2} = z^{-\tilde{\gamma}} , \tag{10.26}$$

to the form

$$Z = \sum_h \kappa^{2h-2} f_h = \kappa^{-2} f_0 + f_1 + \kappa^2 f_2 + \dots . \tag{10.27}$$

In the double scaling limit $N \to \infty$, $g \to g_c$, the 'renormalized' coupling constant κ is kept fixed.

Exercise

X-1 ▷ (*) Consider 'random squarulations' of a Riemann surface with n squares, and introduce the dual to this covering which is a four-valent graph. Show that the MM integral $e^F = \int dM \, \exp[-\text{tr}\, M^2/2 - (g/N)\text{tr}\, M^4]$ is the generating functional for the 'random squarulations'. Check that the identifications in eq. (10.18) apply.

X.2   One-matrix model solution

The confirmation of the claims about the MM's stated above comes from their solution in the large-N limit. It is by now standard to use the *orthogonal polynomials* for this purpose [472, 411]. [9]

The starting point is the MM integral (10.19) specified by a polynomial potential $V(M)$, which is assumed to be *even* for simplicity. It can be represented in the form

$$e^Z = \int dM \, \exp\left[-\text{tr}\, V(M)\right] = \int \prod_{i=1}^{N} d\lambda_i \, \Delta^2(\lambda) \exp\left[-\sum_i V(\lambda_i)\right] , \tag{10.28}$$

---

[9]The first MM solution (in the planar limit) was found by saddle point methods in ref. [473].

where λ_i's are eigenvalues of the hermitian matrix M, and $\Delta(\lambda)$ is the Vandermonde determinant,

$$\Delta(\lambda) = \prod_{i<j}(\lambda_i - \lambda_j) = \det \lambda_i^{j-1} . \qquad (10.29)$$

To prove eq. (10.28), one uses a trick familiar from the quantum theory of gauge fields [474], namely substituting the identity represented by [10]

$$1 = \int dU \, \delta(UMU^\dagger - \Lambda)\Delta^2(\lambda) \qquad (10.30)$$

into the MM integral. In eq. (10.30), Λ is a diagonal matrix with the eigenvalues λ_i, which is related to the matrix M via a unitary matrix U_0 such that $M = U_0^\dagger \Lambda U_0$. The integration over M in eq. (10.28) is then easily performed, and this leaves the integral over the eigenvalues λ_i only. The integration over U now becomes trivial since this matrix actually decouples due to the cyclic invariance of the trace. To determine the remaining FP-determinant, it is enough to consider the infinitesimal neighborhood $U = (1 + T)U_0$ as usual, which yields

$$1 = \int dU \, \delta(UMU^\dagger - \Lambda)\Delta^2(\lambda) = \int dT \, \delta([T, \Lambda])\Delta^2(\lambda) , \qquad (10.31)$$

where $[T, \Lambda] = T_{ij}(\lambda_j - \lambda_i)$. Eqs. (10.28) and (10.29) immediately follow after integrating over the real and imaginary parts of the off-diagonal matrix elements of T.

Let $P_n(\lambda)$ be an infinite set of polynomials of a single real variable λ, which are orthogonal in the sense

$$\int_{-\infty}^{\infty} d\lambda \, e^{-V(\lambda)} P_n(\lambda)P_m(\lambda) = h_n \delta_{mn} , \qquad (10.32)$$

with some constants h_n. The normalization of these polynomials is supposed to be fixed by their leading term as $P_n(\lambda) = \lambda^n + \dots$, and so are the constants h_n. The orthogonal polynomials form a basis in the space of all polynomials of single variable. Hence, in particular, $\lambda P_n(\lambda)$ is expressible as a linear combination of lower P_i's, namely, $\lambda P_n(\lambda) = \sum_{i=0}^{n+1} a_i P_i(\lambda)$, where $a_i = h_i^{-1} \int d\lambda \, e^{-V} \lambda P_n P_i$. This can actually be represented as the simple recursion relation

$$\lambda P_n = P_{n+1} + r_n P_{n-1} , \qquad (10.33)$$

where the coefficient r_n is independent of λ. This happens since any term proportional to P_n on the r.h.s. of eq. (10.33) has to vanish (the potential is even!), whereas the possible

---

[10]The Haar measure on the unitary group is assumed to be normalized as $\int dU = 1$.

terms proportional to P_i with $i < n - 1$ should also vanish since $\int d\lambda \, e^{-V} P_n \lambda P_i = 0$. In addition, the defining equation (10.32) implies the relations

$$\int d\lambda \, e^{-V} P_n \lambda P_{n-1} = r_n h_{n-1} = h_n \ . \tag{10.34}$$

Pairing the λ in $P_n' \lambda P_n$ before and afterwards, and integrating by parts give the identities

$$n h_n = \int d\lambda \, e^{-V} P_n' \lambda P_n = \int d\lambda \, e^{-V} P_n' r_n P_{n-1} = r_n \int d\lambda \, e^{-V} V' P_n P_{n-1} \ . \tag{10.35}$$

The advantage of using the orthogonal polynomials to solve eq. (10.28) becomes clear after noticing that

$$\Delta(\lambda) = \det \lambda_i^{j-1} = \det P_{j-1}(\lambda_i) = \sum (-1)^\pi \prod_k P_{i_k - 1}(\lambda_k) \ , \tag{10.36}$$

where the sum goes over permutations i_k, and $(-1)^\pi$ is the parity of permutation. Substituting eq. (10.36) for each of the two $\Delta(\lambda)$'s in eq. (10.28) factorizes the integrals over individual λ_i's, while the only nonvanishing contributions come from the terms where all $P_i(\lambda_j)$'s are paired, due to the orthogonality condition. There are $N!$ such contributions in total, so that one finds

$$e^Z = N! \prod_{i=0}^{N-1} h_i = N! h_0^N \prod_{k=1}^{N-1} f_k^{N-k} \ , \tag{10.37}$$

where $f_k \equiv h_k / h_{k-1}$. Eq. (10.34) now implies that $r_n = f_n$.

In the planar large-N limit, the rescaled index k/N can usually be substituted by a continuous variable ξ running from 0 to 1, whereas f_k/N can be substituted by a continuous function $f(\xi)$. Correspondingly, the sum representing the partition function becomes a one-dimensional integral having the form

$$\frac{1}{N^2} Z = \frac{1}{N} \sum_k (1 - k/N) \ln f_k \sim \int_0^1 d\xi \, (1 - \xi) \ln f(\xi) \ . \tag{10.38}$$

A functional form of $f(\xi)$ is, of course, dependent on a choice of the potential $V(M)$. As a particular example, let us take the 'regularized' potential [459] [11]

$$V(\lambda) = \frac{1}{2g} \left(\lambda^2 + \frac{\lambda^4}{N} + b \frac{\lambda^6}{N^2} \right) \ , \tag{10.39}$$

---

[11] As can be easily seen, when using, e.g., eq. (10.41), the value of g_c is generically negative. Therefore, it is a good idea to add the λ^6-term with a coefficient $b < 0$ to the 'naive' λ^4-potential, in order to get the convergent MM integral.

so that $gV'(\lambda) = \lambda + 2\lambda^3/N + 3b\lambda^5/N^2$. The corresponding coefficients $f_n = r_n$ are to be determined from eq. (10.35). The r.h.s. of the equation contains terms having the form $\int d\lambda\, e^{-V}\lambda^{2p-1} P_n P_{n-1}$. The recursion relation (10.33) allows us to vizualise them as 'walks' comprising $2p - 1$ steps, namely, $p - 1$ steps up and p steps down [459]. These 'walks' start at n and end at $n - 1$, and each step down from m to $m - 1$ results in a factor of r_m while each step up only produces a factor of unity. There are $\begin{pmatrix} 2p - 1 \\ p \end{pmatrix}$ of such walks in total, and each contributes the factor of h_{n-1} coming from the integral $\int d\lambda\, e^{-V} P_{n-1} P_{n-1}$. Being combined with the corresponding r_n factor whose origin was just explained, it cancels the h_n factor on the l.h.s. of eq. (10.35). Given the potential (10.39), one finds the discrete *string equation* [459, 475]

$$gn = r_n + \frac{2}{N} r_n (r_{n+1} + r_n + r_{n-1}) + \frac{3b}{N^2}(\ldots)\,, \qquad (10.40)$$

where the dots stand for ten r^3-type terms [475].

In the large-N limit, the index n becomes a continuous variable ξ, so that $r_n/N \to r(\xi)$ and $r_{n\pm1}/N \to r(\xi \pm \varepsilon)$, where $\varepsilon \equiv 1/N$. Eq. (10.40) *to the leading order in* $1/N$ gives the continuous 'string equation'

$$g\xi = r + 6r^2 + 30br^3 \equiv W(r) = g_c + \tfrac{1}{2}W''\big|_{r=r_c} (r(\xi) - r_c)^2 + \ldots\,, \qquad (10.41)$$

where the function $W(r)$ has been expanded in a vicinity of the critical point r_c at which $W'|_{r=r_c} = 0$ and $W(r_c) \equiv g_c$. Hence, in this particular case, one finds the relation [459, 475] [12]

$$r(\xi) - r_c \sim (g_c - g\xi)^{1/2}\,. \qquad (10.42)$$

Of course, many details of this calculation, as well as the specific relations among the coefficients, are dependent on the matrix model chosen, and they vary among the different MM's. What turns out to be in common, however, is the specific form of the leading singular behaviour of the function $f(\xi) = r(\xi)$ at $\xi \to 1$, i.e. for g near some g_c in the large-N limit, namely

$$f(\xi) - f_c \sim (g_c - g\xi)^{-\gamma}\,. \qquad (10.43)$$

The critical exponent γ just introduced in eq. (10.43) will be shown to coincide with the string susceptibility in a moment. For the 2d gravity partition function Z in eq. (10.38),

---

[12]For a general even potential of the form $V(\lambda) = \frac{1}{2g}\sum_p a_p \lambda^{2p}$, one finds $W(r) = \sum_p a_p \frac{(2p-1)!}{(p-1)!^2} r^p$ [459, 475].

eq. (10.43) implies the following behaviour at $g \to g_c$:

$$\frac{1}{N^2}Z \sim \int_0^1 d\xi \, (1-\xi)(g_c - g\xi)^{-\gamma} \sim (1-\xi)(g_c - g\xi)^{-\gamma+1}\Big|_0^1 + \int_0^1 d\xi \, (g_c - g\xi)^{-\gamma+1}$$

$$\sim (g_c - g)^{-\gamma+2} \sim \sum_n n^{\gamma-3}(g/g_c)^n \ . \tag{10.44}$$

Comparing eq. (10.44) with eq. (10.21) shows the γ defined by eq. (10.43) is the string susceptibility indeed, while the large-N behaviour of the second derivative of Z (known as the *specific heat*) is given by

$$Z'' \sim (g_c - g)^{-\gamma} \sim f(1) - f_c \ . \tag{10.45}$$

The particular MM result in eq. (10.42) gives $\gamma = -\frac{1}{2}$ for the string susceptibility, which corresponds to pure 2d gravity. The fine-tuning of the parameters in the MM potential allows us to obtain different values of γ [471]. For example, a higher-order critical point for the W-function with $W'|_{r=r_c} = W''|_{r=r_c} = 0$ can be achieved by the fine-tuning of the parameter b in eq. (10.39). This obviously leads to the different leading term on the r.h.s. of eq. (10.41), which is proportional to $(r(\xi) - r_c)^3$. A simple calculation that is very similar to the one just described above then gives $\gamma = -1/3$ for the string susceptibility. Having a general potential $V(M)$ with enough parameters to prepare a critical point of the p-th order for the $W(r)$-function at $r = r_c$, one obtains $r - r_c \sim (g_c - g\xi)^{1/p}$ on the r.h.s. of eq. (10.41), and $\gamma = -1/p$ as the associated critical exponent (the string susceptibility) [471]. The MM interactions resulting in a higher-order critical behaviour can be associated with the additional degrees of freedom due to 2d conformal matter in the continuum limit.

The all-genus results are derivable from the MM in the double scaling limit, when retaining the higher order terms in eqs. (10.40) and (10.41). Then one gets

$$g\xi = W(r) + 2r(\xi)[r(\xi + \varepsilon) + r(\xi - \varepsilon) - 2r(\xi)]$$

$$= g_c + \tfrac{1}{2}W''\Big|_{r=r_c} [r(\xi) - r_c]^2 + 2r(\xi)[r(\xi + \varepsilon) + r(\xi - \varepsilon) - 2r(\xi)] + \dots \ . \tag{10.46}$$

Introducing the new parameter a of dimension [13] [*length*] as [459]

$$g - g_c = \kappa^{-4/5}a^2 \ , \tag{10.47}$$

the coherent double scaling limit $g \to g_c$, $N \to \infty$ at the fixed quantity $\kappa^{-1} = (g - g_c)^{5/4}N$ implies $a \to 0$ and $\varepsilon \equiv 1/N = a^{5/2}$. Since the critical point corresponds to

---

[13] The coupling constant g is of dimension [*length*]^2. We consider the case of $\gamma = -1/2$.

$\xi \to 1$, defining the new variable z by $g_c - g\xi = a^2 z$ is quite appropriate. After all, on dimensional reasons, it is natural to assume $r(\xi) = r_c + au(z)$ as the scaling ansatz. Substituting all this into eq. (10.41) gives the leading-order terms which are proportional to a^2, so that $u^2 \sim z$. One finds, in addition,

$$r(\xi + \varepsilon) + r(\xi - \varepsilon) - 2r(\xi) \sim \varepsilon^2 \frac{\partial^2 r}{\partial \xi^2} = a\frac{\partial^2}{\partial z^2} au(z) \sim a^2 u'' , \qquad (10.48)$$

since $\varepsilon(\partial/\partial\xi) = -ga^{1/2}(\partial/\partial z)$. Eq. (10.46) now implies a non-linear differential 'string equation' having the form [14]

$$z = u^2 - \frac{1}{3}u'' , \qquad (10.49)$$

known as the *Painlevé I* equation. Solutions to this equation characterize critical behaviour of the 2d quantum gravity partition function Z to all orders in the genus h [456, 457, 458]. The leading critical behaviour of the specific heat ($\sim Z''$) is governed by the function $f(\xi)$ at $\xi = 1$, i.e. when $z = (g - g_c)/a^2 = \kappa^{-4/5}$. In particular, as far as the leading term in Z is concerned, one finds $u \sim z^{1/2}$, and, hence, after two integrations, $Z \sim z^{5/2} \sim \kappa^{-2}$, in agreement with eq. (10.27).

A perturbative solution for the specific heat in powers of $z^{-5/2} = \kappa^2$ takes the form

$$u = z^{1/2}\left(1 - \sum_{h=1} u_h z^{-5h/2}\right) , \qquad (10.50)$$

where the coefficients u_h corresponding to the surfaces of genus h are all positive. Therefore, in the double scaling limit, the MM's provide us with the evidence supporting the claims made in the previous section about the scaling behaviour of the partition function, order by order in perturbation theory! Of course, they do not answer the fundamental question whether the non-perturbative MM features really describe the quantum 2d gravity, but, at least, they give some (very incomplete) non-perturbative structure to deal with [476].

An exact solution to the Painlevé I equation (10.49) generically has two integration constants. One of them can be fixed by the perturbative asymptotical behaviour in eq. (10.50), whereas the other constitutes a new parameter, whose influence on a topological perturbation is non-perturbative. Nevertheless, this introduces the ambiguity in a choice of the 'right' non-perturbative solution. The typical behaviour of a real solution $u(z)$ for real values of z is illustrated in Fig. 20. In particular, there are infinitely many double poles in the region $z < 0$. The disconnected regions between each pair of poles

---

[14]Suitable rescalings of the variables u and z have been made.

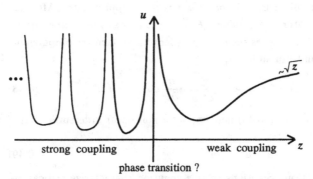

Fig. 20. The schematical behavior of a solution to the Painleve I
           equation for real values of the argument. The detailed shape
           of a solution depends on non-perturbative free parameters.

presumably correspond to strong gravity phases related to a weak gravity phase in the
region $z > 0$ by a kind of a phase transition. The instability of vacuum is signalled by
the fact that any pole-free solution to eq. (10.49) is necessarily complex. [15]

In the cases of higher-order multi-critical points with $W'|_{r=r_c} = W''|_{r=r_c} = \ldots = 0$,
one goes along similar lines by defining $g - g_c = \kappa^{2/(\gamma-2)}a^2$ and $\varepsilon = 1/N = a^{2-\gamma}$, and
considering the limit $a \to \infty$, $\varepsilon \to \infty$ at the fixed 'renormalized' coupling constant
$(g - g_c)^{1-\gamma/2}N = \kappa^{-1}$. The critical value of $\xi = 1$ then corresponds to $z = \kappa^{2/(\gamma-2)}$, or
$\kappa^2 = z^{\gamma-2}$. The scaling ansatz is still dictated by dimensional analysis which implies
$r(\xi) = r_c + a^{-2\gamma}u(z)$. An analogue to the change of variables in eq. (10.48) now takes
the form $\varepsilon(\partial/\partial\xi) = -ga^{-\gamma}(\partial/\partial z)$. In particular, when $\gamma = -1/3$, one gets

$$r(\xi) = r_c + a^{2/3}u(z) , \quad \kappa^2 = z^{-7/3} , \quad \varepsilon(\partial/\partial\xi) = -ga^{1/3}(\partial/\partial z) . \tag{10.51}$$

The 'string equation' (10.40) now implies a non-linear differential equation (after suitable
rescalings of the variables) in the form [459]

$$z = u^3 - uu'' - \tfrac{1}{2}(u')^2 + \alpha u''' , \tag{10.52}$$

where $\alpha = 1/10$. The corresponding perturbative expansion takes the form

$$u = z^{1/3}\left(1 + \sum_{k=1} u_k z^{-7k/3}\right) , \tag{10.53}$$

---

[15] The constraints on admissible solutions to the Painlevé I equation are discussed in ref. [476].

whose coefficients u_k are only positively definite when $\alpha < \frac{1}{12}$. This means that the given theory of 2d gravity with matter fails to be unitary. Although the value of the MM-predicted parameter γ does coincide with its value for the unitary Ising model coupled to 2d gravity, eq. (10.53) actually corresponds to a non-unitary CFT known as the theory of the *Yang-Lee edge singularity* which is obtained by coupling the Ising model to a magnetic field of particular (imaginary) value, in addition to the coupling to 2d gravity [459, 475]. [16] Being coupled to 2d gravity, the (p, q) minimal models have the critical exponent γ given by eq. (9.70a). The p-th order multicritical point of the one-matrix models we consider happens to describe the $(2p - 1, 2)$ minimal model coupled to 2d gravity, whose critical exponent value is also $\gamma = -1/p$, just like that for the unitary discrete series (coupled to 2d gravity). The other (p, q) minimal models appear when considering the multi-matrix models (MMM's) [459, 475] (see sect. 3 also).

Given a potential $W(r)$ with the p-th order critical point, the value of the critical exponent is $\gamma = -1/p$, whereas the scaling ansatz has to be of the form $r(\xi) = r_c + a^{2/p} u(z)$. Hence, one gets $u \sim z^{1/p}$ and $Z \sim z^{2+1/p} = \kappa^{-2}$ for the leading behaviour, which is again consistent with eq. (10.27). The corresponding non-linear differential equation generalizing those in eqs. (10.49) and (10.53) turns out to be the p-th member of the KdV-type hierarchy (see Ch. XI and the next section).

Exercise

X-2 ▷ (*) Consider the Vandermonde determinant and prove the identity in eq. (10.29). When $N = 3$, one gets, in particular

$$(\lambda_3 - \lambda_2)(\lambda_2 - \lambda_1)(\lambda_3 - \lambda_1) = \det \begin{pmatrix} 1 & \lambda_1 & \lambda_1^2 \\ 1 & \lambda_2 & \lambda_2^2 \\ 1 & \lambda_3 & \lambda_3^2 \end{pmatrix} . \qquad (10.54)$$

X.3   Multi-matrix models

The (one-)MM's discussed so far can be generalized to the so-called **multi-matrix models** (MMM's). This generalization was, in fact, inspired by the desire to describe not only the 2d quantum gravity itself, but 2d quantum fields (like CFT matter) also.

---

[16] The specific heat of the conventional critical Ising model coupled to 2d gravity is also determined by eq. (10.53), but with $\alpha = 2/27$ [459, 475].

Once one represents the 2d gravity by the sum over triangulations of a surface, it is quite natural to describe the quantum fields by using statistical mechanics on the lattice associated with a triangulated surface, where the new degrees of freedom are represented by more than one matrix. The tractable (but quite non-trivial) case is given by the 'matrix chain', i.e. an MMM integral in the so-called *Itzykson-Zuber* form [472]

$$
Z = \ln \int \prod_{i=1}^{q-1} dM_i \, \exp\left[-\mathrm{tr}\left(\sum_{i=1}^{q-1} V_i(M_i) - \sum_{i=1}^{q-2} c_i M_i M_{i+1} \right) \right]
$$

$$
= \ln \int \prod_{\substack{i=1,q-1 \\ \alpha=1,N}} d\lambda_i^{(\alpha)} \, \Delta(\lambda_1) \exp\left[-\sum_{i,\alpha} V_i(\lambda_i^{(\alpha)}) + \sum_{i,\alpha} c_i \lambda_i^{(\alpha)} \lambda_{i+1}^{(\alpha)} \right] \Delta(\lambda_{q-1}) , \qquad (10.55)
$$

in terms of $(q-1)$ $N \times N$ hermitian matrices M_i, $i = 1, \ldots, q-1$, with the eigenvalues $\lambda_i^{(\alpha)}$, $\alpha = 1, \ldots, N$, and the associated Vandermonde determinants, $\Delta(\lambda_i) = \prod_{\alpha < \beta}(\lambda_i^{(\alpha)} - \lambda_i^{(\beta)})$. The coefficients $\{c_i\}$ characterize the coupling of the matrices along a line ('chain'), with no closed loops. This allows us to integrate over the relative angular variables w.r.t. M_i, in the defining equation (10.55).

Via its diagrammatic perturbative expansion, the MMM integral (10.55) can be interpreted as the generating function for discretized coverings of Riemann surface by the dual Feynman graphs. The different matrices M_i then represent $q - 1$ different matter states existing at the (dual) faces, where each 'vertex' in the original graph comes from expanding the factor $\exp[-\mathrm{tr}V(M_i)]$. The quantity Z in eq. (10.55) can now be interpreted as the partition function of the 2d gravity coupled with matter. When computing the MMM integral, one should sums not only over all the equivalence classes of graphs as before, but over all the maps of the set of vertices in a given graph to the finite set $(1, 2, \ldots, q-1)$ in addition. A closer inspection of eq. (10.55) shows that these maps are to be summed with some local Boltzmann weights which are not equal, in general.

There are many ways of generalizing eq. (10.55). For instance, letting the chain of matrices to become infinite allows us to consider matter in $D = 1$, or strings propagating on a circle of finite radius; letting the index i to be continuous introduces a field theory of matrices which is connected with the Liouville theory [439, 459]. It is not our purpose here to go far away from the formal definition of the MMM's (see, however, refs. [462, 464, 472]), so, in this section, we confine ourselves to the two examples, each related to different areas. Our first example is the two-matrix model, whose partition function was considered in ref. [459],

$$e^Z = \int dU dV \, \exp\left[-\mathrm{tr}\left(U^2 + V^2 - 2cUV + \frac{g}{N}(e^H U^4 + e^{-H} V^4)\right)\right], \tag{10.56}$$

in terms of two hermitian $N \times N$ matrices U and V, with constant parameters c and H. The Feynman diagrammatic expansion of Z contains two different types of the 4-point vertices (Fig. 19b) corresponding to the insertions of U^4 and V^4, resp. The 'thick' propagator (Fig. 19a) determined by the quadratic term in the exponential of eq. (10.56) reads

$$\begin{pmatrix} 1 & -c \\ -c & 1 \end{pmatrix}^{-1} = \frac{1}{1-c^2}\begin{pmatrix} 1 & c \\ c & 1 \end{pmatrix}. \tag{10.57}$$

Hence, the lines connecting the similar vertices (two U^4 or two V^4) get a factor of $1/(1-c^2)$, whereas the lines connecting the different vertices get a factor of $c/(1-c^2)$. This turns out to be the exact structure needed to realize the Ising model (sect. II.7) on a random lattice! Up to an irrelevant overall normalization factor, the Feynman expansion of the MMM integral (10.56) gives the partition function

$$Z = \sum_{\text{lattices}} \sum_{\substack{\text{spin} \\ \text{configurations}}} \exp\left[\beta \sum_{\langle ij\rangle} \sigma_i \sigma_j + H \sum_i \sigma_i\right], \tag{10.58}$$

where H can now be interpreted as constant magnetic field. The weights for the two allowed values of neighboring spins are $e^{\pm\beta}$, where β is the temperature. Hence, one can identify $e^{2\beta} = 1/c$. Solving the MMM of eq. (10.56) means, therefore, solving the Ising model in the presence of a magnetic field. This illustrates the power of the MMM approach based on the summation over random lattices, when compared to the conventional approach based on the formulation of the theory on a regular lattice, since eq. (10.56) is easier to calculate [477]. Of course, the MMM solution is less informative since one averages over the lattices as well, which results in a higher symmetry.

The basic idea to solve eq. (10.56) is to rewrite it in terms of the eigenvalues x_i of U and y_i of V to the form

$$e^Z = \int \prod_i dx_i dy_i \, \Delta(x)\Delta(y) e^{-W(x_i,y_i)}, \tag{10.59}$$

where

$$W(x_i, y_i) = x_i^2 + y_i^2 - 2c x_i y_i + \frac{g}{N}(e^H x_i^4 + e^{-H} y_i^4). \tag{10.60}$$

The appropriate orthogonal polynomials are to be defined w.r.t. the bilocal measure, viz.

$$\int dx dy \, e^{-W(x,y)} P_n(x) Q_m(y) = h_n \delta_{mn}, \tag{10.61}$$

where $P_n \neq Q_n$ for $H \neq 0$. This results in the partition function $e^Z \propto \prod_i h_i \prod_i f_i^{N-i}$, and the following recursion relations for the polynomials [477]:

$$x P_n(x) = P_{n+1} + r_n P_{n-1} + s_n P_{n-3} ,$$

$$y Q_m(y) = Q_{m+1} + q_m Q_{m-1} + t_m Q_{m-3} , \qquad (10.62)$$

where $f_n \equiv h_n/h_{n-1}$. A calculation of the coefficients f_n is entirely based on these recursion relations, similarly to the previous section. In particular, the specific heat $Z'' \sim u$ is given by a solution to eq. (10.52) with $\alpha = 2/27$, as expected. In order to achieve criticality for lattice *and* spin fluctuations simultaneously, one must tune both parameters to $H_c = 0$ and $c_c = 1/4$, where the free energy undergoes a third-order phase transition. The critical exponents for the gravitationally dressed Ising model differ from the usual ones (*cf* sect. II.7):

	$h[\varepsilon]$	$h[\sigma]$	ν	η	α	β	γ	
Ising	1/2	1/16	1	1/4	0	1/8	7/4	(10.63)
Ising + 2d gravity	2/3	1/6	3/2	2/3	−1	1/2	2	

where α, β, γ follow from ν, η according to the scaling laws in eq. (2.130).

Our second MMM example is relevant for the topological string theories and the topological 2d gravity (sect. IX.3), with a non-trivial ($d \neq 0$) matter sector. To this end, we are going to use the KdV-based description of the MMM's, so that the reader unfamiliar with the KdV approach to the MMM's is invited to consult the next section first, before reading this section any further. In the KdV approach (sect. 4) to the MMM's defined in eq. (10.55) for a chain of $q - 1$ matrices, one starts with the q-order differential operator Q

$$Q = d^q - \frac{1}{q} \sum_{i=0}^{q-2} u_i(z) d^i , \qquad (10.64)$$

where functions $u_i(z)$ are to be determined from eq. (10.97) which replaces the MMM 'string equation'. The physical meaning of functions $u_i(z)$ can be extracted from the generalized KdV flows satisfying (by definition) a Lax equation of the form

$$\frac{\partial Q}{\partial t_{k-1}} = [L_+^k, Q] = [Q, L_-^k] , \qquad (10.65)$$

where $L \equiv Q^{1/q} \equiv \sum_{i=-\infty}^{1} l_i d^i \equiv L_+ + L_-$ is a pseudo-differential operator, and L_+ is its part corresponding to the ordinary differential operator (sect. 4). Let us associate the operators $\mathcal{O}_k$ to the flows $\partial/\partial t_k$, where $\mathcal{O}_0 \equiv \mathcal{P}$ is the puncture operator [445]. Since the puncture-puncture 2-point function is given by $u_{q-2} = Z''_{\text{connected}} = \langle \mathcal{PP} \rangle$, the generalized KdV flows (10.65) can be rewritten to the form

$$\langle \mathcal{PP}\mathcal{O}_{k-1} \rangle = \frac{\partial u_{q-2}}{\partial t_{k-1}} = \frac{\partial}{\partial z} \text{res}\,(L^k)\,, \qquad (10.66)$$

where $\text{res}\,(L^k) \equiv l_{-1}$ is called a *residue* of a pseudo-differential operator. Integrating eq. (10.66) once yields

$$\langle \mathcal{P}\mathcal{O}_{k-1} \rangle = \text{res}\,(L^k)\,, \qquad (10.67)$$

where the r.h.s. is a polynomial in the functions $u_i(z)$ and their derivatives. As far as the first $q-1$ flows $\mathcal{O}_i$, $i = 0, 1, \ldots, q-2$, are concerned, one finds, in particular [445]

$$\langle \mathcal{P}\mathcal{O}_i \rangle \sim u_{q-i-2} + \ldots\,, \qquad (10.68)$$

where the dots stand for some corrections with derivatives, e.g., $\langle \mathcal{P}\mathcal{O}_1 \rangle \sim u_{q-3} + u'_{q-2}$.

The MMM operators $\mathcal{O}_{k-1}$ associated to the flows generated by L^k have, therefore, the 'band' structure, since they are only non-trivial when k is not a multiple of q (when $k = nq$ with some integer n, there are no L^k flows since the operator $L^{nq} = Q^n$ trivially commutes with Q). The first 'band' comprises $q-1$ operators $\mathcal{O}_0, \ldots, \mathcal{O}_{q-2}$, then there is a gap after which the next 'band' appears. This structure has the nice CFT interpretation as a topological string theory [478], after identifying the first 'band' operators $\mathcal{O}_i$ with $q-1$ primaries ϕ_i of the matter sector. The second 'band' would then represent the first-generation of descendants $\sigma_1(\phi_i)$ and so on, $\sigma_k(\phi_i) = \mathcal{O}_{kq+i}$. The corresponding $U(1)$ charges and the matter central charge $c/3 = d$ are given by [478]

$$q_i = \frac{1}{q}\,, \qquad d = 1 - \frac{2}{q}\,. \qquad (10.69)$$

It is not accidental that the central charge formula in eq. (10.69) coincides with that in eq. (5.106) for the $N = 2$ minimal models represented by the $N = 2$ LG theories [479, 480]. To describe topological matter, we actually need the *twisted* $N = 2$ minimal models of central charge $d = c/3 = k/(k + 2)$ [481]. The LG models have the property that their correlators are completely determined by the superpotential $W(X)$, which is a quasi-homogeneous function of chiral primary fields X, whose operator algebra is given by the polynomial ring (5.102). Let us consider the A-series of eq. (5.107) with $\phi_i(t = 0) \equiv X^i$, $i = 0, 1, \ldots, k$, at the conformal point $t = 0$, as an example [479, 480].

The OPE's for products of chiral primary operators from the A_{k+1}-ring at $t = 0$ are

$$\phi_i \cdot \phi_j = \begin{cases} \phi_{i+j} , & i+j \le k , \\ 0 , & \text{otherwise} , \end{cases} \tag{10.70}$$

whereas the general (perturbed) superpotential is

$$W = \frac{X^{k+2}}{k+2} - \sum_{i=0}^{k} g_i(t) X^i . \tag{10.71}$$

where the functions $g_i(t) = t_i + O(t^2)$ describe perturbations. At arbitrary values of t, the functions $\phi_i(t)$ are defined by

$$\phi_i(X) = -\frac{\partial W(X)}{\partial t_i} = \sum_{i=0}^{k} \frac{\partial g_j}{\partial t_i} X^j . \tag{10.72}$$

The OPE's for products of the operators $\phi_i(X)$ also form the polynomial ring,

$$\phi_i(X)\phi_j(X) = \sum c_{ij}{}^l \phi_l(X) \bmod W'(X) . \tag{10.73}$$

Taking the expecation value of both sides of this equation gives

$$\langle \phi_i \phi_j \rangle = c_{ij}{}^k(t) , \tag{10.74}$$

since the field ϕ_k is the only one which has the non-vanishing expectation value at $t = 0$ by charge conservation, and this remains to be true for $t > 0$ because the condensate of $\int \phi_l^{(2)}$ have charges $q_l = \frac{l}{k+2} - 1 < 1$. Therefore, one finds [445, 478]

$$\langle \phi_i \phi_j \rangle = \oint \frac{dx}{2\pi i} \frac{\phi_i(X)\phi_j(X)}{W'(X)} . \tag{10.75}$$

The fields $\phi_i(X)$ are to be orthogonal w.r.t. this inner product, so that we are able to represent them in terms of fractional powers of the superpotential as [445, 478]

$$\phi_i(X) = \frac{\partial}{\partial x} \frac{L_+^{i+1}}{i+1} , \quad \text{where} \quad W = \frac{L^{k+2}}{k+2} . \tag{10.76}$$

One finds indeed,

$$\langle \phi_i \phi_j \rangle = \oint \frac{dx}{2\pi i} \frac{(L^i \partial_x L)_+ (L^j \partial_x L)_+}{L^{k+1} \partial_x L} = \oint \frac{dx}{2\pi i} L^{i+j-k-1} \frac{\partial L}{\partial x} = \delta_{i+j,k} . \tag{10.77}$$

The crucial point comes from the observation that the two definitions of fields $\phi_i(X)$ in eqs. (10.72) and (10.76) are to be compatible. This leads to the compatibility equation

$$\frac{\partial W}{\partial t_i} = const. \frac{\partial}{\partial x} W_+^{\frac{i+1}{k+2}} . \tag{10.78}$$

Eq. (10.78) is quite restrictive, and its solution reads [445, 478]

$$\phi_i(X) = (-1)^i \det \begin{pmatrix} -X & 1 & 0 & \cdots & 0 \\ t_k & -X & 1 & \ddots & \vdots \\ t_{k-1} & t_k & \ddots & \ddots & 0 \\ \vdots & \vdots & \ddots & \ddots & 1 \\ t_{k-i+2} & \cdots & t_{k-1} & t_k & -X \end{pmatrix} \qquad (10.79)$$

The information just given above is enough to determine the correlation functions of the twisted $N = 2$ models. To show that this topological solution agrees with the corresponding MMM solution, one should use the KdV operator Q of the form

$$Q = \frac{d^{k+2}}{k+2} - \sum_{i=0}^{k} u_i(z) d^i \ , \qquad (10.80)$$

which is dictated by the form of LG superpotential in eq. (10.71). Then, one notices that the associated KdV flow equation

$$\frac{\partial Q}{\partial t_i} = [Q_+^{\frac{i+1}{k+2}}, Q] \ , \qquad (10.81)$$

is, in fact, equivalent to eq. (10.78) [450]. Therefore, the twisted $N = 2$ minimal models are equivalent to the MMM's indeed.

X.4 Matrix models and KdV

The purpose of this section is to explain the relevance of the KdV hierarchy for the (M)MM's and 2d gravity [482]. [17] First, let us introduce the orthonormal basis $\{\Pi_n \equiv h_n^{-1/2} P_n\}$, satisfying the orthonormality condition

$$\int_{-\infty}^{+\infty} d\lambda\, e^{-V} \Pi_n \Pi_m = \delta_{nm} \ , \qquad (10.82)$$

and the recursion relations

$$\lambda \Pi_n = \sqrt{\frac{h_{n+1}}{h_n}} \Pi_n + r_n \sqrt{\frac{h_{n-1}}{h_n}} \Pi_{n-1} = \sqrt{r_{n+1}} \Pi_{n+1} + \sqrt{r_n} \Pi_{n-1} \equiv Q_{nm} \Pi_m \ , \qquad (10.83)$$

---

[17]If the reader is not familiar with the KdV theory, he is invited to read sect. XI.1 first.

which are the obvious consequences of eqs. (10.32) and (10.33), resp. In matrix notation, one writes

$$\lambda \Pi = Q \Pi \,, \qquad (10.84)$$

where the matrix Q has elements

$$Q_{nm} = \sqrt{r_m} \delta_{m,n+1} + \sqrt{r_n} \delta_{m+1,n} \,. \qquad (10.85)$$

Eq. (10.82) implies $\int_{-\infty}^{+\infty} d\lambda \, e^{-V} \lambda \Pi_n \Pi_m = Q_{nm} = Q_{mn}$, so that Q is the symmetric Jacobi matrix which should correspond to a hermitian operator in the continuum limit. Fortunately, this can be realized explicitly [482, 483]. Substituting the scaling anzatz $r(\xi) = r_c + a^{2/p} u(z)$ for the p-th multi-critical point (sect. 2) into eq. (10.85) yields the Q-operator [459]

$$Q \rightarrow [r_c + a^{2/p} u(z)]^{1/2} e^{\varepsilon \frac{\partial}{\partial \xi}} + e^{-\varepsilon \frac{\partial}{\partial \xi}} [r_c + a^{2/p} u(z)]^{1/2} \,. \qquad (10.86)$$

Since $\varepsilon \frac{\partial}{\partial \xi} \rightarrow -g a^{1/p} \frac{\partial}{\partial z}$, as far as the *leading* term of Q is concerned, eq. (10.86) gives the hermitian second-order differential operator

$$Q = 2\sqrt{r_c} + \frac{a^{2/p}}{\sqrt{r_c}} (u + r_c \kappa^2 \partial_z^2) \,, \qquad (10.87)$$

where the constant term value $(2r_c^{1/2})$ is not universal, unlike the structure of the operator Q itself.

Given any differential operator Q, another (pseudo-)differential operator P can be introduced as a solution to the differential equation

$$[P, Q] = 1 \,. \qquad (10.88)$$

Eq. (10.88) plays the fundamental role in the theory of 2d integrable equations, where it occurs in their Lax representations. The operator P is called a '*fractional power*' of Q. Differentiating eq. (10.84) gives $[A, Q] = 1$, where the operator A satisfies the equation

$$\frac{\partial}{\partial \lambda} \Pi_n = A_{nm} \Pi_m \,. \qquad (10.89)$$

The operator P defined by eq. (10.88) is not unique, but it can be constrained even further. In the case under consideration, this operator can be made anti-symmetric, or (anti-)hermitian in the continuum limit (while the matrix A does not have any (anti)symmetry properties) by defining

$$P \equiv A - \tfrac{1}{2} V'(Q) = \tfrac{1}{2}(A - A^{\mathrm{T}}) \,. \qquad (10.90)$$

Indeed, differentiating term by term on the r.h.s. of the identity

$$0 = \int_{-\infty}^{+\infty} d\lambda \frac{\partial}{\partial \lambda} \left(\Pi_n \Pi_m e^{-V} \right) , \tag{10.91}$$

and using eqs. (10.82), (10.83) and (10.89), one gets $A + A^{\mathrm{T}} = V'(Q)$.

Given a potential V of order $2l$, $V = \sum_{k=0}^{l} a_k \lambda^{2k}$, the matrix elements $A_{mn} = \int d\lambda\, e^{-V} \Pi_n \partial_\lambda \Pi_m = \int d\lambda\, e^{-V} V' \Pi_n \Pi_m$ at $m > n$ may be nonvanishing for $m - n \leq 2l - 1$. This implies $P_{mn} \neq 0$ for $|m - n| \leq 2l - 1$, and, hence, the highest possible order of differential operator P is $(2l - 1)$. In particular, the only condition $W' = 0$ implies a 3rd-order differential operator P in the continuum limit, while the $l - 1$ multi-critical conditions $W' = W'' = \ldots = W^{(l-1)} = 0$ imply that P is a $(2l - 1)$-th order differential operator in the same limit. After suitable rescalings of the variables, eq. (10.87) reads

$$Q = d^2 - u , \tag{10.92}$$

as far as the universal part of Q is concerned. The anti-hermitian 3rd-order differential operator P corresponding to a simple critical point with $W' = 0$ is given by [18]

$$P = d^3 - \frac{3}{4}\{u, d\} , \tag{10.93}$$

so that

$$1 = [P, Q] = 4R_2' = \left(\frac{3}{4}u^2 - \frac{1}{4}u'' \right)' . \tag{10.94}$$

Integrating once this equation w.r.t. z gives, in essence, the Painlevé I equation, which is simultaneously the first member of the KdV hierarchy (sect. XI.1).

The natural generalization of the KdV connection comes from the MMM's. The operators Q_i and P_i satisfying $[P_i, Q_i] = 1$ represent insertions of λ_i and $\partial/\partial\lambda_i$, resp., into the MMM integral on the r.h.s. of eq. (10.55) [472]. In the double scaling limit, both P and Q become differential operators of some finite orders, say p and q, resp., [19] which still satisfy eq. (10.88) [482]. The generalized structure of Q takes the form

$$Q = d^q + \{v_{q-2}(z), d^{q-2}\} + \ldots + 2v_0(z) , \tag{10.95}$$

where $d \equiv d/dz$. A term with d^{q-1} does not appear in eq. (10.95), since it can be removed by rescaling of the type $Q \rightarrow f^{-1}(z)Qf(z)$. Therefore, the whole problem of solving the MMM in the double scaling limit is reduced to the problem of finding solutions to the

---

[18] The construction of P in the general MMM case will be explained in a moment.

[19] For definiteness, we assume $p > q$.

differential equation (10.88), with the operator Q given by eq. (10.95). When p and q are relatively prime integers, the p-th order *pseudo-differential* operator P is a fractional power of the operator Q,

$$P = Q_+^{p/q} \, , \tag{10.96}$$

where only non-negative powers of d are kept on the r.h.s. Therefore, the differential equations describing the (p, q) minimal models coupled to 2d quantum gravity are given by

$$[Q_+^{p/q}, Q] = 1 \, . \tag{10.97}$$

The formal operations of the *pseudo-differential calculus* just introduced above can be properly defined [250]. An ordinary differential operator d acts on a function $f(z)$ according to the rule $[d, f] = f'$. Pseudo-differential operators D (of order n) generalize ordinary differential operators by allowing (infinitely many) negative powers of the derivative d, but still a finite number (n) of positive powers of d,

$$D = \sum_{i=-\infty}^{n} u_i(z) d^i \, . \tag{10.98}$$

The formal inverse d^{-1} can be introduced as the formal Laurent series [250],

$$d^{-1} f = \sum_{j=0}^{\infty} (-1)^j f^{(j)} d^{-j-1} \, , \tag{10.99}$$

so that

$$dd^{-1} = d^{-1}d = 1 \, . \tag{10.100a}$$

It follows, in particular

$$[f, d^{-1}] = f d^{-1} - d^{-1} f = d^{-1}(df - fd)d^{-1} = d^{-1} f' d^{-1}$$

$$= f' d^{-2} - [f', d^{-1}]d^{-1} = f' d^{-2} - f'' d^{-3} + \dots \, . \tag{10.100b}$$

The definition (10.98) is consistent with the Leibniz rule:

$$[d^p, f] = \sum_{k=1}^{\infty} \binom{p}{k} a^{(k)} d^{p-k} \, , \tag{10.101a}$$

where the binomial coefficients have been extended to arbitrary values of p,

$$\binom{p}{k} = \frac{p(p-1)\cdots(p-k+1)}{k!} \, . \tag{10.101b}$$

This formalism allows us to associate with every linear (pseudo-)differential operator D of order q, normalized so as to have d^q as its leading term,

$$\Delta = d^q + \sum_{j \geq 1} a_j d^{q-j} , \qquad (10.102)$$

its inverse Δ^{-1}, its q-th root $L = \Delta^{1/q}$ and its fractional powers $\Delta^{p/q}$. They are all the uniquely defined differential operators, and they satisfy the properties [250]:

$$\Delta \Delta^{-1} = \Delta^{-1} \Delta = 1 ,$$

$$L^q = \Delta ,$$

$$\Delta^{p/q} = L^p ,$$

$$[\Delta^{p/q}, \Delta^{\tilde{p}/q}] = [L^p, L^{\tilde{p}}] = 0 . \qquad (10.103)$$

Indeed, given an operator Δ of the form (10.102), one writes

$$L = \Delta^{1/q} = d + \sum_{k \geq 0} e_k d^{-k} , \qquad (10.104)$$

whose coefficients e_k are recursively and uniquely determined by identifying L^q and Δ:

$$q e_k + \text{differential polynomial in } (e_1, \ldots, e_{k-1}) = \text{coefficient at } d^{q-1-k} \text{ in } \Delta . \quad (10.105)$$

In particular, when $a_1 = 0$, the first terms read as follows [484]:

$$L = d + \frac{1}{q} a_2 d^{-1} + \left(\frac{a_3}{q} - \frac{q-1}{2q} a_2' \right) d^{-2}$$

$$+ \left(\frac{a_4}{q} + \frac{q^2-1}{12q} a_2'' - \frac{q-1}{2q} [a_2^2 + a_3'] \right) d^{-3} + \ldots . \qquad (10.106)$$

The existence and uniqueness of the inverse Δ^{-1} follows along the similar lines [484]:

$$\Delta^{-1} = d^{-q} - a_1 d^{-q-1} + (q a_1' - a_2 + a_1^2) d^{-q-2} + \ldots . \qquad (10.107)$$

A pseudo-differential operator Δ of the form (10.102) can be decomposed as $\Delta = \Delta_+ + \Delta_-$, where Δ_+ means its ordinary differential part,

$$\Delta_+ = d^q + \sum_{j=1}^{q} a_j d^{q-j} , \qquad (10.108)$$

and Δ_- means the rest. The *residue* of Δ is a coefficient at the d^{-1} term,

$$\text{res } \Delta \equiv a_{n+1}(z) \ .\tag{10.109}$$

The residue of a commutator of two pseudo-differential operators is a total derivative (exersise # X-3).

The power of the KdV-based approach can be illustrated by reproducing the MM results of sect. 2, for the case of $(2l - 1, 2)$ minimal models coupled to 2d gravity [459]. These models correspond to the hermitian $Q \equiv K$ operator in eq. (10.92). [20] The anti-hermitian operator $Q^{l-1/2} = K^{l-1/2}$ now reads

$$K^{l-1/2} = d^{2l-1} - \frac{2l - 1}{4} \left\{ u, d^{2l-3} \right\} + \dots \ ,\tag{10.110}$$

where only the symmetrized odd powers of d appear. Decomposing $K^{l-1/2} = K_+^{l-1/2} + K_-^{l-1/2}$ with all the non-negative powers of d in the $K_+^{l-1/2}$, one finds for the remainder $K_-^{l-1/2} = K^{l-1/2} - K_+^{l-1/2}$ the following expression:

$$K_-^{l-1/2} = \sum_{i=1}^{+\infty} \{ e_{2i-1}, d^{-(2i-1)} \} = \{ R_l, d^{-1} \} + O(d^{-3}) + \dots \ ,\tag{10.111}$$

where the *Gel'fand-Dikii potentials* R_l have been introduced as $R_l \equiv e_1^{(l)}$. When $l = 1$, one gets, in particular

$$K_+^{1/2} = d \ ; \quad R_1 = -u/4 \ .\tag{10.112}$$

Our purpose is to investigate the basic equation (10.97) for the case of $p = 2l - 1$, i.e. when

$$[K_+^{l-1/2}, K] = 1 \ .\tag{10.113a}$$

Since $[K^{l-1/2}, K] = 0$, we have in addition

$$[K_+^{l-1/2}, K] = [K, K_-^{l-1/2}] = 1 \ .\tag{10.113b}$$

The commutator on the l.h.s. of this equation has only positive powers of d. Hence, only the leading term proportional to d^{-1} can contribute to the r.h.s.,

$$[K_+^{l-1/2}, K] = \text{leading term in } [K, 2R_l d^{-1}] = 4R_l' \ ,\tag{10.114}$$

---

[20]More general MMM potentials (with a general Q) correspond to the Toda or Volterra hierarchy generalizing the KdV hierarchy. The KdV hierarchy is the only one we consider here.

where we have used of the fact that the operator K begins with d^2. After one integration, eq. (10.113) takes the form of the Painlevé equation

$$cR_l[u] = z \; , \qquad\qquad (10.115)$$

where the integration constant c is irrelevant, since it can be changed by rescaling z.

The recursion relations for the Gel'fand-Dikii potentials R_l can be found as follows [485, 486]. Because of $K^{l+1/2} = KK^{l-1/2} = K^{l-1/2}K$, we have

$$K_+^{l+1/2} = \tfrac{1}{2}\left(K_+^{l-1/2}K + KK_+^{l-1/2}\right) + \{R_l, d\} \; . \qquad\qquad (10.116)$$

Calculating the commutators of both sides of this equation with K and using eq. (10.114), one easily finds (exercise # X-5)

$$R'_{l+1} = \frac{1}{4}R'''_l - uR'_l - \frac{1}{2}u'R_l \; . \qquad\qquad (10.117)$$

All the potentials R_l at $l \neq 0$ can now be recursively determined from eq. (10.117), demanding them to vanish at $u = 0$. For instance, the first few of them are

$$R_0 = \tfrac{1}{2} \; , \quad R_i = -\frac{1}{4}u \; , \quad R_2 = \frac{3}{16}u^2 - \frac{1}{16}u'' \; ,$$

$$R_3 = -\frac{5}{32}u^3 + \frac{5}{32}\left(uu'' + \tfrac{1}{2}u'^2\right) - \frac{1}{64}u'''' \; . \qquad\qquad (10.118)$$

It follows

$$K_+^{1/2} = d \; , \quad K_+^{3/2} = d^3 - \frac{3}{4}\{u, d\} \; ,$$

$$K_+^{5/2} = d^5 - \frac{5}{4}\{u, d^3\} + \frac{5}{16}\{(3u^2 + u''), d\} \; . \qquad\qquad (10.119)$$

It is the R_3 of eq. (10.118) that appears in the 'string equation' (10.52), after rescalings the variables. It follows from eq. (10.118) that $\alpha = 1/10$ indeed, which corresponds to the $(5, 2)$ minimal model. The other equations following from the hierarchy (10.97) characterize the partition functions of the (p, q) minimal models coupled to 2d gravity. They all can be realized as the *two*-matrix models of the type (10.55) in the continuum limit [439].

The MM *correlation functions* (in the Π-basis) are given by

$$\langle \mathcal{O}_{k_1} \cdots \mathcal{O}_{k_s} \rangle = Z^{-1} \int dM \; \mathcal{O}_{k_1} \cdots \mathcal{O}_{k_s} e^{-V(M)} \; , \qquad\qquad (10.120)$$

where $\mathcal{O}_k \equiv \operatorname{tr} M^k$. The two kinds of operators $\mathcal{O}_k$ are to be distinguished: the ones with a *finite* k contain information about local operators integrated over a surface, and are called *microscopic loops*, whereas the ones with an *infinite k* correspond to extended boundaries on a surface (in the continuum limit), and are called *macroscopic loops* [483]. The operators $\mathcal{O}_k$ actually correspond to the differential operators [458]

$$\mathcal{O}_k \to [\text{const.} - aQ]^k \equiv [\text{const.} + aH]^k \ , \tag{10.121}$$

in the double scaling limit, where $H \equiv -Q = -(d/dz)^2 + u(z)$ is playing the role of a Hamiltonian, and a is a constant. It follows (up to a normalization)

$$\langle \mathcal{O}_k \rangle_c \to \int_z^\infty dz' \, \langle z' \, |[\text{const.} + aH]^k| \, z' \rangle \ . \tag{10.122}$$

Having introduced the singular MM potential

$$V = \sum_k t_k \sigma_k \ , \quad \sigma_k = \text{tr} \, (r_c - M)^{k+1/2} \ , \tag{10.123}$$

with the coupling constants t_k, we find from eq. (10.122) that

$$\frac{\partial}{\partial t_k} \ln Z = \langle \sigma_k \rangle \to \int_z^\infty dz' \, \langle z' \, |H^{k+1/2}| \, z' \rangle \sim \int_z^\infty dz' \, R_{k+1}[u(z')] \ , \tag{10.124}$$

where eq. (11.3) for the diagonal of the resolvent of the operator H has been used.

Having differentiated eq. (10.124) twice w.r.t. z, one finds an equation for the specific heat in the form

$$\frac{\partial}{\partial t_k} u(z) = \frac{\partial}{\partial z} R_{k+1}[u(z)] \ , \tag{10.125}$$

which is just the generalized KdV-flow equation!

In the theory of KdV-type equations, the function $u(z; t_1, t_2, \ldots) = Z''_{\text{connected}} = \langle \sigma_0 \sigma_0 \rangle$, representing the specific heat and satisfying the KdV-flow equation (10.125), is called the *KdV potential*, whereas the partition function Z itself is called the *τ-function* of the KdV-type hierarchy. The σ_k's play the role of physical vertex operators.

Denoting $k = l/a$ and considering the limit $a \to 0$ at the fixed l, one finds that the loops become macroscopic, so that the insertions $\mathcal{O}_k$ above are now to be substituted by $\mathcal{O}_k \to W(l) \equiv \frac{1}{l} \text{tr} \, M^{l/a}$, whereas $[\text{const.} + aH]^{l/a} \to e^{-lH}$. Hence, the macroscopic loop of length l is described by the heat kernel of H. In the case of sphere, one easily finds the explicit formulae for the multi-critical VEV's of the macroscopic loops, *viz.*

$$\langle W(l) \rangle = \int_z^\infty dz' \, \langle z' \, |e^{-lH}| \, z' \rangle \propto l^{-1} z^{\frac{p}{2} - \frac{1}{4}} K_{p - \frac{1}{2}} (l\sqrt{z}) \ ,$$

$$\langle W(l_1) W(l_2) \rangle = \int_z^\infty dz' \int_{-\infty}^{-z} dz'' \, \langle z' \, |e^{-l_1 H}| \, z'' \rangle \langle z'' \, |e^{-l_2 H}| \, z' \rangle$$

$$\propto z\sqrt{l_1 l_2}\frac{e^{-(l_1+l_2)u(z)}}{l_1+l_2}\ ,\tag{10.126}$$

while

$$\langle\sigma_k W(l)\rangle = k!l^{-k-1/2}+\dots\ ,\tag{10.127}$$

where the dots stand for regular terms in the limit $l\to 0$, and K_m denote the standard (modified) Hankel functions.

Redefining the KdV pairs (t_k,σ_k) towards the CFT pairs $(\tau_k,\hat\sigma_k)$ in such a way that the p-th multi-critical point is described by $\tau_0 = \dots = \tau_{p-3} = \tau_{p-2} = z$ and $\tau_{p-1} = 0$ in the new variables, we find, for example, for the 'area operator' $\hat\sigma_{p-2}$,

$$\langle\hat\sigma_{p-2}W(l)\rangle = \frac{\partial}{\partial z}\langle W(l)\rangle = z^{\frac{p}{2}-\frac{3}{4}}K_{p-3/2}(l\sqrt{z})\ ,\tag{10.128}$$

whereas

$$\langle\hat\sigma_k\rangle = 0\ ,\quad\text{when}\ k\ne p,\ p-2\ ,$$

$$\langle\hat\sigma_{p-2}\rangle = -\frac{2\pi}{(2p-1)(2p-3)}u^{2p-1}(z)\ ,\tag{10.129}$$

and

$$\langle\hat\sigma_i\hat\sigma_j\rangle = -\delta_{ij}\left(\frac{\pi}{4}\right)\frac{u^{2j+1}}{2j+1}\ .\tag{10.130}$$

It has to be compared with the continuum Liouville theory described by the action (9.79), $S_{\rm L}[\phi] = \int d^2z\,\mathcal{L}_{\rm L}[\phi]$, under the condition (9.80). The Liouville theory correlators in the presence of matter represented by the $(2p-1,2)$ minimal model take the form

$$Z_p(\tau_j) = \int d\phi\, e^{-S_{\rm L}[\phi]}\left\langle e^{S_{\rm int}}\right\rangle_{(2p-1,2)}\ ,\tag{10.131}$$

where $S_{\rm int}$ can be rewritten in terms of the primaries $\Phi_{p-j-1,1}$, $j = 0,\dots,p-2$, as

$$S_{\rm int} = \sum_{j=2}^{p-2}\tau_j\int_\Sigma e^{\alpha_j\phi}\Phi_{p-j-1,1}\ .\tag{10.132}$$

The correspondence between the MM and the Liouville theory can now be established:

$$\hat\sigma_j \leftrightarrow \int_\Sigma e^{\alpha_j\phi}\Phi_{p-j-1,1}\ ,\quad \alpha_j = \tfrac{1}{2}\gamma(p-j)\ ;\qquad \hat\sigma_{p-1}\leftrightarrow\oint_{\partial\Sigma}e^{\gamma\phi/2}\ ,\tag{10.133}$$

while

$$\frac{p+1/2}{p-3/2}\hat\sigma_p - u^2\hat\sigma_{p-2}\leftrightarrow 2\pi\mathcal{L}_{\rm L}\ .\tag{10.134}$$

In the case of pure 2d gravity, one has $S_{\text{int}} = 0$ and $p = 2$. Hence, the 'area operator' $\hat{\sigma}_0$ corresponds to $\int_\Sigma e^{\gamma\phi}$, the $\hat{\sigma}_1$ corresponds to the 'boundary operator' $\oint_{\partial\Sigma} e^{\gamma\phi/2}$, and

$$\langle \hat{\sigma}_0 \rangle = -\frac{2\pi}{3} u^3(z) , \quad \langle \hat{\sigma}_1 \rangle = 0 , \quad \langle \hat{\sigma}_2 \rangle = \frac{2\pi}{15} u^5(z) . \tag{10.135}$$

By construction, the KdV potential u satisfies the generalized 'string equation' [458]:

$$\sum_k t_k (k + \tfrac{1}{2}) R_k[u] = z , \tag{10.136a}$$

or, equivalently, when using the identity $\frac{\delta}{\delta u} R_{k+1}[u] = -(k + \tfrac{1}{2}) R_k[u]$,

$$\sum_k t_k \frac{\delta}{\delta u} R_{k+1}[u] + z = 0 , \tag{10.136b}$$

where the parameter $t_0 = z$ can be identified with the 'cosmological constant'. Using the relations

$$\langle \sigma_k \sigma_0 \rangle = R_{k+1}[u] , \quad \langle \sigma_k \sigma_0 \sigma_0 \rangle = \frac{\partial u}{\partial t_k} = \frac{\partial}{\partial z} R_{k+1}[u] , \tag{10.137}$$

we can rewrite eq. (10.136) to the form

$$\sum_k (k + \tfrac{1}{2}) t_k \langle \sigma_{k-1} \sigma_0 \rangle - t_0 = 0 , \tag{10.138}$$

where $\sigma_0 = \mathcal{P}$ is known as a *'puncture operator'* [445]. Integrating once this equation yields the first Virasoro equation

$$\left\{ \sum_k (k + \tfrac{1}{2}) t_k \frac{\partial}{\partial t_{k-1}} - \tfrac{1}{2} t_0^2 \right\} Z \equiv L_{-1} Z = 0 . \tag{10.139}$$

Since Z is the τ-function of the KdV-type hierarchy, the other Virasoro constraints

$$L_n Z = 0 , \quad n \geq -1 , \tag{10.140}$$

can be shown to follow recursively [445]. This observation implies an equivalence between the MM solution and the topological solution to 2d gravity [445]. To calculate the gravitational descendants, the so-called W-constraints,

$$W_n^{(s)} Z(t) = 0 , \quad \text{for } n \geq 1 - s , \tag{10.141}$$

are to be considered [444, 464], where linear differential operators $W_n^{(s)}$ form the spin-s W algebra associated with a Lie group labeling the minimal model under consideration.

Exercises

X-3 ▷ Prove that the residue of a commutator $[A, B]$ of two (pseudo-)differential operators A and B is a total derivative. *Hint*: use the Leibniz rule (10.82) and consider the case of two monomials, $A = ad^k$ and $B = bd^l$, first.

X-4 ▷ The *trace* of a pseudo-differential operator A is defined by

$$\text{Tr} \, A = \oint_C \text{res} \, A \, . \tag{10.142}$$

Prove that

$$\text{Tr} \, AB = \text{Tr} \, BA \, . \tag{10.143}$$

X-5 ▷ Verify the recursion relations (10.117).

Chapter XI

CFT and Integrable Models

As was shown in sect. X.4, integrable non-linear differential equations of the KdV-type naturally arise in the theory of (multi-)matrix models in the continuum limit where they correspond to certain CFT's. It is therefore quite natural to expect the existence of a fundamental connection between integrable models, having an infinite number of conservation laws, and CFT in general. In this Chapter, some integrable structures relevant for CFT are illustrated on particular examples in two and four dimensions. A more fundamental underlying theory for the observations collected in this Chapter is yet to be found, if any (see, e.g., ref. [487] to get an idea about how this may happen).

We briefly discuss the Hamiltonian structure of the KdV-type equations and of the classical DS reduction, the origin of conformal and W symmetries there, and review some applications. A bulk of this Chapter is devoted to the hierarchies of (integrable) KdV and W flows, which can be associated with an arbitrary differential operator. Their Hamiltonian structure becomes transparent in the Drinfeld-Sokolov (matrix) representation for a differential operator. In particular, this allows us to interpret classical W transformations as covariance-preserving deformations of differential operators. The conformal reduction of the WZNW theories leading to the Toda theories is considered, and the origin of hidden chiral AKM symmetry in the corresponding 2d CFT's (and in the quantum 2d gravity, in particular) is explained by using Gauss decomposition. One of the possible ways to unify *all* 2d integrable models is through their connection with 4d *self-dual* gauge theories and with $N = 2$ strings. This idea is illustrated on the particular example of an embedding the 2d KdV equation into a 4d self-dual Yang-Mills theory via a 'zero-curvature' representation. Finally, the supersymmetric generalizations of this connection are briefly discussed.

XI.1   KdV-type hierarchies and flows

The **Korteweg-de Vries** (KdV) equation is a 2d non-linear partial differential equation having the form [488]

$$u_t = u''' + 6uu' , \tag{11.1}$$

and satisfied by a function $u(x, t)$ of the two variables x and t. [1] This equation was originally introduced in order to describe the wave propagation in shallow water [488]. The KdV equation is known to possess the soliton-type solutions describing the non-local formations with a finite support in space, which conserve their identity during the scattering [432]. This beautiful feature is based on the existence of an infinite number of conserved quantities associated with the KdV equation, namely [432, 489]

$$I_0 = 2 \int dx\, u , \quad I_1 = \int dx\, u^2 , \quad I_2 = \int dx \left(u^3 - \frac{u'^2}{2} \right) ,$$

$$I_3 = \frac{1}{4} \int dx \left(u''^2 + 5u^2 u'' + 5u^4 \right) , \quad \dots , \tag{11.2}$$

which can be generated through the expansion of the resolvent of the operator $d^2 - f$ as follows [249]

$$\frac{\delta I_l}{\delta u(x)} = 2^{l+2} R_l[-u] , \quad \left\langle x \left| \frac{1}{-d^2 + f + \xi} \right| x \right\rangle = \sum_{l=0}^{\infty} \frac{R_l[f]}{\xi^{l+\frac{1}{2}}} . \tag{11.3}$$

The same polynomials R_l actually appear in the (generalized) Painlevé equations $x = R_l[f]$.

An important property of the KdV equation is the existence of *two* Hamiltonian descriptions. Namely, the equation can be rewritten to the Hamiltonian form ($i = 1, 2$) [484]

$$u_t = \{\mathcal{H}^{(i)}, u\}^{(i)} , \quad \text{or, equivalently} , \quad u_t(x) = \mathcal{D}^{(i)} \frac{\delta \mathcal{H}^{(i)}}{\delta u(x)} , \tag{11.4}$$

in terms of a Poisson bracket, a Hamiltonian $\mathcal{H}$ and a differential operator $\mathcal{D}$,

$$\{u(x, t), u(y, t)\}^{(i)} = -\mathcal{D}_x^{(i)} \delta(x - y) , \tag{11.5}$$

where the two different choices are possible:

$$\mathcal{D}^{(1)} = d , \quad \mathcal{H}^{(1)} = I_2 , \quad \{u(x), u(y)\}^{(1)} = -\delta'(x - y) , \tag{11.6a}$$

---

[1] The prime here means a differentiation w.r.t. x: $u' \equiv du \equiv (d/dx)u$, while the subscript t stands for the derivative $\partial/\partial t \equiv \partial_t$, as usual.

and

$$\mathcal{D}^{(2)} = \tfrac{1}{2}(d^3 + 4ud + 2u') \; , \quad \mathcal{H}^{(2)} = I_1 \; ,$$

$$\{u(x), u(y)\}^{(2)} = -\tfrac{1}{2}\delta'''(x-y) - [u(x) + u(y)]\delta'(x-y) \; . \tag{11.6b}$$

It can be verified (exercise # XI-1) that the integrals of motion I_l are in involution with each other, w.r.t. each of the Poisson brackets, $\{I_j, I_k\} = 0$, whereas

$$\{I_k, u\}^{(1)} = \{I_{k-1}, u\}^{(2)} \; . \tag{11.7}$$

This can be used to derive recursion relations for the conserved quantities I_l in the form

$$d\left(\frac{\delta I_k}{\delta u(x)}\right) = \tfrac{1}{2}\left(d^3 + 4u(x)d + 2u'(x)\right)\frac{\delta I_{k-1}}{\delta u(x)} \; . \tag{11.8}$$

Moreover, the same eq. (11.7) implies that all higher-order partial differential equations having the form

$$\delta_k u \equiv \frac{\partial}{\partial t_k}u = \{I_k, u\}^{(1)} = \{I_{k-1}, u\}^{(2)} \tag{11.9}$$

are also integrable, since they have the same set of conserved quantities. This set of integrable equations is called the **KdV hierarchy**.

The hidden conformal symmetry of the KdV equation is explicit in the second Hamiltonian form of eq. (11.6b) [415, 484]. Assuming a solution $u(x)$ to be periodic on the interval $[0, 2\pi]$, we can expand it into a Fourier series with Fourier coefficients

$$u_n = \int_0^{2\pi}\frac{dx}{2\pi}\,u(x)e^{-inx} - \tfrac{1}{4}\delta_{n,0} \; . \tag{11.10}$$

As a consequence of the last line of eq. (11.6b), they satisfy the Virasoro algebra,

$$-2\pi i\{u_n, u_m\}^{(2)} = -(n-m)u_{n+m} + \tfrac{1}{2}n(n^2 - 1)\delta_{n+m,0} \; . \tag{11.11}$$

The *Lax representation* of an evolution equation has the r.h.s. given by a commutator of two differential operators, say Q and D,

$$\partial_t = [Q, D] \; . \tag{11.12}$$

This representation also exists for the KdV equation provided the D and Q are defined as

$$D = d^2 + u \; , \quad Q = d^3 + \frac{3}{2}ud + \frac{3}{4}u' \; . \tag{11.13}$$

Hence, the KdV equation can be equally viewed as the one describing an *isospectral* deformation (*flow*) of the operator D with the evolution operator $S(t) = T \int^t dt' \exp Q(t')$, i.e. $(\partial_t S) S^{-1} = Q(t)$ and

$$D(t) = S(t) D(0) S^{-1}(t) \ . \tag{11.14}$$

The KdV hierarchy of partial differential equations is therefore associated with the second-order differential operator $Q = d^2 + u(z)$ (*cf* sect. X.4), and it can be generalized as follows [250, 490, 491]. Consider a differential operator of order n,

$$D = d^n + a_1(z) d^{n-1} + a_2(z) d^{n-2} + \ldots + a_n(z) \ , \tag{11.15}$$

and introduce the flows

$$\delta_k D = [D_+^{k/n}, D] \tag{11.16}$$

for any integer k which is not a multiple of n. [2] In particular, the flows associated with functions ($k = 0$), $\delta_\phi D = [\phi, D]$, just mean infinitesimal changes in the functions on which the operator D acts [484]:

$$f \to (1 - \phi) f \ , \quad D \to (1 + \phi) D (1 - \phi) = D + [\phi, D] + \ldots \ . \tag{11.17}$$

This simple observation can be used to get rid of the d^{n-1} term in the general expression (11.15) for D, so that one can always assume $a_1 = 0$. It can then be verified by a straightforward calculation [250, 490, 491] that all the flows (11.16) are (i) *isospectral*, (ii) *commute* with each other, $[\delta_k, \delta_l] D = 0$, and (iii) the traces

$$I_k = \text{Tr} \left(D^{k/n} \right) \tag{11.18}$$

form an infinite set of conserved quantities (the integrals of motion), where the trace (Tr) of a pseudo-differential operator has been defined in eq. (10.142).

Therefore, a hierarchy of integrable flows generalizing the KdV flows can, in fact, be associated with *any* differential operator D. A Hamiltonian interpretation of these hierarchies is of particular interest, since this would provide us with a variety of the *Gel'fand-Dikii* algebras representing classical versions of the chiral algebras, and of the W algebras (Ch. VI) in particular, in CFT. Similarly to the KdV case considered above, there exist two Hamiltonian structures reproducing the flows (11.16) in the general case [250, 490, 491]. Their action has first to be defined on linear functionals of the coefficients

---

[2]When k is a multiple of n, the flows (11.16) are trivial.

$\{a_i\}$ of D, and then extended to arbitrary (polynomial) functions by the differentiation (chain) rule [484]. The two Hamiltonian structures read as follows [485, 486, 492, 493]:

$$\{l_U(D), l_V(D)\}^{(1)} = \text{Tr}\,(D[U, V]) = l_V(DU - UD)\,, \tag{11.19}$$

$$\{l_U(D), L_V(D)\}^{(2)} = \text{Tr}\,((DU)_+(DV) - (VD)(UD)_+) = l_V((DU)_+D - D(UD)_+)\,,$$

where a linear functional $l_U(D)$ has been introduced,

$$l_U(D) = \int dz \sum_{i=2}^{n} u_i(z) a_i(z) \tag{11.20}$$

$$= \text{Tr}\,(d^n + a_2 d^{n-2} + \ldots + a_n)(d^{1-n} u_2 + d^{2-n} u_3 + \ldots + d^{-1} u_n) = \text{Tr}\,(DU)\,.$$

This action of D on $l_U(D)$ can be extended to an arbitrary polynomial functional $\Psi(D)$ as follows:

$$\{\Psi(D), l_U(D)\} = \{l_{V_\Psi}(D), l_U(D)\}\,, \tag{11.21}$$

where

$$V_\Psi \equiv \sum_{i=1}^{n-1} d^{-i} \frac{\delta \Psi(D)}{\delta a_{n-i+1}}\,, \qquad \frac{\delta}{\delta a_j} \equiv \sum_k (-d)^k \frac{\delta}{\delta a_j^{(k)}}, \tag{11.22}$$

and $l_{V_\Psi}(D) = \text{Tr}\,(V_\Psi D)$. The Hamiltonian structures (11.19) are defined to reproduce the flows (11.16) [250, 490, 491], so that

$$\delta_k D = \{I_k, D\}^{(1)} = \{I_{k-1}, D\}^{(2)}\,. \tag{11.23}$$

The origin of these Hamiltonian formulae can be made less mysterious after introducing the *Drinfeld-Sokolov* (matrix) representation for differential operators [250], in which an n-th order differential operator D of the form (11.15) (with $a_1 = 0$) is substituted by a *first-order* differential $n \times n$ *matrix* operator $\hat{D} \equiv d + \mathcal{A}$ having the form

$$\hat{D} = \begin{pmatrix} d & a_2 & a_3 & \cdots & a_n \\ -1 & d & 0 & \cdots & 0 \\ 0 & -1 & d & \cdots & 0 \\ \vdots & \cdots & \ddots & \ddots & \vdots \\ 0 & \cdots & \cdots & -1 & d \end{pmatrix}, \tag{11.24}$$

where a matrix $\mathcal{A}$ is assumed to be valued in the Lie algebra $A_{n-1} \sim sl(n)$. [3] The operator $\hat{D}$ is equivalent to D in the sense that the kernels of both operators are in

---

[3] This construction can be extended to general Lie algebras (see the next sections).

one-to-one correspondence [484]. Indeed, if $\vec{f} \in \text{Ker } \hat{D}$, the last component of $\vec{f}$ belongs to Ker D, whereas, given an element of Ker D, the corresponding element in Ker $\hat{D}$ can be easily picked up from eq. (11.24). Of course, the form of the operator $\hat{D}$, as presented in eq. (11.24), is not unique because of a covariance w.r.t. the gauge transformations

$$\hat{D} \to N^{-1}\hat{D}N \ , \tag{11.25}$$

written down in terms of z-dependent upper-triangular matrices N having units on their diagonal (let $\mathcal{T}$ be the group of such matrices) [250]. The gauge transformations (11.25) do not disturb an isomorphism between Ker $\hat{D}$ and Ker D. Indeed, using the convenient decomposition of $\hat{D}$ as [484]

$$\hat{D} = -J_- + d + \mathcal{A} \ , \tag{11.26}$$

where

$$J_- = \begin{pmatrix} 0 & 0 & \cdots & & 0 \\ 1 & 0 & \cdots & & 0 \\ 0 & 1 & \cdots & & 0 \\ \vdots & & \ddots\ddots & & \vdots \\ 0 & \cdots & & 1 & 0 \end{pmatrix} \ , \quad \mathcal{A} = \begin{pmatrix} 0 & a_2 & a_3 & \cdots & a_n \\ 0 & 0 & \cdots & \cdots & 0 \\ \vdots & \vdots & \vdots & \vdots & \vdots \\ \vdots & \vdots & \vdots & \vdots & \vdots \\ 0 & 0 & \cdots & \cdots & 0 \end{pmatrix} \ . \tag{11.27}$$

a gauge transformation of $\hat{D}$ takes the form [484]

$$\hat{D} \to -J_- + d + \left\{ N^{-1}\mathcal{A}N + N^{-1}(dN + [N, J_-]) \right\} \ , \tag{11.28}$$

where the matrix in curly brackets belongs to $\mathcal{T}$. Conversely, given a matrix operator $\hat{D}$ in the form (11.26) with an upper-triangular matrix $\mathcal{A}$, there exists a unique matrix operator of the form (11.24) in the $\mathcal{T}$-orbit which a chosen $\hat{D}$ belongs to.

The Poisson brackets introducing a Hamiltonian structure are naturally defined on gauge-invariant functionals $s(\mathcal{A})$ of the $sl(n)$-valued matrix $\mathcal{A}$. The specific structure of eq. (11.24) implies certain constraints, which can be imposed by the usual gauge-fixing procedure. In the simplest case of $n = 1$, a natural Poisson bracket is defined on ordinary functions $a(z)$ as

$$\{a(x), a(y)\} = \partial_x \delta(x - y) \ . \tag{11.29}$$

For two functionals $s_1(a)$ and $s_2(a)$, eq. (11.29) yields

$$\{s_1, s_2\} = \oint dx dy \, \frac{\delta s_1}{\delta a(x)} \frac{\delta s_2}{\delta a(y)} \partial_x \delta(x - y)$$

$$= \oint dx \, \frac{\delta s_1}{\delta a(x)} \partial_x \frac{\delta s_2}{\delta a(x)} \; . \tag{11.30}$$

This Poisson bracket is antisymmetric and satisfies the Jacobi identity (exercise # XI-3). In the matrix case ($n > 1$), the generalizations of eq. (11.30) take the form [250, 484]

$$\{s_1, s_2\}^{(1)} = \oint dx \, \mathrm{tr} \left(\frac{\delta s_2}{\delta \mathcal{A}(x)} [\mathcal{B}, \frac{\delta s_2}{\delta \mathcal{A}(x)}] \right) \; , \tag{11.31a}$$

$$\{s_1, s_2\}^{(2)} = \oint dx \, \mathrm{tr} \left(\frac{\delta s_2}{\delta \mathcal{A}(x)} [d + \mathcal{A}(x), \frac{\delta s_2}{\delta \mathcal{A}(x)}] \right) \; , \tag{11.31b}$$

resp., where the functional derivative $\delta s / \delta \mathcal{A}(x)$,

$$s(\mathcal{A} + \delta \mathcal{A}) \approx s(\mathcal{A}) + \oint dx \, \mathrm{tr} \left(\frac{\delta s}{\delta \mathcal{A}(x)} \delta \mathcal{A}(x) \right) \; , \tag{11.32}$$

and the constant matrix $\mathcal{B}$,

$$\mathcal{B} = \begin{pmatrix} 0 & \cdots & 1 \\ 0 & \cdots & 0 \\ & \cdots & \\ 0 & \cdots & 0 \end{pmatrix} \; , \tag{11.33}$$

have been introduced. Being expressed in terms of the original operator D, the Hamiltonian structures introduced in eq. (11.31) were shown to coincide with those introduced earlier in eq. (11.19) [250].

Exercises

XI-1 ▷ Consider the particular case of $n = 3$, $k = 2$ in eqs. (11.15) and (11.16), where one has

$$D = d^3 + ud + v \; , \quad D_+^{2/3} = d^2 + u \; ,$$

$$\delta u = u_t = 2v' - u'' \; , \quad \delta v = v_t = v'' - \frac{2}{3}(uu' + u''') \; . \tag{11.34}$$

Eliminate v from these equations, and prove that this leads to the *Boussinesq equation*

$$u_{tt} = -\frac{1}{3}(4uu'' + 4u'^2 + u'''') \; . \tag{11.35}$$

XI-2 ▷ (*) Prove that the trace of the flow vanishes,

$$\mathrm{Tr}\,(\delta_k D^{k/n}) = 0 \; . \tag{11.36}$$

XI-3 ▷ (*) Verify that the Poisson bracket defined in eq. (11.30) satisfies the Jacobi identity. *Hint*: first use linear functionals.

XI.2 W-flows

The construction of (classical) W algebras via DS reduction was considered in Ch. VII. The other (equivalent) way of their construction naturally follows when one studies deformations of differential operators and their covariance properties [243, 484]. One introduces a linear differential operator

$$D = d^n + \sum_{j=2}^{n} a_j d^{n-j} , \qquad (11.37)$$

acting on functions $f(x)$, and one considers how its coefficients $a_j(x)$, $j = 2, \ldots, n$, vary under changes of variables $x \to \tilde{x}$. First, however, the action of D on f has to be properly represented. Let $\mathcal{F}_h$ be the space of h-differentials, i.e. the space of functions $f(x)$ transforming under $x \to \tilde{x}$ according to the rule

$$\tilde{f}(\tilde{x})d\tilde{x}^h = f(x)dx^h , \qquad (11.38)$$

and let $(f_1, f_2, \ldots, f_n)$ be a basis in the kernel of D. Because of $a_1 = 0$ in eq. (11.37), the logarithmic derivative of the Wronskian of basis functions f_i vanishes. Hence, this Wronskian is a constant, which can be set to one by normalization [484]:

$$W(f_1, f_2, \ldots, f_n) \equiv \det \begin{pmatrix} f_1^{(n-1)} & \cdots & f_n^{(n-1)} \\ f_1^{(n-2)} & \cdots & f_n^{(n-2)} \\ \vdots & \ddots & \vdots \\ f_1 & \cdots & f_n \end{pmatrix} = 1 . \qquad (11.39)$$

The action of D on f can now be conveniently represented as [243, 484]

$$[Df] = \det \begin{pmatrix} f^{(n)} & f_1^{(n)} & \cdots & f_n^{(n)} \\ f^{(n-1)} & f_1^{(n-1)} & \cdots & f_n^{(n-1)} \\ \vdots & \vdots & \ddots & \vdots \\ f & f_1 & \cdots & f_n \end{pmatrix} . \qquad (11.40)$$

When all the functions f_i and f belong to $\mathcal{F}_h$, the Wronskian $W(f_1, f_2, \ldots, f_n)$ belongs to $\mathcal{F}_{nh+\frac{1}{2}n(n-1)}$, while $[Df]$ belongs to $\mathcal{F}_{(n+1)h+\frac{1}{2}n(n+1)}$ [31]. Hence, the condition (11.39) will be preserved if $h = -\frac{1}{2}(n-1)$. Identifying the coefficients a_i with minors of the determinant on the r.h.s. of eq. (11.40), one finds that the operator D maps

$\mathcal{F}_{-\frac{1}{2}(n-1)}$ into $\mathcal{F}_{\frac{1}{2}(n+1)}$. In particular, the transformation law for a_2 has an 'anomaly' given by the Schwartzian derivative,

$$\tilde{a}_2(\tilde{x}) = a_2(x) \left(\frac{dx}{d\tilde{x}}\right)^2 + \frac{n(n^2 - 1)}{12} S\{x, \tilde{x}\} . \tag{11.41}$$

Under $u \to x \to \tilde{x}$, the Schwartzian derivative transforms as (Ch. I)

$$S\{u, \tilde{x}\} = S\{u, x\} \left(\frac{dx}{d\tilde{x}}\right)^2 + S\{x, \tilde{x}\} . \tag{11.42}$$

Hence, the a_2 coefficient transforms as $\frac{n(n^2-1)}{12} S\{u, x\}$, where u is a fixed coordinate for which $a_2(u) = 0$. Transformation laws for the other coefficients a_i, $i \neq 2$, are more complicated. Fortunately, there exists a (non-unique) invertible transformation $a_k \to w_k$ to differential polynomials $w_k \in \mathcal{F}_k$ at $3 \leq k \leq n$ [243, 484]. Taking w_k to be *linear* functionals of a_l, one can split the operator D into separately covariant pieces $\Delta_k^{(n)}$ mapping $\mathcal{F}_{-\frac{1}{2}(n-1)}$ into $\mathcal{F}_{\frac{1}{2}(n+1)}$ and linearly depending on w_k at $3 \leq k \leq n$ [243, 484]:

$$D = \Delta_2^{(n)}(a_2) + \Delta_3^{(n)}(w_3, a_2) + \Delta_4^{(n)}(w_4, a_2) + \ldots . \tag{11.43}$$

The w_k can be computed explicitly, in principle. For instance, using the coordinate choice u in which $a_2 = 0$, one finds [243, 484]

$$w_k(u) = \sum_{l=3}^{k} (-1)^{k-l} \frac{\begin{pmatrix} k-1 \\ k-l \end{pmatrix} \begin{pmatrix} n-l \\ k-l \end{pmatrix}}{\begin{pmatrix} 2k-2 \\ k-l \end{pmatrix}} a_l^{(k-l)}(u) . \tag{11.44}$$

Another basis for the w-differentials can be constructed, when using the matrix form of the differential operator in eq. (11.24), and the gauge freedom (11.25) to put the matrix $\hat{D}$ into the desired form as well [243, 484]. Let $-J_- + \mathcal{A} \equiv A \in A_{n-1}$, and the grading of traceless matrices belonging to the Lie algebra A_{n-1} be defined according to the rule

$$\text{grade} (A_{ij}) = j - i . \tag{11.45}$$

One has grade $(J_-) = -1$, whereas the nilpotent algebra T of upper triangular matrices can be decomposed w.r.t. the grading (11.45) as follows:

$$T = \oplus_{k=1}^{n-1} T^{(k)} , \tag{11.46}$$

where $T^{(k)}$ are of dimension $n - k$. The adjoint operator acts as follows:

$$\text{ad } J_- : \quad T^{(k)} \to T^{(k-1)} , \qquad 1 \le k \le n - 1 , \tag{11.47}$$

where $T^{(0)}$ is the CSA of traceless diagonal matrices. We can now use the gauge transformations (11.28) to bring $\mathcal{A}$ into the form

$$\mathcal{A}(x) = \sum_{k=1}^{n-1} r_k(x) R_k , \tag{11.48}$$

where R_k is a representative of $T^{(k)} \bmod (\text{ad } J_-(T^{(k+1)}))$. To choose the representatives properly, we use the simple observation that any n-dimensional vector space carries, in fact, an $sl(2)$-irrep of spin $j = \frac{1}{2}(n - 1)$ when, in addition to the generator J_- of eq. (11.27), the generators J_+ and J_0 are defined as

$$
J_+ = \begin{pmatrix}
0 & (n-1)\cdot 1 & 0 & \cdots & 0 \\
0 & 0 & (n-2)\cdot 2 & \cdots & 0 \\
\vdots & & & \ddots & \vdots \\
0 & & \cdots & 0 & 1\cdot(n-1) \\
0 & & \cdots & & 0
\end{pmatrix} , \quad
J_0 = \begin{pmatrix}
j & 0 & \cdots & 0 \\
0 & j-1 & \cdots & 0 \\
\vdots & & & \vdots \\
\vdots & & \ddots & \vdots \\
0 & 0 & \cdots & -j
\end{pmatrix} .
$$

$$\tag{11.49}$$

Taken together, they satisfy the commutation relations

$$[J_+, J_-] = 2J_0 , \quad [J_0, J_\pm] = \pm 2J_\pm . \tag{11.50}$$

We note that $J_+^k \in T^{(k)}$, while the sum of entries of J_+^k is nonvanishing [494]. This means that we can choose J_+^k as the coefficients R_k in eq. (11.48), which gives rise to a matrix operator $\hat{D}$ in the form [243, 484, 494]

$$\hat{D} = J_- + d + \sum_{k=1}^{n-1} W_{k+1}(x) J_+^k , \tag{11.51}$$

where the W_k's form a basis of k-differentials, $W_k \in \mathcal{F}_k$ (for $k > 2$). [4] Of course, the two bases, w_k's and W_k's, are related (in a non-linear way), e.g. [494]

$$w_k = \sigma_{k-1} W_k , \text{ for } 2 \le k \le 5 , \quad w_6 = 5!^2 (W_6 + \frac{1}{9} W_3^2) , \text{ etc. }, \tag{11.52}$$

where σ_k is the sum of entries of J_+^k: $\sigma_k = \frac{k!^2}{(2k+1)!} n(n^2 - 1) \cdots (n^2 - k^2)$.

---

[4] We omit details of the proof which can be found in ref. [494].

In more general terms, we just solved the following problem (in a very particular case): find two infinitesimal differential operators X and Y, acting in $\mathcal{F}_{-\frac{1}{2}(n-1)}$ and $\mathcal{F}_{\frac{1}{2}(n+1)}$, resp., such that the equation $F = Df$, after the change of functions as $g = (1 + X)f$ and $G = (1 + Y)F$, takes the form $G = (D + \delta D)g$, where $D + \delta D$ still has the form (11.37) [484]. The variation δD is then given by

$$\delta D = YD - DX .\tag{11.53}$$

This equation is clearly different from the KdV-flows, which correspond to the case $X = Y = D^{k/n}$, and where all the KdV variations δ_k are commuting with each other. Instead, eq. (11.53) describes quite different deformations of a differential operator, namely those which preserve its covariance properties. In particular, the eigenvalues and isospectrality, which are inherent to the KdV flows, are *not* invariant under coordinate changes [243, 484, 494]. The deformations (11.53) are called the *W-flows*. The infinitesimal changes of coordinates give only one specific example of the problem summarized by eq. (11.53), just when the operators X_1 and Y_1, generating changes of the variable $x \to x + \epsilon(x)$ in $\mathcal{F}_{-\frac{1}{2}(n-1)}$ and $\mathcal{F}_{\frac{1}{2}(n+1)}$, resp., are of the first order and read as

$$X_1 = \epsilon d - \tfrac{1}{2}(n-1)\epsilon' , \quad Y_1 = \epsilon d + \tfrac{1}{2}(n+1)\epsilon' .\tag{11.54}$$

Indeed, it follows

$$\delta_1 D = Y_1 D - DX_1 .\tag{11.55}$$

In general, X and Y may be of higher order $n > 1$, and they may depend on $n - 1$ (= rank $sl(n)$) independent functions [484]. Denoting $\delta_X D$ the variation of D in eq. (11.53), it follows from the definition that

$$[\delta_X, \delta_{X'}] = \delta_{[X,X']+\delta_{X'}X-\delta_X X'} ,\tag{11.56}$$

where X' may be a function of D. A connection to the k-differentials $\eta \in \mathcal{F}_k$ is given by the formula [243, 484, 494]

$$[\delta_1(\epsilon), \delta_k(\eta)] = \delta_k(\epsilon \eta' - k\eta\epsilon') ,\tag{11.57}$$

where a basis $\delta_k(\eta) = \delta_{X_k(\eta)}$ has been introduced. In the Hamiltonian formalism, the variations δ_k are generated by Hamiltonians of the from $H_k = \int dx\, \eta(x)w_{k+1}(x)$ via the second Poisson structure introduced in eq. (11.19).

Covariant differential operators X_k and Y_k can be explicitly constructed [243, 484, 494] by comparing the second line of eq. (11.19) with eq. (11.53), which allows us to identify X_k and Y_k from the expression for the w_{k+1}. Choosing $\Psi(D) = H_k = \int dx\, \epsilon w_{k+1}$, one gets

$$\delta_k l_U(D) = \left\{\int dx\, \epsilon w_{k+1}, l_U(D)\right\}^{(2)} = l_U(\delta_k D) \ , \tag{11.58}$$

where

$$\delta_k D = (DV_k)_+ D - D((V_k D)_+ \ , \quad V_k \equiv \hat{V}_{H_k} \ . \tag{11.59}$$

Hence, we can take $X_k = (V_k D)_+$ and $Y_k = (DV_k)_+$. Knowing the w_{k+1}, the full explicit expressions for X_k and Y_k can then be reconstructed [243, 484, 494]. For instance, as far as the $SU(n)$ case is concerned, the generators X_k read as follows [484]:

$$X_1 = \epsilon d - \tfrac{1}{2}(n-1)\epsilon' \ , \quad X_2 = \epsilon d^2 - \tfrac{1}{2}(n-2)\epsilon' d + \left[\frac{2}{n}\epsilon a_2 + \frac{1}{12}(n-1)(n-2)\epsilon''\right] \ ,$$

$$X_3 = \epsilon d^3 - \tfrac{1}{2}(n-3)\epsilon' d^2 + \left[\frac{(n-2)(n-3)}{10}\epsilon'' + \frac{6}{5}\frac{3n^2-7}{n(n^2-1)}a_2\epsilon\right] d \tag{11.60}$$

$$+ \left[\frac{3}{n}w_3\epsilon - \frac{3(n+2)(n-7)}{10n(n+1)}a'_2\epsilon - \frac{(n-3)(4n+7)}{5n(n+1)}a_2\epsilon' - \frac{(n-1)(n-2)(n-3)}{5!}\epsilon'''\right] \ ,$$

whereas $Y_k = (-1)^k X_k^*$. Given an x-independent ϵ and $k = 1, 2$, the operators $X_k = Y_k$ coincide with $D^{k/n}$ and, hence, represent the KdV flows. In general, with an x-dependent ϵ, the W flows can be considered as the covariant local extensions of the KdV flows.

Having obtained the explicit expressions for w_k's, X_k's and Y_k's, the Poisson brackets among them can be computed. In particular, the bracket $\{w_k(x), w_l(y)\}^{(2)}$ is a sum of monomials in the w's and their derivatives (multiplied by a derivative of $\delta(x-y)$). The second set of the Poisson brackets defines a (classical) W algebra of the A_{n-1}-type. This algebra always contains the (classical) Virasoro algebra which is generated by the a_2, in particular. Taking into account that the w_k's transform as k-differentials, one finds

$$\{a_2(y), a_2(x)\}^{(2)} = (a'_2(x) + 2a_2(x)d + c_n d^3)\delta(x-y) \ ,$$

$$\{a_2(y), w_k(x)\}^{(2)} = (w'_k(x) + kw_k(x)d)\delta(x-y) \ . \tag{11.61}$$

The (non-trivial) Poisson brackets among the w's themselves have the general form

$$\{w_k(x), w_l(y)\}^{(2)} = \phi^k \Delta(\phi^{-j} w_j, \phi^{-1} d)\phi^{l-1}\delta(x-y) \ , \tag{11.62}$$

where a differential operator Δ is to be introduced in such a way that the r.h.s. only depends on the Schwartzian derivative associated with the change of coordinates. The explicit formulae for the Poisson brackets can be found in refs. [243, 484, 494]. In particular, the explicit form of a central term is given by [243]:

$$\left.\{w_k(y), w_l(x)\}^{(2)}\right|_{\text{central term}}$$

$$= (n - k + 1)(n - k + 2)\cdots(n + k - 1)\frac{(-1)^k (k - 1)!^2}{(2k - 2)!(2k - 1)!}\delta_{k,l}d^{k+l-1}\delta(x - y) . \quad (11.63)$$

In the next section, we describe the connection which exists between the Liouville-Toda hierarchies and the classical W algebras. The efficient method of refs. [243, 484, 494] just outlined above gives generators of the classical W algebras, as well as their transformations properties, in a very explicit form. In this approach, the W transformations can be recognized as the deformations of differential operators which preserve their covariance, with the simplest transformations corresponding to the coordinate changes. This may lead to a useful *geometrical* interpretation of the W symmetries [484], which is yet to be found.

XI.3 WZNW models and Toda field theories

The origin of the hidden affine $SL(2, \mathbf{R})$ symmetry, which is present in the induced 2d gravity (the Liouville theory) formulated in the light-cone gauge (sect. IX.1), can be explained by regarding this theory as a (conformally) reduced [5] WZNW model [495]. The hidden $SL(2, \mathbf{R})$ symmetry of the quantum 2d gravity can therefore be understood in terms of the chiral AKM symmetry of the WZNW theory. We follow ref. [495] here.

One starts with the WZNW action (4.55) for the non-compact group $SL(2, \mathbf{R})$ in 2d Minkowski spacetime, [6]

$$S(g) = -\frac{k}{8\pi}\int_{S^2}d^2\zeta\eta^{\mu\nu}\,\text{tr}\left[(g^{-1}\partial_\mu g)(g^{-1}\partial_\nu g)\right] + \frac{k}{12\pi}\int_B\text{tr}\left[(g^{-1}dg)^3\right] . \quad (11.64)$$

The left and right AKM symmetries of this theory are generated by the Noether currents (sect. IV.3)

$$J(\lambda) = \kappa\,\text{tr}\left[\lambda\cdot(\partial_+ g)\cdot g^{-1}\right] , \quad \tilde{J}(\lambda) = -\kappa\,\text{tr}\left[\lambda\cdot g^{-1}\cdot(\partial_- g)\right] , \quad (11.65)$$

---

[5] *Conformal reduction* means constraining CFT consistently with conformal invariance.

[6] The Minkowski space-time, where the notion of left-right chirality is well-defined, is more convenient here than the Euclidean space. Our conventions in the 2d Minkowski space-time are: $\eta^{00} = -\eta^{11} = \varepsilon^{01} = -\varepsilon^{10} = 1$, $\partial_\pm\phi = (\partial_0 \pm \partial_1)\phi = \dot{\phi} \pm \phi'$.

resp., where $\kappa \equiv -k/4\pi$, and λ is an element of the Lie algebra $sl(2, \mathbf{R})$. The WZNW equations of motion are known to be equivalent to the current conservation (sect. IV.3)

$$\partial_- J = \partial_+ \tilde{J} = 0 . \tag{11.66}$$

The Gauss decomposition (4.82) of an arbitrary element $g = ABC \in SL(2, \mathbf{R})$ [7]

$$A = \begin{pmatrix} 1 & x \\ 0 & 1 \end{pmatrix} = \exp(x E_+) , \quad C = \begin{pmatrix} 1 & 0 \\ y & 1 \end{pmatrix} = \exp(y E_-) ,$$

$$B = \begin{pmatrix} e^{\frac{1}{2}\phi} & 0 \\ 0 & e^{-\frac{1}{2}\phi} \end{pmatrix} = \exp(\tfrac{1}{2}\phi H) , \tag{11.67}$$

together with the Polyakov-Wiegmann identity (sect. V.3)

$$S(ABC) = S(A) + S(B) + S(C) + \kappa \int d^2\zeta \, \mathrm{tr} \left[(A^{-1}\partial_- A)(\partial_+ B)B^{-1} \right.$$

$$\left. + (B^{-1}\partial_- B)(\partial_+ C)C^{-1} + (A^{-1}\partial_- A)B(\partial_+ C)C^{-1}B^{-1} \right] , \tag{11.68}$$

can be used to bring the WZNW action (11.64) to the form

$$S(g) = S(x, y, \phi) = \frac{\kappa}{2} \int d^2\zeta \left[\tfrac{1}{2}\partial_+\phi\partial_-\phi + 2(\partial_- x)(\partial_+ y)e^{-\phi} \right] . \tag{11.69}$$

The corresponding equations of motion read

$$\partial_-(\partial_+ y e^{-\phi}) = \partial_+(\partial_- x e^{-\phi}) = 0 , \tag{11.70a}$$

and

$$\partial_+\partial_-\phi + 2(\partial_- x)(\partial_+ y)e^{-\phi} = 0 . \tag{11.70b}$$

Eq. (11.70a) has the special solution

$$\partial_+ y = c_1 e^\phi , \quad \partial_- x = c_2 e^\phi , \tag{11.71}$$

where two arbitrary constants $c_{1,2}$ have been introduced. Substituting eq. (11.71) into eq. (11.70b) gives rise to the Liouville equation

$$\partial_+\partial_-\phi + \mu^2 e^\phi = 0 , \quad \text{where} \quad \mu^2 = 2c_1 c_2 . \tag{11.72}$$

This means that the Liouville theory can be regarded as the (conformally) reduced WZNW theory. It should be noticed that this reduction is *canonical* in the sense that

---

[7]We are considering a neighbourhood of the group identity, for definiteness.

the Poisson brackets between the phase-space variables ϕ and $\partial_0\phi$ can be calculated by using either the WZNW Lagrangian or the Liouville one. The relevant constraints can be expressed in terms of currents $J(E_+)$ and $\bar{J}(E_-)$ having the form

$$J(E_+) = \kappa\partial_+ y e^{-\phi} , \quad \bar{J}(E-) = -\kappa\partial_- x e^{-\phi} . \tag{11.73}$$

Choosing the solutions (11.71) is equivalent to imposing the constraints

$$J(E_+) = \kappa c_1 , \quad \bar{J}(E_-) = -\kappa c_2 , \tag{11.74}$$

which are globally defined and conformally invariant.

The conformal symmetry of the Liouville theory is characterized by the improved stress tensor (sect. IX.2)

$$T_{0\pm} = \tfrac{1}{2}\kappa\left[\tfrac{1}{2}(\partial_\pm\phi)^2 + \mu^2 e^\phi \mp 2(\partial_\pm\phi)'\right] \tag{11.75}$$

of (classical) central charge $c = -6k$, which can also be understood in the WZNW context as follows. Though the (left) SS-constructed WZNW Virasoro density (i.e. the T_{++}-component of the (improved) stress tensor)

$$L = -\frac{2\pi}{k}\left[\tfrac{1}{2}J(H)^2 + 2J(E_+)J(E_-)\right] \tag{11.76}$$

does not commute with the Liouville constraints (11.74), there exists a family of the SS-type Virasoro subalgebras having the form

$$l = L + a_\lambda J(\lambda) + b_\lambda J'(\lambda) \tag{11.77}$$

with some real parameters a_λ and b_λ, which actually commute with the constraints. The commuting solution is given by [495]

$$l = L - J'(H) , \quad \tilde{l} = \tilde{L} - \tilde{J}'(H) . \tag{11.78}$$

Due to $\{J(H,\sigma), J(H,\sigma')\} = \frac{k}{2\pi}\mathrm{tr}\, H^2\delta'(\sigma - \sigma')$, the central charge of the Virasoro algebra generated by l or $\tilde{l}$ is the same, $c = -3k\,\mathrm{tr}\, H^2 = -6k$. In its explicit form, the field l reads

$$l = L - J'(H) = \tfrac{1}{2}\kappa\{[\tfrac{1}{2}(\partial_+\phi)^2 + 2(\partial_+ x)(\partial_+ y)e^{-\phi}] - [2\partial_+\phi + 4x(\partial_+ y)e^{-\phi}]'\} . \tag{11.79}$$

Eq. (11.79) reduces to eq. (11.75) after imposing the constraints (11.74).

The connection to the Polyakov approach to 2d quantum gravity (sect. IX.1) can be made by imposing only the right-moving part of the constraints (11.74). The corresponding equation of motion takes the form

$$\partial_-[e^{-\phi}\partial_+\partial_-\phi] = 0 \;, \quad \text{where} \;\; \phi \equiv \ln(b^{-1}\partial_-x) \;. \tag{11.80}$$

It can be derived from the Lagrangian

$$\mathcal{L} = -\frac{k}{16\pi}\frac{(\partial_+\partial_-x)(\partial_-\partial_-x)}{(\partial_-x)^2} \;. \tag{11.81}$$

Being rewritten in terms of Polyakov's variable f which was defined in eq. (9.6), eq. (11.81) just takes the form of eq. (9.7) after the identification $x(\zeta^+, f(\zeta^+, \zeta^-)) = \zeta^-$. By construction, the effective Lagrangian (11.81) is therefore invariant under the *residual* left-moving $SL(2, \mathbf{R})$ transformations

$$\tilde{x}(\zeta^+, \zeta^-) = \frac{a(\zeta^+)x(\zeta^+, \zeta^-) + b(\zeta^+)}{c(\zeta^+)x(\zeta^+, \zeta^-) + d(\zeta^+)} \;, \quad \text{where} \;\; ad - bc = 1 \;. \tag{11.82}$$

The Liouville theory is obtained by imposing the left-moving part of the constraints (11.74) in addition, which clearly breaks down this residual symmetry. The E_+ component of the $SL(2, \mathbf{R})$ Noether current corresponding to the symmetry (11.82) takes the form

$$J(E_+) = -\frac{\kappa}{2b}e^{-\phi}\partial_+\partial_-\phi \;. \tag{11.83}$$

Though the full constraints (11.74) do not respect the AKM symmetry of the WZNW theory, they still preserve at least one member of the family (11.77).

The construction of the Liouville theory from the WZNW theory based on the noncompact group $SL(2, \mathbf{R})$ can be generalized [495] towards the so-called **Toda** theory [496]. Let $\mathcal{G}$ be any complex simple Lie algebra, $\{\alpha\}$ its roots w.r.t. CSA, and Δ a set of simple roots (sect. XII.2). In a Cartan-Weyl basis, all structure constants of $\mathcal{G}$ can be chosen to be real numbers, so that one gets a particular (sometimes called 'maximally non-compact') real form $\mathcal{G}_R$ of $\mathcal{G}$, which is a real Lie algebra of the type $sl(n + 1, \mathbf{R})$, $sp(2n, \mathbf{R})$ or $so(p, q; \mathbf{R})$. This real Lie algebra $\mathcal{G}_R$ allows a locally unique Gauss decomposition $g = ABC$,

$$A = \exp\left(\sum_{\alpha \in \Delta^+} x^\alpha E_\alpha\right), \quad B = \exp\left(\tfrac{1}{2}\sum_{\alpha \in \Delta} \phi^\alpha H_\alpha\right), \quad C = \exp\left(\sum_{\alpha \in \Delta^-} y^\alpha E_\alpha\right) \;, \tag{11.84}$$

where Cartan-Weyl root vectors E_α, CSA generators $H_\alpha \equiv [E_\alpha, E_{-\alpha}]$, and a set of positive (negative) roots $\Delta^\pm$ have been introduced. They have the properties [5]

$$K_{\alpha,\beta} \equiv \alpha(H_\beta) = \frac{2\alpha \cdot \beta}{|\alpha|^2} , \quad \text{where} \quad \alpha, \beta \in \Delta , \quad |\alpha_{\text{long}}|^2 = 2 ,$$

$$\text{tr}(H_\alpha \cdot H_\beta) = \frac{2}{|\alpha|^2} K_{\alpha,\beta} \equiv C_{\alpha,\beta} ,$$

$$\text{tr}(E_\alpha \cdot E_\beta) = \frac{2}{|\alpha|^2} \delta_{\alpha,-\beta} , \quad \text{tr}(E_\alpha \cdot H_\beta) = 0 . \tag{11.85}$$

Eqs. (11.84) and (11.85) can be used to argue that the generalized constraints [495]

$$J(E_\alpha) = \kappa c_1^\alpha , \quad \tilde{J}(E_{-\alpha}) = -\kappa c_2^\alpha , \quad \text{where} \quad \alpha \in \Delta^+ , \tag{11.86}$$

with some real numbers c_1^α and c_2^α whose values do not vanish only for primitive roots $\alpha \in \Delta$, are enough to reduce the $\mathcal{G}_R$-based WZNW theory to the *Toda* field theory defined by the Lagrangian

$$\mathcal{L}_{\text{Toda}} = -\frac{k}{8\pi} \left[\frac{1}{4} C_{\alpha,\beta} \partial_+ \phi^\alpha \partial_- \phi^\beta - \sum_{\alpha \in \Delta} (\mu^2)^\alpha \exp(\tfrac{1}{2} K_{\alpha,\beta} \phi^\beta) \right] , \tag{11.87}$$

where $(\mu^2)^\alpha \equiv |\alpha|^2 c_1^\alpha c_2^\alpha$.

Due to $c_{1,2}^\alpha \neq 0$ for the primitive roots, the constraints (11.86) can be rewritten in terms of the Gauss decomposition (11.84) as follows [495]:

$$A^- \partial_- A = B \left(\sum_{\alpha \in \Delta} \tfrac{1}{2} |\alpha|^2 c_2^\alpha E_\alpha \right) B^{-1} = \sum_{\alpha \in \Delta} \tfrac{1}{2} |\alpha|^2 c_2^\alpha E_\alpha \exp(\tfrac{1}{2} K_{\alpha,\beta} \phi^\beta) ,$$

$$(\partial_+ C) C^{-1} = B^{-1} \left(\sum_{\alpha \in \Delta} \tfrac{1}{2} |\alpha|^2 c_1^\alpha E_{-\alpha} \right) B = \sum_{\alpha \in \Delta} \tfrac{1}{2} |\alpha|^2 c_1^\alpha E_{-\alpha} \exp(\tfrac{1}{2} K_{\alpha,\beta} \phi^\beta) . \tag{11.88}$$

In the WZNW equations of motion, the matrices A and C only occur in the combinations (11.88), so that they can be eliminated in favor of B or ϕ^α. The remaining equation is just the *Toda* equation [496]

$$\partial_+ \partial_- \phi^\alpha + \tfrac{1}{2} |\alpha|^2 (\mu^\alpha)^2 \exp(\tfrac{1}{2} K_{\alpha,\beta} \phi^\beta) = 0 . \tag{11.89}$$

This completes the (canonical) reduction of the $\mathcal{G}_R$-based WZNW theory to the Toda field theory. [8] The conformal symmetry of the Toda theory follows along the lines of

---

[8] See ref. [496], as far as the general solution to the Toda equations of motion is concerned. It can also be deduced from the general solution to the WZNW theory [495].

the Liouville case. It is worthwhile to mention that the conformal reduction used above is totally based on the algebraic structure of AKM symmetry of the WZNW theory, so that it can actually be applied to *any* system with an AKM symmetry [123].

The Toda field theory actually possesses an extended symmetry represented by a (classical) W algebra [238, 498, 513]. This W symmetry can be represented in terms of a Gel'fand-Dikii Hamiltonian structure (sect. VII.3). Regarding the Toda theory as a constrained WZNW theory, this Hamiltonian structure can be obtained by a classical DS reduction from the constrained phase space of AKM algebra [499]. In the Hamiltonian formalism, the AKM symmetry of the WZNW theory is represented by the first-class constraints. The additional conformally invariant constraints (11.86) make the gauge group nilponent. The (classical) W algebra of the Toda theory arises as the Poisson-bracket algebra of gauge-invariant polynomials of the constrained AKM currents and their derivatives [499]. To this end, we summarize the arguments supporting these general statements.

Let $g(z, \bar{z})$ be the $\mathcal{G}_R$-valued WZNW fields and $J(z)$ the corresponding AKM currents, subject to the equations of motion and having the form [9]

$$g(z, \bar{z}) = g(z) \cdot g(\bar{z}) , \quad \partial g(z) = J(z) \cdot g(z) . \tag{11.90}$$

Let $\dim \mathcal{G}_R$ be a dimension of $\mathcal{G}_R$, l its rank, k the level of the associated AKM algebra, g the dual Coxeter number of $\mathcal{G}_R$, ρ the half sum of the positive roots, and $\hat{\rho}$ the half sum of the positive co-roots of $\mathcal{G}_R$. The constrained WZNW theory is specified by (the Euclidean version of) eq. (11.86). After the suitable choice of constants c_i, the currents $J(z)$ can be decomposed as [499]

$$J(z) = I_- + j(z) , \quad I_- = \sum_{i=1}^{l} E_{-\alpha_i} ,$$

$$j(z) = \sum_{i=1}^{l} j^i(z) H_i + \sum_{\phi \in \Delta^+} E_\phi , \tag{11.91}$$

where $\{E_{\alpha_i}\}$ are l simple roots of $\mathcal{G}_R$. The maximal subgroup of $\hat{\mathcal{G}}_R$ leaving this form of currents invariant is the maximal nilpotent subgroup generated by E_ϕ, $\phi \in \Delta^+$, and implemented by the $(\dim \mathcal{G}_R - l)/2$ constrained AKM currents $J^{-\phi}(z)$. This allows us to interpret the constrained WZNW theory as the gauge theory in which all but l of

---

[9] We now consider the left-moving ('holomorphic') sector, and return back to the Euclidean notation.

the $(\dim \mathcal{G}_R + l)/2$ components of J are gauge components [499]. The Virasoro density, generalizing that in eq. (11.76) and having the form

$$L(J) = \tfrac{1}{2} \operatorname{tr} J^2 - \operatorname{tr} H J' , \quad H \equiv \hat{\rho}^i H_i , \tag{11.92}$$

weakly commutes with the constraints (11.86) and, therefore, it is gauge invariant. Under the conformal transformations, the currents transform as

$$\{L(x), J(y)\} = ([H, J(x)] + J(x)) d'(x - y) + [H, J'(x)] \delta(x - y) - H \delta''(x - y) . \tag{11.93}$$

The current $j(z)$ and the gauge transformations corresponding to E_ϕ act on each column of the WZNW field $g(z)$ separately, while each column contains only one (of the highest weight) gauge-invariant component e satisfying $E_\phi e = 0$. The gauge degrees of freedom corresponding to the other elements of each column can be eliminated by a gauge-fixing in favour of e. Because of eq. (11.91), this leads to a linear (pseudo)-differential equation $De = 0$, where D is a (pseudo-)differential operator of the type (11.37), whose coefficients a_j are gauge-invariant polynomials in the currents J. This operator D can now be used to define a classical W algebra, as in sect. 2. A straightforward way to compute the W algebra of the Toda theory is to choose a *Drinfeld-Sokolov gauge*, in which one has

$$j_{\mathrm{DS}}(z) = \sum_{p \geq 2} W^p(z) F_p , \tag{11.94}$$

where p's are orders of l independent Casimir operators of $\mathcal{G}_R$, and F_p are generators with H-weights $(p-1)$, so that the gauge-fixed current (11.94) has only one nonvanishing component in each of the l irreps in a decomposition of the adjoint of $\mathcal{G}_R$ w.r.t. one of its subgroups $SL(2, \mathbf{R})$ [250]. The Poisson brackets between l differential polynomials W^p just define a classical W algebra. As is explained in ref. [499], it is even possible to calculate this W algebra without computing the W^p's themselves, just by choosing a proper gauge *linear* in the currents (a Drinfeld-Sokolov gauge!), and by using the Dirac brackets since the Toda and gauge-fixing constraints together form a second-class system.

As far as a *quantum* version of the WZNW→Toda conformal reduction is concerned, we refer the reader to refs. [500, 501]. The central charge c receives contributions from both the SS term and the improvement term in the (quantized) improved stress tensor (11.92) [499],

$$c_{\mathrm{m}} = \frac{k \dim \mathcal{G}_R}{k + g} - 12 \operatorname{tr} H^2 . \tag{11.95}$$

The BRST ghosts resulting from the $(\dim \mathcal{G}_R - l)/2$ constraints (11.86) form pairs each contributing $12[h_\phi(1 - h_\phi) - \frac{1}{6}]$ to the central charge, since all the constrained currents are primary fields of weights $\Delta(J^{-\phi}) = 1 - h_\phi$, with h_ϕ being the H-weight of E_ϕ. Using the group theory relations (sect. XII.2)

$$\sum_{\Delta^+} h_\phi = 2\hat{\rho} \cdot \rho \,, \quad \sum_{\Delta^+} h_\phi^2 = \frac{1}{2} \operatorname{tr}_{\mathrm{adj}} H^2 \,, \tag{11.96}$$

one finds the total ghost contribution,

$$c_{\mathrm{ghost}} = l - \dim \mathcal{G}_R - 12g \operatorname{tr} H^2 + 24\hat{\rho} \cdot \rho \,, \tag{11.97}$$

where $g \operatorname{tr} H^2 = \frac{1}{2} \operatorname{tr}_{\mathrm{adj}} H^2$. The Freudenthal-de Vries 'strange' formula $g \dim \mathcal{G}_R = 12\rho^2$ now implies that the total central charge $c = c_{\mathrm{m}} + c_{\mathrm{ghost}}$ is given by

$$c = l - 12 \left(\sqrt{k+g}\hat{\rho} - \frac{1}{\sqrt{k+g}}\rho \right)^2 \,, \tag{11.98}$$

which is equivalent to eq. (7.114). For the simply-laced Lie algebras one has $\hat{\rho} = \rho$, so that eq. (11.98) gives [247, 499]

$$c = l - \frac{g \dim \mathcal{G}_R}{k+g}(k+g-1)^2 \,, \tag{11.99a}$$

or, equivalently, [513]

$$c = l \left[1 - g(g+1)\frac{(r-s)^2}{rs} \right] \,, \quad \text{where } k+g \equiv r/s \,. \tag{11.99b}$$

Exercises

XI-4 ▷ (*) Use the $SL(2, \mathbf{R})$ WZNW → Liouville conformal reduction to deduce the general solution (9.88) of the Liouville equation from the general solution of the WZNW theory:

$$g(\zeta^+, \zeta^-) = g_{\mathrm{L}}(\zeta^+) \cdot g_{\mathrm{R}}(\zeta^-) \,, \tag{11.100}$$

where g_{L} and g_{R} are arbitrary $SL(2, \mathbf{R})$-valued functions. *Hint*: use Gauss decompositions for g, g_{L} and g_{R}.

XI-5 ▷ The Liouville equation (11.72) also admits a 'singular' solution having the form

$$e^\phi = \frac{1}{\cos^2 \alpha} \,, \quad \text{where } \alpha = a\zeta^+ - b\zeta^- \,. \tag{11.101}$$

Derive this configuration by the conformal reduction from a regular solution

$$g = \begin{pmatrix} \cos \alpha & -\sin \alpha \\ \sin \alpha & \cos \alpha \end{pmatrix} \tag{11.102}$$

of the $SL(2, \mathbf{R})$ WZNW theory.

XI.4   4d self-duality and 2d integrable models

As was shown in sect. VIII.3, the $N = 2$ strings are closely connected with the 4d self-dual Yang-Mills (SDYM) theory. In its turn, the SDYM theory appears to be a 'master theory' for the whole variety of 2d integrable systems, as we are now going to explain. Though there is no general proof, the statement can be checked on the case-by-case basis. It is the main point here that the 4d self-duality condition admits a '*zero-curvature*' representation [502] underlying a Hamiltonian (or Lax pair) description of SDYM descendants in lower dimensions. This makes it possible to apply the inverse scattering method for integration of the SDYM equations [503]. Simultaneously, it explains the origin of gauge symmetries (like the conformal symmetry, or its extensions) in integrable systems of the KdV-type, since the SDYM theory is both gauge and conformally-invariant in four dimensions (sect. I.1). At last but not least, this connection provides us with a systematic way to associate the KdV-type hierarchy with any simple Lie algebra [504, 505].

SDYM solutions invariant by the action of a subgroup with two conformal generators satisfy a 2d differential equation, since each one-dimensional subgroup reduces by one the number of independent variables. This allows us to describe the invariant SDYM solutions in terms of a 2d integrable system. All known 2d integrable systems seem to be derivable this way, by appropriate truncations and reductions of a 4d self-dual gauge theory. This is true, in particular, for the KdV and non-linear Schrödinger equations [506], the Liouville [507] and Toda equations [496], the Boussinesq equation [504, 505, 508], as well as for many other integrable equations in two and three dimensions [509, 510]. Our presentation in this section is only illustrative: we give one explicit example of embedding of the KdV-equation into the 4d SDYM theory [506], the other examples being referred to the literature.

Let $x^a = (x, y, z, t)$ be coordinates of a flat 4d 'spacetime' of signature $(+, +, -, -)$. The invariant metric reads

$$ds^2 = 2dxdz + 2dydt \ . \tag{11.103}$$

The SDYM equations in $2 + 2$ dimensions ($\varepsilon_{xyzt} = 1$),

$$F_{ab} = \tfrac{1}{2}\varepsilon_{abcd}F^{cd} \ , \tag{11.104}$$

are equivalently represented by three equations having the form

$$\text{(1)} \quad F_{tx} = 0 \ ,$$
$$\text{(2)} \quad F_{yz} = 0 \ , \tag{11.105}$$
$$\text{(3)} \quad F_{ty} + F_{xz} = 0 \ .$$

After a dimensional reduction, which is equivalent to introducing two Killing vectors, of the type

$$\partial_y = \partial_z - \partial_x = 0 \ , \tag{11.106}$$

eq. (11.105) takes the form

$$\text{(1)} \quad [\partial_t - H, \partial_x - Q] = 0 \ ,$$
$$\text{(2)} \quad [P, B] = 0 \ , \tag{11.107}$$
$$\text{(3)} \quad [H, B] = [\partial_x - Q, \partial_x - P] \ ,$$

resp., where the following notation has been introduced:

$$A_t = H \ , \quad A_x = Q \ , \quad A_y = -B \ , \quad A_z = P \ . \tag{11.108}$$

Eq. (11.107) is known as the 'zero-curvature' condition [503].

Let us now choose the non-compact group $SL(2, \mathbf{R})$ as a SDYM gauge group, and an embedding pattern in the form

$$B = \begin{pmatrix} 0 & 0 \\ -1 & 0 \end{pmatrix} \ , \quad Q = \begin{pmatrix} \lambda & 1 \\ -u & -\lambda \end{pmatrix} \ , \tag{11.109}$$

where a constant λ and a function $u = u(t, x + z)$ have been introduced. Since all the SDYM fields are valued in the Lie algebra $sl(2, \mathbf{R})$, it is convenient to introduce a Lie-algebra basis comprising the matrices

$$\tau_+ = \begin{pmatrix} 0 & 1 \\ 0 & 0 \end{pmatrix} \ , \quad \tau_- = \begin{pmatrix} 0 & 0 \\ 1 & 0 \end{pmatrix} \ , \quad \tau_3 = \begin{pmatrix} 1 & 0 \\ 0 & -1 \end{pmatrix} \ . \tag{11.110}$$

They satisfy the commutation relations

$$[\tau_+, \tau_-] = \tau_3 \ , \quad [\tau_3, \tau_\pm] = \pm \tau_\pm \ , \tag{11.111}$$

and allow us to expand the Lie algebra-valued fields H and P as

$$\begin{aligned} H &= H_+\tau_+ + H_-\tau_- + H_3\tau_3 \ , \\ P &= P_+\tau_+ + P_-\tau_- + P_3\tau_3 \ . \end{aligned} \tag{11.112}$$

Substituting all this into the second line of eq. (11.107) gives [10]

$$P_- = P_3 = 0 \ . \tag{11.113}$$

Similarly, the third line of eq. (11.107) gives

$$H_- = -P_+ \ , \quad H_3 = -\tfrac{1}{2}(u + P'_+)' - \lambda P_+ \ . \tag{11.114}$$

Finally, the first line of eq. (11.107) yields three equations:

$$
\begin{aligned}
&(1a) \quad (u + 2P_+)' = 0 \ , \\
&(1b) \quad H_+ = uP_+ - \lambda P'_+ - \tfrac{1}{2}(u + P_+)'' \ , \\
&(1c) \quad \dot u = \tfrac{1}{2}(u + P_+)''' + (u - P_+)u' + 2\lambda^2 P'_+ \ .
\end{aligned}
\tag{11.115}
$$

It follows

$$
\begin{aligned}
&(1a) \ \rightarrow \ P_+ = -\tfrac{1}{2}u \ , \\
&(1b) \ \rightarrow \ H_+ = -\tfrac{1}{2}u^2 + \frac{\lambda}{2}u' - \frac{1}{4}u'' \ , \\
&(1c) \ \rightarrow \ \dot u = \frac{1}{4}u''' + \frac{3}{2}uu' - \lambda^2 u' \ .
\end{aligned}
\tag{11.116}
$$

After changing the notation as

$$u \to u + \frac{2}{3}\lambda^2 \ , \quad t \to 4t \ , \quad x + y \to x \ , \tag{11.117}$$

one gets from eq. (11.116) the KdV equation (sect. 1)

$$u_t = u_{xxx} + 6uu_x \ . \tag{11.118}$$

This particular example together with some other similar patterns mentioned at the beginning of this section (see exercise # XI-6 also) may be relevant towards an ultimate unification of 2d integrable models and 2d CFT's as well, within the 4d SDYM theories or extended fermionic strings, which are known to be closely related (sect. VIII.3).

Exercise

XI-6 ▷ Prove that the $SU(2)$ SDYM equations are solved by the ansatz [506]

$$B = \frac{1}{2i}\begin{pmatrix} 1 & 0 \\ 0 & -1 \end{pmatrix} \ , \quad Q = \begin{pmatrix} 0 & \tfrac{1}{2}\psi \\ -\tfrac{1}{2}\psi^* & 0 \end{pmatrix} \ , \quad H = \begin{pmatrix} i\psi\psi^* & i\psi_x \\ i\psi_x^* & -i\psi\psi^* \end{pmatrix} \ , \tag{11.119}$$

provided a complex field ψ satisfies the *non-linear Schrödinger* equation

$$i\psi_t = -\psi_{xx} - 2|\psi|^2 \psi \ . \tag{11.120}$$

---

[10] The prime and dot stand for differentiations w.r.t. $x + z$ and t, resp.

XI.5   Self-duality and supersymmetry

Extended supersymmetry is compatible with self-duality in $2+2$ dimensions [387, 388, 389]. Therefore, the *supersymmetric* SDYM (= SSDYM) theory should be capable to generate supersymmetric 2d integrable models. A supersymmetrization of the SDYM theory is, however, not unique. One can either replace a gauge group by its graded version, or a $(2+2)$-dimensional 'spacetime' by superspace.

As our first example, consider the $N=1$ *super*-KdV equations in $1+1$ dimensions [511, 512, 513]. They have two dynamical variables, one bosonic $u(x,t)$ and one fermionic $\psi(x,t)$, and read

$$\frac{\partial u}{\partial t} = \tfrac{1}{2}\partial_x^3 u + 3u\partial_x u + \frac{3}{2}(\partial_x^2\psi)\psi \ , \qquad \frac{\partial\psi}{\partial t} = \tfrac{1}{2}\partial_x^3\psi + \frac{3}{2}\partial_x(u\psi) \ . \qquad (11.121)$$

Eqs. (11.121) are invariant under the $N=1$ supersymmetry transformations

$$\delta u = \varepsilon\partial_x\psi \ , \quad \delta\psi = \varepsilon u \ , \qquad (11.122)$$

where ε is a constant Grassmannian (anticommuting) parameter. Eqs. (11.121) are integrable [511, 512], and can be obtained from the 'zero-curavture' condition associated with the *graded* Lie algebra $OSp(2|1)$,

$$\partial_t A_x - \partial_x A_t + [A_t, A_x] = 0 \ , \qquad (11.123)$$

when using the following ansatz for 2d Yang-Mills potentials [514]:

$$A_t(x,t) = \begin{pmatrix} \tfrac{1}{2}\partial_x u & \tfrac{1}{2}\partial_x^2 u + u^2 + \tfrac{1}{2}\partial_x\psi\psi & -\tfrac{i}{2}\partial_x^2\psi - iu\psi \\ -u & -\tfrac{1}{2}\partial_x u & \tfrac{i}{2}\partial_x\psi \\ \tfrac{i}{2}\partial_x\psi & \tfrac{i}{2}\partial_x^2\psi + iu\psi & 0 \end{pmatrix} \ ,$$

$$A_x(x,t) = \begin{pmatrix} 0 & u & -i\psi \\ -1 & 0 & 0 \\ 0 & i\psi & 0 \end{pmatrix} \ . \qquad (11.124)$$

The 2d super-KdV system can be embedded into the self-duality equations (11.107) by choosing the $OSp(2|1)$-valued matrices H, Q, B and P as follows: $H = A_t(x,t)$, $Q = A_x(x,t)$ and

$$B = \begin{pmatrix} 0 & \tfrac{1}{2} & 0 \\ 0 & 0 & 0 \\ 0 & 0 & 0 \end{pmatrix} \ , \quad P = \begin{pmatrix} 0 & \tfrac{1}{2}u & -\tfrac{3i}{4}\psi \\ 0 & 0 & 0 \\ 0 & \tfrac{3i}{4}\psi & 0 \end{pmatrix} \ . \qquad (11.125)$$

The super-KdV *hierarchy* of differential equations generically reads [514, 515]

$$\frac{\partial u}{\partial t} = \frac{1}{2}\left(\partial_x^3 + 2[\partial_x u + u\partial_x]\right)C_n - \frac{3}{2}\partial_x G_n\psi - \frac{1}{2}G_n\partial_x\psi \ ,$$

$$\frac{\partial\psi}{\partial t} = -\frac{1}{2}\partial_x^2 G_n - \frac{1}{2}uG_n + \frac{3}{2}\partial_x C_n\psi + C_n\partial_x\psi \ , \tag{11.126}$$

where some bosonic (C) and fermionic (G) functionals of the dynamical variables $u(x,t)$ and $\psi(x,t)$ have been introduced. The former are, in fact, simply related with conserved charges H_n of the super-KdV hierarchy,

$$C_n = \frac{\delta H_n}{\delta u(x,t)} \ , \tag{11.127}$$

while G_n are related with C_n by supersymmetry. The super-KdV hierarchy (11.126) can be equally formulated as a 'zero-curvature' condition associated with the group $U(1)$, and it is also derivable from the SDYM equations for the graded gauge algebra $OSp(2|1)$ [514].

Our next examples are based on the $N = 2$ SSDYM in $2 + 2$ dimensions, constructed in refs. [388, 389]. We show that the $N = 1$ and $N = 2$ super-KdV equations [515], as well as the $N = 1$ *super-Liouville* and *super-Toda* equations [516], can all be obtained from the $N = 2$ SSDYM theory by dimensional reductions and truncations, in two different ways [517]. After the first type of reduction, some components of the 4d SSDYM superfield strength remain to be non-zero in two dimensions, whereas after the second type of reduction all of them vanish. Both cases are considered below, along the lines of ref. [517].

The $N = 2$ SSDYM theory has the field content $(A_{\underline{a}}{}^I, \tilde{\lambda}_{\dot{\alpha}i}{}^I, T^I)$ in $2+2$ dimensions, where we use $\underline{a}, \underline{b}, \ldots = 1, \ldots, 4$ for the 4d vectorial indices, and $\alpha, \beta, \ldots = 1, 2$ (or $\dot{\alpha}, \dot{\beta}, \ldots = \dot{1}, \dot{2}$) for the chiral (or anti-chiral) spinorial indices. The $N = 2$ SSDYM field equations in $2 + 2$ dimensions read [388, 389]

$$F_{\underline{ab}}{}^I = \frac{1}{2}\epsilon_{\underline{ab}}{}^{\underline{cd}}F_{\underline{cd}}{}^I \ , \tag{11.128a}$$

$$i(\sigma^{\underline{a}})_\alpha{}^{\dot{\beta}}\nabla_{\underline{a}}\tilde{\lambda}_{\dot{\beta}i}{}^I = 0 \ , \tag{11.128b}$$

$$\Box T^I - f^{IJK}(\tilde{\lambda}^{iJ}\tilde{\lambda}_i{}^K) = 0 \ , \tag{11.128c}$$

where the indices $i, j, \cdots = 1, 2$ for the **2**-representation of $Sp(1)$ are raised and lowered by ϵ^{ij} and ϵ_{ij}, while $I, J, \ldots$ stand for the adjoint representation of a Yang-Mills gauge group (they will be normally suppressed in what follows). $\tilde{\lambda}_{\dot{\alpha}i}{}^I$ is an anti-chiral MW (spinor) 'gaugino' field, and T^I is a real scalar field in the adjoint representation. The derivative $\nabla_{\underline{a}}$ is gauge-covariant.

Since there are many options in choosing coordinates, as well as gauge fixing, dimensional reduction is not unique. We are now going to follow the same pattern used in the previous section, with a flat metric (11.103) and self-duality equations in the form (11.105). We regard the coordinates (x, t) as the ones belonging to two dimensions, into which our dimensional reduction is performed. The first eq. (11.105) implies that the 2d gauge fields A_x and A_t should be pure gauge. It allows us to impose the conditions

$$A_x = A_t = 0 \ . \tag{11.129}$$

Requiring further an independence of all quantities on $y-$ and $z-$ coordinates, the remaining equations (11.105) yield

$$[P, B] = 0 \ , \qquad \dot{P} + B' = 0 \ , \tag{11.130}$$

where $P \equiv A_y$, $B \equiv A_z$, while the prime and dot denote $\partial_x \equiv \partial/\partial x$ and $\partial_t \equiv \partial/\partial t$, resp.

Using 2×2 σ-matrices in a representation of 4×4 Dirac matrices, where (i) Γ_5 is diagonal, (ii) metric takes the form (11.103) in $2 + 2$ dimensions [389], viz.

$$\sigma^x = \begin{pmatrix} +i & -i \\ -i & +i \end{pmatrix}, \ \sigma^y = \begin{pmatrix} +1 & -1 \\ +1 & -1 \end{pmatrix}, \ \sigma^z = \begin{pmatrix} -i & -i \\ -i & -i \end{pmatrix}, \ \sigma^t = \begin{pmatrix} -1 & -1 \\ +1 & +1 \end{pmatrix},$$

$$\tilde{\sigma}^x = \begin{pmatrix} +i & +i \\ +i & +i \end{pmatrix}, \ \tilde{\sigma}^y = \begin{pmatrix} -1 & +1 \\ -1 & +1 \end{pmatrix}, \ \tilde{\sigma}^z = \begin{pmatrix} -i & +i \\ +i & -i \end{pmatrix}, \ \tilde{\sigma}^t = \begin{pmatrix} +1 & +1 \\ -1 & -1 \end{pmatrix}, \tag{11.131}$$

and (iii) the following identification of components of the MW spinor $\tilde{\lambda}_{\dot{\alpha}i}$ takes place [11]

$$(\tilde{\lambda}_{\dot{\alpha}i}) = \frac{1}{\sqrt{2}} \begin{pmatrix} \psi_i - i\chi_i \\ \psi_i + i\chi_i \end{pmatrix} \ , \tag{11.132}$$

the 'gaugino' field equation (11.128b) gives rise to the two equations,

$$\dot{\psi}_i = \chi_i{}' \ , \qquad [P, \chi_i] + [B, \psi_i] = 0 \ . \tag{11.133}$$

Applying the dimensional reduction to eq. (11.128c) yields

$$[B, T'] + [P, \dot{T}] + [\psi^i, \chi_i] = 0 \ . \tag{11.134}$$

---

[11]Spinor components of a MW spinor have to be complex conjugated w.r.t. each other [389].

The full set of 2d field equations is given by eqs. (11.130), (11.133) and (11.134).

The $N = 2$ supersymmetry transformation rules of the $N = 2$ SSDYM fields take the form [12]

$$\delta A_{\underline{a}} = -i(\epsilon^i \sigma_{\underline{a}} \tilde{\lambda}_i) , \qquad \delta T = (\tilde{\epsilon}^i \tilde{\lambda}_i) ,$$

$$\delta \tilde{\lambda}_i = \frac{1}{8} \left(\tilde{\sigma}^{\underline{a}} \sigma^{\underline{b}} - \tilde{\sigma}^{\underline{b}} \sigma^{\underline{a}} \right) \tilde{\epsilon}_i F_{\underline{ab}} - i \tilde{\sigma}^{\underline{a}} \epsilon_i \nabla_{\underline{a}} T . \tag{11.135}$$

Substituting the σ-matrices (11.131) into eq. (11.135), one finds

$$\delta P = -\sqrt{2}(\beta^i \psi_i) , \quad \delta B = \sqrt{2}(\beta^i \chi_i) ,$$

$$\delta \psi^i = -\tilde{\beta}^i P' - \tilde{\alpha}^i \dot{P} + \sqrt{2} \beta^i T' ,$$

$$\delta \chi^i = \tilde{\alpha}^i \dot{B} + \tilde{\beta}^i B' + \sqrt{2} \beta^i \dot{T} ,$$

$$\delta T = -(\tilde{\alpha}^i \chi_i) - (\tilde{\beta}^i \psi_i) , \tag{11.136}$$

where the new parameters, α's and β's, have been introduced,

$$\alpha_i \equiv \frac{1}{\sqrt{2}}(\epsilon^1{}_i + \epsilon^2{}_i) , \qquad \beta_i \equiv -\frac{i}{\sqrt{2}}(\epsilon^1{}_i - \epsilon^2{}_i) ,$$

$$\tilde{\alpha}_i \equiv \frac{1}{\sqrt{2}}(\tilde{\epsilon}^1{}_i + \tilde{\epsilon}^2{}_i) , \qquad \tilde{\beta}_i = -\frac{i}{\sqrt{2}}(\tilde{\epsilon}^1{}_i - \tilde{\epsilon}^2{}_i) , \tag{11.137}$$

The α-parameters have to be put to zero by consistency with the gauge condition (11.129). Eqs. (11.136) give the extended supersymmetry transformation laws for the dimensionally reduced 2d theory.

To find an embedding of the $N = 2$ *extended* 2d super-KdV equations into the $N = 2$, 4d SSDYM theory, let us consider the *abelian* SSDYM theory, where all commutators disappear. The following identifications turn out to be consistent for the lowest $N = 2$ super-KdV flow [517]:

$$B \equiv u'' + 3u^2 + 3\xi_i \xi_i{}' + (a+1)w'^2 + (a-2)ww'' - 3auw^2 - 3a\epsilon_{ij}w\xi_i\xi_j ,$$

$$\chi_i \equiv -\xi_i''' + 3(u\xi_i)' + 3a(w^2\xi_i)' + \epsilon_{ij}\left[(a+2)(w\xi_j{}')' + (a-1)(w'\xi_j)'\right] ,$$

$$P \equiv u , \qquad \psi_i \equiv \xi_i' . \tag{11.138}$$

The original $Sp(1)$ symmetry has been lost, and all the repeated indices $i, j, \ldots = 1, 2$ are now contracted by δ_{ij} instead of ϵ_{ij}.

---

[12]The normalization of the 'gaugino' here is different from the one used in refs. [388, 389].

It follows from the second eq. (11.130) that

$$\dot{u} = -u''' + 6uu' - 3\xi_i\xi_i'' - 3aw'w'' - (a-2)ww''' + 3a(uw^2)' + 3a\epsilon_{ij}(w\xi_i\xi_j)' \ , \quad (11.139)$$

while the first eq. (11.133) gives

$$\dot{\xi_i}' = \left[-\xi_i''' + 3(u\xi_i)' + 3a(w^2\xi_i)' + \epsilon_{ij}\left\{(a+2)(w\xi_j')' + (a-1)(w'\xi_j)'\right\} \right]' \ . \quad (11.140)$$

Integrating the last equation once and requiring all fields to vanish at the (space) infinity $|x| \to \infty$, one gets the fermionic part of the $N = 2$ super-KdV equation [512, 515] in the form

$$\dot{\xi_i} = -\xi_i''' + 3(u\xi_i)' + 3a(w^2\xi_i)' + \epsilon_{ij}\left[(a+2)(w\xi_j')' + (a-1)(w'\xi_j)'\right] \ . \quad (11.141)$$

The bosonic w-equation of the $N = 2$ super-KdV system can be derived by applying the supersymmetry transformation law (11.136) with the parameters $\tilde{\alpha}_i \equiv 0$, $\beta_i \equiv \frac{1}{\sqrt{2}}\epsilon_i$, $\tilde{\beta}_i \equiv \epsilon_i$, namely

$$\delta w = -\epsilon_i\xi_i' \ , \quad \delta u = -\epsilon_{ij}\epsilon_i\xi_j' \ , \quad \delta\xi_i = \epsilon_{ij}\epsilon_j u + \epsilon_i w' \ , \quad (11.142)$$

to the fermionic eq. (11.141), and integrating the result over x. This way one gets the standard form of the bosonic $N = 2$ super-KdV equation[515],

$$\dot{w} = -w''' + 6aw^2w' + (a+2)(uw)' + \tfrac{1}{2}(a-1)\epsilon_{ij}(\xi_i\xi_j)' \ . \quad (11.143)$$

The $N = 2$ super-KdV system obviously contains the $N = 1$ super-KdV equations. Hence, the latter could be embedded similarly. This can be actually seen by putting $w(x,t) = \xi_2(x,t) = 0$ and $\xi_1(x,t) \equiv \xi(x,t)$.

The second type of dimensional reduction has manifest supersymmetry, and it allows us to realize an embedding of the $N = 1$ super-Toda theory as well [517]. In the $N = 2$ SSDYM theory in 2+2 dimensions, the $N = 2$ superfield strength F_{AB} maintains only half of its original components, due to the self-duality condition [388, 389], and it is fully consistent to truncate all of its components. To see this, we first introduce the $N = 2$ SSDYM superspace constraints in $2 + 2$ dimensions as [388, 389]

$$F_{\alpha i \underline{b}} = -i(\sigma_{\underline{b}})_{\alpha\dot{\beta}}\tilde{\lambda}^{\dot{\beta}}{}_i \ , \quad F_{\alpha i \beta j} = 2C_{\alpha\beta}\epsilon_{ij}T \ , \quad F_{\dot{\alpha} i \underline{b}} = F_{\alpha i \dot{\beta} j} = F_{\dot{\alpha} i \dot{\beta} j} = 0 \ , \quad (11.144)$$

$$\widetilde{\nabla}_{\dot{\alpha} i}T = -\tilde{\lambda}_{\dot{\alpha} i} \ , \quad \nabla_{\alpha i}T = 0 \ , \quad \nabla_{\alpha i}\tilde{\lambda}_{\dot{\beta} j} = -i\epsilon_{ij}(\sigma^{\underline{c}})_{\alpha\dot{\beta}}\nabla_{\underline{c}}T \ , \quad \widetilde{\nabla}_{\dot{\alpha} i}\tilde{\lambda}_{\dot{\beta} j} = \tfrac{1}{4}\epsilon_{ij}(\sigma^{\underline{cd}})_{\dot{\alpha}\dot{\beta}}F_{\underline{cd}} \ ,$$

where C is the charge conjugation matrix in $2 + 2$ dimensions. Next, we impose the additional conditions [517]

$$\tilde{\lambda}_{\dot{\alpha}i} = 0 \; , \qquad T = 0 \; . \tag{11.145}$$

Accordingly, the equations $F_{xy} = F_{xz} = F_{yz} = F_{yt} = F_{zt} = 0$ are satisfied, and we are left with the purely 2d superfield equations $F_{xt} = 0$, $F_{\underline{\alpha}x} = 0$, $F_{\underline{\alpha}t} = 0$. This amounts to requiring 2d superspace constraints in the very simple form

$$F_{AB} = 0 \; , \tag{11.146}$$

where the 2d superspace indices A, B, ... have been introduced. The superfield strength (11.146) satisfies the Bianchi identities

$$\nabla_{[A}F_{BC)} - T_{[AB|}{}^{D}F_{D|C)} \equiv 0 \; . \tag{11.147}$$

It is now straightforward to apply this type of dimensional reduction, in order to get the $N = 2$ super-KdV equation. Let us consider, for example, an *abelian* superfield strength F_{AB} in the $N = (2,0)$ superspace having super-coordinates $Z^M = (x, t, \theta^1, \theta^2)$, and define

$$\nabla_i \equiv \mathbf{D}_i + \mathbf{D}_i\Phi \; , \quad \mathbf{D}_i \equiv \frac{\partial}{\partial\theta^i} + \theta^i\frac{\partial}{\partial x} \; , \quad (i, j, \ldots = 1, 2) \; , \quad F_{\theta^i\theta^j} = 0 \; ,$$

$$\nabla_x \equiv \tfrac{1}{2}\{\nabla_1, \nabla_1\} = \tfrac{1}{2}\{\nabla_2, \nabla_2\} = \frac{\partial}{\partial x} + A_x \; , \qquad \nabla_t \equiv \frac{\partial}{\partial t} + A_t \; , \tag{11.148}$$

where Φ is a *bosonic* superfield,

$$\Phi(x, t, \theta^1, \theta^2) = w(x,t) + \theta^1\xi_1(x,t) + \theta^2\xi_2(x,t) + \theta^2\theta^1 u(x,t) \; , \tag{11.149}$$

with vanishing boundary conditions at the space infinity.

It follows from eq. (11.148) that $A_x = \Phi'$, whereas identifying

$$A_t = -\Phi''' + 3(\Phi\mathbf{D}_1\mathbf{D}_2\Phi)' + \tfrac{1}{2}(a - 1)(\mathbf{D}_1\mathbf{D}_2\Phi^2)' + 3a\Phi^2\Phi' \; , \tag{11.150}$$

one finds

$$F_{\theta^it} = -\mathbf{D}_i\left[\dot{\Phi} - \left(-\Phi''' + 3(\Phi\mathbf{D}_1\mathbf{D}_2\Phi)' + \tfrac{1}{2}(a - 1)(\mathbf{D}_1\mathbf{D}_2\Phi^2)' + 3a\Phi^2\Phi'\right)\right]$$

$$F_{xt} = -\left[\dot{\Phi} - \left(-\Phi''' + 3(\Phi\mathbf{D}_1\mathbf{D}_2\Phi)' + \tfrac{1}{2}(a - 1)(\mathbf{D}_1\mathbf{D}_2\Phi^2)' + 3a\Phi^2\Phi'\right)\right]' . \tag{11.151}$$

The condition $F_{\theta^it} = 0$ then yields

$$\dot{\Phi} = -\Phi''' + 3(\Phi\mathbf{D}_1\mathbf{D}_2\Phi)' + \tfrac{1}{2}(a - 1)(\mathbf{D}_1\mathbf{D}_2\Phi^2)' + 3a\Phi^2\Phi' \; . \tag{11.152}$$

The constant a introduced above has to be equal to -2, 1 or 4 for the equations to be exactly soluble [512, 515]. The only remaining condition $F_{xt} = 0$ is now automatically satisfied.

The system of differential equations (11.146) has more applications, with the $N = 1$ super-Liouville equation being one of them [517]. Let us choose the supergauge potential in the form [516]

$$A_\theta = 2\beta(\mathbf{D}\Phi)L_0 + G_- \ , \qquad A_{\widetilde\theta} = -e^{\beta\Phi}G_+ \ , \tag{11.153}$$

where Φ is a scalar superfield, $\mathbf{D}^2 = \partial_z$, $\widetilde{\mathbf{D}}^2 = \partial_{\widetilde z}$, $\{\mathbf{D}, \widetilde{\mathbf{D}}\} = 0$, and the $OSp(1,2)$ Lie superalgebra generators (L's and G's) have been introduced. The latter satisfy the algebra

$$[L_0\,,\ L_\pm] = \mp L_\pm \ , \qquad [L_+\,,\ L_-] = 2L_0 \ ,$$

$$[L_0\,,\ G_\pm] = \mp\tfrac{1}{2}G_\pm \ , \qquad [L_\pm\,,\ G_\mp] = \pm G_\pm \ , \tag{11.154}$$

$$\{G_+\,,\ G_-\} = 2L_0 \ , \qquad \{G_\pm\,,\ G_\pm\} = 2L_\pm \ .$$

Inserting all this into the equation $F_{\theta\widetilde\theta} = 0$, and identifying the grading in the Lie superalgebra with the 2d space-time supersymmetry, one obtains

$$F_{\theta\widetilde\theta} = -2\left[\beta(\mathbf{D}\widetilde{\mathbf{D}}\,\Phi) + e^{\beta\Phi}\right]L_0 = 0 \ , \tag{11.155}$$

which is just the super-Liouville equation [516]. The other field equations, $F_{z\theta} = 0$, $F_{z\widetilde\theta} = 0$ and $F_{\underline{ab}} = 0$, are easily satisfied by using

$$\nabla_z \equiv \partial_z + A_z \equiv \nabla^2 \ , \qquad \widetilde\nabla_{\widetilde z} \equiv \partial_{\widetilde z} + A_{\widetilde z} \equiv \widetilde\nabla^2 \ ,$$

$$\nabla \equiv \mathbf{D} + A_\theta \ , \qquad \widetilde\nabla \equiv \widetilde{\mathbf{D}} + A_{\widetilde\theta} \ . \tag{11.156}$$

The case of the $N = 1$ super-Toda theory [516] is quite similar [517]. One chooses a Cartan-Weyl-type basis with commutation relations

$$[H\,,\ H] = 0 \ , \qquad [H\,,\ e_i^\pm] = \pm\alpha_i e_i^\pm \ , \qquad [e_i^+\,,\ e_j^-\} = \delta_{ij}\alpha_i \cdot H \ , \tag{11.157}$$

where α_i are simple roots, and H is an l-vector of CSA (sect. XII.2). CSA vectors are all bosonic, while $e_i^\pm$ can be either bosonic or fermionic. Introducing a real superfield Φ in the l-dimensional representation, and putting

$$A_\theta = \beta(\mathbf{D}\Phi)\cdot H + \sum_{i\in\mathcal{F}} e_i^+ \ , \qquad A_{\widetilde\theta} = -\sum_{i\in\mathcal{F}} \exp(\beta\alpha_i \cdot \Phi)e_i^- \ , \tag{11.158}$$

into the superfield equation $F_{\theta\tilde{\theta}} = 0$, gives the $N = 1$ super-Toda superfield equation [516]

$$\mathbf{D}\widetilde{\mathbf{D}}\,\Phi + \frac{1}{\beta}\sum_{i\in\mathcal{F}}\alpha_i\exp(\beta\alpha_i\cdot\Phi) = 0 \,, \tag{11.159}$$

where $\mathcal{F}$ stands for the *fermionic* grading indices $\mathcal{F} \subset \{1, 2, \cdots, l\}$. Other components in $F_{AB} = 0$ vanish because of eq. (11.156).

The master superfield equation (11.146) is, in fact, connected with the (topological) three-dimensional supersymmetric Chern-Simons theory described by the vanishing superfield strength $F_{AB} = 0$ [180]. The $N = 1$ or $N = 2$ SSDYM theory in $2+2$ dimensions [388, 389] can be reduced to the three-dimensional $N = 1$ or $N = 2$ *supersymmetric* Chern-Simons theory, resp., by dimensional reduction and truncation [518]. Therefore, an embedding of the three-dimensional supersymmetric *Kadomtsev-Petviashvili* (KP) equations [519] into the SSDYM theory in $2 + 2$ dimensions is also possible (exercise # XI-7). The gauged supersymmetric 2d WZNW theories on coset manifolds are generated from the SSDYM theory along the similar lines [520].

Exercise

XI-7 ▷ The *three*-dimensional integrable $N = 1$ super-KP system of equations reads in superspace as [515]

$$\frac{3}{4}\partial_y^2\Psi = -\partial_x\left[\partial_t\Psi + \frac{1}{4}\partial_x^3\Psi + \frac{3}{2}\partial_x(\Psi\mathbf{D}\Psi)\right] \,, \tag{11.160}$$

in terms of an $N = 1$ *fermionic* superfield $\Psi(t, x, y, \theta) = \phi(t, x, y) + \theta u(t, x, y)$ and the supercovariant derivative $\mathbf{D} = \mathbf{D}_x$. Verify that the component field equations following from eq. (11.160) take the form

$$\begin{aligned}
\frac{3}{4}\partial_y^2 u + \partial_x\left[\partial_t u + \frac{1}{4}\partial_x^3 u + 3u\partial_x u + \frac{3}{2}(\partial_x^2\phi)\phi\right] &= 0 \,, \\
\frac{3}{4}\partial_y^2\phi + \partial_x\left[\partial_t\phi + \frac{1}{4}\partial_x^3\phi + \frac{3}{2}\partial_x(u\phi)\right] &= 0 \,,
\end{aligned} \tag{11.161}$$

and show that eq. (11.160) can be embedded into the *three*-dimensional $N = 1$ supersymmetric Chern-Simons theory, whose superfield equations of motion take the form $F_{AB} = 0$ for $A, B = t, x, y, \theta$, in terms of the $U(1)$ superfield strength F_{AB}. *Hint:* use the following ansatz for the superfield potentials [518]

$$A_x = \frac{3}{4}\partial_x\partial_y\mathbf{D}\Psi \,, \quad A_y = -\partial_x\mathbf{D}\left[\partial_t\Psi + \frac{1}{4}\partial_x^3\Psi + \frac{3}{2}\partial_x(\Psi\mathbf{D}\Psi)\right] \,, \quad A_t = \frac{3}{4}\partial_t\partial_y\mathbf{D}\Psi \,,$$

$$\Gamma_\theta = \frac{3}{4}\partial_x\partial_y\Psi \,. \tag{11.162}$$

Chapter XII

Comments

XII.1   About the Literature

The Conformal Field Theory has a long history, but the modern period of its development started from the famous paper [12] of Belavin, Polyakov and Zamolodchikov. Many reviews are now available [7, 9, 13, 14, 15, 16, 17, 18, 19, 20, 21, 32, 76] (see also refs. [28, 77, 70, 276, 292, 293, 303, 521]). [1]

The operator product expansion was introduced by Wilson [25], see also ref. [26]. The Virasoro algebra appeared in the physical literature due to Virasoro [29], while in the mathematical literature this algebra and its representations were extensively studied by Kač [30], Feigin and Fuchs [31]. Conformal anomalies of QFT in curved space are discussed in the book of Birrell and Davies [39]. An introduction to the index theorems can be found in the review of Eguchi, Gilkey and Hanson [40].

Conformal bootstrap in CFT is discussed at length in the Lectures of Dotsenko [13] and Zamolodchikov [14]. The Kač determinant was introduced by Kač [30, 44]. Fundamental contributions to the theory of Verma modules are due to Feigin and Fuchs [31, 45]. The 'Coulomb gas' picture was introduced into CFT by Dotsenko and Fateev [54, 55]. The importance of fusion rules in CFT was pointed out by Verlinde [51], Dijkgraaf and Verlinde [52]. The BRST approach to the minimal models was developed by Felder [66]. The Ising model is a standard issue in statistical physics, which has a lot of references (see, e.g., refs. [6, 7, 8, 9, 10, 11] and references therein).

---

[1] I didn't attempt to make an exhaustive bibliography, but wanted to help the reader to find his own way through the literature about CFT and related topics.

GSO projection was first introduced in string theory by Gliozzi, Scherk and Olive [72]. Spin structures and their relevance in string theory were emphasized by Seiberg and Witten [73]. An evaluation of partition functions for 2d CFT's defined on a torus is due to Itzykson and Zuber [82], and Cardy [7]. The Verlinde formula was conjectured by Verlinde [51], and was proved later by Moore and Seiberg [74, 75]. Fundamentals of chiral bosonization techniques were developed by Friedan, Martinec and Shenker [35, 93], and applied by Kostelecký, Lechtenfeld, Lerche, Samuel and Watamura [91]. Chiral bosonization on Riemann surfaces was considered, in particular, by Verlinde and Verlinde [87, 95]. The BRST approach to the minimal models on a torus was initiated by Felder [66].

A physics-oriented review about affine Lie (or AKM) symmetry structures is due to Goddard and Olive [106, 107]. The theory of AKM algebras from the mathematical viewpont can be found in the book of Kač [44], which is the standard reference on the subject. The SS construction first appeared in the papers of Sugawara [110] and Sommerfeld [111]. The Knizhnik-Zamolodchikov equation was introduced in ref. [109].

The WZNW term and action were discovered by Wess and Zumino [113], Novikov [114, 115], and Witten [112]. From the viewpoint of current algebra, the WZNW theory was investigated by Knizhnik and Zamolodchikov [109]. The relation between topological quantization of the WZ coupling constant and DeRham cohomology theory was explained by Alvarez [306], Braaten, Curtright and Zachos [117]. The WZNW model as a theory of free fields was discussed by Gerasimov, Morozov, Olshanetsky, Marshakov and Shatashvili in ref. [123].

The vertex operator construction is due to Frenkel and Kač [118], and Segal [119]. The parafermions were introduced in CFT by Fateev and Zamolodchikov [120, 151]. The AKM characters and their relation to modular forms were clarified by Kač and Peterson [211], see also ref. [108]. The A-D-E classification of the $SU(2)$ modular invariant partition functions was given by Capelli, Itzykson and Zuber [78, 79]. See refs. [80, 129, 347, 348] for further developments of the CFT classification issues.

Superconformal invariance in two dimensions was investigated by many authors, see, for example, refs. [35, 59, 133, 134, 135] for the $N = 1$ case, refs. [144, 147, 150, 152, 153, 149] for $N = 2$, and refs. [154, 158, 159] for $N = 3$ and $N = 4$.

The $N = 1$ supersymmetric generalization of the WZNW model and the $N = 1$ super-AKM symmetry were introduced by DiVecchia, Knizhnik, Petersen and Rossi [139], see also refs. [140, 141, 160, 162]. The (2, 0) supersymmetric generalization of the

WZNW model and a discussion of the $N = 2$ super-AKM symmetry can be found in the papers of Hull and Spence [174], Gates and Ketov [175].

The chiral ring structure of $N = 2$ SCFT's, and the $N = 2$ LG models were introduced and investigated by Lerche, Vafa and Warner [155]. Catastrophe theory is described in the books of Arnold [182], Arnold, Gusein-Zade and Varchenko [183].

The conditions of quantum equivalence between the SS-constructed bosonic stress tensors and the free fermionic stress tensors were given by Goddard, Nahm and Olive [161]. The GKO construction as a tool for constructing unitary representations of the Virasoro and super-Virasoro algebras was developed by Goddard, Kent and Olive [163], see refs. [106, 107, 165] also. The KS construction was invented by Kazama and Suzuki [191, 193]. The standard reference about symmetric spaces is the book of Helgason [192]. The gauged WZNW theories in connection to CFT coset constructions were investigated in refs. [169, 199, 200, 201, 202, 203, 204, 205]. The BRST approach to coset-based CFT's, and their free-field resolutions are due to Bouwknegt, McCarthy and Pilch [206]. The generalized affine-Virasoro construction is discussed in refs. [213, 214, 215].

The W_3 algebra in CFT was discovered by Zamolodchikov [130]. A review of various generalizations of this algebra can be found in ref. [131]. Free field representations of W algebras and quantum Miura transformations were investigated by Fateev and Lukyanov [238, 239], and Lukyanov [242]. The determinant formulae for the W-symmetric conformal models are due to Mizoguchi [245, 246]. Gel'fand-Dikii algebras were introduced in the theory of integrable models by Gel'fand and Dikii [249]. The classical Hamiltonian reduction is due to Drinfeld and Sokolov [250]. The quantum Hamiltonian (DS) reduction was developed by Belavin [251, 252], Bershadsky and Ooguri [247], Feigin and Frenkel [248]. Fusion rules, characters and coset constructions for the W algebras and the corresponding CFT's are discussed in refs. [131, 186, 216, 253, 255].

The standard reference about strings and superstrings is the book of Green, Schwarz and Witten [128] (see also reviews [257, 258, 259, 260, 261, 262, 265, 266, 267, 269, 270, 271, 272, 273, 274, 276, 277, 278, 279, 280, 281, 282, 283, 285, 286, 287, 288, 290, 291, 293, 300, 302, 303, 343] for more, reprint collections and Conference Proceedings [263, 268, 275, 284, 289, 292, 294, 295, 296, 297, 305, 521] for much more). The importance of conformal symmetry in string theory was first observed by Polyakov [37]. The 2d NLSM conformal invariance conditions were formulated by Friedan [308, 309], Fradkin and Tseytlin [316], Callan, Friedan, Martinec and Perry [317], Tseytlin [312, 314, 327] and many others (see, e.g., refs. [307, 313, 315]). The explicit multi-loop perturbative calculations of the (supersymmetric) NLSM RG β-functions and of the low-energy (super)string effective action can be found, e.g., in the book [320].   The Curci-Paffuti relation was formulated and proved in ref. [321]. The 2d c-theorem is due to Zamolodchikov [323]. The full NSR string action was found by Deser and Zumino [337], Brink, Di Vecchia and Howe [338]. The space-time fermion-vertex operator was found by Friedan, Shenker and Martinec [340], and Knizhnik [341]. The heterotic string theory was invented by Gross, Harvey, Martinec and Rohm [344, 345, 346]. The fundamental connection between space-time supersymmetry of a compactified superstring and $N = 2$ superconformal symmetry on superstring world-sheet is due to Candelas, Horowitz, Strominger and Witten [349], and Gepner [350, 351]. The extended fermionic string theories with $N = 2$ and $N = 4$ superconformal symmetry were invented by Ademollo et al., in refs. [147, 148, 149]. The full $N = 2$ string action was first constructed by Brink and Schwarz [354]. The full $N = 4$ string action is due to Pernici and van Nieuwenhuizen [382]. A BRST quantization of the $N = 2$ string was given by Bilal [357], Mathur and Mukhi [375], Gomis and Suzuki [381], and Ketov [359]. The (untwisted) BRST cohomology of the $N = 2$ string was investigated by Bieńkowska [358], Giveon and Roček [377]. The chiral bosonization techniques were applied to investigation of the $N = 2$ strings by Gomis [379], Li [392], Bischoff, Ketov, Lechtenfeld and Parkes [376, 378]. The connection between $N = 2$ strings and 4d self-dual field theories was discovered by Ooguri and Vafa [362, 363]. The simple free-boson realization of the classical W_3 algebra was found by Hull [395, 396]. See refs. [393, 396, 399] for reviewing the W gravities and the W strings.

The $SL(2, \mathbf{R})$ current algebra in quantum 2d gravity and CFT was discovered by Polyakov [402]. The 2d quantum gravity solution was found by Knizhnik, Polyakov and Zamolodchikov [403], David [404], Distler and Kawai [405]. The classical and quantum Liouville theories were investigated in detail by Gervais and Neveu [414, 415, 416, 417, 418, 419, 420], Braaten, Curtright, Chandour and Thorn [421, 422, 423, 424], see also reviews in refs. [412, 413, 425, 426]. The 'quantum group' structure in the quantum Liouville theory was discovered by Gervais [428, 429, 430, 431]. The Liouville strings and superstrings were introduced by Bilal and Gervais [435, 436, 437, 438]. Topological (conformal) field theories are the subject of reviews [444, 445, 446], see also the original papers of Witten [443, 448, 449].

The exact solutions based on matrix models for strings and 2d quantum gravity were found by Brézin and Kazakov [456], Douglas and Shenker [457], Gross and Migdal [458]. Several reviews are now available [439, 459, 460, 461, 462, 463, 464]. The matrix model technology originated from the earlier resuls of t'Hooft [470], Brézin, Itzykson, Parisi and Zuber [473], Bessis, Itzykson and Zuber [468], and Mehta [472]. The relation between MMM's and topological field theories was clarified by Dijkgraaf and Witten [478]. The recursion relations in topological gravity are due to Li [479, 480]. The relation between MMM's and KdV was discovered by Douglas [482]. The correlation functions in matrix models, the microscopic and macroscopic loops were introduced by Banks, Douglas, Seiberg and Shenker [483]. Many relevant papers about the large-N expansion in spin systems, lattice QCD and 2d quantum gravity are collected in ref. [522].

Integrable models are discussed in the books [432, 489] (see also the reviews [250, 491]). The W flows in connection to the covariant differential equations were introduced by Bauer, Di Francesco, Itzykson and Zuber [484, 494]. The conformal reductions of the WZNW models, leading to the Liouville and Toda theories, as well as the AKM realization of the W algebras, are due to Balog, Fehér, Forgács, O'Raifeartaigh and Wipf [495, 499]. An integrability of the Toda theory was proved by Leznov and Saveliev [496], while its W symmetry was discovered by Babelon [498] (see also ref. [513]). The 4d SDYM theory as the underlying theory for 2d integrable models is discussed in the mathematical literature [510] (see also refs. [502, 509]), and in the physical literature [504, 505, 506, 507, 508] as well. The 'zero-curvature' interpretation of the SDYM equations of motion is due to Belavin and Zakharov [503]. Supersymmetric 2d integrable models are formulated and studied, in particular, in refs. [511, 512, 513, 514, 515, 516]. Their embeddings into the SSDYM theories in $2+2$ dimensions are due to Gates and Nishino [517], and Nishino [518, 520].

The theory of classical Lie algebras and of their representations can be found in the standard textbooks [5, 125, 523, 524, 525, 526] (see ref. [527] for *finite* simple groups). As to the AKM algebras and their highest weight representations, the standard references are ref. [44] and ref. [528], resp. The theory of Riemann surfaces is nicely introduced in ref. [532]. Theta functions are described in detail in refs. [96, 97] (see also refs. [28, 212]). Quantization of field theories with constraints is discussed in the book of Gitman and Tyutin [534]. Braiding and fusion of chiral blocks in (R)CFT's, and the Seiberg-Moore equations were introduced and investigated in refs. [74, 75, 535]. The 'quantum group' structure in RCFT's is explained in refs. [536, 537, 538] (see refs. [76, 77] for a review). The original mathematical references about 'quantum groups' are refs. [541, 542, 543, 544, 545, 546, 547]. See ref. [548] for more about Hopf algebras, and refs. [549, 550] for more about the Yang-Baxter equation and the quantum inverse scattering method.

XII.2   Basic facts about Lie and AKM algebras

In this section, we remind the reader about the structure of Lie and affine Kač-Moody (AKM) algebras. Our discussion may sometimes be oversimplified, so the pedantic reader is invited to consult the original textbooks [5, 125, 523, 524, 525, 526, 527, 528] whenever he would find it necessary.

Given a finite-dimensional Lie algebra $\mathcal{G}$ [529], the total number of linearly independent generators is called its *dimension* $|G|$, whereas the maximal number of simultaneously diagonalizable generators is called its *rank*, $r_G \equiv l$. Lie group G can be associated with a given Lie algebra $\mathcal{G}$ and vice versa, which does not imply, however, a global one-to-one correspondence between them. A *simple* Lie group has no invariant subgroups, except the whole group and the identity; in terms of the associated Lie algebra this means that a simple algebra $\mathcal{G}$ has no proper ideals. A *semi-simple* Lie algebra can be written as a direct sum of simple Lie algebras. Any Lie algebra can be represented as a finite-dimensional matrix algebra [530].

Most aspects of Lie algebras are better considered after choosing a special basis. In the standard *Cartan-Weyl* basis, generators of $\mathcal{G}$ are divided into two sets: the *Cartan subalgebra* (CSA) which is a maximal Abelian subalgebra of $\mathcal{G}$, containing r_G diagonalizable generators H_i:

$$[H_i, H_j] = 0 , \quad i,j = 1,\ldots,r_G , \tag{12.1}$$

and the remaining generators satisfying eigenvalue equations of the form

$$[H_i, E_\alpha] = \alpha_i E_\alpha , \quad i = 1,\ldots,r_G . \tag{12.2}$$

The numbers α_i in eq. (12.2) represent the structure constants of $\mathcal{G}$ in the Cartan-Weyl basis. For each operator E_α, there are r_G numbers α_i that can be used to label a point in an r_G-dimensional Euclidean space called *root space*. A *root vector* $(\alpha_1,\ldots,\alpha_l)$ is a solution to the eigenvalue problem (12.2). The elegant classification of all possible root systems corresponding to the simple Lie algebras can be given in terms of the Dynkin diagrams (see below).

In quantum mechanics, a description of the symmetry G is usually given by an action of H_i and E_α generators on Hilbert space states. A complete set of states that are necessarily interconnected by E_α's forms an irreducible representation (*irrep*) of G. The physical significance of the diagonalizability of H_i's is that the Hilbert space vectors $|\lambda\rangle$ in an irrep can be labeled by the r_G eigenvalues (*quantum numbers*) of

H_i: $H_i |\lambda\rangle = \lambda_i |\lambda\rangle$. Generally speaking, the λ_i's are *not* a complete set of labels to unambiguously identify a state. The set $\{\lambda\}$ is called the *weight* of representation. The weight vector $(\lambda_1, \lambda_2, \ldots, \lambda_l)$ is an element of the *weight space* which is identical to the root space, since roots can be considered as particular weights (for the adjoint representation). A Lie algebra representation space is also called a *module*.

The only (compact real form of) $r_G = 1$ simple Lie algebra is $su(2)$, where the conventional choice of its generators is given by $\{I^{\pm}, I^3\}$, with the I^3 being diagonal:

$$[I^3, I^{\pm}] = \pm I^{\pm} , \quad [I^+, I^-] = 2I^3 , \quad I^3 |m\rangle = m |m\rangle . \tag{12.3}$$

Hence, the roots in this case are ± 1.

E_α's can be recognized as the ladder operators: if $|\lambda\rangle$ is an eigenfunction of H_i with eigenvalue λ_i, then one has

$$H_i (E_\alpha |\lambda\rangle) = E_\alpha H_i |\lambda\rangle + \alpha_i E_\alpha |\lambda\rangle = (\lambda_i + \alpha_i)(E_\alpha |\lambda\rangle) , \tag{12.4}$$

i.e. either $E_\alpha |\lambda\rangle$ is proportional to the eigenvector $|\lambda_i + \alpha_i\rangle$ for all H_i, or it vanishes. If $\langle \lambda | \lambda' \rangle = \delta_{\lambda\lambda'}$, to compute a normalization of $E_\alpha |\lambda\rangle$, one needs more information from the commutation relations. It is a good idea to start a calculation from a 'vacuum state' such that $E_\alpha |\lambda\rangle = 0$ for a proper subset of α_j. Such state is called the *highest weight state* in a given irrep.

Clearly, if α is a root, then so is $-\alpha$. The commutators of E_α and E_β generically have the form

$$[E_\alpha, E_\beta] = N_{\alpha,\beta} E_{\alpha+\beta} , \quad \alpha + \beta \neq 0 , \tag{12.5}$$

if $\alpha + \beta$ is a root, and $N_{\alpha,\beta} = -N_{\beta,\alpha}$. The differences between real forms of a given classical (complex) Lie algebra can be encoded in the choice of the $N_{\alpha,\beta}$-coefficients. Clearly, $N_{\alpha,\beta} = 0$ when $\alpha + \beta$ is not a root. It follows

$$[E_\alpha, E_{-\alpha}] = \alpha^i H_i , \tag{12.6}$$

where the components α^i are related to the α_i by the Cartan-Weyl metric to be discussed below.

Knowing all possible root systems would allow us to classify all simple Lie algebras, since the root vectors completely determine the structure of a Lie algebra. In the Cartan-Weyl basis, there is only one E_α for each α. A derivation of the root systems from the commutation relations and the Jacobi identity results in the celebrated Cartan classification of the simple Lie algebras.

To describe this classification, one should first consider the roots in a Cartesian basis. A half of the non-zero roots will be *positive*, defined by the requirement that the first non-zero component of a positive root in the basis is positive. Next, one should find the positive roots that cannot be written as linear combinations (with positive coefficients) of other positive roots. There are only r_G such roots and they are all linearly independent; this defines a set of *simple roots*. Of course, different selections of coordinate systems will lead to different sets of simple roots; however, all such sets are equivalent in the sense that there is a *Weyl reflection* of the root diagram relating any two sets. The Weyl reflections do not change relative lengths and angles among the roots, and they constitute the so-called *Weyl group* of the root system. The lengths and angles among the simple roots can be shown to completely characterize any simple Lie algebra [5, 125, 523, 524, 525]. The Weyl group is generated by the *fundamental Weyl reflections*, which are the Weyl reflections associated with the simple roots α_i:

$$w_{\alpha_i}(\lambda) = \lambda - \alpha_i \frac{2(\lambda, \alpha_i)}{(\alpha_i, \alpha_i)} \ , \tag{12.7}$$

where the (bilinear, non-degenerate and symmetric) *Killing form* $(x, y) = \text{tr}[ad(x)ad(y)]$, $ad(x)y \equiv [x, y]$, has been introduced. Dynkin was the first who pointed out how this result can be nicely represented by a two-dimensional diagram. The *Dynkin diagram* indicates the relative lengths of the simple roots, and the angles between each pair of them. Each simple root is denoted by a dot on Dynkin diagram. There may be at most two different root lengths, so the longer roots are usually denoted by open dots, and the shorter roots by filled-in dots. The angle between a pair of simple roots is usually denoted by lines connecting the corresponding dots: no line means the angle is $\pi/2$, one line means $2\pi/3$, two lines mean $3\pi/4$, and three lines mean $5\pi/6$. These are all the possibilities. The ratio of lengths of two roots connected by three lines is $\sqrt{3}$ ($\sqrt{2}$ if connected by two lines, and 1 if connected by one line). The root systems of all the simple Lie algebras are shown in Fig. 21. If a Dynkin diagram has several disconnected pieces, the corresponding Lie algebra is semi-simple.

The simple roots do not form an orthonormal basis in the root space. The matrix that keeps track of the non-orthogonality is called the *Cartan matrix*. It has elements

$$A_{ij} = \frac{2(\alpha_i, \alpha_j)}{(\alpha_j, \alpha_j)} \ , \tag{12.8}$$

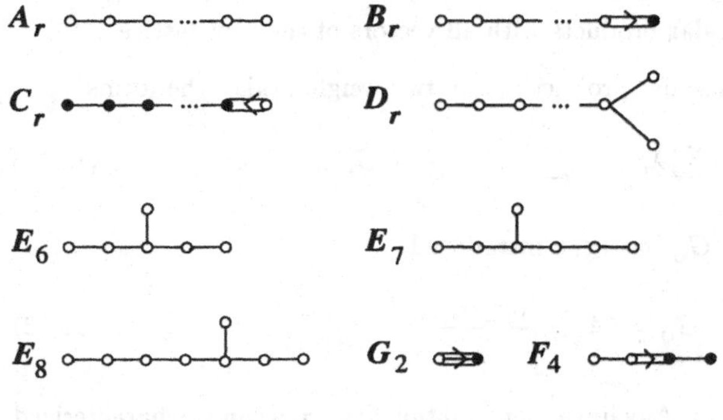

Fig. 21. The Dynkin diagrams for simple Lie algebras.

where the vector α_i represents i-th simple root. Clearly, the Cartan matrix A can be read off from the Dynkin diagram and vice versa. Each element of the Cartan matrix is an *integer*.

A set of roots is called a *root system*. It can be extended to a lattice called a *root lattice*, in the root space. It is convenient to supplement a root lattice with points (or vectors) corresponding to the possible weights of representation vectors. This larger lattice of points is usually referred to as *weight lattice*. Since the simple roots form a basis in the root space, any root or weight vector λ in the root space can be written as a linear combination of the simple roots α_i:

$$\lambda = \sum_i \tilde{\lambda}_i \frac{2}{(\alpha_i, \alpha_i)} \alpha_i \ . \tag{12.9}$$

The longer simple roots are conventionally normalized to length-squared of 2. Therefore, for the *simply-laced* groups ($SU(N)$, $SO(2N)$ and $E_{6,7,8}$), one has $2/(\alpha_i, \alpha_i) = 1$. The coordinates $\{\tilde{\lambda}_i\}$ define the vector λ in the *dual basis*. The dual of a root is called a *co-root*. The celebrated *Dynkin components* a_i of λ are defined by

$$a_i = \frac{2(\lambda, \alpha_i)}{(\alpha_i, \alpha_i)} = \sum_i \tilde{\lambda}_j \frac{2}{(\alpha_j, \alpha_j)} A_{ji} \ . \tag{12.10}$$

The crucial *Dynkin-Weyl theorem* holds: for any weight or root, the *Dynkin labels* a_i in eq. (12.10) are integers. This generalizes the well-known statement about the representations of $su(2)$ algebra, where $2m$ is always an integer.

The weight lattice is exactly the dual of the root lattice, and it is defined by the set of points which have integer scalar products with all vectors of the root lattice.

It is now easy to see that a scalar product of any two weights takes the forms

$$(\lambda, \lambda') = \sum_i \tilde{\lambda}_i a_i' = \sum_i \tilde{\lambda}_i' a_i = \sum_{ij} a_i' G_{ij} a_j \ , \qquad (12.11)$$

where the *Cartan-Weyl metric* G_{ij} has been introduced,

$$G_{ij} = (A^{-1})_{ij} \frac{(\alpha_i, \alpha_j)}{2} \ . \qquad (12.12)$$

CSA members have zero roots. Any linear combination Q of H_i's can be characterized by an axis in root space. If $|\lambda\rangle$ is an eigenstate of Q: $Q|\lambda\rangle = Q(\lambda)|\lambda\rangle$ with eigenvalue $Q(\lambda)$, then one gets

$$Q(\lambda) = (Q, \lambda) \ . \qquad (12.13)$$

From the viewpoint of the representation theory of Lie algebras, the root system is the list of eigenvalues of the CSA generators acting in the adjoint representation. The rules for constructing the root diagram are therefore one particular example of the general rules needed to work out the Dynkin labels of an arbitrary irrep.

Given a Lie algebra irrep, some of its weights may be degenerate, i.e. several vectors in Hilbert space may have the same weight. However, there are always some weights that are not degenerate, while one of these weights may serve to uniquely define the whole irrep. Such weight ('vacuum state') is called the *highest weight state* Λ. When being written in the Dynkin basis, the highest weight is specified by a set of integers. The crucial *Dynkin theorem* reads: the highest weight of an irrep *can* be selected in such a way that the Dynkin labels are *non-negative integers*; each and every irrep is then uniquely identified by the set of integers $(a_1, \ldots, a_l)$ $(a_i \geq 0)$, and each such set is the highest weight of one and only one irrep.

For instance, we can specify all the highest weight representations as arising from the tensor products of the *fundamental* (of lowest dimension) representations, in the same way as any spin-J representation of $SU(2)$ can be constructed by tensoring the spin-1/2 representations. The *fundamental weights* Λ_i are defined by the relation

$$\frac{2(\Lambda_i, \alpha_j)}{(\alpha_j, \alpha_j)} = \delta_{ij} \ , \quad i, j = 1, \ldots, r_G \ . \qquad (12.14)$$

A highest weight of $\mathcal{G}$ is a positive integer combination of the fundamental weights:

$$\Lambda = \sum_{i=1}^{l} n_i \Lambda_i \quad n_i > 0 \; . \tag{12.15}$$

The weights having the form (12.15), with all non-negative integers n_i, are sometimes called the *dominant* weights.

The full set of weights for a given irrep can be derived from the highest weight and the Dynkin diagram by subtracting simple roots. Once the weight system is given, the Dynkin labels can be converted into the eigenvalues (quantum numbers) of a convenient set of the CSA diagonal generators.

For each positive root α, there is an $su(2)$ subalgebra generated by E_α, $E_{-\alpha}$ and their commutator H_α belonging to CSA. Taken all together, they constitute the so-called *Chevalley basis* for a Lie algebra $\mathcal{G}$. Any representation of $\mathcal{G}$ splits into a sum of the highest weight representations w.r.t a given $su(2)$, which is sometimes very useful for applications.

The *level* of an irrep weight is the number of simple roots that must be subtracted from the highest weight to obtain it. The irrep dimensionality is given by the famous *Weyl formula*

$$N(\Lambda) = \prod_{\alpha > 0} \frac{(\Lambda + \delta, \alpha)}{(\delta, \alpha)} \; , \tag{12.16}$$

where $\delta = (1, 1, \ldots, 1, 1)$ in the Dynkin basis, and Λ is the highest weight of the irrep.

Let Λ_i be the fundamental weights of $\mathcal{G}$, and m_i the constants to be introduced via the decomposition of the highest root ψ in terms of its simple roots,

$$\psi/\psi^2 = \sum_{i=1}^{l} m_i \alpha_i / \alpha_i^2 \; , \quad m_0 \equiv 1 \; . \tag{12.17}$$

The sum of m_i's defines the number which is known as the *dual Coxeter number* $\tilde{h}_G$,

$$\tilde{h}_G = \sum_{i=0}^{l} m_i \; . \tag{12.18}$$

In the normalization $\psi^2 = 2$, the dual Coxeter number is just half the second Casimir eigenvalue in the adjoint representation of $\mathcal{G}$. Indeed, since the Casimir eigenvalue can be expressed in terms of the highest weight as [5, 77, 523, 525]

$$C_\Lambda = (\Lambda + 2\rho, \Lambda) \; , \quad \rho = \tfrac{1}{2} \sum_{\alpha > 0} \alpha \; , \tag{12.19}$$

where $2(\rho, \alpha_i)/\alpha_i^2 = 1$ for all simple roots, one has

$$C_A = (\psi + 2\rho, \psi) = \psi^2 \left[1 + \sum_{i=1}^{l} m_i\right] = 2\bar{h}_G \ . \tag{12.20}$$

The *character* of a representation Λ of Lie algebra G is the complex function on the root space, defined by

$$\chi_\Lambda(h) = \sum_\lambda \text{mult}_\Lambda(\lambda) \exp[(\lambda, h)] \ , \tag{12.21}$$

where the sum goes over all weights of the representation. Inspecting the behaviour of weights λ under the Weyl reflections, one finds the formula [5, 77, 523, 525]

$$\chi_\Lambda(h) = \frac{\sum_{\sigma \in W} (\text{sign}\,\sigma) \exp[(\sigma(\Lambda + \rho), h)]}{\sum_{\sigma \in W} (\text{sign}\,\sigma) \exp[(\sigma(\rho), h)]} \ , \tag{12.22}$$

which is known as the *Weyl character formula*, where $\text{sign}\,\sigma = (-1)^{l(\sigma)}$, and $l(\sigma)$ is the length of σ.

The *centralizer* $C_G(\mathcal{K})$ of a subset $\mathcal{K} \subset G$ of Lie algebra G is the set of all elements of G commuting with each element of $\mathcal{K}$. The *normalizer* $\mathcal{N}_G(\mathcal{K})$ of a subalgebra $\mathcal{K}$ of Lie algebra G is the set of all elements of G whose commutators with the elements of $\mathcal{K}$ lie in $\mathcal{K}$.

As our first example, consider the case of $G = A_1$, take $h = \xi J_3$ with some coefficient ξ, and insert it into eq. (12.21). Since the different weights form an arithmetic progression of step 2: $\lambda = -\Lambda, -\Lambda + 2, \ldots, \Lambda - 2, \Lambda$, and each λ has multiplicity 1, one gets [77]

$$\chi_\Lambda(h) = \sum_{n=0}^{\Lambda} q^{\frac{1}{2}\Lambda - n} = \lfloor \Lambda + 1 \rfloor_q \ , \tag{12.23}$$

where $q \equiv e^\xi$, and

$$\lfloor z \rfloor_q \equiv \frac{q^{z/2} - q^{-z/2}}{q^{1/2} - q^{-1/2}} \tag{12.24}$$

is called the *q-number* associated with complex number z (see also sect. 6 for more about q-numbers).

As our next example, consider the Lie algebra $A_{n-1} \sim su(n)$ [523]. Its simple roots e_α can be written in a n-dimensional space as $e_\alpha = (0, \ldots, 0, -1, 1, 0, \ldots, 0)$. The root and weight systems can be introduced in the $(n-1)$-dimensional hyperplane defined by the condition of vanishing for the sum of all n components. In the non-trivial case of $su(3)$, the root and weight diagram is pictured in Fig. 22. There are six non-zero

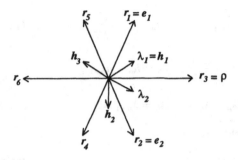

Fig. 22. The weight (and root) diagram for $A_2 \sim su(3)$.

roots r_i, $i = 1, \ldots, 6$, where the r_1, r_2 and r_3 are chosen to be positive. The $r_1 = e_1$ and $r_2 = e_2$ are the simple roots, whereas the r_3 is the highest one. The λ_1 and λ_2 are the two fundamental weights. Given λ_1 as the highest weight (3-representation), the weights of this representation are given by h_1, h_2 and h_3 (see Fig. 22). The roots are supposed to be normalized to have their length-squared equal to 2, so that $e_\alpha^2 = 2$. Let λ_β, $\beta = 1, \ldots, n-1$, be the highest weights corresponding to the totally antisymmetric tensor with i indices (it is normally described by a vertical Young tableaux with i boxes). These weights are dual to the simple roots:

$$\lambda_\alpha \cdot e_\beta = \delta_{\alpha\beta} , \qquad (12.25)$$

while their products $\lambda_\alpha \cdot \lambda_\beta$ are given by the (α, β) elements of the inverse Cartan matrix of A_{n-1}:

$$\lambda_\alpha \cdot \lambda_\beta = \frac{\alpha(n - \beta)}{n}\delta_{\alpha \leq \beta} + \frac{\beta(n - \alpha)}{n}\delta_{\alpha < \beta} . \qquad (12.26)$$

The sum

$$\rho = \sum_{\alpha=1}^{n-1} \lambda_\alpha \qquad (12.27)$$

is the Weyl vector of A_{n-1} [523]. Its definition implies $\rho \cdot e_\alpha = 1$ for all α, and

$$\rho \cdot \lambda_\alpha = \sum_\beta \lambda_\beta \cdot \lambda_\alpha = \frac{\alpha(n - \alpha)}{2} = \frac{n}{2}\lambda_\alpha^2 , \quad \rho^2 = \frac{(n - 1)n(n + 1)}{12} . \qquad (12.28)$$

The n weights λ_μ of the vector representation with the highest weight λ_1 are given by [531]

$$h + 1 = \lambda_1 , \quad h_2 = \lambda_1 - e_1 , \quad \ldots , \quad h_\mu = \lambda_1 - \sum_{\alpha=1}^{\mu-1} e_\alpha . \tag{12.29}$$

It follows

$$\sum_{\mu=1}^{n} h_\mu = 0 , \quad h_\mu \cdot h_\nu = \delta_{\mu\nu} - \frac{1}{n} , \quad e_\alpha = h_\alpha - h_{\alpha+1} ,$$

$$\lambda_\alpha = \sum_{\mu=1}^{\alpha} h_\mu , \quad h_\mu = \lambda_\mu - \lambda_{\mu-1} ; \quad \lambda_0 = \lambda_{n+1} = 0 . \tag{12.30}$$

Eqs. (12.30) imply for the Weyl vector the relations

$$\rho = \sum_\mu (n - \mu) h_\mu = - \sum_\mu \mu h_\mu . \tag{12.31}$$

A permutation of weights corresponds to an element of the Weyl group generated by the reflections w.r.t. the hyperplanes normal to the roots. There is the unique highest (or dominant) weight to which a given weight can be mapped by an element of the Weyl group. In particular, the Weyl vector is always a highest weight.

More information about the specific simple Lie algebras and their irreps can be found in ref. [531]. For the most of our purposes, we only need the characteristics comprised in Table 4 [131]. The so-called *exponents* $\{e_i\}$ in Table 4 mean the orders of the independent Casimir operators of $\mathcal{G}$.

Table 4. Characteristics of simple Lie algebras.

$\mathcal{G}$	$\|G\|$	h	$\tilde{h}$	$\{e_i\}$
A_n	$n(n+2)$	$n+1$	$n+1$	$1, 2, \ldots, n$
B_n	$n(2n+1)$	$2n$	$2n-1$	$1, 3, 5, \ldots, 2n-1$
C_n	$n(2n+1)$	$2n$	$n+1$	$1, 3, 5, \ldots, 2n-1$
D_n	$n(2n-1)$	$2(n-1)$	$2(n-1)$	$1, 3, 5, \ldots, 2n-3, n-1$
E_6	78	12	12	$1, 4, 5, 7, 8, 11$
E_7	133	18	18	$1, 5, 7, 9, 11, 13, 17$
E_8	248	30	30	$1, 7, 11, 13, 17, 19, 23, 29$
F_4	52	12	9	$1, 5, 7, 11$
G_2	14	6	4	$1, 5$

The theory of *affine Kač-Moody* (AKM) algebras [44] and their highest-weight representations [528] can be developed along the lines of the just reviewed case of the simple Lie algebras, as a generalization of Cartan's work. One may think of an AKM algebra as a current (or affine) Lie algebra whose generators, $E^\alpha(z) = \sum_n E_n^\alpha z^{-n-1}$ and $H^i(z) = \sum_n H_n^i z^{-n-1}$, depend on a continuous parameter z. A formal Laurent expansion in powers of z yields an infinite set E_n^α, H_n^i, where $n \in \mathbf{Z}$, for each original ('horizontal') generator of $\mathcal{G}$. In particular, one can write down the defining equations of an AKM algebra $\hat{\mathcal{G}}$ in the generalized Cartan-Weyl basis as a loop algebra with central extensions, namely

$$[E_m^\alpha, E_n^\beta] = \begin{cases} N(\alpha,\beta)E_{m+n}^{\alpha+\beta} & \text{if } \alpha+\beta \text{ a root,} \\ 0 & \text{if } \alpha+\beta \text{ not a root,} \\ 2\alpha^i H_{n+m}^i/\alpha^2 + 2km/\alpha^2 \delta_{n+m,0} & \text{if } \beta = -\alpha, \end{cases}$$

$$[H_n^i, H_m^j] = k\delta^{ij} n\delta_{n+m,0}, \quad [H_n^i, E_m^\alpha] = \alpha^i E_{n+m}^\alpha,$$

$$[k, J_n^a] = 0, \quad [d, J_n^a] = nJ_n^a, \quad [d, k] = 0, \tag{12.32}$$

where the AKM *level* generator $d \equiv -L_0$ has been included to avoid having an infinite degeneracy of the roots. [2] The *horizontal* finite-dimensional Lie algebra $\mathcal{G}$, spanned by the generators J_0^a, is assumed to be simple. In eq. (12.32), the normalization $\psi^2 = 2$ has been chosen, where ψ is a highest root. It follows that the central extension (level) k is then always an integer.

The maximal set of commuting operators in eq. (12.32) is (H_0^i, k, d). The root space is given by the vectors (α, k, n), where α is the root space of the algebra $\mathcal{G}$, whereas the *grade* n and level k are the eigenvalues of d and k, resp. The scalar product in the extended root space is given by

$$(\hat{\alpha}_1, \hat{\alpha}_2) \equiv (\alpha_1, k_1, n_1) \cdot (\alpha_2, k_2, n_2) = (\alpha_1, \alpha_2) + k_1 n_2 + k_2 n_1. \tag{12.33}$$

Some roots of AKM algebra $\hat{\mathcal{G}}$ may have zero norm. They are called *imaginary* roots, and are some integers times $\hat{\delta} = (0,1,1)$. The rest of the roots are called *real*. A root $(\alpha, 0, n)$ is positive if $n > 0$ or if $\alpha > 0$ for $n = 0$. There are $r_G + 1$ simple roots: $(\alpha_i, 0, 0,) \equiv \alpha_i$ where α_i are r_G simple roots of $\mathcal{G}$, and the additional simple root $(-\psi, 0, 1) \equiv \hat{\alpha}_0$. A positive root is a sum of simple roots with positive integer coefficients. The negative roots are minus the positive ones. A real root is either positive or negative. It is due to the existence of imaginary roots that all non-trivial representations of $\hat{\mathcal{G}}$ are of infinite

---

[2]The generator d is sometimes called the *derivation*. It supplies the additional quantum number.

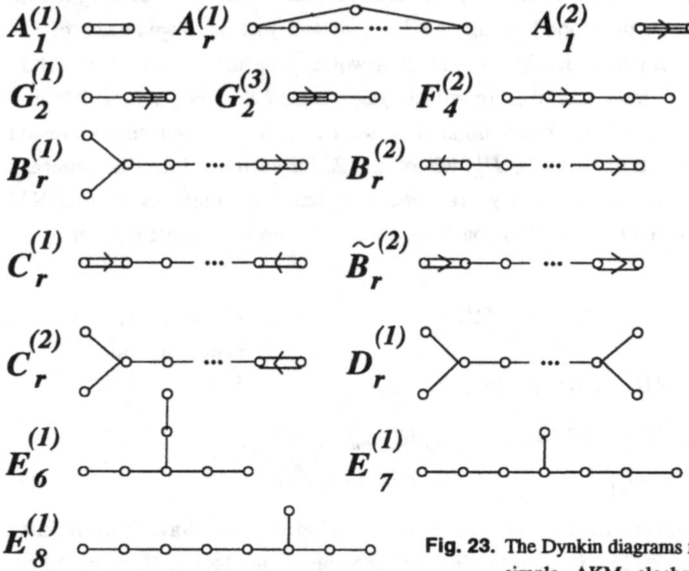

Fig. 23. The Dynkin diagrams for simple AKM algebras.

dimension. The *integrability* of an AKM module means that it is still possible to 'lift' the given AKM-algebra representation up to a representation of the corresponding AKM group via the exponential map, which is not trivial for an infinite-dimensional module.

Affine Dynkin diagrams are in one-to-one correspondence with affine Cartan matrices having the form

$$K_{ij} = \frac{2(\hat{\alpha}_i, \hat{\alpha}_j)}{\hat{\alpha}_j^2} , \quad i,j = 0, 1, \ldots, r_G , \tag{12.34}$$

similarly to the ordinary Lie algebras. The list of affine Dynkin diagrams is given in Fig. 23 [77].

The analogues of eqs. (12.14) and (12.15) still take place for the associated AKM algebras. The fundamental weights dual to the simple co-roots are given by

$$\hat{\Lambda}_i = \left(\Lambda_i, m_i \frac{\psi^2}{2}, 0 \right) , \quad i = 1, \ldots, r_G ,$$

$$\hat{\Lambda}_0 = \left(0, \tfrac{1}{2}\psi^2, 0 \right) , \tag{12.35}$$

where Λ_i are the fundamental weights (12.14) of $\mathcal{G}$, and the constants m_i were defined in eq. (12.17), via the decomposition of the highest root ψ in terms of its simple roots.

Hence, for any highest weight vector ($\hat{\Lambda}$) representation of AKM algebra $\hat{\mathcal{G}}$, we have

$$\hat{\Lambda} = n_0 \hat{\Lambda}_0 + \sum_{i=1}^{l} n_i \hat{\Lambda}_i , \quad n_0, n_i \geq 0 . \tag{12.36}$$

Stated differently, the irreps of $\hat{\mathcal{G}}$ are labeled by the highest weights Λ of the classical (horizontal) Lie algebra and the level k.

The *character* of a representation V of AKM algebra $\hat{\mathcal{G}}$ is the complex function on the root space, which is defined by

$$ch_V = \sum_{\hat{\lambda}} \dim V_{\hat{\lambda}} e^{\hat{\lambda}} , \tag{12.37}$$

where $\dim V_{\hat{\lambda}}$ is the multiplicity of the weight $\hat{\lambda}$ in the representation, and $e^{\hat{\lambda}}$ is the following complex-valued function defined on the root space:

$$\left(e^{\hat{\lambda}} \right) (\hat{\beta}) \equiv e^{(\hat{\lambda}, \hat{\beta})} . \tag{12.38}$$

An arbitrary vector $\hat{\beta}$ of the extended root space can be expanded w.r.t. the basis $\{ \nu_i, \hat{\Lambda}_0, \hat{\delta} \}$, where $\{ \nu_i \}$ define an orthonormal basis for the root space of $\mathcal{G}$, $\hat{\Lambda}_0 = (0, 1, 0)$ and $\hat{\delta} = (0, 0, 1)$,

$$\hat{\beta} = -2\pi i \left(\sum_{s=1}^{l} z_s \nu_s + \tau \hat{\Lambda}_0 + n \hat{\delta} \right) , \tag{12.39}$$

with complex parameters z_s, τ, n. It follows

$$ch_V(z_s, \tau, n) = \sum_{\hat{\lambda}} \dim V_{\hat{\lambda}} \exp \left[-2\pi i \left(\hat{\lambda}, \sum_{s=1}^{l} z_s \nu_s + \tau \hat{\Lambda}_0 + n \hat{\delta} \right) \right] , \tag{12.40}$$

where $\hat{\lambda} = (\lambda, k, n)$. The partition function can now be recognized as

$$ch_V(0, \tau, 0) = \sum_n p_n e^{2\pi i n \tau} = Tr_R e^{2\pi i N \tau} \equiv A_R(\tau) , \tag{12.41}$$

where p_n is the number of vectors at grade $(-n)$ in the representation V, and N is the number operator.

There exists a closed expression for the AKM characters of irreducible integrable highest-weight AKM modules, known as the *Weyl-Kač character formula* [108, 211, 528]:

$$ch_{L(\hat{\Lambda})} = \frac{\sum_{w \in \hat{W}} \epsilon(w) e^{w(\hat{\Lambda} + \hat{\rho}) - \hat{\rho}}}{\prod_{\hat{\alpha} > 0} \left(1 - e^{-\hat{\alpha}} \right)^{\text{mult } \hat{\alpha}}} , \tag{12.42}$$

where $L(\hat{\Lambda})$ is an irreducible (integrable) highest-weight representation with the highest weight $\hat{\Lambda}$, $\epsilon(w)$ is the sign of an element w of the *affine* Weyl group $\hat{W}$: $\epsilon(w) = 1$ or $\epsilon(w) = -1$ depending on whether w comprises even or odd number of Weyl reflections resp., mult $\hat{\alpha}$ is the multiplicity of $\hat{\alpha}$ in the set of roots of the algebra, and $\hat{\rho} = (\rho, \check{h}_G, 0)$.

The *affine* Weyl group $\hat{W}$ of AKM algebra $\hat{\mathcal{G}}$ is the semi-direct product of the finite Weyl group W with certain lattice group T. To see this, consider an action of the reflection $w_{\hat{\alpha}}$, w.r.t. the root $\hat{\alpha} = (\alpha, 0, 1)$ on $\hat{\lambda} = (\lambda, k, n,)$:

$$w_{\hat{\alpha}}(\hat{\lambda}) = \left(w_\alpha \left(\lambda + 2k\alpha/\alpha^2 \right), k, n + \frac{1}{2k} \left[\lambda^2 - \left(\lambda + 2k\alpha/\alpha^2 \right)^2 \right] \right) . \tag{12.43}$$

Eq. (12.43) tells us that $w_{\hat{\alpha}}$ is composed of a reflection from the finite Weyl group W and the certain translation $t_{\hat{\beta}}$ defined by

$$t_{\hat{\beta}}(\hat{\lambda}) = \left(\lambda + k\beta, k, n + \frac{1}{2k} \left[\lambda^2 - (\lambda + k\beta)^2 \right] \right) , \tag{12.44}$$

where β is dual to α: $\beta \equiv 2\alpha/\alpha^2$. Since $t_{\hat{\beta}} t_{\hat{\gamma}} = t_{\hat{\beta}+\hat{\gamma}}$, these translations generate the subgroup of the affine Weyl group, called the *translation subgroup* T. Associated with the translation subgroup T is the lattice $\mathcal{Z}$ generated by the long roots of the algebra $\mathcal{G}$. For simply-laced algebras, $\mathcal{Z}$ is just a root lattice. Since T is a normal subgroup in $\hat{W}$ ($t_{w(\hat{\beta})} = wt_{\hat{\beta}}w^{-1}$), it is easy to see that $\hat{W}$ is the semi-direct product of W and T indeed. This simple observation naturally gives rise to the AKM characters in terms of theta functions, with nice transformation properties of the latter under the modular group $SL(2, \mathbf{R})$, as we are now going to show.

Substituting $\hat{\Lambda} = 0$ into eq. (12.42) results in the nontrivial identity

$$\prod_{\hat{\alpha}>0} \left(1 - e^{-\hat{\alpha}} \right)^{\text{mult } \hat{\alpha}} = \sum_{w\in\hat{W}} \epsilon(w) e^{w(\hat{\rho})-\hat{\rho}} . \tag{12.45}$$

To establish a relation between AKM characters and theta functions, let us introduce the following theta function on the root space:

$$\vartheta_{\hat{\lambda}} \equiv e^{-|\hat{\lambda}|^2\delta/2k} \sum_{t\in T} e^{t(\hat{\lambda})} = e^{k\Lambda_0} \sum_{\gamma\in\mathcal{Z}+k^{-1}\lambda} e^{-k|\gamma|^2\delta+k\gamma/2} , \tag{12.46}$$

where eq. (12.44) has been used. In the basis (12.39), we get a classical theta function of level k:

$$\vartheta_{\hat{\lambda}}(z, \tau, u) = e^{-2\pi i k u} \sum_{\gamma\in\mathcal{Z}+k^{-1}\lambda} \exp\left[\pi i k \tau |\gamma|^2 - 2\pi i k (\gamma_1 z_1 + \ldots + \gamma_{r_G} z_{r_G}) \right] . \tag{12.47}$$

A sum over the affine Weyl group $\hat{W}$ can always be decomposed into a sum over the finite Weyl group W and that over the translation group T. As far as the denominator is concerned, there is the representation (12.45) which can be rewritten as

$$\sum_{w\in\hat{W}} \epsilon(w)e^{w(\hat{\rho})-\hat{\rho}} = e^{-\hat{\rho}}\sum_{w\in W}\epsilon(w)\sum_{\alpha\in\mathcal{Z}}e^{t_\alpha(w(\hat{\rho}))} = e^{-\hat{\rho}+\rho^2\hat{\delta}/2\hbar_G}\sum_{w\in W}\epsilon(w)\vartheta_{w(\hat{\rho})} \ . \qquad (12.48)$$

As far as the numerator in eq. (12.42) is concerned, one finds

$$\sum_{w\in\hat{W}}\epsilon(w)e^{w(\hat{\Lambda}+\hat{\rho})-\hat{\rho}} = \exp\left(-\hat{\rho}+\frac{|\hat{\Lambda}+\hat{\rho}|^2}{2(k+\bar{h}_G)}\hat{\delta}\right)\sum_{w\in W}\epsilon(w)\vartheta_{w(\hat{\Lambda}+\hat{\rho})} \ . \qquad (12.49)$$

Eqs. (12.48) and (12.49) allow us to rewrite the character formula as

$$\exp\left(-s_{\hat{\Lambda}}\hat{\delta}\right)ch_{L(\hat{\Lambda})} = \frac{\sum_{w\in W}\epsilon(w)\vartheta_{w(\hat{\Lambda}+\hat{\rho})}}{\sum_{w\in W}\epsilon(w)\vartheta_{w(\hat{\rho})}} \ , \qquad (12.50)$$

where the number $s_{\hat{\Lambda}}$ (known as the *modular anomaly*) has been introduced,

$$s_{\hat{\Lambda}} \equiv \frac{|\hat{\Lambda}+\hat{\rho}|^2}{2(k+\bar{h}_G)} - \frac{|\hat{\rho}|^2}{2\bar{h}_G} \ . \qquad (12.51)$$

When using the Casimir eigenvalue C_r of the finite-dimensional representation (r) associated with the highest weight Λ according to eq. (12.19): $C_r = (\Lambda + 2\rho, \Lambda) = |\hat{\Lambda}+\hat{\rho}|^2 - |\hat{\rho}|^2$, [3] the *Freudenthal-de Vries 'strange' formula* [44]

$$\frac{|\rho|^2}{2\bar{h}_G} = \frac{1}{24}|G| \ , \qquad (12.52)$$

and eq. (12.20) imply for the modular anomaly

$$s_{\hat{\Lambda}} = \frac{C_r}{C_A+2k} - \frac{1}{24}\frac{k|G|}{k+C_A/2} \ . \qquad (12.53)$$

The r.h.s. of this equation is just the difference between the dimension of the field $\phi_{(r)}$ and the trace anomaly.

The transformation properties of theta functions under the modular group (see Ch. III) are the standard issue in the classical theory of theta functions [96, 97, 532] (see also the next section). Let $\mathcal{Z}^*$ be the lattice dual to $\mathcal{Z}$ (for a simply-laced algebra, $\mathcal{Z}$ is the root lattice, and $\mathcal{Z}^*$ is the weight lattice), and $|\mathcal{Z}|$ (or $|\mathcal{Z}^*|$) the volume of the basic cell of the lattice $\mathcal{Z}$ (or that of the dual one $\mathcal{Z}^*$). Let us denote by $\vartheta_{\hat{\lambda}}\big|_A$ the action of

---

[3]This becomes obvious after choosing the representation first grade to be zero, while $(\hat{\Lambda}, \hat{\Lambda}_0) = 0$.

an element A of the modular group Γ on the theta function $\vartheta_{\hat{\lambda}}$. The action of Γ on the root basis (12.39) is defined by

$$\begin{pmatrix} a & b \\ c & d \end{pmatrix} (\hat{z} + \tau \hat{\Lambda}_0 + u\hat{\delta}) = \frac{1}{c\tau + d}\hat{z} + \frac{a\tau + b}{c\tau + d}\hat{\Lambda}_0 + \left(u + \frac{c(z,z)}{2(c\tau + d)} \right) \hat{\delta} , \qquad (12.54)$$

where $\hat{z} = (z,0,0)$ and $z = \sum_{s=1}^{l} z_s \nu_s$.

Let $\mathcal{Y}$ be the weight lattice of the algebra $\mathcal{G}$, which is defined to contain all vectors λ such that $2(\lambda, \alpha_i)/\alpha_i^2$ is an integer for all the simple roots α_i, and $\mathcal{N}_k$ be the weights which are the Weyl images of the dominant weights appearing at level k, i.e. $\lambda = w(\Lambda)$ and $(\hat{\Lambda}, \hat{\delta}) = k$. The transformation law of $\vartheta_{\hat{\lambda}}$ under the S transformation is given by

$$\vartheta_{\hat{\lambda}} |_S = (-i\tau)^{1/2} |\mathcal{Z}^*/k\mathcal{Z}|^{-1/2} \sum_{\mu \in \mathcal{N}_k \bmod k\mathcal{Z}} \exp\left[-\frac{2\pi i}{k}(\lambda, \mu) \right] \vartheta_{\hat{\mu}} . \qquad (12.55)$$

Notably, the sum in eq. (12.55) goes over all theta functions of the *same* level.

The transformation properties of the theta functions under the T transformation can be read off directly from their definition (12.47):

$$\vartheta_{\hat{\lambda}} |_T = \exp\left[\frac{\pi i}{k}(\lambda, \lambda) \right] \vartheta_{\hat{\lambda}} . \qquad (12.56)$$

The $SU(2)$ case is particularly simple, since there is only one root α, which obeys $\alpha^2 = 2$. Thus, the lattice $\mathcal{Z}$ is given by $j\alpha$ with integer j. The dual lattice $\mathcal{Z}^*$ is the weight lattice $\mathcal{Y}$ and $\mathcal{Z}^* = \frac{1}{2}\mathcal{Z}$. Let n be twice the spin for a weight λ: $\lambda = \frac{1}{2}n\alpha$. Eq. (12.47) now simplifies to

$$\vartheta_{n,k}(z, \tau, u) = e^{-2\pi i k u} \sum_{j \in \mathbf{Z} + n/2k} \exp\left[2\pi i k(j^2\tau - jz) \right] , \quad j \in \mathbf{Z} \bmod 2k\mathbf{Z} , \qquad (12.57)$$

while the transformation law of these functions under the S transformation takes the form

$$\vartheta_{n,k}\left(\frac{z}{\tau}, -\frac{1}{\tau}, u + \frac{z^2}{2\tau} \right) = (-i\tau)^{1/2} \sum_{n' \in \mathbf{Z} \bmod 2k\mathbf{Z}} e^{-\pi i nn'/k} \vartheta_{n',k}(z, \tau, u) . \qquad (12.58)$$

Eq. (12.58) could also be proven in a more straightforward way, by using the Poisson resummation formula (12.71).

Similarly, eq. (12.56) tells us that

$$\vartheta_{n,k}(z, \tau + 1, u) = e^{i\pi n^2/2k} \vartheta_{n,k}(z, \tau, u) . \qquad (12.59)$$

The Weyl-Kač character formula suggests us to introduce another theta function having the form

$$C_{\hat{\lambda}} = \sum_{w \in W} \epsilon(w) \vartheta_{w(\lambda)} , \tag{12.60}$$

since we can then rewrite the characters $\chi_R(\tau)$ to the form

$$\chi_R(\tau) = \frac{C_{\hat{\Lambda}+\hat{\rho}}}{C_{\hat{\rho}}}(0, \tau, 0) , \tag{12.61}$$

where $\hat{\Lambda}$ is the highest weight corresponding to the representation R at level k.

The Weyl group in the $SU(2)$ case is generated by the single reflection w_α, which is isomorphic to $\mathbf{Z}_2$. The dual Coxeter number is $\tilde{h}_{SU(2)} = 2$, and the Weyl vector is $\rho = \frac{1}{2}\alpha$. Eq. (12.60), in the $SU(2)$ case, takes the form

$$C_{n,k} = \vartheta_{n,k} - \vartheta_{-n,k} , \tag{12.62}$$

since the Weyl reflection maps n to $-n$.

XII.3 Basic facts about ϑ-functions

To illustrate some of the properties of the Jacobi theta functions introduced in eq. (3.22), and to develop some 'physical' tools for understanding the sophisticated identities of number theory, we begin with a proof of the *Jacobi triple product identity*

$$\prod_{n=1}^{+\infty}(1 - q^n)(1 + q^{n-1/2}w)(1 + q^{n-1/2}w^{-1}) = \sum_{n=-\infty}^{+\infty} q^{\frac{1}{2}n^2}w^n \tag{12.63}$$

for $|q| < 1$ (this guarantees absolute convergence of relevant series) and $w \neq 0$. We follow ref. [17] here.

Let us consider the partition function of a free electron-positron system with linearly spaced energy levels $E = \varepsilon_0(n - 1/2)$, $n \in \mathbf{Z}$, and total fermion number $N = N_e - N_{\bar{e}}$. Rewriting the energy E and fugacity μ in terms of $q = e^{-\varepsilon_0/T}$ and $w = e^{\mu/T}$, resp., we can introduce the grand canonical partition function as

$$Z(w, q) = \sum_{\substack{\text{fermion} \\ \text{occupations}}} e^{-E/T + \mu N/T} = \sum_{N=-\infty}^{+\infty} w^N Z_N(q)$$

$$= \prod_{n=1}^{+\infty}(1 + q^{n-1/2}w)(1 + q^{n-1/2}w^{-1}) , \tag{12.64}$$

where $Z_N(q)$ is the canonical partition function at the fixed total fermion number N. The factorization property of the *free* partition function (sect. III.1) has been used to write down the second line of eq. (12.64). The lowest energy state contributing to Z_0 has all negative energy levels filled and energy $E = 0$ by definition. Excited states are described by non-negative integers $k_1 \geq k_2 \geq \ldots \geq k_l \geq 0$, with $\sum_{i=1}^{l} k_i = M$ and energy $E = M\varepsilon_0$. The total number of such states is just the number of partitions $P(M)$, so that

$$Z_0 = \sum_{M=0}^{+\infty} P(M)q^M = \frac{1}{\prod_{n=1}^{+\infty}(1 - q^n)} \ . \tag{12.65}$$

The lowest energy state in the sector with fermion number N has the first N positive levels occupied, and it contributes the factor

$$q^{1/2} \ldots q^{N-3/2} q^{N-1/2} = q^{\sum_{n=1}^{N}(j-1/2)} = q^{N^2/2} \ . \tag{12.66}$$

Excitations of this state are described in exactly the same way as that for Z_0, so that $Z_N = q^{N^2/2} Z_0$. Hence, we have

$$Z(w,q) = \sum_{N=-\infty}^{+\infty} w^N Z_N(q) = \sum_{N=-\infty}^{+\infty} w^N \frac{q^{N^2/2}}{\prod_{n=1}^{+\infty}(1 - q^n)} \ , \tag{12.67}$$

which implies eq. (12.63) in full.

The basic result (12.63) can be used to derive a number of subsidiary identities. Substituting $w = \pm 1$ or $w = \pm q^{-1/2}$ into eq. (12.63) allows us to express the ϑ-functions in eq. (3.22) as

$$\vartheta_1 = i \sum_{n=-\infty}^{+\infty} (-1)^n q^{\frac{1}{2}(n-1/2)^2} \ (= 0), \quad \vartheta_2 = \sum_{n=-\infty}^{+\infty} q^{\frac{1}{2}(n-1/2)^2} \ ,$$

$$\vartheta_3 = \sum_{n=-\infty}^{+\infty} q^{n^2/2} \ , \qquad \vartheta_4 = \sum_{n=-\infty}^{+\infty} (-1)^n q^{n^2/2} \ , \tag{12.68}$$

while substituting $q \to q^3$, $w \to -q^{-1/2}$ yields

$$\prod_{n=1}^{+\infty}(1 - q^{3n})(1 - q^{3n-2})(1 - q^{3n-1}) = \sum_{n=-\infty}^{+\infty} q^{3n^2/2}(-1)^n q^{-n/2} \ , \tag{12.69a}$$

or, equivalently,

$$\prod_{n=1}^{+\infty}(1 - q^n) = \sum_{n=-\infty}^{+\infty} (-1)^n q^{\frac{1}{2}(3n^2-n)} \ . \tag{12.69b}$$

The last identity is known as the *Euler pentagonal number theorem*. Multiplying it by $q^{1/24}$ gives, in particular

$$\eta(q) \equiv q^{1/24} \prod_{n=1}^{+\infty} (1 - q^n) = \sum_{n=-\infty}^{+\infty} (-1)^n q^{\frac{1}{2}(n-1/6)^2} \ . \tag{12.70}$$

In addition, the *Poisson resummation formula* having the form

$$\sum_{n=-\infty}^{+\infty} f(nr) = \frac{1}{r} \sum_{m=-\infty}^{+\infty} \tilde{f}\left(\frac{m}{r}\right) \ , \tag{12.71}$$

where the Fourier transform $\tilde{f}$ has been defined as

$$\tilde{f}(y) = \int_{-\infty}^{+\infty} dx\, e^{-2\pi i x y} f(x) \ , \tag{12.72}$$

is quite useful to treat modular transformation properties of ϑ's and η under $\tau \to -1/\tau$. Eq. (12.71) itself is easily checked after substituting eq. (12.72) into it. The natural generalization of eq. (12.71) to higher dimensions takes the form

$$\sum_{v \in \Gamma} f(v) = \frac{1}{V} \sum_{w \in \Gamma^*} \tilde{f}(w) \ , \tag{12.73}$$

where Γ is a lattice, Γ^* is its dual, and V is the volume of its unit cell.

Applying eq. (12.71) to the η-function yields

$$\eta\left(q(-1/\tau)\right) = (-i\tau)^{-1/2} \eta\left(q(\tau)\right) \ . \tag{12.74}$$

Similarly, under $\tau \to -1/\tau$, one has

$$\vartheta_2 \to (-i\tau)^{1/2} \vartheta_4 \ , \quad \vartheta_4 \to (-i\tau)^{1/2} \vartheta_2 \ , \quad \vartheta_3 \to (-i\tau)^{1/2} \vartheta_3 \ , \tag{12.75}$$

whereas, under $\tau \to \tau + 1$, one gets

$$\vartheta_3 \leftrightarrow \vartheta_4 \ , \quad \vartheta_2 \to \sqrt{i}\,\vartheta_2 \ , \quad \eta \to e^{i\pi/12}\eta \ . \tag{12.76}$$

We are now in a good position (after substituting $w = e^{2\pi i z}$ into the r.h.s. of eq. (12.63)) to introduce the 'fundamental' theta function $\vartheta_3(z, \tau)$, as a function of two variables, in the form

$$\vartheta_3(z, \tau) \equiv \sum_{n=-\infty}^{+\infty} q^{n^2/2} e^{2\pi i n z} \ . \tag{12.77}$$

The similar generalizations of all the other ϑ-functions introduced in eq. (3.22) can be defined in the following way:

$$\vartheta_4(z,\tau) = \vartheta_3(z + \tfrac{1}{2}, \tau) = \sum_{\infty}^{+\infty} (-1)^n q^{n^2/2} e^{2\pi i n z} \ ,$$

$$\vartheta_1(z,\tau) = -ie^{iz} q^{1/8} \vartheta_4(z + \frac{\tau}{2}, \tau) = i \sum_{n=-\infty}^{+\infty} (-1)^n q^{\frac{1}{2}(n-1/2)^2} e^{i\pi(2n-1)z} \ ,$$

$$\vartheta_2(z,\tau) = \vartheta_1(z + \tfrac{1}{2}, \tau) = \sum_{n=-\infty}^{+\infty} q^{\frac{1}{2}(n-1/2)^2} e^{i\pi(2n-1)z} \ , \qquad (12.78)$$

and $\vartheta_i(\tau) \equiv \vartheta_i(0, \tau)$.

In string theory (Ch. VIII), where space-time gauge symmetries are realized as affine symmetries on the string world-sheet, the additional z-dependence introduced in eqs. (12.77) and (12.78) is responsible for the dependence of the string partition function on background gauge fields. The space-time gauge and gravitational anomalies in string theory can then be probed via the (anomalous) modular transformation properties of the string partition functions.

XII.4   Basic facts about Riemann surfaces

A *Riemann surface* Σ is a connected (analytic) orientable manifold of real dimension 2 (complex dimension 1). It is therefore assumed that (i) Σ can be covered by coordinate charts with holomorphic transition functions, and (ii) Σ is a *paracompact* manifold , i.e. there exists its locally finite covering by the coordinate charts (*the partition of unity*). The latter ensures the global existence of a Riemannian metric $g_{\alpha\beta}$ defined on the whole Riemann surface Σ, otherwise a metric would exist only chart-wise. Let Diff(Σ) be the group of all diffeomorphisms of a Riemann surface Σ, where Diff$_0(\Sigma)$ is its component of unity, comprising the diffeomorphisms homotopic to the identity map.

The existence of a Riemannian metric $g_{\alpha\beta}(\zeta)$ on an orientable surface Σ allows us to consistently introduce the *complex structure* tensor $J_\alpha{}^\beta$,

$$J_\alpha{}^\beta \equiv \sqrt{g}\,\epsilon_{\alpha\gamma} g^{\gamma\beta} \ , \qquad \epsilon_{\alpha\beta} = -\epsilon_{\beta\alpha} \ , \qquad \epsilon_{12} = 1 \ , \qquad (12.79)$$

having the properties

$$J_\alpha{}^\beta J_\beta{}^\gamma = -\delta_\alpha^\gamma \ , \qquad \nabla_\gamma J_\alpha{}^\beta = 0 \ . \qquad (12.80)$$

After changing the variables, $x^\alpha = x^\alpha(\zeta^1, \zeta^2)$, $z = \frac{1}{\sqrt{2}}(x^1 + x^2)$, in accordance with the *Beltrami* or *Cauchy-Riemann* equations

$$J_\alpha{}^\beta \frac{\partial z}{\partial \zeta^\beta} = i \frac{\partial z}{\partial \zeta^\alpha} , \qquad J_\alpha{}^\beta \frac{\partial \bar{z}}{\partial \zeta^\beta} = -i \frac{\partial \bar{z}}{\partial \zeta^\alpha} , \tag{12.81}$$

we can render the metric (locally) conformally flat, [4]

$$ds^2 = g_{\alpha\beta}(\zeta)d\zeta_\alpha d\zeta^\beta = 2 \left| \frac{\partial z}{\partial \zeta} \right|^{-2} dz d\bar{z} \equiv \rho(z, \bar{z}) dz d\bar{z} . \tag{12.82}$$

The choice of coordinates in eq. (12.82) has the residual invariance under the analytic (or conformal) transformations $z \to f(z)$, which are just the transition functions for a Riemann surface. In other words, a Riemann surface is a pair (Σ, J) comprising a 2d connected oriented manifold Σ and a complex structure J defined on Σ.

The simplest example of a non-compact Riemann surface is a complex plane **C**. The extended complex plane $\mathbf{C} \cup \{\infty\}$ is an example of a closed compact Riemann surface, which is called the *Riemann sphere* and has genus $h = 0$. A compact surface of genus h (without boundary) is topologically a 'sphere' with h handles, a genus-1 surface being topologically equivalent to a *torus*.

Any metric $g_{\alpha\beta}$ on Σ can be diffeomorphism- and Weyl-transformed to the new metric $\hat{g}_{\alpha\beta} = e^{-2\sigma} g_{\alpha\beta}$ which has a *constant* curvature everywhere on Σ. This important observation is equivalent to the statement that the *Liouville* equation

$$\Delta_g \sigma = R_g - R_{\hat{g}} e^{-2\sigma} , \tag{12.83}$$

to which the Weyl-transformation parameter σ is then supposed to satisfy, admits a unique solution, which is clearly the case. The sign of the constant curvature is dictated by the Gauss-Bonnet theorem (1.140): $R_{\hat{g}} = 1$ when $h = 0$, $R_{\hat{g}} = 0$ when $h = 1$, and $R_{\hat{g}} = -1$ when $h \geq 2$. It is one of the central results (called the *uniformization theorem*) in the theory of Riemann surfaces that there are only three simply connected Riemann surfaces, which are homologically trivial ($h = 0$): the Riemann sphere **S**, the plane **C**,

---

[4]Of course, this is not possible globally (except for $h = 0$), because of the Gauss-Bonnet theorem (see below).

and the upper half-plane C_+. [5] The constant-curvature metrics on them are given by

$$ds^2 = \frac{4dzd\bar{z}}{(1+|z|^2)^2} , \qquad R = 1 ,$$

$$ds^2 = 2dzd\bar{z} , \qquad R = 0 , \qquad (12.84)$$

$$ds^2 = -\frac{dzd\bar{z}}{(z-\bar{z})^2} , \qquad R = -1 ,$$

resp. The general uniformization theorem claims that an arbitrary Riemann surface is conformally equivalent to a quotient Υ/G, where Υ is either S, C, or C_+, and G is a freely acting discrete group of (Möbius) transformations that preserve Υ [532]. In fact, G is isomorphic to the fundamental group of Σ (see its definition below).

Given an arbitrary Riemann surface Σ of genus h, the $\mathcal{M}_{\text{const}} = \{\bar{g};\ R_{\bar{g}} = \text{const}\}$ gives the well-defined global slice for the Weyl transformation group in the space of all metrics on Σ. The additional modding by the diffeomorphism group yields the spaces

$$\mathcal{T}_h = \frac{\mathcal{M}_{\text{const}}}{\text{Diff}_0(\Sigma)} ,$$

$$\mathcal{M}_h = \frac{\mathcal{M}_{\text{const}}}{\text{Diff}(\Sigma)} \equiv \frac{\mathcal{T}_h}{\text{MCG}_h} , \qquad (12.85)$$

called the *Teichmüller* and *moduli* space, respectively. The discrete group $\text{MCG}_h = \text{Diff}(\Sigma)/\text{Diff}_0(\Sigma)$ is known as the *mapping class group*. It is the theorem in the theory of Riemann surfaces [532] that the (real) dimension of the Teichmüller (or moduli) space is *finite*, and its value (for compact surfaces) is given by

$$\dim \mathcal{T}_h = \dim \mathcal{M}_h = \begin{cases} 0 , & h = 0 , \\ 2 , & h = 1 , \\ 6h - 6 , & h \geq 2 . \end{cases} \qquad (12.86)$$

The theory of Riemann surfaces naturally contains a topological part (topology of Riemann surfaces), an algebraic part (algebraic geometry of Riemann surfaces), and an analytic part (functions on Riemann surfaces). As far as the topological aspects are concerned, we confine ourselves here to the *compact* Riemann surfaces Σ without boundary or punctures (they correspond to the closed strings of Ch. VIII), whose topology is completely characterized by their genus h via the Gauss-Bonnet theorem (1.140). The

---

[5]The latter is the same as the *Poincaré disc*: $|z| \leq 1$ with the metric $ds^2 = \frac{4dzd\bar{z}}{(1-|z|^2)^2}$ (*cf* exercise # IX-7).

Fig. 24. The canonical homological basis
on a Riemann surface .

fundamental group $\pi_1(\Sigma)$ is defined as the set of equivalence classes of closed curves passing through a fixed point of Σ. Two curves are defined to be equivalent, when they are *homotopic*, i.e. when they can be continuously transformed into each other. For example, a *simply connected* manifold has $\pi_1 = 0$. For any compact Riemann surface of genus h, there are exactly $2h$ non-contractable independent closed curves (*cycles*) (Fig. 24), which are subject to the single relation $\prod_{i=1}^{h} a_i b_i a_i^{-1} b_i^{-1} = 1$. They form the *homological* or *canonical* basis in the linear space of all cycles $H_1(\mathcal{M}) \cong \mathbf{Z}^{2g}$ called the 1st homology group. [6] Therefore, these $2h$ cycles (a_i, b_j) are generators of the fundamental group $\pi_1(\Sigma)$. Their intersections (with the orientation accounted for, see Fig. 24) define the intersection matrix $\#(c_i, c_j)$,

$$\#(a_i, a_j) = \#(b_i, b_j) = 0 \ ,$$
$$\#(a_i, b_j) = -\#(b_i, a_j) = \delta_{ij} \ , \tag{12.87}$$

The canonical homology basis is not unique since the intersection matrix remains invariant under the *symplectic* transformations of the canonical basis with integer entries. The associated symplectic group $Sp(2h, \mathbf{Z})$ is called the *modular group*. This group can be vizualized as generated by 2π-twists around a- and b-cycles called *Dehn twists*. The modular group is always a subgroup of the mapping class group MCG_h, the quotient $MCG_h/Sp(2h, \mathbf{Z})$ being known as the *Torelli group* which does not act on the homology basis.

Given complex analytic coordinates and a metric in the form (12.82), one can introduce the complex derivatives $(z = \zeta_1 + i\zeta_2)$

$$\partial \equiv \partial_z = \tfrac{1}{2}(\partial_1 - i\partial_2) \ , \qquad \bar{\partial} \equiv \partial_{\bar{z}} = \tfrac{1}{2}(\partial_1 + i\partial_2) \ , \tag{12.88}$$

---

[6]Dimensions of the homology groups are called *Betti numbers* β_i. In particular, the Euler characteristic can always be represented as the alternating sum of the Betti numbers [532]: $\chi(\mathcal{M}) = 2 - 2h = \beta_0 - \beta_1 + \beta_2$. As far as the Riemann surfaces are concerned, one has $\beta_0 = \beta_2 = 1$, $\beta_1 = 2h$. The (abelian) homology groups are defined w.r.t. addition.

the tangent space t_P (vectors) and the co-tangent space t^P (1-forms) at a point $P \in \Sigma$. Varying the point $P = P(z, \bar{z})$ allows us to introduce fields as usual. Because of the complex structure on Σ, they can be represented as

$$t_P = t_P^{(1,0)} \oplus t_P^{(0,1)} ,$$
$$t^P = t_{(1,0)}^P \oplus t_{(0,1)}^P , \tag{12.89}$$

where the $(1, 0)$ is for z direction, and the $(0, 1)$ is for $\bar{z}$. More general forms,

$$t_{(n,k)} = t_{\underbrace{zz\ldots z}_{n \text{ times}} \underbrace{\bar{z}\bar{z}\ldots\bar{z}}_{k \text{ times}}}(dz)^n (d\bar{z})^k , \tag{12.90a}$$

as well as tensors,

$$t_{(-n,-k)} \equiv t^{(n,k)} = t^{\overbrace{zz\ldots z}^{n \text{ times}}\overbrace{\bar{z}\bar{z}\ldots\bar{z}}^{k \text{ times}}}(dz)^{-n}(d\bar{z})^{-k} , \tag{12.90b}$$

can be introduced as the generalizations of eq. (12.89). They are all invariant under the analytic or conformal transformations $z \to f(z)$, $\bar{z} \to \bar{f}(\bar{z})$. The invariant scalar products read

$$\left\langle t^{(n,k)} | \tilde{t}^{(1-n,1-k)} \right\rangle = \int t^{(n,k)} \wedge \tilde{t}^{(1-n,1-k)} , \tag{12.91}$$

and zero otherwise. The rank of $\tilde{t}^{(k,n)}$ can be easily changed by multiplying with ρ-powers: $\tilde{t}^{(k,n)} \to \tilde{t}^{(k,n)} \rho^{-(n+k)+1}(z, \bar{z}) \sim \tilde{t}^{(1-n,1-k)}$. This introduces the Hilbert structure in the space of all forms and tensors. The compactness of Σ is needed to ensure the convergence of these products.

Because of eq. (12.81), the components of the metric in the conformal gauge are given by

$$g_{z\bar{z}} = g_{\bar{z}z} = \tfrac{1}{2}\rho , \qquad g_{zz} = g_{\bar{z}\bar{z}} = 0 ,$$
$$g^{z\bar{z}} = g^{\bar{z}z} = 2\rho^{-1} , \qquad g^{zz} = g^{\bar{z}\bar{z}} = 0 . \tag{12.92}$$

This metric can be used for contracting z and $\bar{z}$ indices, like in the usual Riemannian geometry. In particular, a tensor or a form can be reduced to their canonical forms, e.g.

$$t_n = t_{\underbrace{zz\ldots z}_{n \text{ times}}}(dz)^n , \tag{12.93}$$

where

$$t_n = t_{\underbrace{zz\ldots z}_{n \text{ times}}} = (g^{z\bar{z}})^k \, t_{\underbrace{zz\ldots z}_{n \text{ times}} \underbrace{\bar{z}\bar{z}\ldots\bar{z}}_{k \text{ times}}} . \tag{12.94}$$

Hence, it is enough to consider the case of $k = 0$. Number n in eq. (12.93) is called *rank* of t_n. The space of all tensors (or forms) of rank n is denoted by T^n.

Associated with the metric are the Christoffel symbols, the covariant derivatives and the Riemann curvature tensor, whose definitions are quite standard. In particular, when acting on the objects (12.93), one gets

$$\nabla^z t_{z\ldots} = g^{z\bar{z}} \bar{\partial} t_{z\ldots} = g^{z\bar{z}} \bar{\nabla} t_{z\ldots} , \tag{12.95}$$

where the derivative in the $\bar{z}$-direction remains to be unchanged: $\bar{\partial}_{(n)} = \nabla_{\bar{z}}^{(n)}$. Similarly, it is straightforward to check that

$$\nabla_z t_{\underbrace{zz\ldots z}_{n\ \text{times}}} = (\partial - n\partial \log \rho)\, t_{\underbrace{zz\ldots z}_{n\ \text{times}}} , \tag{12.96}$$

and

$$[\nabla^z, \nabla_z]\, t_{\underbrace{zz\ldots z}_{n\ \text{times}}} = \tfrac{1}{2} n R\, t_{\underbrace{zz\ldots z}_{n\ \text{times}}} , \quad R = \rho^{-1}\left(-4\partial\bar{\partial} \log \rho\right) . \tag{12.97}$$

W.r.t. the scalar product introduced in eq. (12.91), we find

$$\left(\nabla_{(n)}^z\right)^\dagger = -\nabla_z^{(n-1)} , \tag{12.98}$$

and, hence,

$$\nabla_z : T^n \to T^{n+1} , \qquad \nabla^z : T^n \to T^{n-1} ,$$

$$\bar{\partial} = \nabla_{\bar{z}} : T^n \to \{t^{(n,1)}\} . \tag{12.99}$$

The *Laplace* operators $\Delta_n^{(\pm)} : T^n \to T^n$,

$$\Delta_n^{(+)} = -2\nabla_{(n+1)}^z \nabla_z^{(n)} , \qquad \Delta_n^{(-)} = -2\nabla_z^{(n-1)} \nabla_{(n)}^z , \tag{12.100}$$

are self-conjugated. In particular, one has $\Delta_n^{(-)} = 2\bar{\partial}_{(n)}^\dagger \bar{\partial}_{(n)}$ and $\Delta_g = \Delta_0^{(+)} = \Delta_0^{(-)}$, where the operator $\Delta_g = -\frac{1}{\sqrt{g}}\partial_\alpha \sqrt{g} g^{\alpha\beta} \partial_\beta$ has been introduced.

The connection with the real notation and with the corresponding real differential operators is provided by the formulae $(T^n \oplus T^{-n} \equiv S^n)$ [28]:

$$P_n : S^n \to S^{n+1} , \qquad P_n = \nabla_z^{(n)} \oplus \nabla_{\bar{z}}^{(-n)} ,$$

$$P_n^\dagger : S^{n+1} \to S^n , \qquad P_n^\dagger = -\left(\nabla_{(n+1)}^z \oplus \nabla_{\bar{z}}^{(-n-1)}\right) . \tag{12.101}$$

Zero modes of these operators are of special interest. Given a compact orientable Riemann surface of genus h, and the Cauchy-Riemann operator $\bar{\partial}$ acting on tensors of rank n, let us consider solutions to the equation

$$\bar{\partial} t_{\underbrace{zz\ldots z}_{n \text{ times}}} = 0 \ . \tag{12.102}$$

A partial information about the number of independent solutions of this equation is provided by the *Riemann-Roch* theorem :

$$\dim \text{Ker } \nabla_z^{(n)} - \dim \text{Ker } \nabla_{(n+1)}^z = \tfrac{1}{2}(2n+1)\chi(\mathcal{M}) \ . \tag{12.103}$$

This index theorems and its various generalizations are the standard topics in the mathematical literature (see, e.g., refs. [28, 532]). One has, in particular

$$\dim \text{Ker } \nabla_z^{(n)} = 0 \quad \text{for} \quad n \geq 1 \ \text{ and } \ h \geq 2 \ , \quad \dim \text{Ker } \nabla_z^{(0)} = 1 \ . \tag{12.104}$$

The last equation implies that there are no non-constant holomorphic functions on compact surfaces, which is the consequence of the maximum modulus principle known in complex analysis. [7] The first equation (12.104) can be proved by choosing a metric with the negative constant curvature ($R = -1$) on the Riemann surface (the dimensions we consider are topologically invariant!). Taking $n = 1$ for simplicity, let us consider the equation $(P_1\delta v)_{\alpha\beta} = 0$ defining the *conformal Killing vector* vector δv^α. Upon differentiating

$$\nabla^\beta \nabla_\beta \delta v^\alpha = -R\delta v^\alpha = \delta v^\alpha \ , \tag{12.105}$$

and integrating it versus δv_α, we find

$$||\nabla^\beta \delta v^\alpha||^2 + ||\delta v^\alpha||^2 = 0 \ . \tag{12.106}$$

Both terms on the l.h.s. of this equation have to vanish, and, hence, $\delta v^\alpha = 0$. The proof is quite similar in the other cases ($n > 1$) of eq. (12.104). Eqs. (12.103) and (12.104) now yield for $h \geq 2$ the desired relations

$$\dim \text{Ker } \nabla_{(n)}^z = \dim \text{Ker } \bar{\partial}_{(n)} = (2n-1)(h-1) \ , \quad \text{for} \ n \geq 2 \ ,$$

$$\dim \text{Ker } \nabla_{(1)}^z = \dim \text{Ker } \bar{\partial}_{(1)} = h \ . \tag{12.107}$$

---

[7]It is the non-trivial theorem [532] that a Riemann surface carries non-constant *meromorphic* functions. For example, in the case of Riemann sphere, the meromorphic functions are just the rational functions.

Notably, there are exactly $3h - 3$ complex independent holomorphic quadratic differentials $\phi = \phi_{zz}(dz)^2$ or Teichmüller deformations, in accordance with eq. (12.86). The space *dual* to the holomorphic quadratic differentials is called the space of *Beltrami differentials* $\mu = \mu_{\bar{z}}{}^z d\bar{z}(dz)^{-1}$ having $\mu_z{}^z = 0$ and

$$\langle \mu | \phi \rangle = \int_{\mathcal{M}} d^2 z \, \mu_{\bar{z}}{}^z \phi_{zz} \ . \tag{12.108}$$

Given a reference metric $ds^2 = \rho(z, \bar{z}) \, |dz|^2$ on a Riemann surface, any other metric can be written in terms of the Beltrami differential μ as

$$ds^2 = \rho(z, \bar{z}) \, |dz + \mu_{\bar{z}}{}^z d\bar{z}|^2 \ . \tag{12.109}$$

Therefore, the cohomology of Beltrami differentials w.r.t. reparametrizations

$$\mu_{\bar{z}}{}^z \to \mu_{\bar{z}}{}^z + \bar{\partial}\varepsilon(z, \bar{z}) \tag{12.110}$$

naturally characterizes the conformal classes of the metric too. This provides the equivalent description of the Teichmüller space as the quotient

$$\mathcal{T}_h = \frac{\{\text{Beltrami differentials}\}}{\{\text{Range } \bar{\partial} \text{ on vectors}\}} \ . \tag{12.111}$$

For superstrings (Ch.VIII), the Teichmüller and moduli spaces are to be extended to the *super-Teichmüller* and *super-moduli* spaces, resp., after taking into account the additional $2h - 2$ complex holomorphic 3/2-differentials associated with the gravitino field. Eq. (12.107) is still valid for $n = 3/2$, and it tells us about the complex dimension ($= 2h - 2$) of odd variables of the supermoduli space.

In the case of Riemann sphere with $\chi = 2$, one has $\dim \operatorname{Ker} P_1 = 6$ since there are 3 complex conformal Killing vectors due to the Möbius automorphism symmetry $SL(2, \mathbf{C})$ the sphere has. Therefore, $\dim \operatorname{Ker} P_1^{\dagger} = 0$ in this case. For a torus with $\chi = 0$, there is only one complex Killing vector, so that $\dim \operatorname{Ker} = 2$ due to the two-dimensional translational invariance the torus has. Hence, one gets

$$h = 1 : \quad \dim \operatorname{Ker} \nabla_z^{(n)} = \dim \operatorname{Ker} \nabla_{(n)}^z = 1 \ , \tag{12.112}$$

and

$$h = 0 : \quad \dim \operatorname{Ker} \nabla_z^{(n)} = 2n + 1 \ , \quad \text{for } \ n \geq -\tfrac{1}{2} \ ,$$
$$\dim \operatorname{Ker} \nabla_{(n)}^z = 0 \ , \quad \text{for } \ n \geq \tfrac{1}{2} \ . \tag{12.113}$$

In particular, there are exactly h holomorhic 1-forms (*abelian differentials of the first kind*) generating the first cohomology group $H^1(\Sigma)$ when $h \geq 1$. [8] Given two holomorphic 1-forms ω and η, the following two identities, known as the *Riemannian bilinear relations*, are valid:

$$(1):\quad \sum_{i=1}^{h} \left(\int_{a_i} \omega \cdot \int_{b_i} \eta - \int_{b_i} \omega \cdot \int_{a_i} \eta \right) = 0 \,, \tag{12.114}$$

and

$$(2):\quad \mathrm{Im}\left(\sum_{i=1}^{h} \overline{\int_{a_i} \omega} \cdot \int_{b_i} \omega \right) > 0 \,, \tag{12.115}$$

where $\{a_i, b_i\}_{i=1}^{i=h}$ is the canonical basis on Σ.

Let $\{\omega_i\}$, $i = 1, \ldots, h$, be a basis in the space of all holomorphic 1-forms, so that for any holomorphic 1-form ω we have $\omega = \sum_i c_i \omega_i$. It follows from eq. (12.115) that

$$\sum_{k,l,i} \bar{c}_k c_l \overline{\int_{a_i} \omega_k} \cdot \int_{b_i} \omega_l \neq 0 \,, \tag{12.116}$$

for any $\omega \neq 0$. Hence, the matrix $A_{kl} \equiv \sum_i \overline{\int_{a_i} \omega_k} \int_{b_i} \omega_l$ is not degenerate, which, in its turn, implies a non-degeneracy of the matrix $\int_{a_i} \omega_k$. Therefore, there exists a basis in which

$$\int_{a_i} \omega_k = \delta_{ik} \,. \tag{12.117}$$

Given this, the matrix

$$\Omega_{ij} = \int_{b_i} \omega_j \tag{12.118}$$

is determined, and it is called the *period matrix*. Choosing $\omega = \omega_k$ and $\eta = \omega_l$ in the Riemannian bilinear relation (12.114) proves that the matrix Ω is symmetric:

$$\Omega_{kl} = \Omega_{lk} \,. \tag{12.119}$$

Choosing in eq. (12.115) $\omega = \sum_{i=1}^{h} c_l \omega_l$, with some *real* coefficents c_l, yields

$$\mathrm{Im}\left(\sum_i c_i \int_{b_i} \omega \right) = \mathrm{Im}\left(\sum_{i,j} c_i \Omega_{ij} c_j \right) = \sum_{ij} c_i c_j \, \mathrm{Im}\Omega_{ij} > 0 \,. \tag{12.120}$$

Therefore, the imaginary part of the period matrix is positively definite. When the complex structure on Σ varies (by a Beltrami differential μ) the period matrix deforms as [28]

$$\delta\Omega_{ij} = -i \int d^2z \, \mu_{\bar{z}}{}^z \omega_i \omega_j \,. \tag{12.121}$$

---

[8] Abelian differentials of the *second* kind are meromorphic with one double pole, whereas the ones of the *third* kind have two simple poles of opposite residues, by definition.

Associated to the period matrix Ω_{ij} of a Riemann surface is the $2h$-dimensional lattice

$$L_\Omega = \{m_i + \Omega_{ij} n_j\} , \qquad m, n \in \mathbf{Z} . \tag{12.122}$$

The complex torus $J(\Sigma) = \mathbf{C}^h / L_\Omega$ is called the *Jacobian variety*. By the modular transformations

$$\begin{pmatrix} A & B \\ C & D \end{pmatrix} \in Sp(2h, \mathbf{Z}) , \tag{12.123}$$

the period matrix transforms as

$$\Omega \to \Omega' = (A\Omega + B)(C\Omega + D)^{-1} , \tag{12.124}$$

so that the lattice defined be eq. (12.122) does not change.

Let $\{P_j\}$ be a set of points on Σ, and $\{\alpha_j\}$ a set of integers. A formal sum

$$D = \sum_{j=1}^{k} \alpha_j P_j \tag{12.125}$$

is called a *divisor*. The sum

$$\sum_{j=1}^{k} \alpha_j \equiv \deg D \tag{12.126}$$

is called the *degree* of a divisor D. Given a non-vanishing meromorphic q-differential ω on Σ, one can associate the natural divisor (ω) with it, namely

$$(\omega) = \sum_{P \in \Sigma} \operatorname{ord}_P \omega \, P = \sum_{j \in \{\text{poles, zeroes}\}} \operatorname{ord}_{P_j} \omega \, P_j , \tag{12.127}$$

where the *order*, $\operatorname{ord}_P \omega$, of the q-differential ω at point $P \in \Sigma$ is just the order of a pole or of a zero, when P is such a point. The divisor (12.127) is called a *q-canonical* divisor, or a *canonical* divisor when $q = 1$.

Given an h-vector of Abelian differentials $\vec{\omega} = (\omega_1, \dots, \omega_h)$ and a fixed point $z_0 \in \Sigma$, the *Abelian map*

$$I(z_1 + \cdots + z_M) \equiv \int_{z_0}^{z_1} \omega + \cdots + \int_{z_0}^{z_M} \omega \tag{12.128}$$

going from Σ to $J(\Sigma)$ is single-valued on Σ. In eq. (12.128), coordinates $\{z_i\}$ specify a set of points $\{P_i\}$ on Σ, while the addition signs on the. l.h.s. have to be understood in the divisor sense.

Given an arbitrary divisor Υ on Σ, it is the *Abel theorem* [532] that the necessary and sufficient conditions for D to be a divisor of a meromorphic function defined on Σ are [532]

$$I(D) = 0 \bmod L_\Omega , \quad \text{and} \quad \deg D = 0 . \tag{12.129}$$

For D to be a divisor of a meromorphic q-differential on a compact Riemann surface Σ of genus h, the necessary and sufficient conditions are [532]

$$I(D) = -2\Delta \bmod L_\Omega \ , \quad \deg D = 2h - 2 \ , \tag{12.130}$$

where Δ is known as the *vector of Riemann constants*:

$$\Delta_i = \tfrac{1}{2} - \tfrac{1}{2}\Omega_{ii} + \sum_{k \neq j} \oint_{a_k} \omega_k(z) \int_{z_0}^{z} \omega_j \ . \tag{12.131}$$

The Jacobian variety and the Abelian map play the important roles in constructing the theta functions associated with Riemann surfaces. The definition of the generalized theta function parametrized by a symmetric complex matrix $\hat\Omega$, [9] having an argument $\mathbf{z} \in \mathbf{C}^h$ and generalizing the Jacobi theta functions considered in the previous section for the case of the torus to arbitrary Riemann surfaces, is given by [97]

$$\vartheta(\mathbf{z}, \hat\Omega) = \sum_{\mathbf{n} \in \mathbf{Z}^h} \exp\left[i\pi \mathbf{n}^T \cdot \hat\Omega \cdot \mathbf{n} + 2\pi i \mathbf{n}^T \cdot \mathbf{z} \right] \ . \tag{12.132}$$

This function satisfies the generalized heat equation having the form

$$\left[4\pi i \frac{\partial}{\partial \hat\Omega_{ij}} + \frac{\partial^2}{\partial z_i \partial z_j} \right] \vartheta(\mathbf{z}, \hat\Omega) = 0 \ , \tag{12.133}$$

and it has the transformation property $(\mathbf{a}, \mathbf{b} \in \mathbf{Z^h})$:

$$\vartheta(\mathbf{z} + \mathbf{a} + \hat\Omega \cdot \mathbf{b}, \hat\Omega) = \exp\left[-i\pi \mathbf{b}^T \cdot \hat\Omega \cdot \mathbf{b} - 2\pi i \mathbf{b}^T \cdot \mathbf{z} \right] \vartheta(\mathbf{z}, \hat\Omega) \ . \tag{12.134}$$

It is due to the last property (12.134) that a Jacobi theta function naturally 'lives' on a Jacobian variety (obviously, it can be defined for an arbitrary $\hat\Omega$) since it is 'almost' (up to a phase) single-valued there. [10] The function (12.132) absolutely and uniformly converges on any compact subset $\mathbf{z} \in \mathbf{C}^h$, and it is holomorphic there.

The theta function associated with a Riemann surface Σ is parametrized by its period matrix $\Omega(\Sigma)$. The detailed information about the zero set of this function (the so-called ϑ *divisor*) is provided by the *Riemann vanishing theorem* known in the theory of Riemann surfaces [532].

---

[9]It is not assumed here that $\hat\Omega$ is the period matrix of a Riemann surface!

[10]The function $\vartheta(\mathbf{z}, \hat\Omega)$ is known to be the holomorphic section of a line bundle over the Jacobian variety, the so-called ϑ *line bundle* .

Slightly more general theta functions are dependent on additional parameters (called *characteristics*) $\vec{\delta} = \begin{bmatrix} \vec{\delta}' \\ \vec{\delta}'' \end{bmatrix}$ valued in $[0,1]^{2h}$. Their relevance in string theory (Ch. VIII) originates from the fact that they describe *spin structures* associated with spinors defined on a Riemann surface. A spin structure is a choice of phases spinors acquire under parallel transport around homology cycles. The definition of the *Jacobi theta function with characteristics* $\vec{\delta}$ is [97]

$$\vartheta[\vec{\delta}](\mathbf{z}, \hat{\Omega}) = \sum_{\mathbf{n} \in \mathbf{Z}_h} \exp\left[i\pi(\mathbf{n} + \vec{\delta}')^T \cdot \hat{\Omega} \cdot (\mathbf{n} + \vec{\delta}') + 2\pi i(\mathbf{n} + \vec{\delta}')^T \cdot (\mathbf{z} + \vec{\delta}'') \right]$$

$$= \exp\left[i\pi\vec{\delta}'^T \cdot \hat{\Omega} \cdot \vec{\delta}' + 2\pi i\vec{\delta}'^T \cdot (\mathbf{z} + \vec{\delta}'') \right] \vartheta(\mathbf{z} + \hat{\Omega} \cdot \vec{\delta}' + \vec{\delta}'', \hat{\Omega}) . \qquad (12.135)$$

Its transformation laws take the form [97]

$$\vartheta[\vec{\delta}](\mathbf{z} + \mathbf{a} + \hat{\Omega} \cdot \mathbf{b}, \hat{\Omega}) = \exp\left[-i\pi\mathbf{b}^T \cdot \hat{\Omega} \cdot \mathbf{b} - 2\pi i\mathbf{b}^T \cdot (\mathbf{z} + \vec{\delta}'') + 2\pi i\vec{\delta}'^T \cdot \mathbf{a} \right] \vartheta[\vec{\delta}](\mathbf{z}, \hat{\Omega}) ,$$

$$\vartheta\begin{bmatrix} \vec{\delta}' + \mathbf{a} \\ \vec{\delta}'' + \mathbf{b} \end{bmatrix}(\mathbf{z}, \hat{\Omega}) = \exp\left[2\pi i\vec{\delta}'^T \cdot \mathbf{b} \right] \vartheta\begin{bmatrix} \vec{\delta}' \\ \vec{\delta}'' \end{bmatrix}(\mathbf{z}, \hat{\Omega}) . \qquad (12.136)$$

Let us consider the holomorphic Abelian differential on a Riemann surface Σ,

$$\omega_{\vec{\delta}}(w) \equiv \sum_{i=1}^{h} \frac{\partial \vartheta}{\partial z^i}[\vec{\delta}](0, \Omega)\omega_i(w) , \qquad (12.137)$$

which has only zeroes of *second* order (this statement is one of the consequences of the Riemann vanishing theorem [97, 532]). Hence, the field

$$h_{\vec{\delta}}(w) = \sqrt{\omega_{\vec{\delta}}(w)} \qquad (12.138)$$

is a well-defined holomorphic spinor on Σ. The spinor $h_{\vec{\delta}}$ is used to define the **prime form**

$$E(z, w) = \frac{\vartheta[\vec{\delta}]\left(\int_w^z \omega, \Omega\right)}{h_{\vec{\delta}}(z)h_{\vec{\delta}}(w)} , \qquad (12.139)$$

which is of fundamental importance in the construction of various holomorphic fields on Riemann surfaces. The prime form is a holomorphic $(-\frac{1}{2}, 0)$ form in z and w, with a single zero at $z = w$. This form is dependent on a choice of the homology basis, and it transforms under the modular transformations (12.123) as follows [97, 532]

$$E(z, w) \rightarrow \exp\left[i\pi \int_z^w \omega(C\Omega + D)^{-1}C \int_z^w \omega \right] E(z, w) . \qquad (12.140)$$

The relevance of the prime form for CFT's defined on Riemann surfaces follows from the observation that it naturally appears in the scalar Green function and, hence, in the correlation functions as well [28]. The full (non-chiral) scalar Green function $G(z, \bar{z}; w, \bar{w}) \equiv \langle \Phi(z, \bar{z}) \Phi(w, \bar{w}) \rangle$ in locally conformal coordinates (12.82) satisfies the equations [87]

$$\int d^2 z \sqrt{g}\, G(z, \bar{z}; w, \bar{w}) = 0 \ ,$$

$$\partial_z \partial_{\bar{z}} G(z, \bar{z}; w, \bar{w}) = -2\pi \delta^2(z - w) + \frac{\pi \rho(z, \bar{z})}{\int d^2 z \sqrt{g}} \ ,$$

$$\partial_z \partial_{\bar{w}} G(z, \bar{z}; w, \bar{w}) = 2\pi \delta^2(z - w) - \pi \sum_{i,j} \omega_i(z)(\mathrm{Im}\Omega)^{-1}_{ij} \overline{\omega_j(w)} \ . \tag{12.141}$$

Although $G(z, \bar{z}; w, \bar{w})$ is *not* Weyl-invariant, it actually appears in the (reqularized and non-chiral) correlation functions of the operators $V_\alpha(z, \bar{z}) =: e^{i\alpha\Phi(z,\bar{z})} := \rho^{\alpha^2/2} e^{i\alpha\Phi(z,\bar{z})}$ (*cf* exercise # II-7),

$$\langle V_{\alpha_1}(z, \bar{z}) \cdots V_{\alpha_M}(z_M, \bar{z}_M) \rangle \sim \delta \left(\sum_j \alpha_j \right) \prod_{i<j} F(z_i, \bar{z}_i; z_j, \bar{z}_j)^{\alpha_i \alpha_j} \ , \tag{12.142}$$

only through the other function $F(z, \bar{z}; w, \bar{w})$ given by

$$F(z, \bar{z}; w, \bar{w}) = [\rho(z, \bar{z})\rho(w, \bar{w})]^{-1/2}$$

$$\times \exp \left[-G(z, \bar{z}; w, \bar{w}) + \tfrac{1}{2} G_{\mathrm{reg}}(z, \bar{z}; z, \bar{z}) + \tfrac{1}{2} G_{\mathrm{reg}}(w, \bar{w}; w, \bar{w}) \right] \ . \tag{12.143}$$

It the function $F(z, \bar{z}; w, \bar{w})$ that can actually be written in terms of the prime form as follows [87]:

$$F(z, \bar{z}; w, \bar{w}) = \exp \left[-2\pi \mathrm{Im} \int_w^z \omega (\mathrm{Im}\Omega)^{-1} \mathrm{Im} \int_w^z \omega \right] |E(z, w)|^2 \ . \tag{12.144}$$

Eqs. (12.143) and (12.144) solve eq. (12.141) for the Green function G in terms of the prime form E.

The prime form has the following short-distance expansion [28]

$$E(z, w) = z - w + (z - w)^3 T(z) + O((z - w)^5) \ , \tag{12.145}$$

where T stands for the stress tensor of a free scalar field on Riemann surface.

The Green function for chiral fermionic fields on Riemann surface is given by the so-called *Szegö kernel* having the form

$$S_{\vec{\delta}}(z, w) = \langle b(z) c(w) \rangle = \frac{1}{E(z, w)} \frac{\vartheta[\vec{\delta}](z - w, \Omega)}{\vartheta[\vec{\delta}](0, \Omega)} \ , \tag{12.146a}$$

for an *even* spin structure, and

$$S_{\vec{\delta}}(z,w) = \langle b(z)c(w)\rangle = \frac{1}{E(z,w)} \frac{\sum_i \partial_i \vartheta[\vec{\delta}](z-w,\Omega)\omega_i(y)}{\sum_i \partial_i \vartheta[\vec{\delta}](0,\Omega)\omega_i(y)} , \qquad (12.146b)$$

for an *odd* spin structure, [11] depending on whether $4\delta'\delta''$ is an even or an odd integer, resp. This is related to the transformation property of the Jacobi theta function with characteristics under reflections:

$$\vartheta[\vec{\delta}](-z,\Omega) = (-1)^{4\delta'\delta''}\vartheta[\vec{\delta}](z,\Omega) . \qquad (12.147)$$

The Szegö kernel $S_{\vec{\delta}}(z,w)$ is meromorphic in z and w, analytic in Ω, and satisfies the equation (for even spin structures)

$$\nabla^z S_{\vec{\delta}}(z,w) = 2\pi\delta^2(z,w) . \qquad (12.148)$$

It can be shown that all the above-mentioned results for the Szegö kernels do not depend on the reference point y at all. The explicit expressions for Green's functions are particularly relevant for chiral bosonization on higher-genus ($h \geq 2$) Riemann surfaces (sect. III.4).

XII.5 Basic Facts about BRST and BFV

It is the standard *Dirac theorem* [533, 534] that gives us the way to quantize a constrained classical system described in the phase space of canonical variables (q^i, p_i), $i = 1, 2, \ldots, n$, by a Hamiltonian $H_0(q,p)$ and the constraints $T_\alpha(q,p)$, $\alpha = 1, 2, \ldots, m$, $m < n$, which are of the *first class*. The relevant Poisson brackets take the form [12]

$$\{T_\alpha, T_\beta\}_{\text{PB}} = T_\gamma U^\gamma_{\alpha\beta} , \quad \{H_0, T_\alpha\}_{\text{PB}} = T_\beta V^\beta_\alpha . \qquad (12.149)$$

The Dirac theorem claims that the dynamics of this theory is equivalent to that of the $(n - m)$ independent physical degrees of freedom (q^*, p^*) to be singled out by additional constraints (called *gauges*), $\Phi^\alpha(q,p) = 0$, satisfying the condition

$$\det\{\Phi^\alpha, T_\beta\}_{\text{PB}} \neq 0 . \qquad (12.150)$$

---

[11] Here, y is an arbitrary reference point on Σ where $h_{\vec{\delta}}$ does not vanish.

[12] The indices used in this section have the meaning different from that in the rest of the book.

It follows

$$\int dt \left[p_i \, \dot{q}^i - H_0 \right]\Big|_{T=\Phi=0} = \int dt \left[p^* \, \dot{q}^* - H_{\text{phys}}(q^*, p^*) \right] \equiv S_{\text{phys}}(q^*, p^*) \,, \qquad (12.151)$$

with a change of gauges being equivalent to a *canonical* transformation in the space of the physical variables (q^*, p^*). The quantum generating functional in the physical subspace takes the standard form

$$Z = \int dq^* dp^* \exp[i S_{\text{phys}}(q^*, p^*)] \,. \qquad (12.152)$$

Therefore, Dirac's quantization first introduces the independent canonically conjugated variables arising from a solution to the constraints subject to admissible gauges, then canonically quantizes these physical variables, and ultimately rewrites the result back to the initial phase space. Dirac's quantization is, therefore, obviously consistent with the canonical one, it is unitary but may not be covariant.

Given a covariant 'Lorentz-like' gauge, the problem arises how to rewrite the correct answer (12.152) in terms of the original known quantities in a covariant form. The corresponding prescription is given by the *Becchi-Rouet-Stora-Tyutin* (BRST) quantization procedure [67, 68]. The BRST prescription is nothing but the Dirac quantization method in the covariant form, when a covariant (w.r.t. space-time Lorentz transformations) gauge is chosen. The covariance is maintained in the BRST quantization via extending the initial phase space by ghosts, so that the extended (fields + ghosts) system can be quantized 'naively', in terms of the BRST-invariant Hamiltonian and the constraints, by integrating over the ghosts in the (extended) quantum generating functional.

To be specific, given bosonic (B) and fermionic (F) constraints satisfying the *closed* algebra

$$[B^a, B^b] = f^{ab}{}_c B^c \,, \quad [B^a, F^\beta] = f^{a\beta}{}_\gamma F^\gamma \,, \quad \{F^\alpha, F^\beta\} = f^{\alpha\beta}{}_c B^c \,, \qquad (12.153)$$

where $B^0 \equiv H_0$ is the initial Hamiltonian, and f's are the constraint algebra structure *constants*, the BRST-invariant quantities are defined by [385]

$$B^a = \{\rho^a, Q\} \,, \qquad \mathcal{F}^\alpha = [\xi^\alpha, Q] \,. \qquad (12.154)$$

In eq. (12.154), the canonically conjugated ghosts for each constraint,

$$\text{B}: \ \rho^a, \eta_b \,, \qquad \text{F}: \ \xi^\alpha, \lambda_\beta \,, \qquad (12.155)$$

have been introduced. By definition, they have statistics opposite to that of the constraints, and satisfy the (anti)commutation relations:

$$\{\rho^a, \eta_b\} = \delta_{ab} , \qquad [\xi^\alpha, \lambda_\beta] = \delta_{\alpha\beta} . \tag{12.156}$$

An operator Q introduced in eq. (12.154) is known as a *BRST charge* or a *BRST operator*, and it takes the form

$$Q = B^a \eta_a + F^\alpha \lambda_\alpha - \tfrac{1}{2} f^{ab}{}_c \rho^c \eta_a \eta_b - f^{\alpha\beta}{}_\gamma \xi^\gamma \eta_a \lambda_\beta - \tfrac{1}{2} f^{\alpha\beta}{}_c \rho^c \lambda_\alpha \lambda_\beta . \tag{12.157}$$

Classically, one has $\{Q, Q\}_{\text{PB}} = 0$ w.r.t. the Poisson bracket. Quantum mechanically, a consistent quantization requires that the BRST operator has to be *nilpotent*, $Q^2 = 0$, and *hermitian*, $Q^\dagger = Q$. In the extended space of all states of the BRST quantized theory, the physical states are the ones which are annihilated by the BRST operator *modulo* BRST-trivial states,

$$Q \left|\text{phys}\right\rangle = 0 , \quad \left|\text{phys}\right\rangle \sim \left|\text{phys}'\right\rangle + Q \left|\text{anything}\right\rangle , \tag{12.158}$$

where more gauge conditions are still needed to fix total ghost number (and 'picture', if any). Given the total ghost number (and the picture), the physical states can be viewed as the BRST cohomology classes Ker$Q$/Image$Q$.

If a classical gauge theory to be quantized has constraints which are either *not* of the first-class or *reducible*, the BRST quantization prescription has to be modified. The proper covariant generalization is known as the *Batalin-Fradkin-Vilkovisky* (BFV) quantization prescription [384, 386]. The BFV-prescription introduces more ghosts, beyond those needed in the BRST framework, when dealing with reducilbe constraints. The original derivation of the generalized BFV-BRST rules goes back and forth: first one chooses an irreducible subset of constraints, then one applies the BRST rules, and ultimately rewrites the quantized theory back, in terms of the original variables and constraints by using '*ghosts for ghosts*' [386].

Given an arbitrary gauge-invariant classical action $S_{\text{cl}}[\Phi^i]$ depending on fields Φ^i, let $\delta \Phi^i = R^i_a[\Phi] \varepsilon^a$ be the infinitesimal gauge transformations, which are now assumed to be reducible, $R^i_\alpha Z^\alpha_a = 0$, $Z^\alpha_a \neq 0$. The gauge generators form, in general, an open gauge algebra, or they may become reducible on-shell. This means that they have zero modes (with eigenvectors Z) and, therefore, the naive FP-determinant would vanish since the FP-ghost action would have the residual gauge symmetry. In its turn, the ghost zero modes may have their own zero modes, etc. Further gauge fixing just leads to the 'ghosts for ghosts'. The structure of the 'ghosts for ghosts' generations does *not* naively combine the new generations of 'FP-ghosts for FP-ghosts', but it takes the form of the BFV triangle: the number of ghosts grows as $1 + 2 + 3 + 4 + \ldots$, instead of $1 + 2 + 4 + 8 + \ldots$ [384, 386].

The formal BFV-procedure can be briefly described as follows [290, 446]. One assigns a ghost field C^α to each local symmetry infinitesimal parameter ε^α, with the statistics of C^α being opposite to that of ε^α. The so-called 'minimal' set of fields comprises $\Phi^A_{\min} = (\Phi^i, C^\alpha, \eta^a)$, where the Grassmann-even ghost fields η^a associated with the first-order zero modes have been added. The so-called *anti-fields* $\Phi^*_{A\ \min} = (\Phi^*_i, C^*_\alpha, \eta^*_a)$ are then introduced, in addition to the already defined fields. The (gauge-unfixed) quantum action S is defined to be a solution to the BFV *'master equation'*

$$(S, S) \equiv \frac{\partial_r S}{\partial \Phi^A} \frac{\partial_l S}{\partial \Phi^*_A} - \frac{\partial_r S}{\partial \Phi^*_A} \frac{\partial_l S}{\partial \Phi^A} = 0 \,, \tag{12.159}$$

where ∂_r and ∂_l denote the right and left derivatives, resp. The unique minimal solution can be found as the power series in anti-fields,

$$S_{\min} \equiv S(\Phi_{\min}, \Phi^*_{\min}) = S_{cl} + \Phi^*_i R^i_\alpha C^\alpha + C^*_\alpha (Z^\alpha_a \eta^a + T^\alpha_{\beta\gamma} C^\gamma C^\beta) + \ldots \,, \tag{12.160}$$

whose coefficients are to be determined from the master equation (12.159). The *anti-ghosts*, $\bar{C}_\alpha$ and $\bar{\eta}_a$, as well as the Lagrange multipliers, Π_α and π_a, are now added to the minimal set of fields, and the minimal solution is completed as follows:

$$S = S_{\min} + \bar{C}^{\alpha*} \Pi_\alpha + \bar{\eta}^{a*} \pi_a \,, \tag{12.161}$$

where the anti-fields for the anti-ghosts, $\bar{C}^{\alpha*}$ and $\bar{\eta}^{a*}$, have also been introduced. The BFV gauge-fixing is performed by choosing a *'gauge fermion'*

$$\Psi = \bar{C}_\alpha F^\alpha[\Phi] + \bar{\eta}_a \omega^a_\alpha C^\alpha \,, \tag{12.162}$$

which specifies gauge constaints $F^\alpha[\Phi] = \omega^a_\alpha C^\alpha = 0$. The total gauge-fixed quantum action is then given by

$$S_q = S\left(\Phi, \Phi^* = \frac{\partial \Psi}{\partial \Phi}\right) \,. \tag{12.163}$$

The key property of this quantum action is its invariance w.r.t. the BRST transformations

$$\delta \Phi^A = \epsilon \left. \frac{\partial_r S}{\partial \Phi^*_A} \right|_{\Phi^* = \partial \Psi / \partial \Phi} \,, \tag{12.164}$$

where the constant BRST parameter ϵ is Grassmann-odd.

To this end, we are going to illustrate the *naive* BRST approach by quantizing a pure Yang-Mills theory in $1 + 3$ dimensions. This theory is characterized by the action [13]

$$S_{\text{cl}} = -\frac{1}{4g^2} \int d^4x \, F^a_{\mu\nu} F^{a\mu\nu} = \frac{1}{8g^2} \int d^4x \, \text{tr} \left(F_{\mu\nu} F^{\mu\nu} \right) , \qquad (12.165)$$

where the matrix-valued gauge field $A_\mu(x) = g A^a_\mu(x) t_a$, the covariant derivative $D_\mu = \partial_\mu + [A_\mu, \ ]$, and the gauge field strength $F_{\mu\nu} = g F^a_{\mu\nu} t_a = [D_\mu, D_\nu]$ have been introduced. A quantization of this theory is usually performed by factorizing the gauge group volume in the naive generating functional $\int \exp[iS_{\text{cl}}]$ and then disregarding this volume, in order to avoid overcounting the gauge field configurations. Then, the gauge field configurations which are related by gauge transformations are not accounted but one representative. This results in the total gauge-fixed quantum action S_{tot}, which is then used as the input instead of S_{cl} in the naive expression $\int \exp[iS]$. The same result follows from the canonical quantization [474]. The total action reads

$$S_{\text{tot}} = S_{\text{cl}} + S_{\text{g.-f.}} + S_{\text{gh}} , \qquad (12.166)$$

and it includes the *gauge-fixing* and *ghost* terms, in addition to the classical action [474]. It is usually the point when one notices that the total action (12.166) (in an explicitly chosen gauge) has a special rigid Grassmannian symmetry, known as the *BRST symmetry* now [67, 68]. The BRST symmetry plays the fundamental role in the quantum theory of gauge fields since it implies the quantum gauge Ward identities and ensures unitarity of the theory [474]. It is also possible to reverse this argument by requiring the BRST symmetry of S_{tot} from the outset. As far as the Yang-Mills theory is concerned, it results in the same total action, but the idea is much more general. We are now going to reformulate the standard Yang-Mills theory quantization in the BRST terms, in order to make it applicable for *any* gauge theory with a closed gauge algebra and independent gauge generators.

A Yang-Mills gauge condition $G(A_\mu) = 0$ can be implemented by adding to S_{cl} a gauge-fixing term in the form

$$S_{\text{g.-f.}} = - \int d^4x \, \text{tr} \left(BG \left[A \right] \right) , \qquad (12.167)$$

---

[13]Let t_a be (anti-hermitian) generators of a gauge group. They satisfy the relations: $[t_a, t_b] = f_{ab}{}^c t_c$, $\text{tr}(t_a t_b) = -2\delta_{ab}$.

where the Lagrange multiplier field B has been introduced. To construct the ghost action and ensure the BRST symmetry of the total construction, the BRST procedure goes as follows. First, the local infinitesimal parameter $\omega = \omega^a t_a$ in the gauge transformation law $\delta A_\mu^a = (D_\mu \omega)^a$ has to be replaced by $\Lambda c^a t_a$, where Λ is a Grassmannian-odd rigid infinitesimal BRST parameter and c^a are ghost fields. Second, anti-ghost fields $b = b^a t_a$ have to be introduced. The nilpotent BRST transformation laws are defined in the form

$$\delta_{\text{BRST}} A_\mu = \Lambda D_\mu c = \Lambda \left(\partial_\mu c + [A_\mu, c] \right) \ ,$$

$$\delta_{\text{BRST}} c = -\tfrac{\Lambda}{2} \{c, c\} \ , \quad \text{or} \quad \delta_{\text{BRST}} c^c = -\tfrac{\Lambda}{2} f_{ab}{}^c c^a c^b \ ,$$

$$\delta_{\text{BRST}} b = \Lambda B \ , \quad \delta_{\text{BRST}} B = 0 \ . \tag{12.168}$$

The BRST prescription for the quantum gauge-fixed action S_{tot} gives

$$\Lambda \left(S_{\text{g.-f.}} + S_{\text{gh}} \right) = - \int d^4 x \, \delta_{\text{BRST}} \text{tr} \left(b G \left[A \right] \right) \ , \tag{12.169}$$

and its BRST invariance immediately follows from the nilpotency property of the BRST transformation, $\delta_{\text{BRST}}^2 = 0$. The classical action is obviously BRST-invariant, since the BRST transformation of the gauge field takes the form of the gauge transformation.

In particular, when $G = \partial^\mu A_\mu$, one easily finds

$$S_{\text{gh}} = - \int d^4 x \, \text{tr} \left(b \frac{\delta G}{\delta A_\mu} D^\mu c \right) = - \int d^4 x \, \text{tr} \left(b \partial_\mu D^\mu c \right) \ , \tag{12.170}$$

as it should.

The general BRST rules can be easily extracted from this very simple pattern, without any reference to a specific gauge theory. Their justification comes from the fact that they allow us to quantize a gauge field theory in the unitary and gauge-independent way.

XII.6   CFT and quantum groups

The origin of '*quantum groups*' in CFT is related with the structure of chiral (or conformal) blocks (sect. II.3) satisfying the conformal bootstrap or duality equation (2.43) and having non-trivial monodromies [74, 75, 76, 535, 536, 537, 538]. Being multi-valued functions, the chiral blocks $\mathcal{F}_{\vec{p}}^{i_1 \cdots i_M} (z_1, \ldots, z_M)$ are supposed to be defined first for a given order of their arguments $\{z_i\}$ and then analytically continued to different domains. Since the analytic continuation depends on the path chosen to make it, it is the *braid* group

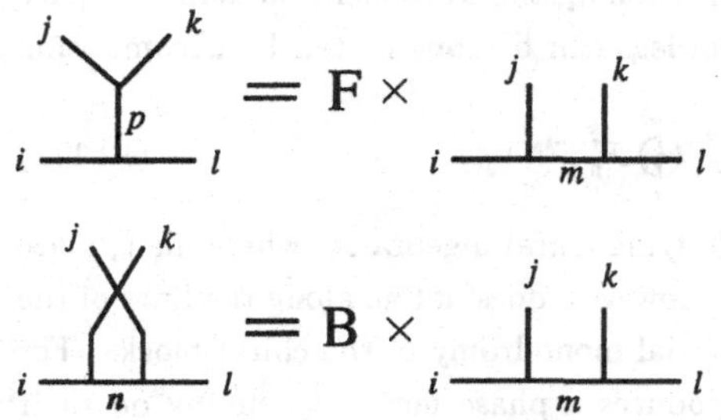

Fig. 25. Fusion (F) and braiding (B)
of a chiral block in CFT.

(not the permutation group!) that actually acts on these blocks. There are two basic moves in the space of chiral blocks, interchanging their external legs and called *braiding* (B) and *fusion* (F). They are both pictured in Fig. 25.

Chiral blocks themselves form a 'representation' of the braid group, in the sense to be described below. The bootstrap (or duality) equations can be reformulated as a set of the *polynomial* constraints in terms of the corresponding braiding and fusion matrices [74, 75, 535]. For a RCFT, this gives the finite set of conditions (known as the *Seiberg-Moore equations* for chiral blocks), which, together with the modular invariance condition for the partition function, represent all the relevant equations to determine the RCFT chiral blocks. [14] Some explicit examples are given in sect. II.8. In this context, the Verlinde formula (3.18), relating fusion rules and modular transformation properties in the non-trivial way, plays the important role since it puts the strong constraint on the possible fusion rules in CFT. In the case of RCFT's, it opens the way for their complete classification by solving the Seiberg-Moore equations. Representations of 'quantum groups' just arise on the way of finding their solutions. In particular, RCFT fusion rules turn out to be dictated by Clebsch-Gordan coefficients for 'quantum groups', when the 'quantum' deformation parameter is a root of unity. The representation theory of Lie and AKM symmetry algebras alone cannot explain the OPE coefficients of CFT primary fields. However, given the additional structure due to a Hopf algebra or a 'quantum group' in CFT, one can do it, in principle [77].

---

[14]This amounts to finding multi-valued functions on a Riemann surface with known transformation properties under analytic continuation and given singular points — the so-called *Riemann monodromy problem*, which is very difficult in general [77].

For a RCFT with primary fields ϕ_i of dimension h_i, the convenient basis in the space of chiral blocks [15] having ϕ_i on their external legs can be constructed by decomposing the tensor product

$$V_{i_1} \otimes \cdots \otimes V_{i_M} = \bigoplus_j \mathcal{F}^{i_1 \cdots i_M}_{\vec{p}, j} V_j \tag{12.171}$$

according to the fusion rules w.r.t. the underlying chiral algebra $\mathcal{A}$, where all V_i's are highest-weight modules. This decomposition, however, does not go along the lines of the conventional group theory, because of non-trivial monodromy of the chiral blocks. For instance, an exchange (braiding) of fields produces a phase factor depending on their conformal dimensions. This happens since the braid group 'representation' described by braiding matrices B acts on the chiral block arguments which are subject to analytic continuation and, therefore, are *dependent* on a local coordinate system. This is mathematically described by Dehn twists, i.e. local rotations around the origin of the local coordinate system in question, which are generated by $L_0 - \bar{L}_0$, and, hence, produce a phase in the finite expression [77]. If σ_{ij} is the operator of braiding of i and j, its effect on a 3-point block F^{ij}_k is given by

$$\sigma_{ij}(F^{ij}_k) = \epsilon^k_{ij} e^{\iota \pi (h_k - h_i - h_j)} F^{ji}_k , \tag{12.172}$$

where ϵ^k_{ij} represents a 'classical' sign contribution depending on whether the primary fields ϕ_i and ϕ_j couple symmetrically or anti-symmetrically to the field ϕ_k. The two basic motions in the space of chiral blocks are related by duality as [75, 535]

$$B_{pp'} \begin{pmatrix} j & k \\ i & l \end{pmatrix} = \epsilon^p_{kl} \epsilon^i_{p'k} e^{\iota \pi (h_p + h_{p'} - h_i - h_l)} F_{pp'} \begin{pmatrix} j & l \\ i & k \end{pmatrix} , \tag{12.173}$$

Duality transformations correspond to essentially non-local operations in terms of the chiral blocks, but they are the same for both primary and secondary fields (sect. II.3). It is, therefore, enough to consider the duality action w.r.t. the zero-mode subalgebra A of the chiral algebra $\mathcal{A}$. Let $\tilde{A}$ be the extension of A by duality operations. We are not going to say anything about an explicit construction of the algebra $\tilde{A}$ in terms of the original chiral algebra $\mathcal{A}$, which may be difficult. [16] The spaces V_i can now be considered as the representations spaces of $\tilde{A}$. If the algebra $\tilde{A}$ acts on V_k via matrices $\rho^k(a)$, $a \in \tilde{A}$, and $K^{ij}_k : V_i \otimes V_j \to V_k$ is an intertwinner (a 3-point conformal block)

---

[15] In the RCFT represented by a WZNW model, the space of chiral blocks can be identified with the physical space of the three-dimensional Chern-Simons theory [539].

[16] See however ref. [77], as far as the attempts to relate generators of $\mathcal{A}$ and $\tilde{A}$ are concerned.

defining the composition rule, we can use eq. (12.171) to define the action of $\tilde{A}$ on the tensor product $V_i \otimes V_j$ via the diagram

$$
\begin{array}{ccc}
V_i \otimes V_j & \xrightarrow{K_k^{ij}} & V_k \\
\downarrow{\scriptstyle \rho_i \otimes \rho_j(\Delta(a))} & & \downarrow{\scriptstyle \rho^k(a)} \\
V_i \otimes V_j & \xrightarrow{K_k^{ij}} & V_k
\end{array}
\qquad (12.174)
$$

This looks formally the same as the introduction of a *co-multiplication* Δ in the algebra $\tilde{A}$, namely, $\Delta : \tilde{A} \to \tilde{A} \otimes \tilde{A}$ and $\rho^i \otimes \rho^j(\Delta(a)) \equiv \Delta^{ij}(a)$. When $\tilde{A}$ and $\otimes$ are associative, the co-multiplication is also associative, the same being true for the monodromy. Having defined Δ, the action of $\tilde{A}$ can be extended to arbitrary tensor products by iterating the co-multiplication.

One can identify $\tilde{A}$ by looking for the *centralizer* (sect. 2) of the braid group action on chiral blocks [77]. Indeed, the space of chiral blocks can always be decomposed into the irreps w.r.t. the braid group, while knowing the centralizer gives the full information about the irreps appearing in the tensor product (12.171) since, due to the same equation, the multiplicity of an irrep V_j in the product is given by the number of linearly independent blocks $\mathcal{F}_{\vec{p},j}^{i_1 \cdots i_M}$. On general grounds, the centralizer is expected to be semisimple and to provide a representation of the braid group. Going backwards, this implies the existence of the matrix R acting on contiguous spaces in the tensor product and commuting with the co-multiplication. In addition, one can construct a second co-multiplication Δ' which is related with the first one by the relation

$$
\Delta'(a) = \mathcal{R}\Delta(a)\mathcal{R}^{-1} \; . \qquad (12.175)
$$

Some extra requirements come from compatibility of braiding and fusion in RCFT and eq. (12.173). Putting all together gives rise to the set of restrictions which actually coincide with the 'quantum group' axioms to be formulated below in terms of the matrix $\mathcal{R}$! This implies that the algebra $\tilde{A}$ encoding the duality properties of the RCFT in question should be a 'quantum' algebra.

We conclude that the Moore-Seiberg equations contain the topological information about the chiral blocks in RCFT. All the information they contain can be translated into the definting properties of 'quantum groups'. More specifically, the 'quantum group' arises in a RCFT as the centralizer of the braid group action on the RCFT chiral blocks. The 'quantum group' representation theory determines braiding and fusion matrices B and F as solutions to the Moore-Seiberg equations. For the RCFT based on a WZNW model, this gives the way to bypass solving the Knizhnik-Zamolodchikov equation (4.47), in order to get the same infromation [540]. The explicit form of B and F matrices in a particular RCFT is given below.

The 'quantum groups' can be formally introduced as follows [541, 542, 543, 544, 545, 546, 547]. Given an associative algebra $\tilde{A}$ (with unity) over complex numbers $\mathbf{C}$, it is called a *Hopf algebra* when the additional structure comprising three operations on $\tilde{A}$ is introduced [548]: (i) the *co-multiplication* $\Delta : \tilde{A} \to \tilde{A} \otimes \tilde{A}$, (ii) the *antipodal map* $\gamma : \tilde{A} \to \tilde{A}$, and (iii) the *co-unit* $\varepsilon : \tilde{A} \to \mathbf{C}$, which are subject to the following axioms (for any $a, b \in \tilde{A}$):

(associativity of Δ) $(id \otimes \Delta)\Delta(a) = (\Delta \otimes id)\Delta(a)$,

(definition of γ) $m(id \otimes \gamma)\Delta(a) = m(\gamma \otimes id)\Delta(a) = \varepsilon(a)1$,   (12.176)

(definition of ε) $(\varepsilon \otimes id)\Delta(a) = (id \otimes \varepsilon)\Delta(a) = a$,

where m is the multiplication in $\tilde{A}$, $m : \tilde{A} \otimes \tilde{A} \to \tilde{A}$, $m(a \otimes b) = a \cdot b$. The operations Δ and ε are homomorphisms, while γ is an anti-homomorphism of $\tilde{A}$,

$$\Delta(a \cdot b) = \Delta(a)\Delta(b) , \quad \varepsilon(a \cdot b) = \varepsilon(a)\varepsilon(b) , \quad \gamma(a \cdot b) = \gamma(b)\gamma(a) .$$ (12.177)

Given this structure, another co-multiplication $\Delta' = \sigma \circ \Delta$ can be introduced with the antipode $\gamma' = \gamma^{-1}$, where $\sigma : \tilde{A} \otimes \tilde{A} \to \tilde{A}$ is the permutation map, $\sigma(a \otimes b) = b \otimes a$. If the two co-multiplications are related by conjugation,

$$\Delta'(a) = \sigma \circ \Delta(a) = \mathcal{R}\Delta(a)\mathcal{R}^{-1} , \quad \mathcal{R} \in \tilde{A} \otimes \tilde{A} ,$$ (12.178)

then a Hopf algebra $\tilde{A}$ is called a (quasi-triangular) *Yang-Baxter algebra*, provided the additional conditions are satisfied:

$$(id \otimes \Delta)(\mathcal{R}) = \mathcal{R}_{13}\mathcal{R}_{12} , \quad (\Delta \otimes id)(\mathcal{R}) = \mathcal{R}_{13}\mathcal{R}_{23} , \quad (\gamma \otimes id)(\mathcal{R}) = \mathcal{R}^{-1} .$$ (12.179)

The axioms above originated from the theory of exactly solvable models [541, 549]. From the viewpoint of CFT, these axioms exactly correspond to the definition of the centralizer and the compatibility between fusion and braiding, which were discussed above. The matrix $\mathcal{R}$ is known as the *universal R-matrix* of the Yang-Baxter algebra $\tilde{A}$. It satisfies the *Yang-Baxter equation* [541, 549, 550]

$$\mathcal{R}_{12}\mathcal{R}_{13}\mathcal{R}_{23} = \mathcal{R}_{23}\mathcal{R}_{13}\mathcal{R}_{12} ,$$ (12.180)

which follows from the axioms. A solution to the Yang-Baxter equation gives a representation of the braid group. The term 'quantum group' [542] is actually the substitute for a (quasi-triangular) Yang-Baxter algebra. The meaning of word 'quantum' in this context has nothing to do with the meaning of words 'classical' and 'quantum' in physics, but the fact that the 'classical' Lie algebra structure is reproduced from the 'quantum group' one in the 'classical' limit $q \to 1$. Defining $q \equiv e^{\hbar}$ makes it even more similar. The formal parameter $\hbar$ here should not, however, be confused with the Planck constant.

It is possible to associate 'quantum groups' with certain deformations of the universal enveloping algebra [17] for a classical Lie algebra. Let us consider the Lie algebra $sl(2)$ with generators $(H, X^{\pm})$ as the simplest example, and define the *quantum* universal enveloping algebra as the ('quantum') q-deformation of its enveloping algebra, i.e. as the algebra of power series in the generators $(H, X^{\pm})$ *modulo* the relations

$$[X^+, X^-] = \lfloor H \rfloor \,, \quad [H, X^{\pm}] = \pm 2X^{\pm} \,, \quad [X^{\pm}, X^{\pm}] = [H, H] = 0 \,. \tag{12.181}$$

The q-numbers have already been introduced in eq. (12.24) as

$$\lfloor X \rfloor \equiv \lfloor X \rfloor_q = \frac{q^{X/2} - q^{-X/2}}{q^{1/2} - q^{-1/2}} \,. \tag{12.182}$$

The other useful definitions are

$$\begin{pmatrix} n \\ m \end{pmatrix} = \frac{\lfloor n \rfloor!}{\lfloor m \rfloor! \lfloor n - m \rfloor!} \,, \quad \lfloor n \rfloor! = \lfloor n \rfloor \lfloor n-1 \rfloor \cdots \lfloor 1 \rfloor \,. \tag{12.183}$$

The Hopf algebra operations $\Delta, \gamma, \varepsilon$ in the $sl(2)$ case are defined by

$$\Delta(H) = H \otimes 1 + 1 \otimes H \,, \quad \Delta(X^{\pm}) = X^{\pm} \otimes q^{H/4} + q^{-H/4} \otimes X^{\pm} \,,$$

$$\gamma(X^{\pm}) = -q^{\pm 1/2} X^{\pm} \,, \quad \gamma(H) = -H \,,$$

$$\varepsilon(X^{\pm}) = \varepsilon(H) = 0 \,, \quad \varepsilon(1) = 1 \,. \tag{12.184}$$

It is now straightforward to verify that they actually fulfill the axioms of a quasi-triangular Hopf algebra. Most remarkably, there exists an explicit solution to the Yang-Baxter equation in this case, having the form

$$\mathcal{R} = q^{H \otimes H/4} \sum_{n \geq 0} \frac{(1 - q^{-1})^n}{\lfloor n \rfloor!} q^{-n(n-1)/4} q^{nH/4} (X^+)^n \otimes q^{-nH/4} (X^-)^n \,, \tag{12.185}$$

---

[17]The *universal enveloping algebra* of a Lie algebra is the algebra of formal power series in Lie algebra generators.

and known as the $U_q(sl(2))$ universal R matrix. There exists the efficient general method of *deriving* the R matrix, known as the *'quantum double'* construction [542].

The simple realization of the $U_q(sl(2))$ 'quantum group' is provided by the deformation of the classical *Borel-Weyl* construction, which we are now going to describe. The fundamental $sl(2)$ irrep has two states, which can be put to a correspondence with two variables u, v. An arbitrary $sl(2)$ irrep of spin J can then be represented by a homogeneous polynomial of degree $2J$, with $sl(2)$ generators having the form

$$X^+ = u\frac{\partial}{\partial v} \ , \quad X^- = v\frac{\partial}{\partial u} \ , \quad H = u\frac{\partial}{\partial u} - v\frac{\partial}{\partial v} \ . \tag{12.186}$$

The spin-J representation basis reads

$$|J, M\rangle = \frac{u^{J+M}v^{J-M}}{\sqrt{(J+M)!(J-M)!}} \ , \quad -J \le M \le J \ . \tag{12.187}$$

The variables u, v can be thought of as the $\mathbf{CP}^1$ coordinates parametrizing the Riemann sphere. Given this interpretation, $2J + 1$ vectors (12.187) are in one-to-one correspondence with holomorphic sections in the space of $(-J)$-differentials on the sphere, according to the Riemann-Roch theorem (sect. 4).

The 'quantum' deformation of the classical Borel-Weyl construction can be described in terms of the q-derivatives defined by

$$D_u f(u) = \frac{f(q^{1/2}u) - f(q^{-1/2}u)}{(q^{1/2} - q^{-1/2})u} \ , \tag{12.188}$$

and similarly for v. It is straightforward to check that the operators

$$X^+ = uD_v \ , \quad X^- = vD_u \ , \quad H = u\frac{\partial}{\partial u} - v\frac{\partial}{\partial v} \ , \tag{12.189}$$

satisfy the defining relations (12.181) of the $U_q(sl(2))$ 'quantum group'. The q-deformed spin-J 'quantum group' representation basis is

$$|J, M\rangle = \frac{u^{J+M}v^{J-M}}{\sqrt{\lfloor J+M \rfloor! \lfloor J-M \rfloor!}} \ , \tag{12.190}$$

where an action of the 'quantum group' generators takes the form

$$(X^\pm)^m|J, M\rangle = \sqrt{\frac{\lfloor J \mp M \rfloor! \lfloor J \pm M + m \rfloor!}{\lfloor J \mp M - m \rfloor! \lfloor J \pm M \rfloor!}}|J, M \pm m\rangle \ . \tag{12.191}$$

Another standard procedure to obtain the 'quantum group' structures is to quantize the so-called *Poisson-Hopf algebras* associated with the classical Hamiltonian systems, where the Poisson algebra has the structure of a Hopf algebra.

As far as the representation theory of 'guantum groups' is concerned, it goes in parallel with the classical (Lie) case when q is *not* a root of unity, by replacing ordinary numbers by q-numbers at appropriate places. The co-multiplication is analogous to the rule for addition of angular momenta. However, when q *is* a root of unity, the 'quantum group' and the centralizer algebra become not semisimple. For instance, when $q = e^{2\pi i/p}$, $p \in \mathbf{Z}$, one gets $(X^\pm)^p = 0$ which generates null vectors in some representations. The 'quantum group' representations introduced in sect. IX.2 for the quantum Liouville theory with $q = \exp[2i\hat{h}] = \exp\left[\frac{2\pi i}{\kappa+2}\right]$ do not share the last property since they are labeled by the parameter $\hat{h}$ or, equally, the $SL(2, \mathbf{R})$ 'level' κ which are not integer, as is easily seen from eqs. (9.40) and (9.105).

In the tractable case of the $\widehat{SU}(2)_k$ WZNW theory, where a 'quantum group' structure appears in the prosess of solving the corresponding Moore-Seiberg equations, all the 'quantum group' quantities introduced above can be explicitly calculated [76, 537]. In particular, the braiding and fusion matrices take the form

$$F_{jj'}\begin{pmatrix} j_2 & j_3 \\ j_1 & j_4 \end{pmatrix} = \left\{ \begin{matrix} j_1 & j_2 & j \\ j_3 & j_4 & j' \end{matrix} \right\}, \tag{12.192a}$$

$$B_{jj'}\begin{pmatrix} j_2 & j_3 \\ j_1 & j_4 \end{pmatrix} = (-)^{j+j'-j_1-j_4} q^{(c_{j_1}+c_{j_4}-c_j-c_{j'})/2} \left\{ \begin{matrix} j_2 & j_1 & j \\ j_3 & j_4 & j' \end{matrix} \right\}, \tag{12.192b}$$

where the 'quantum' (Clebsch-Gordan) $6j$-symbols have been introduced as follows:

$$\left\{ \begin{matrix} j_1 & j_2 & j_{12} \\ j_3 & j & j_{23} \end{matrix} \right\} = \Delta(j_1, j_2, j_{12})\Delta(j_3, j, j_{12})\Delta(j_1, j, j_{23}) \sum_{k \geq 0} (-)^k \lfloor k+1 \rfloor!$$

$$\times \frac{1}{\lfloor k-j_1-j_2-j_{12} \rfloor! \lfloor k-j_3-j-j_{12} \rfloor! \lfloor k-j_1-j-j_{23} \rfloor! \lfloor k-j_3-j_2-j_{23} \rfloor!}$$

$$\frac{1}{\lfloor j_1+j_2+j_3+j-k \rfloor! \lfloor j_1+j_3+j_{12}+j_{23}-k \rfloor! \lfloor j_2+j+j_{12}+j_{23}-k \rfloor!},$$

$$\Delta(a, b, c) \equiv \sqrt{\frac{\lfloor -a+b+c \rfloor! \lfloor a-b+c \rfloor! \lfloor a+b-c \rfloor!}{\lfloor a+b+c+1 \rfloor!}}. \tag{12.193}$$

The modular transformation matrices S and T of the $\widehat{SU}(2)_k$ WZNW theory can also be computed this way [76, 537]. In the DF representation for the correlation functions

(Ch. II), the RCFT chiral blocks appear as the linear combinations of tensor products of the screened vertex operators, with the 'quantum' $6j$-symbols as the coefficients. A fundamental relation, if any, between the 'quantum groups' and the AKM algebras is yet to be found.

Bibliography

[1] C. Itzykson and J.-B. Zuber, *Quantum Field Theory*, McGraw-Hill, N.-Y., 1980.

[2] P. Ramond, *Field Theory: a Modern Primer*, Benjamin-Cummings, Reading, MA, 1981.

[3] N. N. Bogolyubov and D. V. Shirkov, *Quantum Fields*, Benjamin-Cummings, Reading, MA, 1983.

[4] L. H. Ryder, *Quantum Field Theory*, Cambridge Univ. Press, Cambridge, 1985.

[5] A. Barut and R. Raczka, *Theory of Group Representations and Applications*, World Scientific, Singapore, 1986.

[6] R. J. Baxter, *Exactly Solved Models in Statistical Mechanics*, Academic Press, 1982.

[7] J. L. Cardy, *Conformal Invariance and Statistical Mechanics*, in "Les Houches Summer School Proceedings", 1988, p. 169.

[8] G. Parisi, *Statistical Field Theory*, Addison-Wesley, 1988.

[9] C. Itzykson, H. Saleur and J.-B. Zuber, eds., *Conformal Invariance and Applications to Statistical Mechanics*, World Scientific, Singapore, 1988.

[10] M. Martellini and M. Rasetti, eds., *Topological and Quantum Group Methods in Field Theory and Condensed Matter Physics*, Int. J. Mod. Phys. **B6** (1992) #'s 11 and 12.

[11] A. M. Polyakov, *Conformal symmetry of critical fluctuations*, JETP Lett. **12** (1970) 381.

[12] A. Belavin, A. Polyakov and A. Zamolodchikov, *Infinite conformal symmetry in two-dimensional quantum field theory*, Nucl. Phys. **B241** (1984) 333.

[13] V. L. Dotsenko, *Lectures on Conformal Field Theory*, in "Conformal Field Theory and Solvable Lattice Models", edited by M. Jimbo et al., World Scientific, Singapore, 1986, p. 123.

[14] A. B. Zamolodchikov, *Exact solutions of conformal field theory in two dimensions and critical phenomena*, Kiev preprint IMP-87-65P, 1987.

[15] M. Peskin, *Introduction to string and superstring theory*, preprint SLAC-PUB-4251, in "From the Plank Scale to the Weak Scale", edited by H. Haber, World Scientific, Singapore, 1987.

[16] T. Banks, *Lectures on conformal field theory*, Stanford preprint SLAC-PUB-4251, 1986; in "The Santa Fe TASI'87", edited by R. Slansky and G. West, World Scientific, Singapore, 1988.

[17] P. Ginsparg, *Applied conformal field theory*, Harvard preprint HUTP-AO54, 1988; in "Les Houches, 1988", edited by E. Brézin and J. Zinn-Justin, Elsevier Publishers, Amsterdam, 1989.

[18] P. Furlan, G. M. Sotkov and I. T. Todorov, *Two-dimensional conformal field theories*, Boll. Soc. Ital. Fis., Nuova Ser. **12** (1989) # 6, 1.

[19] A. Kato, *Introduction to conformal field theories*, KEK Rep. **88** (1989) 1.

[20] P. Goddard, *Conformal symmetry and its extensions*, in "Proceedings of the 9-*th* International Congress on Mathematical Physics", Bristol, 1989.

[21] J. Bagger, *Basic conformal field theory*, Harvard preprint HUTP-A006, 1989; in "Particles and Fields", Proc. Banff Summer Inst., 1988, edited by A. N. Kamal and F. C. Khanna, World Scienfific, Singapore, 1989, p. 556.

[22] P. Dirac, *Forms of relativistic dynamics*, Rev. Mod. Phys. **21** (1949) 392.

[23] S. Fubini, A. Hansen and R. Jackiw, *New approach to field theory*, Phys. Rev. **D7** (1973) 1732.

[24] C. Lovelace, *Adiabatic dilations and essential Regge spectrum*, Nucl. Phys. **B99** (1975) 109.

[25] K. Wilson, *Broken scale invariance and anomalous dimensions*, in "Proceedings of Midwest Conference on Theoretical Physics", Notre Dame, 1970, p. 131.

[26] W. Zimmermann, *Lectures on Field Theory and Elementary Particles*, MIT Press, Cambridge, 1970.

[27] J. C. Collins, *Renormalization: An Introduction to Renormalization Group, and the Operator-Product Expansion*, Cambridge Univ. Press, Cambridge, 1984.

[28] E. D'Hoker and D. Phong, *The geometry of string perturbation theory*, Rev. Mod. Phys. **60** (1988) 917.

[29] M. Virasoro, *Subsidiary conditions and ghosts in dual-resonance models*, Phys. Rev. **D1** (1970) 2933.

[30] V. G. Kač, *Contravariant form for infinite-dimensional Lie algebras and superalgebras*, Lect. Notes in Phys. **94** (1979) 441.

[31] B. L. Feigin and D. B. Fuchs, *Invariant skew-symmetric differential operators on the Line and Verma modules over the Virasoro algebra*, Funct. Anal. Appl. **16** (1982) 114.

[32] P. West, *An introduction to string theory*, CERN Preprint TH-5165, 1988.

[33] Al. B. Zamolodchikov, *Conformal scalar field on the hyperelliptic curve and critical Ashkin-Teller multipoint correlation functions*, Nucl. Phys. **B285** (1987) 481.

[34] S. A. Apikyan and A. B. Zamolodchikov, *Conformal blocks related to conformally invariant Ramond states of a free scalar field*, Pis'ma ZhETF **65** (1987) 34; [JETP Lett.65 (1987) 19].

[35] D. Friedan, E. Martinec and S. Shenker, *Conformal invariance, supersymmetry and string theory*, Nucl. Phys. **B271** (1986) 93.

[36] L. Dixon, D. Friedan, E. Martinec and S. Shenker, *The conformal field theory of orbifolds*, Nucl. Phys. **B282** (1987) 13.

[37] A. Polyakov, *Quantum geometry of bosonic strings*, Phys. Lett. **103B** (1981) 207.

[38] D. Friedan, *Introduction to Polyakov's string theory*, in "Recent Advances in Field Theory and Statistical Mechanics", Elsevier North-Holland, Amsterdam, 1984, p. 839.

[39] N. D. Birrell and P. C. W. Davies, *Quantum Fields in Curved Space*, Cambridge University Press, Cambridge, 1982.

[40] T. Eguchi, P. Gilkey and A. Hanson, *Gravitation, Gauge Theories and Differential Geometry*, Phys. Rep. **66** (1980) 213.

[41] J. Gomez, *The triviality of representations of the Virasoro algebra with vanishing central element and L_0 positive*, Phys. Lett. **171B** (1986) 75.

[42] J. Serre, *A Course in Arithmetic*, Springer-Verlag, Berlin, 1973.

[43] P. D. B. Collins, *An Introduction to Regge Theory and High Energy Physics*, Cambridge Univ. Press, Cambridge, 1977.

[44] V. Kač, *Infinite Dimensional Lie Algebras*, Cambridge University Press, Cambridge, 1985.

[45] B. Feigin and D. Fuchs, *Verma modules over the Virasoro algebra*, in "Topology: Proceedings of Leningrad Conference 1982", edited by L. Faddeev and A. Mal'tsev, Springer-Verlag, N.-Y., 1985.

[46] C. Thorn, *Computing the Kac determinant using dual model techniques and more about the no-ghost theorem*, Nucl. Phys. **B248** (1984) 551.

[47] M. B. Halpern, *Recent progress in irrational conformal field theory*, Talk presented at the Conference "Strings 1993", Berkeley, USA, May 23–29, 1993.

[48] D. Friedan, Z. Qui and S. Shenker, *Conformal invariance, unitarity, and critical exponents in two dimensions*, Phys. Rev. Lett. **52** (1984) 1575.

[49] D. Friedan, Z. Qui and S. Shenker, *Conformal invariance, unitarity, and two-dimensional critical exponents*, in "Vertex Operators in Mathematics and Physics", Springer-Verlag, N.-Y., 1985, p. 419.

[50] D. Friedan, Z. Qui and S. Shenker, *Details of the non-unitarity proof for highest weight representations of the Virasoro algebra*, Commun. Math. Phys. **107** (1986) 535.

[51] E. Verlinde, *Fusion rules and modular transformations in 2-D conformal field theory*, Nucl. Phys. **B300** (1988) 360.

[52] R. Dijkgraaf and E. Verlinde, *Modular invariance and the fusion algebra*, Utrecht preprint ThU-88-25, 1988; Nucl. Phys. [Proc. Suppl.] **5B** (1988) 87.

[53] C. Vafa, *Towards classification of conformal fields*, Phys. Lett. **206B** (1988) 421.

[54] V. Dotsenko and V. Fateev, *Conformal algebra and multipoint correlation functions in 2d statictical models*, Nucl. Phys. **B240 FS** (1984) 312.

[55] V. Dotsenko and V. Fateev, *Four-point correlation functions and the operator algebra in 2d conformal invariant theories with central charge c < 1*, Nucl. Phys. **B251 FS** (1985) 691.

[56] M. Kato and S. Matsuda, *Construction of singular vertex operators as degenerate primary conformal fields*, Phys. Lett. **172B** (1986) 216.

[57] V. Fateev and A. Zamolodchikov, *Conformal quantum field theory models in two dimensions having Z_3 symmetry*, Nucl. Phys. **B280 FS** (1987) 644.

[58] V. S. Dotsenko, *Critical behaviour and associated conformal algebra of the Z_3 Potts model*, J. Stat. Phys. **34** (1984) 781.

[59] D. Friedan, Z.Qiu and S.Shenker, *Superconformal invariance in two dimensions and the tricritical Ising model*, Phys. Lett. **151** (1985) 37.

[60] R. J. Baxter, *Hard hexagons: exact solution*, J. Phys. A: Math. Gen. **13** (1980) L61.

[61] R. B. Potts, *Some generalized order-disorder transformations*, Proc. Cambridge Phil. Soc. **48** (1952) 106.

[62] R. J. Baxter, *Potts model at the critical temperature*, J. Phys. C: Solid State Phys. **6** (1973) L445.

[63] G. E. Andrews, R. J. Baxter and J. P. Forrester, *Eight-vertex SOS model and generalized Rogers-Ramanujan-type identities*, J. Stat. Phys. **35** (1984) 193.

[64] D. A. Huse, *Exact exponents for infinitely many new multicritical points*, Phys. Rev. **B30** (1984) 3908.

[65] V. Dotsenko and V. Fateev, *Operator algebra of two-dimensional conformal theories with central charge $c \leq 1$*, Phys. Lett. **154B** (1985) 291.

[66] G. Felder, *BRST approach to minimal models*, Nucl. Phys. **B317** (1989) 21 [Erratum: **B324** (1989) 548].

[67] C. Becchi, A. Rouet and R. Stora, *Renormalization of gauge theories*, Ann. Phys. **98** (1976) 287.

[68] I. V. Tyutin, *Gauge invariance in field theory and statistical physics, in operatorial formulation*, Lebedev preprint N 39, Moscow, 1975 (in Russian, unpublished).

[69] P. Bouwknegt, J. Mc'Carthy and K. Pilch, *Free field appproach to 2-dimensional conformal field theories*, Progr. Theor. Phys. **102** (1990) 67.

[70] J. Lepowsky, S. Mandelstam and I. Singer, eds., *Vertex Operators in Mathematics and Physics*, Springer-Verlag, New York, 1985.

[71] A. Rocha-Caridi, *Vacuum vector representations of the Virasoro algebra*, in "Vertex Operators in Mathematics and Physics", Springer-Verlag, New York, 1985, p. 451.

[72] F. Gliozzi, J. Scherk and D. Olive, *Supersymmetry, supergravity and the dual spinor model*, Nucl. Phys. **B122** (1977) 253.

[73] N. Seiberg and E. Witten, *Spin structures in string theory*, Nucl. Phys. **B276** (1986) 272.

[74] G. Moore and N. Seiberg, *Polynomial equations for rational conformal field theories*, Phys. Lett. **212B** (1988) 454.

[75] G. Moore and N. Seiberg, *Naturality in conformal field theory*, Nucl. Phys. **B313** (1989) 16.

[76] L. Alvarez-Gaumé, G. Sierra and C. Gomez, *Topics in conformal field theory*, CERN preprint TH-5540/89, in "Physics and Mathematics of strings", Contribution to the Knizhnik Memorial Volume, edited by L. Brink et al., World Scientific, Singapore, 1990, p. 16.

[77] J. Fuchs, *Affine Lie Algebras and Quantum Groups*, Cambridge University Press, Cambridge, 1992.

[78] A. Capelli, C. Itzykson and J.-B. Zuber, *Modular invariant partition functions in two dimensions*, Nucl. Phys. **B280 FS** (1987) 445.

[79] A. Capelli, C. Itzykson and J.-B. Zuber, *The A-D-E classification of minimal and $A_1^{(1)}$ conformal invariant theories*, Commun. Math. Phys. **113** (1987) 1.

[80] D. Gepner and Z. Qui, *Modular invariant partition functions for parafermionic field theories*, Nucl. Phys. **B285 FS** (1987) 423.

[81] J. Polchinski, *Evaluation of the one loop string path integral*, Commun. Math. Phys. **104** (1986) 37.

[82] C. Itzykson and J.-B. Zuber, *Two-dimensional conformal invariant theories on a torus*, Nucl. Phys. **B275 FS** (1986) 580.

[83] R. Narain, *New heterotic string theories in uncompactified dimensions < 10*, Phys. Lett. **169B** (1986) 49.

[84] J. Cardy, *Operator content of two-dimensional conformally invariant theories*, Nucl. Phys. **B270 FS** (1986) 186.

[85] L. Alvarez-Gaumé, J.-B. Bost, G. Moore, P. Nelson and C. Vafa, *Bosonization in arbitrary genus*, Phys. Lett. **178B** (1986) 41.

[86] L. Alvarez-Gaumé, J.-B. Bost, G. Moore, P. Nelson and C. Vafa, *Bosonization on higher genus Riemann surfaces*, Commun. Math. Phys. **112** (1987) 503.

[87] E. Verlinde and H. Verlinde, *Chiral bosonization, determinants, and the string partition function*, Nucl. Phys. **B288** (1988) 357.

[88] T. Eguchi and H. Ooguri, *Chiral bosonization on Riemann surface*, Phys. Lett. **187B** (1987) 127.

[89] M. Dugan and H. Sonoda, *Functional determinants on Riemann surfaces*, Nucl. Phys. **B289** (1987) 227.

[90] V. Knizhnik, *Analytic fields on Riemann surfaces*, Commun. Math. Phys. **112** (1987) 567.

[91] V. A. Kostelecký, O. Lechtenfeld, W. Lerche, S. Samuel and S. Watamura, *Conformal techniques, bosonization and tree level string amplitudes*, Nucl. Phys. **B288** (1987) 173.

[92] O. Lechtenfeld, *Superconformal ghost correlations on Riemann surfaces*, Phys. Lett. **232B** (1989) 193.

[93] D. Friedan, E. Martinec and S. Shenker, *Covariant quantization of superstrings*, Phys. Lett. **160** (1985) 55.

[94] V. Knizhnik, *Covariant fermionic vertex in superstrings*, Phys. Lett. **160** (1985) 403.

[95] E. Verlinde and H. Verlinde, *Multiloop calculations in covariant superstring theory*, Phys. Lett. **192B** (1987) 95.

[96] J. Fay, *Theta Functions on Riemann Surfaces*, Springer-Verlag, Berlin, 1973.

[97] D. Mumford, *Theta Lectures on Theta*, Birkhäuser, Basel, 1983.

[98] L. Alvarez-Gaumé, G. Moore and C. Vafa, *Theta functions, modular invariance and strings*, Commun. Math. Phys. **106** (1986) 40.

[99] G. Felder and R. Silvotti, *Free field representation of minimal models on a Riemann surface*, Phys. Lett. **231B** (1989) 411.

[100] M. Frau, A. Lerda, J. McCarthy and S. Sciuto, *Minimal models on Riemann surfaces*, Phys. Lett. **228B** (1989) 205.

[101] M. Frau, A. Lerda, J. McCarthy and S. Sciuto, *Operator formalism and free field representation for minimal models on Riemann surfaces*, Nucl. Phys. **B338** (1990) 415.

[102] S.-J. Sin, *Feigin-Fuchs construction on Riemann surfaces*, Berkeley preprint UCB-PTH-89/20, 1989.

[103] M. Frau, A. Lerda, J. McCarthy, S. Sciuto and J. Sidenius, *Free field representation for $SU(2)_k$ WZNW models on Riemann surfaces*, Phys. Lett. **245B** (1990) 453.

[104] M. Frau, A. Lerda, J. McCarthy, S. Sciuto and J. Sidenius, $N = 2$ *minimal models on Riemann surfaces*, Phys. Lett. **254B** (1990) 381.

[105] D. Friedan and S. Shenker, *The analytic geometry of two-dimensional conformal field theory*, Nucl. Phys. **B281** (1987) 509.

[106] P. Goddard and D. Olive, *An introduction to Kač-Moody algebras and their physical applications*, in "Unified String Theories", World Scientific, Singapore, 1986, p. 214.

[107] P. Goggard and D. Olive, *Kač-Moody and Virasoro algebras in relation to quantum physics*, Int. J. Mod. Phys. **A1** (1986) 303.

[108] D. Gepner and E. Witten, *String theory on group manifolds*, Nucl. Phys. **B278** (1986) 493.

[109] V. Knizhnik and A. Zamolodchikov, *Current algebra and Wess-Zumino model in two dimensions*, Nucl. Phys. **B247** (1984) 83.

[110] H. Sugawara, *A field theory of currents*, Phys. Rev. **170** (1968) 1659.

[111] C. Sommerfeld, *Currents as dynamical variables*, Phys. Rev. **176** (1968) 2019.

[112] E. Witten, *Non-abelian bosonization in two dimensions*, Commun. Math. Phys. **92** (1984) 455.

[113] J. Wess and B. Zumino, *Consequences of anomalous Ward identities*, Phys. Lett. **37B** (1971) 95.

[114] S. Novikov, *Multi-valued functions and functionals. An analogue of the Morse theory*, Sov. Math. Dokl. **24** (1981) 222.

[115] S. Novikov, *Hamiltonian formalism and multi-valued analogue of Morse theory*, Usp. Mat. Nauk **37** (1982) 3.

[116] S. Mukhi, *The geometric background-field method, renormalization and the Wess-Zumino term in non-linear σ-models*, Nucl. Phys. **B264** (1986) 640.

[117] E. Braaten, T. Curtright and K. Zachos, *Torsion and geometrostasis in non-linear sigma models*, Nucl. Phys. **B260** (1985) 630.

[118] I. Frenkel and V. Kač, *Basic representations of affine Lie algebras and dual resonance models*, Inv. Math. **62** (1980) 23.

[119] G. Segal, *Unitary representations of some infinite dimensional groups*, Commun. Math. Phys. **80** (1981) 301.

[120] A. Zamolodchikov and V. Fateev, *Non-local (parafermion) currents in two-dimensional conformal quantum field theory and self-dual critical points in Z_N-symmetric statistical systems*, ZhETF **89** (1985) 380 [JETP **62** (1985) 215].

[121] D. Gepner and Z. Qiu, *Modular-invariant partition functions for parafermionic field theories* Nucl. Phys. **B285** (1987) 423.

[122] D. Gepner, *New conformal field theories associated with Lie algebras and their partition functions*, Nucl. Phys. **B290** (1987) 10.

[123] A. Gerasimov, A. Morozov, M. Olshanetsky, A. Marshakov and S. Shatashvili, *Wess-Zumino-Witten model as a theory of free fields*, Int. J. Mod. Phys. **A5** (1990) 2495.

[124] M. Wakimoto, *Fock representations of the affine Lie algebra $A_1^{(1)}$*, Commun. Math. Phys. **104** (1986) 605.

[125] M. Naimark, *Theory of Group Representations*, Nauka, Moscow, 1976.

[126] A. Morozov and A. Perelomov, *Complex geometry and string theory*, VINITI: Sovr. Probl. Mat. **54** (1989) 197.

[127] V. Knizhnik, *Analytic fields on Riemannian surfaces*, Phys. Lett. **180B** (1986) 247.

[128] M. Green, J. Schwarz and E. Witten, *Superstring Theory*, Vols. 1 and 2, World Scientific, Singapore, 1987.

[129] A. Schellekens and S. Yankielowicz, *New modular-invariants for $N = 2$ tensor products and four-dimensional strings*, Nucl. Phys. **B330** (1990) 103.

[130] A. Zamolodchikov, *Infinite additional symmetries in two-dimensional conformal quantum field theory*, Teor. Mat. Fiz. **65** (1986) 347 [Theor. Math. Phys. **65** (1986) 1205].

[131] P. Bouwknegt and K. Schoutens, *W-symmetry in conformal field theory*, Phys. Rep. **223** (1993) 183.

[132] E. Martinec, *Conformal field theory on a (super)-Riemann surface*, Nucl. Phys. **B281** (1987) 157.

[133] M. Baranov, I. Frolov and A. Schwarz, *Geometry of superconformal moduli space*, Teor. Mat. Fiz. **79** (1989) 241 [Theor. Math. Phys. **79** (1989) 509].

[134] M. Bershadsky, V. Knizhnik and M. Teitelman, *Superconformal symmetry in two dimensions*, Phys. Lett. **151B** (1985) 31.

[135] S. Shenker, *Introduction to two-dimensional conformal and superconformal field theory*, in "Unified String Theories", World Scientific, Singapore, 1986, p. 141.

[136] C. Montonen, *Multi-loop amplitudes in additive dual resonance models*, Nuovo Cim. **19** (1974) 69.

[137] E. Martinec, *Superspace geometry of fermionic strings*, Phys. Rev. **D28** (1983) 2604.

[138] J. Schwarz, *Superconformal symmetry and superstring compactification*, Int. J. Mod. Phys. **A4** (1989) 2653.

[139] P. Di Vecchia, V. Knizhnik, J. Peterson and P. Rossi, *A supersymmetric Wess-Zumino lagrangian in two dimensions*, Nucl. Phys. **B253** (1985) 701.

[140] E. Kiritsis and G. Siopsis, *Operator algebra of the $N = 1$ super–Wess-Zumino model*, Phys. Lett. **184B** (1987) 353.

[141] S. Nahm, *Superconformal and super–Kač-Moody invariant quantum field theories in two dimensions*, Phys. Lett. **187B** (1987) 340.

[142] K. Schoutens, *$O(N)$-extended superconformal field theory in superspace*, Nucl. Phys. **B295** (1988) 634.

[143] P. Binetruy, P. Sorba and R. Stora, eds., *Conformal Field Theories and Related Topics*, Elsevier North-Holland, Amsterdam, 1990.

[144] S. Nam, *The Kač formula for the $N = 1$ and $N = 2$ superconformal algebras*, Phys. Lett. **172B** (1986) 323.

[145] Z. Qiu, *Supersymmetry, two-dimensional critical phenomena, and the tricritical Ising model*, Nucl. Phys. **B270** (1986) 205.

[146] M. J. Tejwani, O. Ferreira and O. E. Vilches, *Possible Ising Transition in a ^4He monolayer adsorbed on Kr-plated Graphite*, Phys. Rev. Lett. **44** (1980) 152.

[147] M. Ademollo, L. Brink, A. D'Adda, R. D'Auria, E. Napolitano, S. Sciuto, E. Del Giudice, P. Di Vecchia, S. Ferrara, F. Gliozzi, R. Musto and R. Pettorino, *Supersymmetric strings and colour confinement*, Phys. Lett. **62B** (1976) 105.

[148] M. Ademollo, L. Brink, A. D'Adda, R.D'Auria, E. Napolitano, S. Sciuto, E. Del Guidice, P. Di Vecchia, S. Ferrara, F. Gliozzi, R. Musto, R. Pettorino and J. Schwarz, *Dual string with $U(1)$ colour symmetry*, Nucl. Phys. **B111** (1976) 77.

[149] M. Ademollo, L. Brink, A. D'Adda, R.D'Auria, E. Napolitano, S. Sciuto, E. Del Guidice, P. Di Vecchia, S. Ferrara, F. Gliozzi, R. Musto and R. Pettorino, *Dual string models with non-Abelian colour and flavour symmetries*, Nucl. Phys. **B114** (1976) 297.

[150] P. Di Vecchia, J. Petersen, M. Yu and H. Zheng, *$N = 2$ extended superconformal theories in two dimensions*, Phys. Lett. **162B** (1985) 327.

[151] A. Zamolodchikov and V. Fateev, *Disorder fields in two-dimensional conformal quantum field theory and $N = 2$ extended supersymmetry*, ZhETF **90** (1986) 1533 [JETP **63** (1986) 913].

[152] P. Di Vecchia, J. Petersen and M. Yu, *On the unitary representations of $N = 2$ superconformal theory*, Phys. Lett. **172B** (1986) 211.

[153] W. Boucher, D. Friedan and A. Kent, *Determinant formulae and unitarity for the $N = 2$ superconformal algebras in two dimensions or exact results on string compactification*, Phys. Lett. **172B** (1986) 316.

[154] T. Eguchi and A. Taormina, *On the unitary representations of $N = 2$ and $N = 4$ superconformal algebras*, Phys. Lett. **210B** (1988) 125.

[155] W. Lerche, C. Vafa and N. Warner, *Chiral rings in $N = 2$ superconformal theories*, Nucl. Phys. **B324** (1989) 427.

[156] E. Witten, *Constraints on supersymmetry breaking*, Nucl. Phys. **B202** (1982) 253.

[157] T. Eguchi and A. Taormina, *The unitary representations of the $N = 4$ superconformal algebra*, Phys. Lett. **196B** (1986) 75.

[158] T. Eguchi and A. Taormina, *Character formulas for the $N = 4$ superconformal algebra*, Phys. Lett. **200B** (1988) 315.

[159] A. Schwimmer and N. Seiberg, *Comments on the $N = 2, 3, 4$ superconformal algebras in two dimensions*, Phys. Lett. **184B** (1987) 191.

[160] P. Di Vecchia, *The Wess-Zumino action in two dimensions and non-abelian bosonization*, Phys. Lett. **144B** (1984) 245.

[161] P. Goddard, W. Nahm and D. Olive, *Symmetric spaces, Sugawara's energy-momentum tensor in two dimensions and free fermions*, Phys. Lett. **160B** (1985) 111.

[162] P. Windey, *Super Kač-Moody algebras and supersymmetric two-dimensional free fermions*, Commun. Math. Phys. **105** (1986) 511.

[163] P. Goddard, A. Kent and D. Olive, *Unitary representations of the Virasoro and super-Virasoro algebras*, Commun. Math. Phys. **103** (1986) 105.

[164] I. Antoniadis, C. Bachas, C. Kounnas and P. Windey, *Supersymmetry among free fermions and superstrings*, Phys. Lett. **171B** (1985) 51.

[165] V. Kač and I. Todorov, *Superconformal current algebras and their unitary representations*, Commun. Math. Phys. **102** (1985) 337.

[166] C. Hull and E. Witten, *Supersymmetric sigma models and the heterotic string*, Phys. Lett. **160B** (1985) 398.

[167] M. Sakamoto, $N = 1/2$ *supersymmetry in two dimensions*, Phys. Lett. **151B** (1985) 115.

[168] R. Brooks, F. Muhammad and S. Gates, Jr., *Unidexterious $D = 2$ supersymmetry in superspace*, Nucl. Phys. **B268** (1986) 599.

[169] S. J. Gates Jr., S. V. Ketov, S. M. Kuzenko and O. A. Soloviev, *Lagrangian chiral coset construction of heterotic string theories in $(1, 0)$ superspace*, Nucl. Phys. **B362** (1991) 199.

[170] A. M. Polyakov and P. B. Wiegmann, *Theory of nonabelian Goldstone bosons in two dimensions*, Phys. Lett. **131B** (1983) 121.

[171] A. M. Polyakov and P. B. Wiegmann, *Goldstone fields in two dimensions*, Phys. Lett. **141B** (1984) 223.

[172] A. Kirillov, *Elements of the Theory of Representations*, Springer-Verlag, New York, 1975.

[173] G. Delius, P. van Nieuwenhuizen and V. Rodgers, *The method of coadjoint orbits: an algorithm for the construction of invariant actions*, Int. J. Mod. Phys. **A5** (1990) 3943.

[174] C. Hull and B. Spence, *The $(2,0)$ supersymmetric Wess-Zumino-Witten model*, Nucl. Phys. **B345** (1990) 493.

[175] S. J. Gates Jr. and S. V. Ketov, *More about the $(2,0)$ supersymmetric WZNW model in $(2,0)$ superspace*, Phys. Lett. **271B** (1991) 355.

[176] G. W. Delius, *The $N = 2$ super Kač-Moody algebra and the WZW model in $(2,0)$ superspace*, Int. J. Mod. Phys. **A5** (1990) 4753.

[177] P. S. Howe and G. Papadopoulos, *Ultraviolet behaviour of two-dimensional supersymmetric nonlinear sigma models*, Nucl. Phys. **B289** (1987) 264.

[178] C. Vafa and N. Warner, *Catastrophes and the classification of conformal theories*, Phys. Lett. **218B** (1989) 51.

[179] A. B. Zamolodchikov, *Conformal symmetry and multi-critical points in two-dimensional quantum field theory*, Sov. J. Nucl. Phys. **44** (1986) 529.

[180] S. J. Gates, Jr., M. T. Grisaru, M. Roček and W. Siegel, *Superspace, or One Thousand and One Lessons in Supersymmetry*, Benjamin/Cummings, 1983.

[181] A. Marshakov and A. Morozov, *Landau-Ginzburg models with $N = 2$ supersymmetry as conventional conformal theories*, Phys. Lett. **235B** (1990) 97.

[182] V. I. Arnold, *Singularity Theory*, Cambridge Univ. Press, Cambridge, 1981.

[183] V. I. Arnold, S. M. Gusein-Zade and A. N. Varchenko, *Singularities of Differentiable Maps*, Birkhäuser, Basel, 1985.

[184] A. N. Schellekens and N. P. Warner, *Conformal subalgebras of Kač-Moody algebras* Phys. Rev. **D34** (1986) 3092.

[185] F. A. Bais and P. G. Bouwknegt, *A classification of subgroup truncations of the bosonic string*, Nucl. Phys. **B279** (1987) 561.

[186] F. A. Bais, P. Bouwknegt, M. Surridge and K. Schoutens, *Coset constructions for extended Virasoro algebras*, Nucl. Phys. **B304** (1988) 371.

[187] P. Goddard and A. Schwimmer, *Unitary construction of extended conformal algebras*, Phys. Lett. **206B** (1988) 62.

[188] J. Bagger, D. Nemeschansky and S. Yankielowicz, *Virasoro algebras with central charge $c > 1$*, Phys. Rev. Lett. **60** (1988) 389.

[189] S. Elitzur, E. Gross, E. Rabinivici and N. Seiberg, *Aspects of bosonization in string theory*, Nucl. Phys. **B283** (1987) 413.

[190] Y. Kazama and H. Suzuki, *Characterization of $N = 2$ superconformal models generated by coset space method*, Phys. Lett. **216B** (1989) 112.

[191] Y. Kazama and H. Suzuki, *New $N = 2$ superconformal field theories and superstring compactification*, Nucl. Phys. **B321** (1989) 232.

[192] S. Helgason, *Differential Geometry, Lie Groups, and Symmetric Spaces*, Academic Press, New York, 1978.

[193] Y. Kazama and H. Suzuki, *Bosonic construction of conformal field theories with extended supersymmetry*, Mod. Phys. Lett. **A4** (1989) 235.

[194] T. Nakatsu, *Supersymmetric gauged Wess-Zumino-Witten models*, Progr. Theor. Phys. **87** (1992) 795.

[195] C. C. Chevalley and S. Eilenberg, *Cohomology theory of Lie groups and Lie algebras*, Trans. Amer. Math. Soc. **63** (1948) 85.

[196] B. Kostant, *Lie algebra cohomology and the generalized Borel-Weyl theorem*, Ann. Math. **74** (1961) 329.

[197] P. Griffiths and W. Schmid, *Locally homogeneous complex manifolds*, Acta Math. **123** (1969) 253.

[198] R. Bott, *An application of the Morse theory to the topology of Lie groups*, Bull. Soc. Math. France, **84** (1956) 251.

[199] K. Bardakci, E. Rabinovici and B. Säring, *String models with $c < 1$ components*, Nucl. Phys. **B299** (1988) 151.

[200] K. Gawedzki and A. Kupiainen, G/H conformal field theory from gauged WZW model, Phys. Lett. **215B** (1988) 119.

[201] K. Gawedzki and A. Kupiainen, Coset construction from functional integrals, Nucl. Phys. **B320** (1989) 625.

[202] Q.-H. Park, Lagrangian formulation of coset conformal field theory, Phys. Lett. **223B** (1989) 422.

[203] D. Karabali, Q.-H. Park, H. J. Schnitzer and Z. Yang, A GKO construction based on a path integral formulation of gauged Wess-Zumino-Witten actions, Phys. Lett. **216B** (1989) 307.

[204] H. J. Schnitzer, A path integral construction of superconformal field theories from a gauged supersymmetric Wess-Zumino-Witten model, Nucl. Phys. **B324** (1989) 412.

[205] D. Karabali and H. J. Schnitzer, BRST quantization of the gauged WZW action and coset conformal field theories, Nucl. Phys. **B329** (1990) 649.

[206] P. Bouwknegt, J. McCarthy and K. Pilch, On the free field resolutions for coset conformal field theories, Nucl. Phys. **B352** (1991) 139.

[207] B. L. Feigin and E. V. Frenkel, Representations of affine Kač-Moody algebras, bosonization and resolutions, Lett. Math. Phys. **19** (1990) 307.

[208] B. L. Feigin and E. V. Frenkel, Affine Kač-Moody algebras and semi-infinite flag manifolds, Commun. Math. Phys. **128** (1990) 161.

[209] P. Bouwknegt, J. McCarthy and K. Pilch, Free field realizations of the WZNW models: the BRST complex and its quantum group structure, Phys. Lett. **234B** (1990) 297.

[210] P. Bouwknegt, J. McCarthy and K. Pilch, Quantum group structure in the Fock space resolutions of $SL(N)$ representations, Commun. Math. Phys. **131** (1990) 125.

[211] V. Kač and D. Peterson, Infinite-dimensional Lie algebras, theta functions and modular forms, Adv. Math. **53** (1984) 125.

[212] B. Schoenberg, Elliptic Modular Functions, Springer-Verlag, New York, 1974.

[213] M. B. Halpern and E. Kiritsis, *General Virasoro construction of affine G*, Mod. Phys. Lett. **A4** (1989) 1373 [Erratum: **A4** (1989) 1797].

[214] M. B. Halpern and J. P. Yamron, *Geometry of the general affine-Virasoro construction*, Nucl. Phys. **B332** (1990) 411.

[215] A. Giveon, M. B. Halpern, E.B. Kiritsis and N. A. Obers, *The superconformal master equation*, Int. J. Mod. Phys. **A7** (1992) 947.

[216] F. A. Bais, P. Bouwknegt, K. Schoutens and M. Surridge, *Extensions of the Virasoro algebra constructed from Kac-Moody algebras using higher order Casimir invariants*, Nucl. Phys. **B304** (1988) 348.

[217] T. Inami, Y. Matsuo and I. Yamanaka, *Extended conformal algebras with $N = 1$ supersymmetry*, Phys. Lett. **215B** (1988) 701.

[218] L. J. Romans, *The $N = 2$ super-W_3 algebra*, Nucl. Phys. **B369** (1992) 403.

[219] P. Bouwknegt, *Extended conformal algebras*, Phys. Lett. **207B** (1988) 295.

[220] H. G. Kausch and G. M. T. Watts, *A study of W-algebras using Jacobi identities*, Nucl. Phys. **B354** (1991) 740.

[221] R. Blumenhagen, M. Flohr, A. Kliem, W. Nahm, A. Recknagel and R. Varnhagen, *W-algebras with two and three generators*, Nucl. Phys. **B361** (1991) 255.

[222] S. Komata, K. Mohri and H. Nohara, *Classical and quantum extended superconformal algebras*, Nucl. Phys. **B359** (1991) 168.

[223] J. M. Figueroa-O'Farrill and S. Schrans, *Extended superconformal algebras*, Phys. Lett. **257B** (1991) 69.

[224] I. Bakas, *The large-N limit of extended conformal systems*, Phys. Lett. **228B** (1989) 57.

[225] A. Bilal, *A note on super W-algebras*, Phys. Lett. **238B** (1990) 239.

[226] C. N. Pope, L. J. Romans and X. Shen, *The complete structure of W_∞*, Phys. Lett. **236B** (1990) 173.

[227] C. N. Pope, L. J. Romans and X. Shen, *W_∞ and the Racah-Wigner algebra*, Nucl. Phys. **B339** (1990) 191.

[228] C. N. Pope, L. J. Romans and X. Shen, *A new higher-spin algebra and the lone-star product*, Phys. Lett. **242B** (1990) 401.

[229] E. Bergshoeff, B. de Wit and M. Vasiliev, *The structure of the super-$W_\infty(\lambda)$ algebra*, Nucl. Phys. **B366** (1991) 315.

[230] E. Bergshoeff, C. N. Pope, L. J. Romans, E. Sezgin and X. Shen, *The super-W_∞ algebra*, Phys. Lett. **245B** (1990) 447.

[231] Q. Ho-Kim and H. B. Zheng, *Twisted conformal field theories with Z_3 invariance*, Phys. Lett. **212B** (1988) 71.

[232] Q. Ho-Kim and H. B. Zheng, *Twisted structures of extended Virasoro algebras*, Phys. Lett. **223B** (1989) 57.

[233] Q. Ho-Kim and H. B. Zheng, *Twisted characters and patrition functions in extended Virasoro algebras*, Mod. Phys. Lett. **A5** (1990) 1181.

[234] A. M. Morozov, *On the concept of universal W-algebra*, Nucl. Phys. **B357** (1991) 619.

[235] H. Lu, C. N. Pope, S. Shen and K.-J. Wang, *The complete structure of W_n from W_∞*, Phys. Lett. **267B** (1991) 356.

[236] I. Bakas and E. Kiritsis, *Beyond the large-N limit: non-linear W_∞ as symmetry of the $SL(2, \mathbf{R})/U(1)$ model*, Int. J. Mod. Phys. **A7** (1992) 55.

[237] K. Schoutens and A. Sevrin, *Minimal super W_n algebras in coset conformal field theories*, Phys. Lett. **258B** (1991) 134.

[238] V. A. Fateev and S. L. Lukyanov, *The models of two-dimensional conformal quantum field theory with Z_n symmetry*, Int. J. Mod. Phys. **A3** (1988) 507.

[239] V. A. Fateev and S. L. Lukyanov, *Conformal invariant models of two-dimensional quantum field theory with Z_n symmetry*, JETP **67** (1988) 447.

[240] A. Bilal, *Introduction to W-algebras*, in "String Theory and Quantum Gravity '91", edited by J. Harvey, R. Iengo, K. S. Narain, S. Randjbar-Daemi and H. Verlinde, World Scientific, Singapore, 1992, p. 245.

[241] R. M. Miura, *Korteweg-de Vries equation and generalizations*, J. Math. Phys. **9** (1968) 1202.

[242] S. L. Lukyanov, *Quantization of the Gel'fand-Dikii brackets*, Funct. Anal. Appl. **22** (1990) 1.

[243] P. Di Francesco, C. Itzykson and J.-B. Zuber, *Classical W-algebras*, Commun. Math. Phys. **140** (1991) 543.

[244] A. Bilal, V. V. Fock and I. I. Kogan, *On the origin of W-algebras*, Nucl. Phys. **B359** (1991) 635.

[245] S. Mizoguchi, *Determinant formula und unitarity for the W_3-algebras*, Phys. Lett. **222B** (1989) 226.

[246] S. Mizoguchi, *Non-unitarity theorem for the A-type W_n-algebras*, Phys. Lett. **231B** (1989) 112.

[247] M. Berschadsky and H. Ooguri, *Hidden $SL(n)$ symmetry of in conformal field theories*, Commun. Math. Phys. **126** (1989) 49.

[248] B. L. Feigin and E. Frenkel, *Quantization of the Drinfeld-Sokolov reduction*, Phys. Lett. **246B** (1990) 75.

[249] I. M. Gel'fand and L. A. Dikii, *A family of Hamiltonian structures connected with integrable non-linear differential equations*, IPM AN USSR preprint, Moscow, 1978; in "Gel'fand Collected Papers", edited by Gindinkin et al., Springer-Verlag, N.-Y., 1987, p. 625.

[250] V. Drinfeld and V. Sokolov, *Lie algebras and equations of Korteweg-de Vries type*, J. Sov. Math. **30** (1984) 1975.

[251] A. Belavin, *On the connection between Zamolodchikov's W algebras and Kač-Moody algberas*, in "Quantum Field Theory", edited by N. Kawamoto and T. Kugo, Proc. Second Yukawa Memorial Symposium, Nishinomiya, Japan, 1987; Proceedings in Physics, Springer, Berlin, **31** (1989) p. 132.

[252] A. Belavin, *KdV-type equations and W-algebras*, Adv. Stud. in Pure Math. **19** (1989) 117.

[253] E. Frenkel, V. G. Kač and M. Wakimoto, *Characters and fusion rules for W-algebras via quantized Drinfeld-Sokolov reductions*, Commun. Math. Phys. **147** (1992) 295.

[254] F. A. Bais, T. Tjin and P. van Driel, *Coupled chiral algebras obtained by reduction of WZNW models*, Nucl. Phys. **B357** (1991) 632.

[255] P. Bouwknegt, J. McCarthy and K. Pilch, *Some aspects of free field resolutions in 2D CFT with application to the quantum Drinfeld-Sokolov reduction*, in "Strings and Symmetries 1991", World Scientific, Singapore, 1992, p. 407.

[256] I. Bakas ans E. Kiritsis, *Beyond the large N limit: non-linear W_∞ as symmetry of the $SL(2, \mathbf{R})/U(1)$ coset model*, in "Infinite Symmetries", edited by A. Tsuchiya, T. Eguchi and M. Jimbo, World Scientific, Singapore, Int. J. Mod. Phys. **A7** (1992) Suppl. 1A, p. 55.

[257] M. Jacob, ed., *Dual Theory*, Physics Reports reprints, North-Holland, Amsterdam, 1974.

[258] J. Scherk, *An introduction to the theory of dual models and strings*, Rev. Mod. Phys. **47** (1975) 123.

[259] J. Schwarz, *Superstring theory*, Phys. Rep. **89** (1982) 223.

[260] M. Green, *Supersymmetrical dual string theories and their field theory limits — a review*, Surveys in High Energy Physics **3** (1983) 127.

[261] D. Friedan, *Introduction to Polyakov's string theory*, in "Recent Advances in Field Theory and Statistical Mechanics", Elsevier North-Holland, Amsterdam, 1984, p. 839.

[262] M. Green, *Unification of forces and particles in superstring theories*, Nature **314** (1985) 409.

[263] J. Schwarz, ed., *"Superstrings: The First 15 Years of Superstring Theory"*, Vols. 1 and 2, World Scientific, Singapore, 1985.

[264] J. Schwarz, *Topics in superstring theory*, in "Supersymmetry and Its Applications: Superstrings, Anomalies and Supergravity", Cambridge Univ. Press, Cambridge, 1985, p. 109.

[265] L. Clavelli, *Historical overview of superstring theory*, in "Lewes String Theory Workshop Proceedings", World Scientific, Singapore, 1986, p. 3.

[266] P. Frampton, *Introduction to superstrings*, in "Lewes String Theory Workshop Proceedings", World Scientific, Singapore, 1986, p. 21.

[267] D. Gross, *The heterotic string*, in "Unified String Theories", World Scientific, Singapore, 1986, p. 357.

[268] M. Green and J. Schwarz, eds., *Unified String Theories*, World Scientific, Singapore, 1986.

[269] M. Perry, *Strings and spacetime physics*, in "Supersymmetry and Its Appications: Superstrings, Anomalies and Supergravity", Cambridge Univ. Press, Cambridge, 1986, p. 269.

[270] M. Green, *Superstrings*, Sci. Amer. **255** (1986) 44.

[271] P. Frampton, *Dual Resonance Models and Superstrings*, World Scientific, Singapore, 1986.

[272] J. Schwarz, *Introduction to superstrings*, CALTECH preprint 68–1290, 1986.

[273] G. Horowitz, *Introduction to string theories*, in "Topological Properties and Global Structure of Space-Time", Plenum, N.-Y., 1986.

[274] M. Dine, *Superstring primer*, in "New Frontiers in Particle Physics", World Scientific, Singapore, 1986.

[275] B. de Wit, P. Fayet and M. Grisaru, eds., *Supersymmetry, Supergravity and Superstrings '86*, World Scientific, Singapore, 1986.

[276] A. Polyakov, *Gauge Fields and Strings*, Harcourt Brace Jovanovich, New York, 1987.

[277] P. Nelson, *Lectures on strings and moduli space*, Phys. Rep. **149** (1987) 337.

[278] J. Schwarz, *Review of recent developments in superstring theory*, Int. J. Mod. Phys. **A2** (1987) 593.

[279] W. Siegel, *Introduction to String Field Theory*, World Scientific, Singapore, 1988.

[280] L. Brink, *Superstrings — towards a unification of all interactions*, Göteborg preprint ITP-28, 1988.

[281] G. Veneziano, *Topics in string theory*, CERN preprint TH-5019, 1988.

[282] M. Green, *Introduction to string and superstring theory*, in "Jerusalem Winter School on Theoretical Physics", Vol. 3, World Scientific, Singapore, 1988, p. 1.

[283] M. Kaku, *Introduction to Superstrings*, Springer-Verlag, 1988.

[284] S. J. Gates Jr., C. R. Preitschopf and W. Siegel, eds., *Strings '88*, World Scientific, Singapore, 1989.

[285] A. Morozov and A. Perelomov, *Complex geometry and string theory*, VINITI: Sovr. Probl. Mat. **54** (1989) 197.

[286] J. Polchinski, *Superstring theory*, Austin preprint UTTG-1-89, 1989.

[287] J. Schwarz, *Superconformal symmetry and superstring compactification*, Int. J. Mod. Phys. **A4** (1989) 2653.

[288] D. Lüst and S. Theisen, *Lectures in String Theory*, Lecture Notes in Physics **346** (1989).

[289] R. Arnowitt, R. Bryan, M. J. Duff, D. Nanopoulos and C. N. Pope, eds., *Strings '89*, World Scientific, Singapore, 1990.

[290] S. V. Ketov, *An Introduction to the Quantum Theory of Strings and Superstrings*, Nauka, Novosibirsk, 1990.

[291] B. Greene, *Lectures on string theory in four dimensions*, Cornell preprint CLNS 91-1046, 1990.

[292] L. Brink, D. Friedan and A. Polyakov, eds., *Physics and Mathematics of Strings*, World Scientific, Singapore, 1990.

[293] M. Kaku, *Strings, Conformal Fields, and Topology. An Introduction*, Springer-Verlag, New York, 1991.

[294] M. Kaku, A. Jevicki and K. Kikkawa, eds., *Quarks, Symmetries and Strings*, World Scientific, Singapore, 1991.

[295] M. Green, R. Iengo, S. Randjbar-Daemi, E. Sezgin and H. Verlinde, eds., *String Theory and Quantum Gravity*, World Scientific, Singapore, 1991.

[296] N. Berkovitz, H. Itoyama, K. Schoutens, A. Sevrin, W. Siegel, P. van Nieuwenhuizen and J. Yamron, eds., *Strings and Symmetries 1991*, World Scientific, Singapore, 1992.

[297] J. Harvey, R. Iengo, K. S. Narain, S. Randjbar-Daemi and H. Verlinde, eds., *String Theory and Quantum Gravity '91*, World Scientific, Singapore, 1992.

[298] C. B. Thorn, *String field theory*, Phys. Rep. **175** (1989) 1.

[299] R. Narain, *New heterotic string theories in uncompactified dimensions* < 10, Phys. Lett. **169B** (1986) 49.

[300] A. Schellekens, *Four-dimensional superstrings*, CERN preprint TH-4807, 1987.

[301] I. Antoniadis, C. Bachas and C. Kounnas, *Four-dimensional superstrings*, Nucl. Phys. **B289** (1987) 87.

[302] A. Schellekens, *Conformal field theory for 4-dimensional strings*, CERN preprint TH-5515, 1989.

[303] C. Lütken, *String theory of Calabi-Yau compactifications*, Nordita preprint 51P, 1989.

[304] A. Schellekens and S. Yankielowicz, *New modular-invariants for N = 2 tensor products and four-dimensional strings*, Nucl. Phys. **B330** (1990) 103.

[305] A. Schellekens, ed., *Superstring Construction*, Elsevier North-Holland, Amsterdam, 1991.

[306] O. Alvarez, *Theory of strings with boundaries*, Nucl. Phys. **B216** (1983) 125.

[307] C. Lovelace, *Strings in curved space*, Phys. Lett. **135B** (1984) 75.

[308] D. Friedan, *Non-linear sigma models in 2+ε dimensions*, Phys. Rev. Lett. **45** (1980) 1057.

[309] D. Friedan, *Non-linear sigma models in 2 + ε dimensions*, Ann. Phys. **163** (1985) 318.

[310] A. Blasi, F. Delduc and S. P. Sorella, *The background-quantum split symmetry in two-dimensional σ-models, a regularization independent proof of its renormalizability*, Nucl. Phys. **B314** (1989) 409.

[311] C. Becchi and O. Piguet, *On the renormalization of two-dimensional chiral models*, Nucl. Phys. **B315** (1989) 153.

[312] A. A. Tseytlin, *Conformal anomaly in a two-dimensional sigma model on a curved background and strings*, Phys. Lett. **178B** (1986) 34.

[313] G. M. Shore, *A local renormalization group equations, diffeomorphisms and conformal invariance in sigma models*, Nucl. Phys. **B286** (1987) 349.

[314] A. A. Tseytlin, *Sigma-model Weyl invariance conditions and string equations of motion*, Nucl. Phys. **B286** (1987) 383.

[315] H. Osborn, *String theory effective actions from bosonic σ-models*, Nucl. Phys. **B308** (1988) 629.

[316] E. S. Fradkin and A. A. Tseytlin, *Quantum string theory effective action*, Nucl. Phys. **B261** (1985) 1.

[317] C. G. Callan, D. Friedan, E. J. Martinec and M. J. Perry, *Strings in background fields*, Nucl. Phys. **B262** (1985) 593.

[318] S. V. Ketov, *Two-loop calculations in the non-linear sigma-model with torsion*, Nucl. Phys. **B294** (1987) 813.

[319] S. V. Ketov, A.A. Deriglazov and Ya. S. Prager, *Three-loop β-function for the two-dimensional non-linear σ-model with a Wess-Zumino-Witten term*, Nucl. Phys. **B332** (1990) 447.

[320] S. V. Ketov, *Non-Linear Sigma-Models in Supersymmetry, Supergravity and Strings*, Nauka, Novosibirsk, 1992.

[321] G. Curci and G. Paffuti, *Consistency between the string background field equations of motion and the vanishing of the conformal anomaly*, Nucl. Phys. **B286** (1987) 399.

[322] G. F. Chapline and N. S. Manton, *Unification of Yang-Mills theory and supergravity in ten dimensions*, Phys. Lett. **120B** (1983) 105.

[323] A. B. Zamolodchikov, *On 'irreversability' of the renormalization group flow in two-dimensional conformal field theory*, Pis'ma ZhETF, **43** (1986) 565 [JETP Lett. **43** (1986) 730].

[324] N. E. Mavromatos and J. L. Miramontes, *Zamolodchikov's c-theorem and string effective action*, Phys. Lett. **212B** (1988) 33.

[325] J. L. Cardy, *Is there a c-theorem in four dimensions?*, Phys. Lett. **215B** (1988) 749.

[326] B. P. Dolan, *Integrability conditions for potential flow of the renormalization group*, Hannover and DESY preprint ITP–UH–03/93 and DESY 93–040, March 1993.

[327] A. A. Tseytlin, *Vector field effective action in the open superstring theory*, Nucl. Phys. **B276** (1986) 391.

[328] I. Jack and D. R. T. Jones, *σ-model β-functions and ghost-free string effective actions*, Nucl. Phys. **B303** (1988) 260.

[329] I. Jack, D. R. T. Jones and D. A. Ross, *The four-loop dilaton β-function*, Nucl. Phys. **B307** (1988) 531.

[330] Y. Nambu, *Lectures at the Copenhagen Symposium on Symmetries and Quark Models*, Gordon and Breach, N.-Y., 1970, p. 269.

[331] T. Goto, *Relativistic quantum mechanics of one-dimensional mechanical continuum and subsidiary condition of dual resonance model*, Progr. Theor. Phys. **46** (1971) 1560.

[332] P. Ramond *Dual theory for free fermions*, Phys. Rev. **D3** (1971) 2415.

[333] A. Neveu and J. Schwarz, *Factorizable dual model of pions*, Nucl. Phys. **B31** (1971) 86.

[334] A. Neveu and J. Schwarz, *Quark model of dual pions*, Phys. Rev. **D4** (1971) 1109.

[335] A. Neveu, J. Schwarz and C. Thorn, *Reformulation of the dual pion model*, Phys. Lett. **35B** (1971) 529.

[336] J. Schwarz, *Physical states and pomeron poles in the dual pion model*, Nucl. Phys. **B46** (1972) 61.

[337] S. Deser and B. Zumino, *A complete action for the spinning string*, Phys. Lett. **65B** (1976) 369.

[338] L. Brink, P. Di Vecchia and P. Howe, *A locally supersymmetric and reparametrization invariant action for the spinning string*, Phys. Lett. **65B** (1976) 471.

[339] P. Howe, *Superspace and spinning string*, Phys. Lett. **70B** (1977) 453.

[340] D. Friedan, S. Shenker and E. Martinec, *Covariant quantization of superstrings*, Phys. Lett. **160B** (1985) 55.

[341] V. G. Knizhnik, *Covariant fermionic vertex in superstrings*, Phys. Lett. **160B** (1985) 403.

[342] S. V. Ketov, *Self-dual lattices and strings in $D < 10$ dimensions*, Izv. VUZov USSR, Matematika N 3 (1991) 14 (in Russian).

[343] J. H. Schwarz, *Superconformal symmetry in String Theory*, Caltech preprint 68–1503, 1988; in "Particles and Fields", Proc. Banff Summer Inst., 1988, edited by A. N. Kamal and F. C. Khanna, World Scientific, Singapore, 1989, p. 55.

[344] D. Gross, J. Harvey, E. Martinec and R. Rohm, *Heterotic string*, Phys. Rev. Lett. **54** (1985) 502.

[345] D. Gross, J. Harvey, E. Martinec and R. Rohm, *Heterotic string theory (I). The free heterotic string*, Nucl. Phys. **B256** (1985) 253.

[346] D. Gross, J. Harvey, E. Martinec and R. Rohm, *Heterotic string theory (II). The interacting heterotic string*, Nucl. Phys. **B267** (1986) 75.

[347] A. Capelli, *Modular invariant partition functions of superconformal theories*, Phys. Lett. **185B** (1987) 82.

[348] D. Kastor, *Modular invariance in superconformal models*, Nucl. Phys. **B280 FS** (1987) 304.

[349] P. Candelas, H. T. Horowitz, A. Strominger and E. Witten, *Vacuum configurations for superstrings*, Nucl. Phys. **B258** (1985) 46.

[350] D. Gepner, *Space-time supersymmetry in compactified string theory and superconformal models*, Nucl. Phys. **B296** (1988) 757.

[351] D. Gepner, *Exactly solvable string compactifications on manifolds of $SU(N)$ holonomy*, Phys. Lett. **199B** (1987) 380.

[352] A. Sevrin, W. Troost and A. van Proeyen, *Superconformal algebras in two dimensions with $N = 4$*, Phys. Lett. **208B** (1988) 447.

[353] S. V. Ketov, *How many $N = 4$ strings exist ?*, Hannover preprint ITP–UH–13/94, hep-th/9409020; to appear in Class. and Quantum Gravity (1995).

[354] L. Brink and J. Schwarz, *Local complex supersymmetry in two dimensions*, Nucl. Phys. **B121** (1977) 285.

[355] R. Penrose and W. Rindler, *Spinors and Space-Time*, Cambridge University Press, Cambridge, 1987.

[356] E. Fradkin and A. Tseytlin, *Quantization of two-dimensional supergravity and critical dimensions for string models*, Phys. Lett. **106B** (1981) 63.

[357] A. Bilal, *BRST approach to the $N = 2$ superconformal algebra*, Phys. Lett. **180B** (1986) 255.

[358] J. Bieńkowska, *The generalized no-ghost theorem for $vN = 2$ SUSY critical strings*, Phys. Lett. **281B** (1992) 59.

[359] S. V. Ketov, *Space-time supersymmetry of extended fermionic strings in $2 + 2$ dimensions*, Class. and Quantum Grav. **10** (1993) 1689.

[360] A. D'Adda and F. Lizzi, *Space dimensions from supersymmetry for the $N = 2$ spinning string: a four-dimensional model*, Phys. Lett. **191B** (1987) 85.

[361] A. R. Bogojevic and Z. Hlousek, *The BRST quantization of the $O(2)$ string*, Phys. Lett. **179B** (1986) 69.

[362] H. Ooguri and C. Vafa, *Self-duality and $N = 2$ string magic*, Mod. Phys. Lett. **A5** (1990) 1389.

[363] H. Ooguri and C. Vafa, *Geometry of $N = 2$ strings*, Nucl. Phys. **B361** (1991) 469.

[364] J. F. Plebanski, *Some solutions of complex Einstein equations*, J. Math. Phys. **16** (1975) 2395.

[365] S. Kalitzin and E. Sokatchev, *An action principle for self-dual Yang-Mills and Einstein equations*, Phys. Lett. **257B** (1991) 151.

[366] N. Marcus, Y. Oz and S. Yankielowicz, *Harmonic space, self-dual Yang-Mills and N = 2 string*, Nucl. Phys. **B379** (1992) 121.

[367] Ch. Devchand and V. I. Ogievetsky, *Super-self-duality as analyticity in harmonic superspace*, Phys. Lett. **297B** (1992) 93.

[368] W. Siegel, Self-dual $N = 8$ supergravity as closed $N = 2$ ($N = 4$) strings, Phys. Rev. **D47** (1993) 2504.

[369] N. Marcus, *The N = 2 open string*, Nucl. Phys. **B387** (1992) 263.

[370] W. Siegel, *The N = 4 string is the same as the N = 2 string*, Phys. Rev. Lett. **69** (1992) 1493.

[371] W. Siegel, $N = 2$, ($N = 4$) *string is self-dual Yang-Mills theory*, Phys. Rev. **D46** (1992) 3235.

[372] A. Parkes, *A cubic action for self-dual Yang-Mills*, Phys. Lett. **286B** (1992) 265.

[373] H. Nishino and S. J. Gates Jr., $N = (2,0)$ *superstring as the underlying theory of self-dual Yang-Mills theory*, Mod. Phys. Lett. **A7** (1992) 2543.

[374] S. V. Ketov, *Supersymmetric σ-models with torsion in supergravity background and critical dimensions for string theories*, Class. and Quantum Grav. **4** (1987) 1163.

[375] S. D. Mathur and S. Mukhi, *Becchi-Rouet-Stora-Tyutin quantization of twisted extended fermionic strings*, Phys. Rev. **D36** (1987) 465.

[376] S. V. Ketov, O. Lechtenfeld and A. Parkes, *Twisting the N = 2 string*, Hannover and DESY preprint, ITP–UH–24/93 and DESY 93–191, hep-th/9312150; to appear in Phys. Rev. **D** (1995).

[377] A. Giveon and M. Roček, *On the BRST operator structure of the N = 2 string*, Nucl. Phys. **B400** (1993) 145.

[378] J. Bischoff, S. V. Ketov and O. Lechtenfeld, *The GSO projection, BRST cohomology and picture-changing in N = 2 string theory*, Hannover preprint ITP–UH–05/94, hep-th/9406101; to appear in Nucl. Phys. **B** (1995).

[379] J. Gomis, *Some aspects of the $N = 2$ superstring*, Phys. Rev. **D40** (1989) 408.

[380] N. Berkovits and C. Vafa, $N = 4$ *topological strings*, Harvard and King's College preprint, HUTP–94/A018 and KCL–TH–94–12, hep-th/9407191.

[381] J. Gomis and H. Suzuki, $N = 2$ *string as a topological conformal theory*, Phys. Lett. **278B** (1992) 266.

[382] M. Pernici and P. van Nieuwenhuizen, *A covariant action for the $SU(2)$ spinning string as a hyper-Kähler or quaternionic non-linear sigma model*, Phys. Lett. **169B** (1986) 381.

[383] S. J. Gates Jr., L. Lu and R. Oerter, *Simplified $SU(2)$ spinning string superspace supergravity*, Phys. Lett. **218B** (1989) 33.

[384] E. S. Fradkin and G. A. Vilkovisky, *Quantization of relativistic systems with constraints*, Phys. Lett. **55B** (1975) 224.

[385] I. A. Batalin and G. A. Vilkovisky, *Relativistic S-matrix of dynamical systems with boson and fermion constraints*, Phys. Lett. **69B** (1977) 309.

[386] I. A. Batalin and G. A. Vilkovisky, *Feynman rules for reducible gauge theories*, Phys. Lett. **120B** (1983) 166.

[387] S. J. Gates, Jr., H. Nishino and S. V. Ketov, *Extended supersymmetry and self-duality in $2 + 2$ dimensions*, Phys. Lett. **297B** (1992) 99.

[388] S. V. Ketov, H. Nishino and S. J. Gates, Jr., *Self-dual supersymmetry and supergravity in Atiyah-Ward space-time*, Nucl. Phys. **B393** (1993) 149.

[389] S. V. Ketov, H. Nishino and S. J. Gates, Jr., *Majorana-Weyl spinors and self-dual gauge fields in $2 + 2$ dimensions*, Phys. Lett. **307B** (1993) 323.

[390] E. Fradkin and A. Tseytlin, *Conformal supergravity*, Phys. Rep. **119** (1985) 233.

[391] P. van Nieuwenhuizen, *Supergravity*, Phys. Rep. **68** (1981) 189.

[392] M. Li, *Gauge symmetries and amplitudes in $N = 2$ strings*, Nucl. Phys. **B395** (1993) 129.

[393] E. Bergshoeff, A. Bilal and K. S. Stelle, *W-symmetries: gauging and geometry*, Int. J. Mod. Phys. **A6** (1991) 4959.

[394] K. Schoutens, A. Sevrin and P. van Nieuwenhuizen, *Covariant formulation of classical W-gravity*, Nucl. Phys. **B349** (1991) 791.

[395] C. M. Hull, *Gauging the Zamolodchikov's W-algebra*, Phys. Lett. **240B** (1990) 110.

[396] C. M. Hull, *Classical and Quantum W-gravity*, in "Strings and Symmetries 1991", edited by N. Berkovitz et al., World Scientific, Singapore, 1991, p. 495.

[397] J. Thierry-Mieg, *BRS-analysis of Zamolodchikov's spin 2 and 3 current algebra*, Phys. Lett. **197B** (1987) 368.

[398] K. Schoutens, A. Sevrin and P. van Nieuwenhuizen, *Quantum BRST charge for quadratically non-linear Lie algebras*, Commun. Math. Phys. **124** (1989) 87.

[399] C. N. Pope, *Lectures on W-algebras and W-strings*, in the Proceedings of the 1991 Trieste Summer School in High Energy Physics, and the 1993 Trieste Spring School in Gravity and Strings, World Scientific, Singapore, 1992 and 1994.

[400] J. de Boer and J. Goeree, *The covariant W_3 action*, Phys. Lett. **274B** (1992) 289.

[401] H. Ooguri, K. Schoutens, A. Sevrin and P. van Nieuwenhuizen, *The induced action of W_3-gravity*, Commun. Math. Phys. **145** (1992) 515.

[402] A. M. Polyakov, *Quantum gravity in two dimensions*, Mod. Phys. Lett. **A2** (1987) 893.

[403] V. G. Knizhnik, A. M. Polyakov and A. B. Zamolodchikov, *Fractal structure of 2d quantum gravity*, Mod. Phys. Lett. **A3** (1988) 819.

[404] F. David, *Conformal field theories coupled to 2-D gravity in the conformal gauge*, Mod. Phys. Lett. **A3** (1988) 1651.

[405] J. Distler and H. Kawai, *Conformal field theory and 2D quantum gravity*, Nucl. Phys. **B321** (1989) 509.

[406] A. Alekseev and S. Shatashvili, *Path integral quantization of the coadjoint orbits of the Virasoro group and 2d gravity*, Nucl. Phys. **B323** (1989) 719.

[407] A. B. Zamolodchikov, *On the entropy of random surfaces*, Phys. Lett. **117B** (1982) 87.

[408] K. Fujikawa, *Path integral of relativistic strings*, Phys. Rev. **D25** (1982) 1584.

[409] E. S. Fradkin and A. A. Tseytlin, *Quantized string actions*, Ann. Phys. **143** (1982) 413.

[410] D. V. Boulatov, V. A. Kazakov, I. K. Kostov and A. A. Migdal, *Analytical and numerical study of the model of dynamically triangulated random surfaces*, Nucl. Phys. **B275** (1986) 641.

[411] V. A. Kazakov and A. A. Migdal, *Recent progress in the theory of non-critical strings*, Nucl. Phys. **B311** (1988) 171.

[412] N. Seiberg, *Notes on Quantum Liouville theory and quantum gravity*, Progr. Theor. Phys. Suppl. **102** (1990) 319.

[413] J. Polchinski, *Remarks on the Liouville field theory*, in "Strings '90", edited by R. Arnowitt et al, World Scientific, Singapore, 1991, p. 62.

[414] J.-L. Gervais and A. Neveu, *The Dual string spectrum in Polyakov's quantization (I)*, Nucl. Phys. **B199** (1982) 59.

[415] J.-L. Gervais and A. Neveu, *The Dual string spectrum in Polyakov's quantization (II). Mode separation*, Nucl. Phys. **B209** (1982) 125.

[416] J.-L. Gervais and A. Neveu, *New quantum treatment of Liouville field theory*, Nucl. Phys. **B224** (1983) 329.

[417] J.-L. Gervais and A. Neveu, *Novel triangle relation and absence of tachyons in Liouville string field theory*, Nucl. Phys. **B238** (1984) 125.

[418] J.-L. Gervais and A. Neveu, *Green functions and scattering amplitudes in Liouville string field theory*, Nucl. Phys. **B238** (1984) 396.

[419] J.-L. Gervais and A. Neveu, *Non-standard 2D critical statistical models from Liouville theory*, Nucl. Phys. **B257** (1985) 59.

[420] J.-L. Gervais, *Critical dimensions for non-critical strings*, Phys. Lett. **243B** (1990) 85.

[421] T. L. Curtright and C. B. Thorn, *Conformally invariant quantization of the Liouville theory*, Phys. Rev. Lett. **48** (1982) 1309.

[422] E. Braaten, T. L. Curtright and C. B. Thorn, *Quantum Bäcklund transformation for the Liouville theory*, Phys. Lett. **118B** (1982) 115.

[423] E. Braaten, T. L. Curtright and C. B. Thorn, *An exact operator solution of the quantum Liouville field theory*, Ann. Phys. **147** (1983) 365.

[424] E. Braaten, T. L. Curtright, G. Chandour and C. B. Thorn, *Nonperturbative weak-coupling analysis of the Liouville quantum field theory*, Phys. Rev. Lett. **51** (1983) 19; Ann. Phys. **153** (1984) 147.

[425] E. D'Hoker, *Lecture Notes on 2D Quantum Gravity and Liouville Theory*, UCLA preprint UCLA/91/TEP/35, 1991.

[426] Y. Kazama and H. Nicolai, *On the exact operator formalism of two-dimensional Liouville quantum gravity in Minkowski spacetime*, DESY and Tokyo preprint, DESY 93-043 and UT Komaba 93-6, March 1993.

[427] L. A. Takhtajan, *Topics in quantum geometry of Riemann surfaces: two-dimensional quantum gravity*, Lectures given at Intern. School of Physics "Enrico Fermi", Varenna, 28 June – 8 July 1994, hep-th/9409088.

[428] J.-L. Gervais, *The quantum group structure of 2d gravity and minimal models*, Commun. Math. Phys. **130** (1990) 257.

[429] J.-L. Gervais, *On the algebraic structure of quantum gravity in two dimensions*, Int. J. Mod. Phys. **A6** (1991) 2805.

[430] J.-L. Gervais, *Solving the strongly-coupled 2d gravity: unitary truncation and quantum group structure*, Commun. Math. Phys. **138** (1991) 301.

[431] J.-L. Gervais, *Gravity-matter couplings from Liouville theory*, Nucl. Phys. **B391** (1993) 287.

[432] L. D. Faddeev and L. A. Takhtadzhian. *Hamiltonian Methods in the Theory of Solitons*, Springer-Verlag, Berlin, 1987.

[433] A. V. Bäcklund, *Zur Theorie der Flächentransformationen*, Math. Ann. **19** (1882) 387.

[434] J. Ambjørn, *Barriers in quantum gravity*, Invited Talk presented at the ICTP Spring School and Workshop on String Theory, Trieste, Italy, April 1993, to appear in the Proceedings; Copenhagen preprint NBI–HE–93–31, July 1993.

[435] A. Bilal and J.-L. Gervais, *New critical dimensions for string theories*, Nucl. Phys. **B284** (1987) 397.

[436] A. Bilal and J.-L. Gervais, *Modular invariance for closed strings at the new critical dimensions*, Phys. Lett. **187B** (1987) 39.

[437] A. Bilal and J.-L. Gervais, *The five-dimensional open Liouville superstring*, Nucl. Phys. **B293** (1987) 1.

[438] A. Bilal and J.-L. Gervais, *Liouville superstring and Ising model in three dimensions*, Nucl. Phys. **B295** (1987) 277.

[439] P. Ginsparg and G. Moore, *Lectures on 2D Gravity and 2D String Theory*, Los Alamos and Yale preprint LA-UR-92-3479 and YCTP-P23-92, 1992/1993; to appear in Cambridge University Press.

[440] J. Labastida, M. Pernici and E. Witten, *Topological gravity in two dimensions*, Nucl. Phys. **B310** (1988) 611.

[441] D. Montano and J. Sonnenschein, *The topology of moduli space and quantum field theory*, Nucl. Phys. **B324** (1989) 348.

[442] R. Myers and V. Periwal, *Topological gravity and moduli space*, Nucl. Phys. **B333** (1990) 536.

[443] E. Witten, *On the structure of the topological phase of two-dimensional gravity*, Nucl. Phys. **B340** (1990) 281.

[444] R. Dijkgraaf, E. Verlinde and H. Verlinde, *Notes on topological string theory and 2d quantum gravity*, in "String Theory and Quantum Gravity", edited by M. B. Green et al., World-Scientific, Singapore, 1990, p. 91.

[445] R. Dijkgraaf, *Topological field theory and 2d quantum gravity*, in "Two-Dimensional Quantum Gravity and Random Surfaces", edited by D. Gross et al., World Scientific, Singapore, 1992, p. 191.

[446] D. Birmingham, M. Blau, M. Rakowski and G. Thompson, *Topological field theory*, Phys. Rep. **209** (1991) 129.

[447] A. S. Schwartz, *The partition function of a degenerate quadratic functional and the Ray-Singer invariants*, Lett. Math. Phys. **2** (1978) 247.

[448] E. Witten, *Topological quantum field theory*, Commun. Math. Phys. **117** (1988) 353.

[449] E. Witten, *Topological sigma models*, Commun. Math. Phys. **118** (1988) 411.

[450] R. Dijkgraaf, E. Verlinde and H. Verlinde, *Topological strings in $D < 1$*, Nucl. Phys. **B352** (1991) 59.

[451] E. Verlinde and H. Verlinde, *A solution of two-dimensional topological quantum gravity*, Nucl. Phys. **B348** (1991) 457.

[452] S. Donaldson, *An application of gauge theory to four-dimensional topology*, J. Diff. Jeom. **18** (1983) 279.

[453] S. Donaldson, *Polynomial invariants of smooth four-manifolds*, Topology **29** (1990) 257.

[454] E. Witten, *Surprises with topological field theories*, in "Strings '90", edited by R. Arnowitt et al., World Scientific, 1991, p. 50.

[455] H. Nishino, *Self-dual supersymmetric Yang-Mills theory generates Witten's topological field theory*, Phys. Lett. **309B** (1993) 68.

[456] E. Brézin and V. A. Kazakov, *Exactly solvable field theories of closed strings*, Phys. Lett. **236B** (1990) 144.

[457] M. Douglas and S. Shenker, *Strings in less than one dimension*, Nucl. Phys. **B335** (1990) 635.

[458] D. Gross and A. A. Migdal, *Nonperturbative two-dimensional quantum gravity*, Phys. Rev. Lett. **64** (1990) 127.

[459] P. Ginsparg, *Matrix models of 2d gravity*, Los Alamos preprint LA-UR-91-4101, July 1991.

[460] V. Kazakov, *Bosonic strings and string field theories in one-dimensional target space*, Paris preprint LPTENS 30/90, 1990; in "Random Surfaces and Quantum Gravity", edited by O. Alvarez et al., p. 269.

[461] E. Brézin, *Large N limit and discretized two-dimensional quantum gravity*, in "Two-Dimensional Quantum Gravity and Random Surfaces", edited by D. Gross et al., World Scientific, Singapore, 1992, p. 1.

[462] D. Gross, *The c = 1 matrix models*, in "Two-Dimensional Quantum Gravity and Random Surfaces", edited by D. Gross et al., World Scientific, Singapore, 1992, p. 143.

[463] A. Bilal, *2d gravity from matrix models*, Johns Hopkins Lectures, CERN preprint TH.-5867, 1990.

[464] A. Marshakov, *Integrable structures in matrix models and physics of 2D gravity*, Int. J. Mod. Phys. **A8** (1993) 3831.

[465] A. M. Polyakov, Lecture at Northeastern University, in Spring 1990.

[466] F. David, *A model of random surfaces with nontrivial critical behavior*, Nucl. Phys. **B257** (1985) 573.

[467] D. V. Boulatov, V. A. Kazakov and A. A. Migdal, *Possible types of critical behavior and the mean size of dynamically triangulated random surfaces*, Phys. Lett. **174B** (1986) 87.

[468] D. Bessis, C. Itzykson and J.-B. Zuber, *Integral methods in combinatorial topology*, Adv. Appl. Math. **1** (1980) 109.

[469] P. Ginsparg and J. Zinn-Justin, *Large order behaviour of non-perturbative gravity*, Phys. Lett. **255B** (1991) 189.

[470] G. t'Hooft, *A planar diagram theory for strong interactions*, Nucl. Phys. **B72** (1974) 461.

[471] V. A. Kazakov, *The appearance of matter fields from quantum fluctuations of 2d gravity*, Mod. Phys. Lett. **A4** (1989) 2125.

[472] M. L. Mehta, *Random Matrices*, Academic Press, 1991.

[473] E. Brézin, C. Itzykson, G. Parisi and J.-B. Zuber, *Planar diagrams*, Commun. Math. Phys. **59** (1978) 35.

[474] L. D. Faddeev and A. A. Slavnov, *Gauge Fields: Introduction to Quantum Theory*, Benjamin/Cummings, Reading, MA, 1982.

[475] C. Crnković, P. Ginsparg and G. Moore, *Multicritical multicut matrix models*, Phys. Lett. **237B** (1990) 196.

[476] F. David, *Nonperturbative effects in 2d gravity and matrix models*, Saclay preprint SPHT-90-178, 1990.

[477] D. Boulatov and V. Kazakov, *The Ising model on random planar lattice: the structure of phase transition and the exact critical exponents*, Phys. Lett. **186B** (1986) 379.

[478] R. Dijkgraaf and E. Witten, *Mean field theory, topological field theory and multi-matrix models*, Nucl. Phys. **B342** (1990) 486.

[479] K. Li, *Topological gravity with minimal matter*, Nucl. Phys. **B354** (1991) 711.

[480] K. Li, *Recursion relations in topological gravity with minimal matter*, Nucl. Phys. **B354** (1991) 725.

[481] T. Eguchi and S.-K. Yang, *N=2 superconformal models as topological field theories*, Mod. Phys. Lett. **A5** (1990) 1693.

[482] M. R. Douglas, *Strings in less than one dimension and the generalized KdV hierarchies*, Phys. Lett. **238B** (1990) 176.

[483] T. Banks, M. Douglas, N. Seiberg and S. Shenker, *Microscopic and macroscopic loops in nonperturbative two-dimensional gravity*, Phys. Lett. **238** (1990) 279.

[484] J.-B. Zuber, *KdV and W - Flows*, Saclay preprint SPhT/91-052, 1991.

[485] I. M. Gel'fand and L. A. Dikii, *Asymptotic behaviour of the resolvent of Sturm-Liouville equations and the algebra of the KdV equations*, Sov. Math. Surveys **30** (1975) 77.

[486] I. M. Gel'fand and L. A.Dikii, *Fractional powers of operators and Hamiltonian systems*, Funct. Anal. Appl. **10** (1976) 259.

[487] F. A. Smirnov, *On the symmetries of quantum field theory in two dimensions*, Progr. Theor. Phys. Suppl. **110** (1992) 329.

[488] D. J. Korteweg and G. de Vries, *On the change of form of long waves advancing in a rectangular canal, and on a new type of long stationary waves*, Philos. Mag. **39** (1895) 422.

[489] A. Das, *Integrable Models*, World Scientific, Singapore, 1989.

[490] V. G. Drinfeld and V. V. Sokolov, *Equations of Korteweg -de Vries type and simple Lie algebras*, Dokl. Akad. Nauk USSR **258** (1981) 11 [Sov. Math. Dokl. **23** (1981) 457].

[491] V. G. Drinfeld and V. V. Sokolov, *Lie algebras and equations of Korteweg-de Vries type*, Itogi Nauki i Tekhn., Ser.: Sovrem. Probl. Mat. **24** (1984) VINITI, Moscow, p. 81.

[492] M. Adler, *On a trace functional for formal pseudo-differential operators and symplectic structure of the Korteweg-de Vries type equations*, Invent. Math. **50** (1979) 219.

[493] B. A. Kupershmidt and G. Wilson, *Conservation laws and symmetries of generalized sine-Gordon equations*, Commun. Math. Phys. **81** (1981) 189.

[494] M. Bauer, P. Di Francesco, C. Itzykson and J.-B. Zuber, *Covariant differential equations and singular vectors in Virasoro representations*, Nucl. Phys. **B362** (1990) 515.

[495] P. Forgács, A. Wipf, J. Balog, L. Fehér and L. O'Raifeartaigh, *Liouville and Toda theories as conformally reduced WZNW models*, Phys. Lett. **237B** (1989) 214.

[496] A. N. Leznov and M. V. Saveliev, *Representation theory and integration of nonlinear spherically symmetric equations of gauge theories*, Commun. Math. Phys. **74** (1980) 111.

[497] A. Bilal and J.-L. Gervais, *Systematic approach to conformal systems with extended Virasoro symmetries*, Phys. Lett. **206B** (1988) 412.

[498] O. Babelon, *Extended conformal algebra and the Yang-Baxter equatuon*, Phys. Lett. **215B** (1988) 523.

[499] J. Balog, L. Fehér, P. Forgács, L. O'Raifeartaigh and A. Wipf, *Kac-Moody realization of W algebras*, Phys. Lett. **244B** (1990) 435.

[500] L. Fehér, L. O'Raifeartaigh P. Ruelle and , I. Tsutsui, *On the Hamiltonian reduction of the Wess-Zumino-Novikov-Witten models*, Phys. Rep. **222** (1992) 1.

[501] L. O'Raifeartaigh, P. Ruelle and I. Tsutsui, *Quantum equivalence of constrained WZNW and Toda theories*, Phys. Lett. **258B** (1991) 359.

[502] R. Ward and R. Wells, *Twistor Geometry and Field Theory*, Cambridge Univ. Press, Cambridge, 1990.

[503] A. A. Belavin and V. E. Zakharov, *Yang-Mills equations as inverse scattering problem*, Phys. Lett. **73B** (1978) 53.

[504] I. Bakas and D. Depireux, *Self-duality and generalized KdV flows*, Mod. Phys. Lett. **A6** (1991) 399.

[505] I. Bakas and D. Depireux, *The origin of gauge symmetries in integrable systems of KdV type*, Int. J. Mod. Phys. **A7** (1992) 1767.

[506] L. J. Mason and G. A. J. Sparling, *Non-linear Schrödinger and Korteweg-de Vries are reductions of self-dual Yang-Mills*, Phys. Lett. **137A** (1989) 29.

[507] E. Witten, *Some exact multi-instanton solutions of classical Yang-Mills theory*, Phys. Rev. Lett. **38** (1977) 121.

[508] D. Depireux and I. Bakas, *Self-duality, KdV flows and W algebras*, University of Maryland preprint UMDEPP-91-0426, June 1991.

[509] R. S. Ward, *Integrable and solvable systems, and relations among them*, Phil. Trans. R. Soc. Lond. **A315** (1985) 451–457.

[510] M. A. Ablowitz and P. Clarkson, *Solitons, Nonlinear Evolution Equations and Inverse Scattering*, Cambridge Univ. Press, Cambridge, 1989.

[511] B. A. Kupershmidt, *A super Korteweg-de Vries equation: an integrable system*, Phys. Lett. **102A** (1984) 213.

[512] P. Mathieu, *Superconformal algbera and supersymmetric Korteweg-de Vries equation*, Phys. Lett. **203B** (1988) 287.

[513] A. Bilal and J.-L. Gervais, *Super-conformal algebra and super KdV equation: two infinite families of polynomial functions with vanishing Poisson brackets*, Phys. Lett. **211B** (1988) 85.

[514] A. Das and C. A. P. Galvao, *Self-duality and the supersymmetric KdV hierarchy*, Mod. Phys. Lett. **A8** (1993) 1399.

[515] Y. Manin and O. A. Radul, *A supersymmetric extension of the Kadomtsev-Petviashvili hierarchy*, Commun. Math. Phys. **98** (1985) 65.

[516] L. A. Leites, M. V. Saveliev and V. V. Serpukhov, *Embeddings of OSp(N|2) and associated non-linear supersymmetric equations*, in Proceedings of the Third Intern. Seminar on Group-Theory Methods in Physics, Yurmala, 20-24 May 1985; Nauka, Moscow, 1986.

[517] S. J. Gates Jr. and H. Nishino, *Supersymmetric soluble systems embedded in supersymmetric self-dual Yang-Mills theory*, Phys. Lett. **299B** (1993) 255.

[518] H. Nishino, *Supersymmetric KP systems embedded in supersymmetric self-dual Yang-Mills theory*, Phys. Lett. **318B** (1993) 107.

[519] B. B. Kadomtsev and V. I. Petviashvili, *On the stability of solitary waves in weakly dispersing media*, Dokl. Akad. Nauk USSR **192** (1970) 753 [Sov. Phys. Dokl. **15** (1970) 539].

[520] H. Nishino, *Self-dual supersymmetric Yang-Mills theory generates two-dimensional supersymmetric WZNW models*, Univ. of Maryland preprint UMDEPP 93-213, June 1993.

[521] A. Tsuchiya, T. Eguchi and M. Jimbo, eds., *Infinite Symmetries*, Proc. RIMS Research Project, June-August 1991, Kyoto University; World Scientific, Singapore, Int. J. Mod. Phys. **A7** (1992) Suppl. 1A and 1B.

[522] E. Brézin and S. R. Wadia, eds., *The Large N Expansion in Quantum Field Theory and Statistical Physics. From Spin Systems to Two-Dimensional Gravity*, World Scientific, Singapore, 1993.

[523] R. Gilmore, *Lie Algebras and Some of Their Applications*, John Wiley & Sons, New York, 1974.

[524] N. Bourbaki, *Groups et Algebras de Lie*, Masson, Paris, 1982.

[525] J. E. Humphreys, *Introduction to Lie Algebras and Representation Theory*, Springer-Verlag, Berlin, 1970.

[526] D. Zhelobenko, *Compact Lie Groups and Their Representations*, American Mathematical Society, Providence, 1973.

[527] D. Gorenstein, *Finite Simple Groups*, Plenum Press, New York, 1982.

[528] V. G. Kač and A. K. Raina, *The Highest Weight Representations of Infinite-Dimensional Lie Algebras*, World Scientific, Singapore, 1987.

[529] S. Lie, *Begründung einer Invariantentheorie der Berührungstransformationen*, Math. Ann. **8** (1874) 215.

[530] I. Ado, *The representation of Lie algebras by matrices*, Usp. Mat. Nauk **2** (1947) 159 [Transl. Amer. Math. Soc. **9** (1962) 308].

[531] R. Slansky, *Group theory for unified model building*, Phys. Rep. **79** (1981) 1.

[532] H. Farkas and I. Kra, *Riemann Surfaces*, Springer-Verlag, New York, 1980.

[533] P. A. M. Dirac, *Lectures on Quantum Mechanics*, Yeshiva University, New York, 1964.

[534] D. M. Gitman and I. V. Tyutin, *Quantization of Fields with Constraints*, Springer-Verlag, Berlin, 1990.

[535] G. Moore and N. Seiberg, *Classical and quantum conformal field theory*, Commun. Math. Phys. **123** (1989) 177.

[536] N. Yu. Reshetikhin and G. Moore, *A comment on quantum group symmetry in conformal field theory*, Nucl. Phys. **B328** (1989) 557.

[537] L. Alvarez-Gaumé, G. Sierra and C. Gomez, *Duality and quantum groups*, Nucl. Phys. **B330** (1990) 347.

[538] N. Yu. Reshetikhin and F. Smirnov, *Hidden quantum group symmetry and integrable perturbations of conformal field theories*, Commun. Math. Phys. **131** (1990) 157.

[539] E. Witten, *Quantum field theory and Jones polynomials*, Commun. Math. Phys. **121** (1989) 351.

[540] V. G. Drinfeld, *Quasi-Hopf algebras and Knizhnik-Zamolodchikov equations*, in "Problems of Modern Quantum Field Theory", edited by A. A. Belavin et al., Springer-Verlag, Berlin, 1989, p. 1.

[541] V. G. Drinfeld, *Hamiltonian structure on Lie groups, Lie bialgebras and the geometrical meaning of classical Yang-Baxter equation*, Sov. Math. Dokl. **27** (1983) 68 [Dokl. Akad. Nauk USSR, Ser. Mat. **268** (1982) 285; *ibid.* **283** (1985) 1060].

[542] V. G. Drinfeld, *Quantum Groups*, in Proc. Intern. Congress of Mathematicians, Berkeley 1986, edited by A. M. Gleason; American Math. Soc., Providence, 1987, Vol. 1, p. 798.

[543] M. Jimbo, *A q-difference analogue of $U(q)$ and the Yang-Baxter equation*, Lett. Math. Phys. **10** (1985) 63.

[544] M. Jimbo, *A Q analog of $U(GL(N+1))$ Hecke algebra and the Yang-Baxter equation*, Lett. Math. Phys. **11** (1986) 247.

[545] M. Jimbo, *Quantum R Matrix Related to the Generalized Toda System: an Algebraic Approach*, Springer Lecture Notes in Physics **246** (1986) 335.

[546] N. Yu. Reshetikhin, *Quantized universal enveloping algebras and invariants of links*, parts 1 and 2, LOMI preprints E-4-87 and E-17-87, Leningrad, 1987.

[547] A. N. Kirillov and N. Yu. Reshetikhin, *Representations of the algebra $U_q(SL(2))$, q-orthogonal polynomials and invariants of links*, LOMI preprint E-9-88, in "New Developments in the Theory of Knots", edited by T. Kohno, 1991, p. 202.

[548] E. Abe, *Hopf Algebras*, Cambridge Univ. Press, Cambridge, 1977.

[549] M. Jimbo, *Yang-Baxter Equation in Integrable Systems*, World Scientific, Singapore, 1989.

[550] V. E. Korepin, N. M. Bogoliubov and A. G. Izergin, *Quantum Inverse Scattering Method and Correlation Functions*, Cambridge Univ. Press, Cambridge, 1993.

Text Abbreviations

ASD	anti-self-dual
AKM	affine Kač-Moody
BFV	Batalin-Fradkin-Vilkovisky
BPZ	Belavin-Polyakov-Zamolodchikov
BRST	Becchi-Rouet-Stora-Tyutin
BS	Brink-Schwarz
c.c.	complex conjugation
cf	compare
CFT	conformal field theory
CSA	Cartan sub-algebra
2d (4d)	2-dimensional (4-dimensional)
DDK	David-Distler-Kawai
DF	Dotsenko-Fateev
DS	Drinfeld-Sokolov
ed(s).	editor(s)
e.g.	for example
FP	Faddeev-Popov
GKO	Goddard-Kent-Olive
GSO	Gliozzi-Scherk-Olive
ICFT	irrational conformal field theory
i.e.	that is
IR	infra-red
irrep	irreducible representation
h.c.	hermitian conjugation
KP	Kadomtsev-Petviashvili
KPZ	Knizhnik-Polyakov-Zamolodchikov
KS	Kazama-Suzuki

LG	Landau-Ginzburg
l.h.s. (r.h.s.)	left-hand-side (right-hand-side)
LM	left-loving
M	Majorana
MM	matrix model
M-M	Mathur-Mukhi
MMM	multi-matrix model
MW	Majorana-Weyl
NLSM	non-linear sigma-model
NS	Neveu-Schwarz
NSR	Neveu-Schwarz-Ramond
OPE	operator product expansion
PN	Pernici-Nieuwenhuizen
QFT	quantum field theory
R	Ramond
RCFT	rational conformal field theory
resp.	respectively
RG	renormalization group
RM	right-moving
RSOS	restricted solid-on-solid
SCA	super-conformal algebra
SCFT	super-conformal field theory
SD	self-dual
SDG	self-dual gravity
SDSG	self-dual supergravity
SDYM	self-dual Yang-Mills
SSDYM	supersymmetric self-dual Yang-Mills
SS	Sugawara-Sommerfeld
TCFT	topological conformal field theory
TFT	topological field theory
UV	ultra-violet
VEV	vacuum expectation value
viz.	vizibly
w.r.t.	with respect to
WZNW	Wess-Zumino-Novikov-Witten

Index